TABLE B Sex and Sex-condition Terminology for Several Livestock and Poultry Species

Species	Female Young[a]	Female Mature[b]	Uncastrated Male Young[a]	Uncastrated Male Mature[b]	Castrated Male Young[a]	Castrated Male Mature[b]
Cattle	Heifer[c]	Cow	Bull[c]	Bull	Steer	Stag
Chicken	Chick or pullet	Hen	Chick or cockerel	Cock/rooster	Capon	—
Goat	Doe	Doe	Buck	Buck	Wether	—
Horse[d]	Filly	Mare	Colt[e]	Stallion	Gelding	—
Sheep	Ewe[f]	Ewe	Ram[f]	Ram/buck	Wether	Stag
Swine	Gilt	Sow	Boar[g]	Boar	Barrow	Stag
Turkey	Young hen or poult[h]	Hen	Young tom or poult[h]	Tom	—	—

[a] *Young:* generally prior to puberty or sexual maturity and before the development of secondary sex characteristics.

[b] *Mature:* generally after puberty and before the development of secondary sex characteristics.

[c] Referred to as a *bull calf* (under 1 year of age); *heifer* can be a heifer calf, yearling heifer, or first calf heifer (after the first calf is born and prior to the birth of the second calf).

[d] A close relative of the horse is the *donkey,* also known as an *ass* or a *burro.* The male ass is referred to as a *jack*; the female is known as a *jennet.* A *mule* is produced by crossing a *jackass* with a mare. A *hinny* is produced by crossing a stallion with a jennet. Mules and hinnys are reproductively sterile, although their visual sexual characteristics appear to be normal.

[e] Under 3 years of age.

[f] Referred to as a *ewe lamb* or a *ram lamb* (under 1 year of age).

[g] Referred to as a *boar pig* (under 6 months of age).

[h] Under 10 weeks of age. Chicks or poults are newly hatched or a few days old.

Scientific Farm Animal Production

SIXTH EDITION

Scientific Farm Animal Production

AN INTRODUCTION TO ANIMAL SCIENCE

ROBERT E. TAYLOR
THOMAS G. FIELD
Colorado State University

Prentice Hall, Upper Saddle River, N.J. 07458

Library of Congress Cataloging-in-Publication Data

Taylor, Robert E. (Robert Ellis),
    Scientific farm animal production : an introduction to animal
science / Robert E. Taylor, Thomas G. Field. — 6th ed.
      p.   cm.
    Includes bibliographical references and index.
    ISBN 0-13-456591-6
    1. Livestock.   I. Field, Thomas G. (Thomas Gordon)   II. Title.
SF61.T39   1998
636—dc21   97-12212
                                          CIP

Acquisitions Editor: *Charles Stewart*
Production Editor: *Christina Palaia*
Developmental Editor: *Carol Robison*
Marketing Manager: *Debbie Yarnell*
Director of Manufacturing & Production: *Bruce Johnson*
Managing Editor: *Mary Carnis*
Design Director: *Marianne Frasco*
Manufacturing Buyer: *Marc Bove*
Editorial Assistant: *Kate Linsner*
Formatting/page make-up: *North Market Street Graphics*
Art Director: *Cathy Dunn*
Cover Designer: *Miguel Ortiz*

Printed in the United States of America

10 9 8 7 6 5 4 3 2 1

ISBN 0-13-456591-6

Prentice-Hall International (UK) Limited, *London*
Prentice-Hall of Australia Pty. Limited, *Sydney*
Prentice-Hall Canada Inc., *Toronto*
Prentice-Hall Hispanoamericana, S.A., *Mexico*
Prentice-Hall of India Private Limited, *New Delhi*
Prentice-Hall of Japan, Inc., *Tokyo*
Simon & Schuster Asia Pte. Ltd., *Singapore*
Editora Prentice-Hall do Brasil, Ltda., *Rio de Janeiro*

Dedication

This book is dedicated to the hundreds of students the authors have taught. The positive and rewarding interactions with these students have provided the stimulus to write this book. Hopefully, the reading and studying of *Scientific Farm Animal Production* will be a motivational influence to current and future students as they seek to enhance their education.

Contents

Preface

Scientific Farm Animal Production is distinguished by an appropriate combination of both breadth and depth of livestock and poultry production and their respective industries. The book gives an overview of the biological principles applicable to the Animal Sciences with chapters on reproduction, genetics, nutrition, lactation, consumer products, and others. The book also covers the breeding, feeding, and management of beef cattle, dairy cattle, horses, sheep, swine, poultry, and goats. Although books have been written on each of these separate chapters, the authors have highlighted the significant biological principles, scientific relationships, and management practices in a condensed but informative manner.

Target Audience

This book is designed as a text for the introductory Animal Science course typically taught at universities and junior or community colleges. It is also a valuable reference book for livestock producers, vocational agriculture instructors, and others desiring an overview of livestock production principles and management. The book is basic and sufficiently simple for urban students with limited livestock experience, yet challenging for students who have a livestock production background.

Key Features

Chapters 1–9 cover animal products and give an overview of the livestock and poultry industries, Chapters 10–21 discuss the biological principles, while livestock and poultry management practices are presented in Chapters 22–33.

The glossary of the terms used throughout the book has been expanded so students can readily become familiar with animal science terminology. The bold-lettered words in the text are included in the glossary.

Many illustrations in the form of photographs and line drawings are used throughout the book to communicate key points and major relationships. If "a picture is worth a thousand words," the numerous photographs and drawings expand the usefulness of the book beyond its pages.

Selected references are provided for each chapter to direct students into greater depth and breadth as they become intrigued with certain topics. Instructors can also use the references to expand their knowledge in current background material. Also included in the selected-references section are references to visuals that relate to the specific chapter. Instructors are encouraged to review these visuals and use those which will enrich their course.

Using The Book

The book is designed to accommodate several instructional approaches to teaching the introductory course: (1) the life-cycle biological principles approach, including such areas as consumer products, reproduction, breeding, nutrition, and animal health; (2) the species approach (teaching the course primarily in reference to the various species); or (3) a combination of the previous two. The latter appears to be the most popular teaching approach, covering principles in lecture and combining principles and species into laboratory exercises.

Some instructors will assign one or more papers on a topic selected by them or the students. The references at the end of each chapter are designed for students who want or need to explore certain topics in more depth.

Most instructors will not have sufficient time in their course to assign all the chapters. Course outlines can be developed to include the chapters assigned and put them in the sequence that meets their preference.

Changes In This Edition

This edition has been updated with current technical and applied information. Numerous tables and figures have been revised with current data, especially Chapters 1–5. Major changes have been made to Chapter 26, Swine Breeds and Breeding, and Chapter 34, Animal Behavior. Chapter 35, Issues in the Animal Industries, has been rewritten to update an increased number of current issues. Students should understand the issues in animal agriculture because many of them will be involved in finding solutions to these issues.

The management emphasis, where bioeconomics focuses on combining biology with economics, is expanded. The global dimensions of the animal industries are also emphasized.

Acknowledgments

Appreciation is expressed to those individuals and organizations who have reviewed all or part of the fifth and sixth editions and offered suggestions for the revision. The following individuals and organizations made a major contribution to the sixth edition: Allen Christian (Iowa State University), Dr. Ted Friend (Texas A&M University) and Dr. Temple Grandin (Colorado State University). A special thanks to Barbara Holst and Pat Beebe who typed the sixth edition.

Many instructors who teach the introductory animal science course completed a questionnaire and made numerous helpful suggestions. Those individuals who offered helpful comments that were incorporated into the sixth edition are:

W. R. Backus, University of Tennessee
Russ Danielson, North Dakota State University
Keith Gilster, University of Nebraska
Tim Marshall, University of Florida
Mark Mirando, Washington State University
Doug Parrett, University of Illinois
Tommy Perkins, Southwest Texas State University
Tim Ross, New Mexico State University
Cliff Stokes, California Polytechnic State University
Lyle Westrom, University of Minnesota

Robert E. Taylor
Thomas G. Field

About The Authors

Dr. Taylor was raised on an Idaho livestock operation where several livestock species were produced. He received his B.S. and M.S. degrees from Utah State University. This background, combined with his Ph.D. work in animal breeding and physiology from Oklahoma State University, has provided much depth to his knowledge of livestock production. He has had practical production experience with beef cattle, dairy cattle, horses, poultry, sheep, and swine.

Dr. Taylor received teaching awards at Iowa State University (where he also managed a swine herd) and at Colorado State University. He also received the Distinguished Teaching Award from the American Society of Animal Science and the USDA National Excellence in Teaching Award. Many of his concepts for effective teaching are used in this book.

Dr. Field was raised on a Colorado cow–calf and seedstock enterprise. He managed a seedstock herd after completing his B.S. degree. A competitive horseman as a youth, he has had practical experience with seedstock cattle, commercial cow–calf, stockers, and horses. He has an M.S. and Ph.D. in animal science from Colorado State University.

Dr. Field is highly respected by the hundreds of students he has taught. He has received numerous teaching awards including the NACTA National Teaching Fellow and the USDA National Excellence in Teaching Award.

CHAPTER 1

Animal Contributions to Human Needs

Our basic human needs are food, shelter, clothing, fuel, and emotional well-being. Animals and animal products supply many of these basic physical and emotional needs and contribute to a high standard of living which is associated with a high consumption of animal products.

Since the domestication of dogs, horses, cattle, sheep, and other animals, some 6,000 to 10,000 years ago, wide differences have developed among people in various regions of the world in using agricultural technology to improve their standard of living. But in all societies, domestic animals are a source of food, other consumer products, and companionship for people.

Of particular importance among the multitude of benefits that domestic animals provide for humans are food; clothing; by-products used for consumer goods and animal feeds; power; manure for fuel (Fig. 1.1), buildings, and fertilizer; information on human disease from research using experimental animals; and pleasure for those who keep animals. Table 1.1 shows the major domesticated animal species, their approximate numbers, and how they are used by people throughout the world. Chickens (12.0 billion) are most numerous, followed by cattle (1.3 billion), sheep (1.1 billion), and swine (875 million).

CONTRIBUTIONS TO FOOD NEEDS

When opportunity exists, most humans consume both plant and animal products (Fig. 1.2). Meat is nearly always consumed in quantity when it is available. Its availability in most countries is closely related to the economic status of the people and their agricultural technology. Vegetarianism in countries such as India may be the long-term result of intense population pressures and scarcity

FIGURE 1.1 A load of cow-dung cakes en route to a market in India. Courtesy of R. E. McDowell, Cornell University.

of feed for animals because of competition between humans and animals for food. Rising population pressures, particularly in developing regions (Southeast Asia, Africa, and Latin America), force people to consume foods primarily of plant origin. Some major groups in human society practice vegetarianism for ethical reasons. In the Buddhist philosophy and some religions of India, for example, all animal life is considered sacred.

The contribution of animal products to the per-capita calorie and protein supply in food is shown in Table 1.2. Animal products comprise approximately 16% of the calories and 35% of the protein in the total world food supply. Large differences exist between developed countries and developing countries in both total daily supply of calories and protein. For example, the contribution of animal products in calories is more than 30% in some developed countries and less than 10% in several developing countries. Some developed countries have more than 50% of their daily per-capita protein supply from animal products, while several developing countries are less than 25%. The United States ranks high compared to other countries in the contribution of animal products to the available calories and protein.

Changes in per-capita calorie supply and protein supply during the past 25–30 years are shown in Table 1.3. Both per-capita caloric and protein supply have increased in most areas of the world. The contribution of animal products to the per-capita protein supply has increased in most of the world.

Selected recommended daily intakes (recommended daily allowance) of calories and protein are given in Table 1.4. Although the data in Table 1.4 do not represent an average of the U.S. population, a comparison of these data with those in Table 1.2 is interesting. Some countries have a larger supply of calories and protein than needed, where other countries have an inadequate calorie and protein supply. This assumes an equal distribution of the available supply, which in reality does not occur.

The large differences among countries in the importance of animal products in their food supply can be partially explained by available resources and development of those resources. Most countries with only a small percentage of their population involved in agriculture have higher standards of living and a higher per-capita consumption of animal products. When comparing Table 1.5

TABLE 1.1 Major Domesticated Animal Species—Their Numbers and Uses in the World

Animal Species	World Numbers (mil)	Leading Countries or Areas with Numbers[a] (mil)	Primary Uses
Ruminants			
Cattle	1,288	India (193), Brazil (152), United States (101), China (91), Russian Fed. (49)	Meat, milk, hides
Sheep	1,087	Australia (133), China (112), New Zealand (51), Iran (45), United States (10)	Wool, meat, milk, hides
Goats	609	India (118), China (106), Pakistan (41), Nigeria (26), United States (2)	Milk, meat, hair, hides
Buffalo	149	India (79), China (22), Pakistan (19), Thailand (4)	Draft, milk, meat, hides, bones
Camels	19	Somalia (6.0), Sudan (2.8), India (1.5), Pakistan (1.1)	Packing, riding, draft, meat, milk, hides
Yaks	13	Russian Federation,[b] Tibet[b]	Packing, riding, draft,
Llamas	13	South America[b]	meat, milk, hides
Nonruminants			
Chickens	12,002	China (2,692), United States (1,530), Brazil (680), Indonesia (640)	Meat, eggs, feathers
Swine	875	China (403), United States (58), Germany (26)	Meat
Turkeys	247	United States (88), France (33), Italy (23)	Meat, eggs, feathers
Ducks	681	China (443), Vietnam (30), Indonesia (27), United States (4)	Meat, eggs, feathers
Horses	58	China (10), Mexico (6), Brazil (6), United States (4), Ethiopa (3)	Draft, packing, riding,
Asses	44	China (11), Ethiopa (5), Mexico (3), United States (0.05)	meat, companion animals
Mules	15	China (5.6), Mexico (3.2), Brazil (2.1), United States (0.03)	

[a] U.S. numbers are given for comparison; may not always be among the leading countries.
[b] Data not available.
Source: Adapted from several sources, including the 1994 *FAO Production Yearbook.*

with Table 1.2, note that the countries in Table 1.5 are listed by percentage of their population involved in agriculture.

Agriculture mechanization (note tractor numbers in Table 1.5) has been largely responsible for increased food production and allowing many people to work in other industries. This facilitates the provision of many goods and services and thus raises the standard of living in a country.

The tremendous increase in the productivity of U.S. agriculture and the relative cost of food are vividly demonstrated in Table 1.6. From 1820 it took 100 years to double productivity. Then productivity doubled in shorter, successive time periods—30 years (1920–50), 15 years (1950–65), and 10 years (1965–75). A dramatic change occurred after World War II, when productivity increased more than fivefold in 30 years. During that time, the abundant production of feed grains provided a marked stimulus in increasing livestock production, thus providing large amounts of animal products for the human population.

FIGURE 1.2 Animal food products, such as meat, milk, and eggs, are the preferred foods in countries with high standards of living. Courtesy of The American Egg Board.

Releasing people from producing their own food in the United States has given them the opportunity to improve their per-capita incomes. The increased per-capita income associated with an abundance of animal products has resulted in a decrease in relative costs of some animal products with time (Table 1.7).

United States consumers allocate a smaller share (12%) of their disposable income for food than do people in other countries. In contrast, people of India and China spend 55–65% of their incomes for food.

Table 1.8 shows that cereal grains are the most important source of energy in world diets. The energy derived from cereal grains, however, is twice as important in developing countries (as a group—there are exceptions) as in developed countries. Table 1.8 also shows that meat and milk are the major animal products contributing to the world supply of calories and protein.

Most of the world meat supply comes from cattle, buffalo, swine, sheep, goats, and horses. There are, however, 20 or more additional species, unfamiliar to most Americans, that collectively contribute about 6.5 billion lb of edible protein per year or approximately 10% of the estimated total protein from all meats. These include the alpaca, llama, yak, deer, elk, antelope, kangaroo, rabbit, guinea pig, capybara, fowl other than chicken (duck, turkey, goose, guinea fowl, pigeon), and wild game exclusive of birds. For example, the former Soviet Union cans more than 110 million lb of reindeer meat per year, and in West Germany the annual sales of local venison exceed $1 million. Peru derives more than 5% of its meat from the guinea pig.

Meat is important as a food for two scientifically based reasons. The first is that the assortment of amino acids in animal protein more closely matches the needs of the human body than does the assortment of amino acids in plant pro-

TABLE 1.2   Animal Product Contribution to Per-Capita and Protein Supply

Country	Per-Capita Kilocalorie Supply (kilocalories per day) Total Kilo calories	From Animal Products Kilo calories	Percent	Per-Capita Protein Supply Total Protein (g/day)	From Animal Products Grams	Percent
Denmark	3,664	1,596	44	99	63	64
Greece	3,815	961	25	114	59	52
Iceland	3,005	1,218	40	123	95	77
United States	3,732	1,228	33	113	74	66
United Kingdom	3,317	1,075	32	91	52	57
Germany	3,344	1,163	35	100	64	64
Australia	3,179	1,214	38	100	68	68
Japan	2,903	629	22	98	56	57
Mexico	3,146	543	17	78	31	40
Brazil	2,824	470	17	52	18	35
Kenya	2,075	255	12	54	16	30
China	2,727	345	13	67	16	24
Egypt	3,335	210	6	87	13	15
Nigeria	2,124	64	3	43	5	12
India	2,395	162	7	58	10	17
Bangladesh	2,019	58	3	42	5	12
Afghanistan	1,523	173	11	43	10	23
Rwanda	1,821	49	3	44	3	7
World Total	2,718	428	16	71	25	35

Source: 1994 *FAO Production Yearbook.*

TABLE 1.3   Changes in Per-Capita Calorie and Protein Supply in the World

Area	Year	Per-Capita Calorie Supply (calories per day) Total Calories	From Animal Products Calories	Percent	Per-Capita Protein Supply Total Protein (g/day)	From Animal Products Grams	Percent
World	1961–63	2,287	359	16	63	20	32
	1988–90	2,697	424	16	71	25	35
	1994	2,718	428	16	71	25	35

Source: 1994 *FAO Production Yearbook.*

TABLE 1.4   Recommended Daily Caloric and Protein Intake for Selected Males and Females in the United States

Sex	Age (years)	Weight (lb)	Height	Average Daily Calories	Protein (g/day)
Female	25–50	120	5 ft 4 in	2,000	44
Male	25–50	155	5 ft 10 in	2,700	56

TABLE 1.5   Population Involved in Agriculture in Selected Countries (countries ranked by percent of population in agriculture)

Country	1994 Population (mil)	1994 Population in Agriculture[a] (mil)	Percent of Economically Active Population in Agriculture (1994)[b]	Number of Tractors (1993) (thou)
United States	261	6	2	4,800
United Kingdom	58	1	2	500
Germany	81	3	4	1,300
Australia	18	0.8	4	315
Denmark	5	0.2	4	156
Japan	125	6	5	2,041
Iceland	0.3	0.01	6	11
Ireland	4	.04	12	168
Greece	10	2	22	216
Brazil	159	35	22	735
Mexico	92	25	28	172
Egypt	62	24	39	61
Turkey	61	25	44	746
Afghanistan	19	10	52	<1
Nigeria	109	69	63	12
China	1,209	779	64	737
India	919	564	65	1,195
Bangladesh	118	78	66	5
Kenya	27	21	75	14
Nepal	21	20	91	5
Rwanda	8	7	91	<1
World Total	5,630	2,443	45	25,704

[a] *Agricultural population* is defined as all persons depending for their livelihood on agriculture. This comprises all persons actively engaged in agriculture and their nonworking dependents.
[b] Includes all economically active persons engaged principally in agriculture, forestry, hunting, or fishing.
Source: 1994 *FAO Production Yearbook.*

TABLE 1.6   Persons Supplied by One Farm Worker in the United States and the Percent of Income Spent on Food

Year	No. of Persons in the U.S. Supplied Per Farm Worker	Annual Personal Disposable Income[a]	Amount of Personal Disposable Income Spent on Food		
			At Home	Away From Home	Percent of Total Income
1820	4	NA	NA	NA	NA
1850	4	NA	NA	NA	NA
1880	6	NA	NA	NA	NA
1910	7	NA	NA	NA	NA
1930	9	691	NA	NA	23.9
1940	11	586	NA	NA	21.9
1950	16	1,502	NA	NA	22.1
1960	26	2,265	301	109	17.8
1970	48	3,489	380	194	14.1
1980	76	8,421	818	530	13.8
1985	78	11,861	991	717	12.6
1990	80	16,236	1,242	1,013	11.7
1995	94[b]	19,703	1,320	1,165	11.2

[a] Current dollars. Based on constant (1987) dollars per capita income for 1970, 1980, and 1990 is $9,875, $12,005, and $14,101, respectively.
[b] The number increases to 128 persons when export markets are considered.
Source: USDA.

TABLE 1.7 Amount of Food Purchased by an Hour's Pay (average U.S. worker)

Food Item	1950	1970	1980	1990	1995
Frying chicken (lb)	2.3	7.9	9.4	11.1	12.5
Milk (half gal)	3.5	9.0	11.6	7.0	8.0
Eggs (doz)	2.2	5.3	7.9	9.9	12.4
Pork (lb)	2.5	4.2	4.8	4.7	5.9
White bread (lb)	9.3	13.3	13.1	14.4	14.5
Ground beef (lb)	2.4	5.0	4.2	6.3	8.4

Note: Average hourly pay $1.34 (1950), $3.23 (1970), $6.66 (1980), $10.02 (1990), and $11.46 (1995).
Source: USDA.

TABLE 1.8 Contributions of Various Food Groups to the World Food Supply

Food Group	Calories (%)	Protein (%)
Cereals	49	43
Roots, tubers, pulses	10	10
Nuts, oils, vegetable fats	8	4
Sugar and sugar products	9	2
Vegetables and fruits	8	7
All animal products	16	34
Meat	(7)	(15)
Eggs	(1)	(2)
Fish	(1)	(5)
Milk	(5)	(11)
Other	(2)	(1)

Source: Adapted from several FAO world food surveys.

tein. The second is that vitamin B_{12}, which is required in human nutrition, may be obtained in adequate quantities from consumption of meat or other animal products but not from consumption of plants.

Milk is one of the largest single sources of food from animals. In the United States, 99% of the milk comes from cattle, but on a worldwide basis, milk from other species is important, too; the domestic buffalo, sheep, goat, alpaca, camel, reindeer, and yak supply significant amounts of milk in certain countries. Milk and products made from milk contribute protein, energy, vitamins, and minerals for humans.

Besides the nutritional advantages, a major reason for human use of animals for food is that most countries have land areas unsuitable for growing cultivated crops. Approximately two-thirds of the world's agricultural land is permanent pasture, range, and meadow; of this, about 60% is unsuitable for producing cultivated crops that would be consumed directly by humans. This land, however, can produce feed in the form of grass and other vegetation that is digestible by grazing ruminant animals, the most important of which are cattle and sheep (Fig. 1.3). These animals can harvest and convert the vegetation, which is for the most part undigestible by humans, to high-quality protein food. In the United States, about 385 million acres of range land and forest, representing 44% of the total land area, are used for grazing. Although this acreage now supports only about 40% of the total cattle population, it could carry twice this amount if developed and managed intensively.

FIGURE 1.3 Animals produce food for humans by utilizing grass, crop residues, and other forages from land that cannot produce crops to be consumed directly by humans. (A) Sheep grazing a steep hillside. Courtesy of *California Agriculture Magazine*. University of California. (B) Cattle grazing a mountain valley in Switzerland. Courtesy of the American Simmental Association. (C) Cattle produce meat from the mountains and Plains areas of the western United States. Courtesy of the American Simmental Association. (D) Sheep utilizing the crop residue remaining after harvesting the corn grain. Courtesy of Winrock International.

Animal agriculture therefore does not compete with human use for production of most land used as permanent pasture, range, and meadow. On the contrary, the use of animals as intermediaries provides a means by which land that is otherwise unproductive for humans can be made productive (Fig. 1.4).

People today are concerned about energy, protein, population pressures (Fig. 1.5), and land resources as they relate to animal agriculture. Quantities of energy and protein present in foods from animals are smaller than quantities consumed by animals in their feed because animals are inefficient in the ratio of nutrients used to nutrients produced. More acres of cropland are required per person for diets high in foods from animals than for diets including only plant products. As a consequence, animal agriculture has been criticized for wasting food and land resources that could otherwise be used to provide persons with adequate diets. Consideration must be given to economic systems and consumer preferences to understand why agriculture perpetuates what critics perceive as resource-inefficient practices. These practices relate primar-

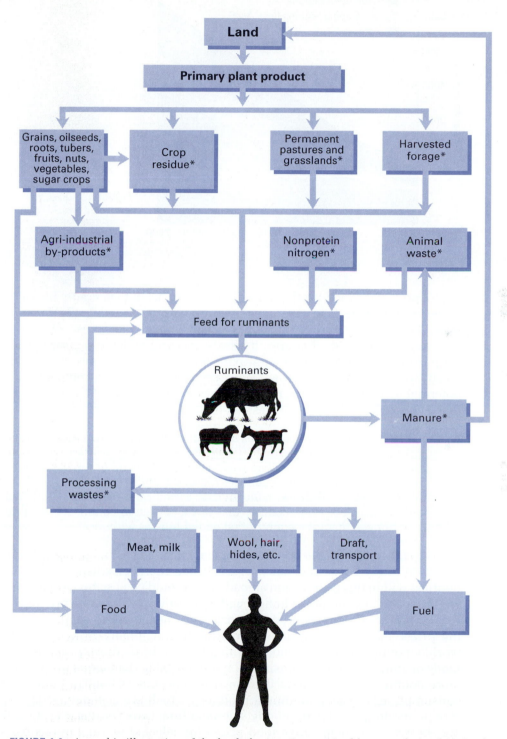

FIGURE 1.4 A graphic illustration of the land-plant-ruminant-animal-human relationship. Products marked with an asterisk (*) are not normally consumed by humans, but ruminants convert many of these products into useful products for humans. Courtesy of Winrock International.

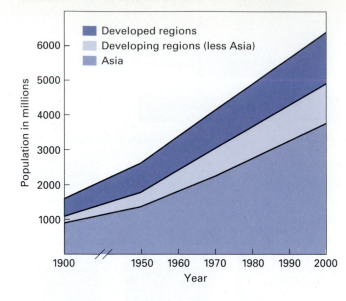

FIGURE 1.5 Past, present, and projected world population, 1900–2000.

ily to providing food-producing animals with feed that could be eaten by humans and using land resources to produce crops specifically for animals instead of producing crops that could be consumed by humans.

Hunger does exist, of course. It affects, in varying degrees, nearly 1 billion people in the world. This is approximately 20% of the world's population, mostly in developing countries. However, hunger is being reduced, and the potential exists to make even more improvement. The book *Ending Hunger: An Idea Whose Time Has Come* states:

■ Since 1900, seventy-five countries have ended hunger within their borders as a basic, society-wide issue. It is both important and heartening to note that there is no single prescribed way to achieve the end of the persistence of hunger in a society. Some countries focused on land reform, while others emphasized food subsidies, collectivized agriculture, or privately owned "family farms." For every country that saw a particular action as crucial to ending hunger, there is a country that ended hunger without it.

Ending Hunger quotes the National Academy of Sciences in its World Food and Nutrition Study, for which 1,500 scientists had been consulted: "If there is the political will in this country and abroad . . . it should be possible to overcome the worst aspects of widespread hunger and malnutrition within one generation."

Agriculture producers produce what consumers want to eat as reflected in the prices consumers can and are willing to pay. Eighty-five percent of the world's population desires food of animal origin in its diets, perhaps because foods of animal origin are considered more palatable than foods from plants. In more countries, as per-capita income rises, consumers tend to increase their consumption of meat and animal products, which are generally more expensive pound for pound than products derived from cereal grains (Fig. 1.6). Table 1.9 shows some comparative food prices for selected world capitals by comparing the time (earnings) required to purchase a specified market basket. The range in time is from a low of 2 hours, 35 minutes in Washington, D.C. to a high of 12 hours, 37 minutes in Brasilia, Brazil.

FIGURE 1.6 Income and supply of animal protein (per person) in selected countries. Source: USDA and FAO.

If many consumers in countries where animal products are consumed at a high rate were to decide to eat only food of plant origin, consumption and price of foods from plants would increase, and consumption and price of foods from animals would decrease. Agriculture would then adjust to produce greater quantities of food from plants and lesser quantities of food from animals. Ruminant animals can produce large amounts of meat without grain feeding. The amount of grain feeding in the future will be determined by cost of grain and the price consumers are willing to pay for meat.

Some people in the United States advocate shifting from the consumption of foods from animals to foods from plants. They see this primarily as a moral issue, believing it is unethical to let people elsewhere in the world starve when our own food needs could be met by eating foods from plants rather than feeding plants to animals. The balance of plant-derived foods could then be sent abroad. These people believe that grain can be shipped with comparative ease because a surplus of grain exists in the United States and because any surplus we have should be provided at no cost. Providing free food to other countries has met with limited success in the past. In some situations, it upsets their own agricultural production, and in many cases the food cannot be adequately distributed in the country because transportation and marketing systems are poor.

There are strong feelings that the United States has a moral obligation to share its abundance with other people in the world, particularly those in developing countries. It appears that this can best be done by sharing our time and sharing our basic, but not necessarily advanced technology. People need to have self-motivation to improve, need to be shown how to help themselves step by step, and need to develop fully the agricultural resources in their own

TABLE 1.9   Time Spent to Earn the Retail Value of Selected Food Items, by Country

Item in Market Basket	Bonn, Germany	Brasilia, Brazil	London, United Kingdom	Mexico City, Mexico	Paris, France	Pretoria, South Africa	Rome, Italy	Tokyo, Japan	Metro. Washington, DC, USA
Total time[a]	**2:42**	**12:37**	**3:52**	**7:19**	**5:20**	**7:53**	**4:42**	**4:57**	**2:35**
Sirloin steak, boneless	0:27	1:04	0:37	0:53	0:51	0:43	0:42	1:14	0:21
Pork roast, boneless	0:14	1:16	0:25	0:55	0:30	0:26	0:24	0:21	0:24
Broilers, whole	0:05	0:29	0:08	0:17	0:23	0:17	0:14	0:09	0:04
Large eggs, dozen	0:05	0:30	0:13	0:19	0:14	0:18	0:14	0:05	0:05
Butter	0:09	0:47	0:11	0:27	0:18	0:31	0:21	0:14	0:09
Cheddar cheese, Emmenthaler	0:18	1:08	0:18	0:57	0:23	0:43	0:27	0:22	0:18
Whole milk (quart)	0:03	0:31	0:05	0:10	0:08	0:14	0:08	0:05	0:04
Cooking oil (quart)	0:05	0:39	0:10	0:17	0:19	0:27	0:08	0:09	0:07
Potatoes	0:02	0:32	0:04	0:23	0:07	0:11	0:05	0:08	0:04
Apples	0:07	1:21	0:11	0:23	0:10	0:22	0:10	0:10	0:08
Oranges	0:08	0:12	0:12	0:11	0:12	0:09	0:13	0:12	0:07
Flour	0:02	0:25	0:05	0:09	0:10	0:14	0:04	0:05	0:03
Rice	0:11	0:28	0:10	0:18	0:18	0:20	0:15	0:09	0:06
Sugar	0:05	0:24	0:06	0:11	0:11	0:13	0:08	0:06	0:05
Coffee	0:40	2:49	0:58	1:27	1:07	2:45	1:09	1:28	0:30

[a] Time is shown in hours and minutes per pound (exceptions noted) of product using the countries' or cities' average wages. For example, in Tokyo, Japan, it would take 4 hours and 57 minutes to purchase the total market basket and 1 hour and 14 minutes to purchase a pound of sirloin steak.
Source: USDA (*Food Review*, Sept./Dec. 1994).

countries. Another challenge in effectively distributing food is the black market created by political and economic instability. The United Nations reports that direct cash contributions are more effective in addressing world hunger than are contributions of food.

Increases in the efficiency of animal production in the United States have been remarkable during the past half century (Table 1.10). These improvements in efficiency have occurred primarily because people had an incentive to progress under a free-enterprise system. They learned how to improve their standard of living by effectively using available resources. This progress has taken time under the environmental conditions that motivated people to be successful.

Citizens of the United States should not only share their technology with other countries but also share how this technology was developed (Fig. 1.7). These achievements have been built on knowledge developed through experience and research, the extension of knowledge to producers, and the development of an industry to provide transportation, processing, and marketing in addition to production. Dwindling dollars currently being spent to support agricultural research and extension of knowledge in the United States may not provide the technology needed for future food demands.

TABLE 1.10 Productivity Changes in Several Farm Animal Species in the United States

Species and Measure of Productivity	1925	1950	1975	1990	1995
Beef cattle					
Carcass weight (per year) weight marketed per breeding female (lb)	220	310	482	524	540
Sheep					
Liveweight marketed per breeding female (lb)	60	90	130	145	145
Dairy cattle					
Milk marketed per breeding female (lb)	4,189	5,313	10,500	14,000	16,400
Swine					
Liveweight marketed per breeding female (lb)	1,600	2,430	2,850	3,500	4,590
Broiler chickens					
Age to market weight (weeks)	15.0	12.0	7.5	7.4	7.3
Feed per pound of gain (lb)	4.0	3.3	2.1	1.9	1.8
Liveweight at marketing (lb)	2.8	3.1	3.8	4.5	5.0
Turkeys					
Age to market weight (weeks)	34	24	19	16	16
Feed per pound of gain (lb)	5.5	4.5	3.1	2.6	2.5
Liveweight at marketing (lb)	13.0	18.6	18.4	21.1	23.1
Laying hens					
Eggs per hen per year (no.)	112	174	232	250	254
Feed per dozen eggs (lb)	8.0	5.8	4.2	4.0	3.2

Source: Adapted from *Food from Animals,* CAST Report 82.

About 30% of the world human population and 32% of the ruminant animal population live in developed regions of the world, but ruminants of these same regions produce two-thirds of the world's meat and 80% of the world's milk. In developed regions, a higher percentage of animals are used as food producers, and these animals are more productive on a per-animal basis than animals in developing regions. This is the primary reason for the higher level of human nutrition in developed countries of the world.

Possibly many developing regions of the world could achieve levels of plant and animal food productivity similar to those of developed regions. Except perhaps in India, abundant world supplies of animal feed resources that do not compete with production of food for people are available to support expansion of animal populations and production. It has been estimated that through changes in resource allocation, an additional 8 billion acres of arable land (twice what is now being used) and 9.2 billion acres of permanent pasture and meadow (23% more than is now being used) could be put into production in the world. These estimates, plus the potential increase in productivity per acre and per animal in developed countries, demonstrate the magnitude of world food-production potential. This potential cannot be realized, however, without proper government planning and increased incentive to individual producers.

In the long run, each nation must assume the responsibility of producing its own food supply by efficient production, barter, or purchase and by keeping future food-production technology ahead of population increases and demand. Extensive untapped resources that can greatly enhance food production exist throughout the world, including an ample supply of animal products. The greatest resource is the human being, who can, through self-motivation, become more productive and self-reliant.

A

B

C

FIGURE 1.7 Mechanization has increased plant and animal production in the United States. (A) In the late 1800s and early 1900s, a team of horses, one worker, and a single moldboard plow could plow about 2 acres per day. Courtesy of Charles L. Benn, Iowa State University. (B) An early tractor in 1918 increased agricultural production on a per-worker basis. Courtesy of Michigan State University. (C) In the 1980s, one worker with a powerful modern tractor pulling three plows, each with five moldboards, could plow 110 acres in a 10-hour day, accomplishing the work that once required 55 workers and 110 horses. Courtesy of the USDA and CAST.

CONTRIBUTIONS TO CLOTHING AND OTHER NONFOOD PRODUCTS

Products other than food from ruminants include wool, hair, hides, and pelts. Although synthetic materials have made some inroads into markets for these products, world wool production has been decreasing in the past few years. However, it is higher than it was 10 years ago. It is important to note that in more than 100 countries, ruminant fibers are used in domestic production and cottage industries for clothing, bedding, housing, and carpets.

Annual production of animal wastes from ruminants contains millions of tons of nitrogen, phosphorus, and potassium. The annual value of these wastes for fertilizer is estimated at $1 billion.

Inedible tallow and greases are animal by-products used primarily in soaps and animal feeds and as sources of fatty acids for lubricants and industrial use. Additional tallow and grease by-products are used in the manufacture of pharmaceuticals, candles, cosmetics, leather goods, woolen fabrics, and tin plating. The individual fatty acids can be used to produce synthetic rubber, food emulsifiers, plasticizers, floor waxes, candles, paints, varnishes, printing inks, and pharmaceuticals.

Gelatin is obtained from hides, skins, and bones and can be used in foods, films, and glues. Collagen, obtained primarily from hides, is used to make sausage casings.

CONTRIBUTIONS TO WORK AND POWER NEEDS

The early history of the developed world abounds with examples of the importance of animals as a source of work energy through draft work, packing, and human transport. The horse made significant contributions to winning wars and exploration of the unknown regions of the world.

In the United States during the 1920s, approximately 25 million horses and mules were used, primarily for draft purposes. The tractor has replaced all but a few of these draft animals. In parts of the developing world, however, animals provide as much as 99% of the power for agriculture even today.

In more than half the countries of the world, animals—mostly buffalo and cattle but also horses, mules, camels, and llamas—are kept primarily for work and draft purposes (Fig. 1.8). About 20% of the world's human population depends largely or entirely on animals for moving goods. Animal draft power makes a significant contribution to the production of major foods (rice and other cereal grains) in several heavily populated areas of the world. India, for example, has more than 200 million cattle and buffalo, the largest number of any country. Although not slaughtered because they are considered sacred, cattle of India contribute significantly to the food supply by providing power for field work and milk for people. It is estimated that India alone would have to spend more than $1 billion annually for gasoline to replace the animal energy it uses in agriculture.

ANIMALS FOR COMPANIONSHIP, RECREATION, AND CREATIVITY

Estimates of the number of companion animals in the world are unavailable. There are an estimated 26 million family-owned dogs and 21 million family-owned cats in the United States in addition to the animals identified in Table 1.1. The U.S. pet food industry processes more than 3 million tons of cat and

FIGURE 1.8   Animals provide significant contributions to the draft and transportation needs of countries lacking mechanization in their agricultural technology. (A) Cattle used for draft in Honduras. Courtesy of Winrock International. (B) Water buffalos are used to cultivate many of the rice patties in the Far East. Courtesy of Dr. Budi S. Nara. (C) Camels being used to transfer fertilizer to farming areas in Ethiopia. FAO photo, courtesy of C. N. Coombes.

dog food annually valued at more than $1 billion. Many species of animals would qualify as companions where people derive pleasure from them. The contribution of animals as companions, especially to the young and elderly, is meaningful, even though it is difficult to quantify the emotional value.

Animals used in rodeos, bullfighting, and other sports provide income for thousands of people and recreational entertainment for millions (Fig. 1.9). Numerous people who have made money from nonagricultural businesses have invested in land and animals for recreational and emotional fulfillment away from the urban environment.

Our livestock heritage involves the interactions of humans with animals over the centuries. Historically, animals have been highly respected, revered, and even worshiped by humans. Early humans expressed the sacred, mysterious images for some animals through art on cave walls some 15,000–30,000 B.C.

A

B

C

FIGURE 1.9   (A) The dog serves humans in a variety of ways. A guide dog leads the way for a visually impaired man. Courtesy of Guide Dogs for the Blind, San Rafael, Calif. (B) Many people enjoy riding horses. Courtesy of the *Western Horseman.* (C) Horse racing is one of the most highly attended spectator sports in the United States. Courtesy of the Kentucky Derby.

D

E

F

FIGURE 1.9 (continued)   (D) Steer wrestling is exciting for the spectator but dangerous for the participants. Fast, well-trained horses and strong participants are needed to wrestle the steers to the ground in less than 5 seconds. Courtesy of Larry Thomas. (E) Calf roping is a favorite rodeo event. Courtesy of Larry Thomas. (F) Bullfighting originated in Spain centuries ago. It is popular in Mexico and several South American countries. Courtesy of José Rafael Cortes, editor of *La Nación* newspaper, San Cristobal, Venezuela.

Thus, through these early paintings and sculptures, animals found their way into an expression of the things of humans—the humanities. Several historians have noted that an extremely high form of art is the intelligent manipulation of animal life—the modeling and molding of different types through the application of breeding principles. They say that there are boundless possibilities of genetic expression that have godlike creative opportunities. Many seedstock animal breeders today are inspired with those same creative challenges.

HUMAN HEALTH RESEARCH

Laboratory animals are commonly used to provide valuable information for improving human life. Larger, domestic farm animals are used less frequently because the initial cost and maintenance costs are high. Farm animals serve as research models for approximately 200 human diseases. Cattle and sheep have been used to test artificial organs before these organs have been implanted into humans.

Iowa State University has done an extensive study using swine as the model to evaluate how irradiation might increase the frequency of genetic defects. Colorado State University has studied the long-term effects of low levels of irradiation on the incidence of cancer in a beagle dog colony.

Miniature pigs (Yucatan swine whose ancestors came from the wilds of southern Mexico and central South America) have been used as laboratory animals because their pulmonary, cardiac, dental, and even prenatal brain development closely resembles that of humans. The utilization of miniature Yucatan pigs in biomedical research has increased dramatically since 1978. Miniature pigs are considered ideal in studying human aging, disease resistance, and the effect of diet on diabetes and atherosclerosis (Fig. 1.10). In one research study, for example, it was discovered that genetically predisposed pigs became susceptible to diabetes after consuming a low-fiber diet with 40% of the calories coming from saturated fat. There is less cost in maintaining the miniature pigs because their mature weights are approximately 120 lb, compared to 600–800 lb for typical swine. A new breed of Yucatan swine, the micropig, may prove even more useful as an experimental animal; its mature weight range is 50–70 lb.

During the 1990s, several press releases identified promising human health research projects using animals—such as a gene from a bacterium in cattle can be used to detect colon cancer in humans; a protein defense system preventing organ rejection has been developed making pig kidney and heart transplants to humans much safer (could be routine transplants in a decade); a company has successfully developed genetically engineered swine that express approximately 25% authentic human hemoglobin; a patient was the first to receive a pig's liver; the Food and Drug Administration gave clearance to treat liver

FIGURE 1.10 Yucatan miniature pigs are used as experimental animals to help solve human problems such as diabetes, obesity, and atherosclerosis. Courtesy of Colorado State University.

failure by routing patients' blood through livers from a strain of pigs developed to be compatible with the human immune system; scientists create pig with humanlike heart; and cow tongue tissue produces a natural antibiotic.

OTHER ANIMALS

This book gives major attention to cattle, sheep, swine, horses, poultry, and goats. Several other useful species are briefly mentioned in this chapter. In addition, there are other animals that provide useful products in specific areas of the world. There are also several domesticated animals about which little is known, and undomesticated animals that have potential for long-term world agricultural development. Some of these domesticated animals, relatives to cattle, are the banteng of Indonesia and the mithan of India, Burma, and Bangladesh. Wild bovines include the kouprey in Thailand and the gaur in India and Southeast Asia.

Among the undomesticated Asian pigs is the babirusa of eastern Indonesia (Fig. 1.11). The stomach of these unusual animals has an extra sac, suggesting they may have the ability to break down cellulose. These "ruminant pigs" browse leaves, a behavior more similar to deer than to pigs. In the 1990s, Israeli pig breeders were considering importing the babirusa with an objective of breeding a "kosher pig."

In recent years, new species of mammals have been discovered—such as, a small horse in Tibet and the Vu Quang ox near the Vietnamese-Laos border. It is intriguing to contemplate that in this modern world there may be unknown species of animals that are genetically related to our domestic farm animals.

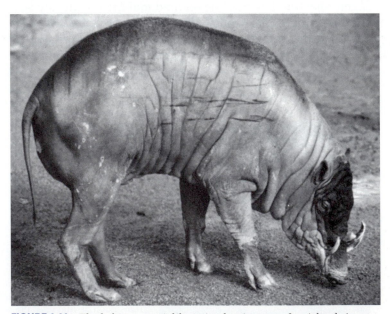

FIGURE 1.11   The babirusa, a piglike animal unique to a few islands in eastern Indonesia. Courtesy of Phillip Coffey, Jersey Wildlife Preservation Trust.

There are feral animals that have previously been domesticated but have returned to nature to survive and reproduce. Examples are (1) today's wild horses that escaped from the Spanish, Native Americans, U.S. Cavalry, and livestock producers; and (2) the Cracker cattle (sometimes called Florida Scrub or Piney Woods cattle) that are descendants of the Spanish cattle brought to Florida.

Several of these rare animals may have disease-resistance and other genetic material that might be incorporated into breeding programs of our common farm animals. A realistic question could be posed: Are we overlooking animals that could benefit humans?

CHAPTER SUMMARY

- Domesticated animals (>16 billion) contribute to the well-being of 5.6 billion humans throughout the world in providing food, clothing, shelter, power, recreation, and companionship.

- Animal products contribute significantly to the world's human protein needs and energy supply.

- As people increase their standard of living, the per capita consumption of animal products also increases.

- Human health is improved from the continuing research utilizing domestic animals.

REVIEW QUESTIONS

1. *True or False:* Developed nations have less of their population economically involved in agriculture than do developing nations.

2. *True or False:* People in developed nations consume more of their daily supply of protein and calories from animal products than do people from developing nations.

3. Consumers from which country allocate the smallest share of their disposable income toward the purchase of food?

4. What are the two scientifically based reasons that meat is important food for humans?

5. What is the single most important reason for increased food production in the twentieth century?

6. What percentage of the world's agricultural land is unsuitable for cultivation of crops and is therefore used to pasture or graze livestock?

7. What are the basic human needs to which animals and animal products contribute?

SELECTED REFERENCES

Baldwin, R. L. 1980. *Animals, Feed, Food and People: An Analysis of the Role of Animals in Food Production.* Boulder, CO: Westview Press.

Devendra, C. 1980. Potential of sheep and goats in less developed countries. *J. Anim. Sci.* 51:461.

Ending Hunger: An Idea Whose Time Has Come. 1985. Sparks, NV: Praeger.

FAO Production Yearbook. Vol. 48, 1994. Rome: FAO.

Kunkel, H. O. 1990. *World Hunger—Grain Versus Meat Production.* College Station, TX: Texas A&M University.

National Research Council. 1983. *Little-Known Asian Animals with a Promising Economic Future.* Washington: National Academy Press.

Pearson, R. A., 1994. Draft Animal Power. *Encyclopedia of Agriculture Science.* San Diego: Academic Press, Inc.

Reid, J. T., White, O. D., Anrique, R., and Fortin, A. 1980. Nutritional energetics of livestock: Some present boundaries of knowledge and future research needs. *J. Anim. Sci.* 51:1393.

Willham, R. L. 1985. *The Legacy of the Stockman.* Morrilton, AR: Winrock International.

CHAPTER 2

An Overview of the Animal Industries

It is important to see the broad picture of the animal industries before studying the specific biological and economic principles that explain animal function and production. These industries are typically described with numbers of animals, pounds produced, production systems, prices, products, people (producers and consumers), and profitability. Products and consumers are covered in the next several chapters.

An understanding of the animal industries begins with basic terminology, especially the various species and sex classifications (see Table B at the beginning of this book). Refer to the glossary, which follows Chapter 37, for definitions of many commonly used animal terms.

U.S. LIVESTOCK AND POULTRY: AN OVERVIEW

The livestock and poultry industries in the United States are big businesses—they must be large to meet the high animal-product preference of more than 263 million U.S. consumers and also to supply the export market. Table 2.1 shows the number of livestock and poultry producers, the inventory number, and value of animals at one point in time during the year.

Cash Receipts

An evaluation of farm cash receipts from the sale of animals and animal products provides another perspective of U.S. animal industries. Table 2.2 shows the cash receipts for animal commodities ranked against all agricultural commodities. The top five states for each commodity are also shown in this table.

TABLE 2.1   Numbers of Producers and Animals in the U.S. Livestock and Poultry Industries

Species	Number of Producers (thousands)	Number of Animals (mil head)	Inventory Value ($mil)
Beef cattle	909.1	103.8[a]	$52,160
Swine	182.7	60.2	4,256
Dairy cattle	140.1	9.4[b]	—
Chickens	179.2	384.2	916
Sheep	82.1	8.5	731
Horses (for sale)	88.4	—	—
Ducks and geese	39.6	—	—
Milk goats	15.4	—	—
Angora goats[c]	5.4	1.9	89

[a] All cattle.
[b] Cows only.
[c] Texas only.
Source: USDA (Agricultural Statistics).

TABLE 2.2   Leading U.S. States for Farm Cash Receipts

Commodity	Rank[a]	Value ($mil)	Five Leading States ($mil)				
			1	2	3	4	5
All Commodities	—	$185,750	Calif. $22,261	Tex. $13,288	Iowa $10,959	Nebr. $8,690	Ill. $7,887
Livestock/products	—	98,843	Tex. 8,454	Calif. 5,549	Nebr. 5,189	Iowa 5,068	Kans. 4,693
Cattle and calves	1	33,983	Tex. 6,296	Nebr. 5,158	Kans. 4,235	Colo. 2,081	Okla. 1,759
Dairy products	2	19,923	Calif. 3,078	Wis. 2,916	N.Y. 1,494	Pa. 1,458	Minn. 1,186
Broilers	5	11,760	Ga. 1,772	Ark. 1,769	Ala. 1,438	N.C. 1,162	Miss. 992
Hogs	7	10,073	Iowa 2,550	N.C. 1,274	Minn. 865	Nebr. 739	Ind. 720
Chicken eggs	10	3,958	Ark. 294	Ga. 290	Calif. 288	Pa. 265	Ohio 253
Turkeys	12	2,774	N.C. 582	Minn. 299	Ark. 241	Mo. 232	Calif. 213
Aquaculture	a	657	Miss. 272				
Sheep/lambs	a	507	Tex. 64	Calif.	Colo.	S. Dak.	Wyo. 44
Wool	a	52	Tex. 14	Wyo.	Calif.	Mont.	Colo.
Horses	a	—	Ky. 559	—	—	—	—
Mohair	a	30	Tex. 21	—	—	—	—
Other livestock		1,575					

[a] Ranking is in comparison to all agricultural commodities. Ranking of sheep/lambs, wool, horses, and mohair is below 25th.
Source: USDA (*Ranking of States and Commodities by Cash Receipts,* 1995).

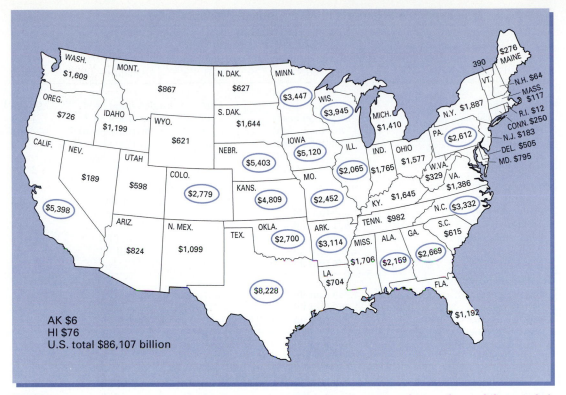

FIGURE 2.1 Farm cash receipts for livestock and products, 1994 ($ mil). States with more than $2 bil are circled. Source: USDA (Statistical Highlights of U.S. Agriculture, 1995/96).

Note the cash receipts for all livestock and livestock products comprises approximately 50% of all agricultural commodities in the United States.

Figure 2.1 shows the farm cash receipts from livestock and poultry products for each state. Note that 16 states each have annual cash receipts exceeding $2 billion.

World Trade

The economic well-being of U.S. animal industries is influenced by world trade. Table 2.3 shows the export and import markets for several animal commodities. While world trade for all U.S. products shows a deficit, agricultural and animal products show a positive trade balance. The economics of the U.S. export market for animal products is significantly influenced by cattle hides, meat (beef), fat/tallow, and dairy products. The highest import expenditures are for beef, dairy products, pork, and live cattle.

Of course, world trade affects the economics of some states more than others. Table 2.4 shows the export values of animal products for the five leading states in each product area. Live animals and meat ranks fourth in all agricultural commodities. Feed grains, wheat, and soybeans are the leading agricultural exports.

Commodity Prices

The profitability of U.S. animal industries is partly influenced by the prices paid to producers for animals and animal products. Prices can fluctuate

TABLE 2.3   U.S. Exports and Imports of Major Animal Products, 1995

Commodity	Exports Quantity	Value ($mil)	Imports Quantity	Value ($mil)
Live Animals				
(excluding poultry)	NA	$397	4,665 (thou head)	$1,678
Cattle and calves	95 (thou head)	86	2,786 (thou head)	1,413
Hogs	16 (thou head)	4	1,750 (thou head)	138
Sheep and lambs	288 (thou head)	12	41 (thou head)	4
Horses, mules, burros	61 (thou head)	262	43 (thou head)	88
Baby chicks	52,350 (thou head)	106	NA	NA
Meats and Meat products				
(excluding poultry)	3,704 (mil lb)	4,522	2,299 (mil lb)	2,305
Beef and veal	1,309 (mil lb)	2,647	1,555 (mil lb)	1,447
Pork	581 (mil lb)	847	592 (mil lb)	686
Lamb and mutton	7 (mil lb)	8	66 (mil lb)	86
Poultry	4,490 (mil lb)	2,026	11 (mil lb)	14
Horse meat	40 (mil lb)	67	NA	22
Other products				
Tallow, grease, fat, and lard	3,749 (mil lb)	827	95 (mil lb)	31
Variety meats	1,030 (mil lb)	732	57 (mil lb)	42
Sausage casings	24 (mil lb)	37	29 (mil lb)	64
Hides and skins[a]	NA	1,748[b]	NA	199
Wool and mohair	11 (mil lb)	35	88 (mil lb)	163
Dairy products	NA	711	NA	1,089
Eggs	NA	171	NA	20
Feathers and down	7 (mil lb)	27	44 (mil lb)	120
Bull semen	NA	67	NA	4
Total for Animals and Products		10,933		5,990
Total for Agricultural Products		55,814[c]		29,993

[a] All hides and skins (including fur skins).
[b] Number of cattle hides (valued at $1,388 mil).
[c] Grains ($18.5 bil) are most important.
Source: USDA (Foreign Agricultural Trade of the United States; U.S. Dairy, Livestock and Poultry Trade).

TABLE 2.4   Leading U.S. States in Animal and Animal Product Export Values, 1995

Commodity	Rank[a]	Total U.S. ($mil)	Five Leading States ($mil) 1	2	3	4	5
Live animals and meat	4	$4,798	Nebr. 764	Kans. 731	Tex. 659	Iowa 529	Colo. 269
Poultry	8	2,208	Ark. 311	Ga. 286	N.C. 252	Ala. 245	Miss. 158
Hides and skins	9	1,738	Kans. 323	Nebr. 316	Tex. 273	Colo. 128	Iowa 103
Animal fats	14	841	Nebr. 148	Kans. 145	Tex. 121	Iowa 83	Colo. 52
Dairy products	15	714	Wis. 165	Calif. 127	Minn. 59	Pa. 42	N.Y. 37
Total for All Agricultural Commodities $54,160							

[a] Ranking is in comparison to export of all agricultural commodities—feed grains are 1st at $8,833 mil.
Source: USDA (Foreign Agricultural Trade of the United States).

monthly, weekly, even daily. These price changes are influenced primarily by supply and demand.

Those individuals interested in animal profitability should know the average prices of animal products. In addition, an understanding of the prices for specific classes and grades of animals, of what causes prices to fluctuate, and of how high prices are obtained is also necessary. Figure 2.2 compares market

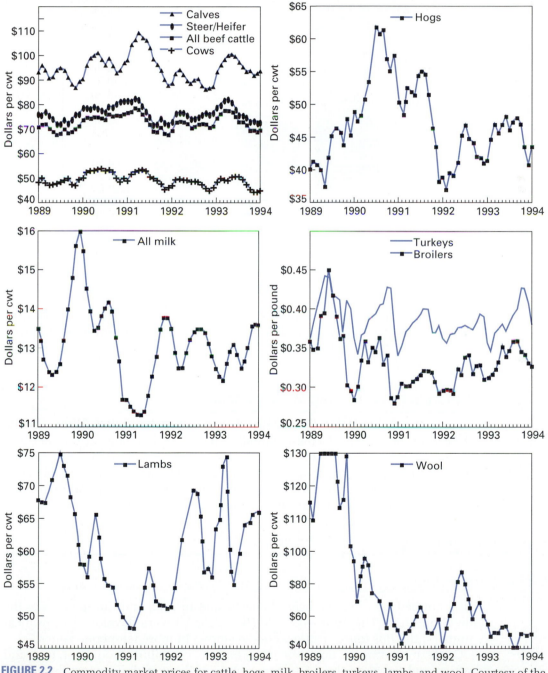

FIGURE 2.2 Commodity market prices for cattle, hogs, milk, broilers, turkeys, lambs, and wool. Courtesy of the USDA (Agricultural Prices).

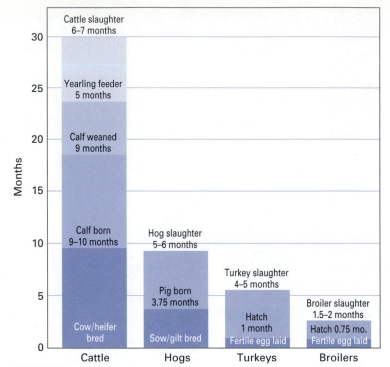

FIGURE 2.3 Biological time-lag in changing numbers of several species of farm animals. Courtesy of the USDA.

prices of several livestock and poultry commodities. This overview shows average prices and variability in prices producers receive for the animals and products they sell. Details of how some factors influence prices are discussed in later chapters.

Biological Differences in Meeting Market Demand

Changes in consumer demand, feed prices, weather, and other factors dictate the need to increase or decrease animal numbers and amount of product produced. Figure 2.3 shows the large differences between some farm animal species and how quickly or slowly numbers can change. Broiler numbers can be increased or decreased in a couple of months, while several years are needed to make significant changes in cattle numbers.

THE BEEF INDUSTRY

Global Perspective

Cattle are believed to have been domesticated in Asia and Europe during the New Stone Age. The humped cattle (*Bos indicus*) were developed in tropical countries; the *Bos taurus* cattle were developed in more temperate zones.

Cattle, including the domestic water buffalo, contribute food, fiber, fuel, and draft animal power to the 5.5 billion people of the world. For most developed countries, beef (meat) is a primary product. For developing countries, beef is a secondary product, as draft animal power and milk are the primary products. In some countries, cattle are still a mode of currency or a focus of religious beliefs and customs.

For 30 years, cattle numbers have continued to increase due to (1) greater demand for beef in developing countries, and (2) increased export demand (the result of more liberal trade policies in some countries). In recent years, drought in the Southern Hemisphere has slowed the growth in cattle numbers.

Table 2.5 shows the leading countries for cattle numbers, beef production, and beef consumption. India has the largest cattle population; however, its per-capita consumption is low because religious customs forbid cattle (considered sacred) from being slaughtered. The United States produces the most beef, but Argentina and Uruguay are higher in per-capita consumption.

Countries with a high cattle population relative to their human population typically have a high per-capita consumption of beef and high export tonnage. For example, the cattle versus human population in Australia is 24 million versus 16 million, and in Argentina it is 51 million versus 32 million. Both countries rank high in per-capita consumption of beef. Japan, with its 5 million cattle, 123 million people, and rapidly expanding economy, has an increasing demand for beef (Table 2.6).

United States

The U.S. beef industry is made up of a series of producing, processing, and consuming segments that relate to each other but that operate independently. Table 2.7 identifies the various segments and the products they produce.

Figure 2.4 shows total cattle and amount of carcass beef produced in the United States from 1950–96. Interestingly, 100 million head of cattle currently produce as much carcass beef as 120 million head produced in the 1970s. There are several reasons for this relatively high production of beef from fewer numbers of cattle: (1) the average carcass weight has increased from 579 lb in 1975 to 703 lb in 1995, (2) an increased number of cattle are fed per feedlot (2.4 times the feedlot capacity), (3) the market age of fed cattle has decreased, and (4) more crossbreeding and faster-gaining Continental breeds (e.g., Simmental and Charolais) are being used in commercial breeding programs.

TABLE 2.5   World Cattle Numbers, Production, and Consumption

Country	No. Cattle (mil head)	Country	Production (bil lb)[a,b]	Country	Per-capita Consumption (lb)[b]
1. India	193	1. United States	25	1. Argentina	152
2. Brazil	152	2. USSR (former)	7	2. Uruguay	133
3. United States	101	3. Brazil	7	3. United States	96
4. China	91	4. Argentina	6	4. Canada	78
5. Russian Fed.	49	5. China	5	5. Australia	77
World total	1,288	**World total**	111	**World average**[c]	23

[a] Does not include buffalo meat.
[b] Carcass weight.
[c] Estimated by dividing the annual beef production by the world population.
Sources: USDA, FAS (*World Livestock Situation*); 1994 *FAO Production Yearbook*.

TABLE 2.6   World Beef Trade

Exports		Imports	
Country	(bil lb)	Country	(bil lb)
1. Australia	2.6	1. United States	2.4
2. Germany	1.4	2. Japan	1.3
3. United States	1.3	3. USSR (former)	1.2
4. France	1.2	4. Italy	1.1
5. Ireland	1.0	5. Germany	1.0
World total	14.3	**World total**	12.1

NOTE: Carcass weight.
Source: USDA, FAS (*World Livestock Situation*).

TABLE 2.7   An Overview of the Beef Industry Segments in the United States

Segment of the Beef Industry	Products Produced or Utilized	Approximate Number	Investment ($bil)
Seedstock Producers	Breeding stock—primarily bulls (14–30 months old) and some heifers and cows. Some steers and heifers for feeding. Slaughter cows and bulls.	120,000 breeders	$ 15.0
↓ ↑ (a) ↓ ↑ (b)			
Commercial Cow-Calf Producers	Calves (6–10 months old), weighing 300–700 lb. Slaughter cows and bulls with majority over 5 years of age, weighing 800–1,500 lb (cows) and 1,000–2,500 lb (bulls).	909,000 producers[c]	180.0
↓ ↑			
Yearling or Stocker Operator	Feeder steers and heifers (most of them 12–20 months old, weighing 500–900 lb).		
↓ ↑			
Feeders	Market steers, heifers, cows, and bulls (mostly steers and heifers 16–30 months old, weighing 900–1,400 lb).	44,000 feedlots	7.5
↓ ↑			
Packers	Carcasses (approx. 600–800 lb). Boxed beef (carcasses into subprimal cuts).	1250 packers	3.8
↓ ↑			
Retailers	Retail cuts. By-products.	250 food chains	50.0
↓ ↑			
Consumers	Cooked products. By-products (leather, pharmaceuticals, variety meats, etc.).	263 million consumers	

[a] Animal and/or product flow.
[b] Demands and expectations.
[c] In addition, there are approximately 200,000 dairy farms that produce about 20% of U.S. beef.

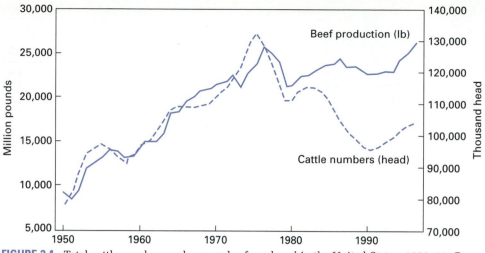

FIGURE 2.4 Total cattle numbers and carcass beef produced in the United States, 1950–96. Courtesy of USDA.

Cattle Production

Most commercial beef cattle production occurs in three phases: the **cow–calf, stocker–yearling,** and **feedlot** operations. The cow–calf operator raises the young calf from birth to 6–10 months of age (400–650 lb). The stocker–yearling operator then grows the calf to 600–850 lb, primarily on roughage. Finally, the feedlot operator uses high-energy rations to finish the cattle to a desirable slaughter weight, approximately 900–1,300 lb. Most fed steers and slaughter heifers are between 15 and 24 months of age when marketed.

However, there are alternatives to the typical three-phase operation. In an **integrated** operation, for instance, the cattle may have a single owner from cow–calf to feedlot, or ownership may change several times before the cattle are ready for slaughter. Alternative production and marketing strategies are diagrammed in Fig. 2.5.

Cow–Calf Production. U.S. cow–calf production involves some 35 million head of beef cows that are distributed throughout the country. Most of the cows are concentrated in areas where forage is abundant. As Fig. 2.6 shows, 11 states each have over 1 million head of cows (51% of the U.S. total), most of them located in the Plains, Corn Belt, and southeastern states. Approximately 80% of the 909,000 beef cow operations have less than 50 cows per operation. However, nearly 50% of the beef cow inventory is in operations with more than 100 cows (Table 2.8). Cow numbers fluctuate over the years, depending on drought, beef prices, and land prices.

There are two kinds of cow–calf producers. **Commercial** cow–calf producers raise most of the potential slaughter steers and heifers. **Seedstock breeders,** specialized cow–calf producers, produce primarily breeding cattle and semen.

Stocker–Yearling Production. Stocker–yearling producers feed cattle for growth prior to their going into a feedlot for finishing. Replacement heifers intended for the breeding herd are typically included in the stocker–yearling category. Our focus here, however, is on steers and heifers grown for later feedlot finishing.

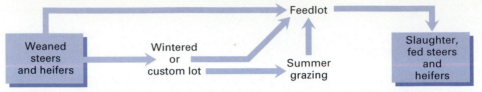

FIGURE 2.5 Alternative feeding and marketing pathways for weaned calves. Courtesy of Colorado State University.

Several alternate stocker–yearling production programs are identified in Fig. 2.5. In some programs, a single producer owns the calves from birth through the feedlot finishing phase, and the cattle are raised on the same farm or ranch. In other programs, one operator retains ownership, but the cattle are custom-fed during the growth and finishing phases. In still other programs, the cattle are bought and sold once or several times.

The primary basis of the stocker–yearling operation is to market available forage and high-roughage feeds, such as grass, crop residues (e.g., corn stalks, grain stubble, and beet tops), wheat pasture, and silage. Stocker–yearling operations also make use of summer-only grazing areas that are not suitable for the production of supplemental winter feed.

Stocker–yearling operations are desirable for early-maturing cattle. These cattle need slower gains to achieve heavier slaughter weights without being

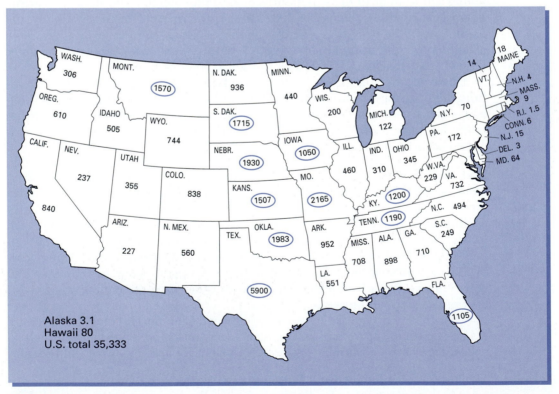

FIGURE 2.6 U.S. beef cows that have calved, January 1, 1996 (1,000 head). States with more than 1 million cows are circled. Courtesy of the USDA.

TABLE 2.8   **U.S. Beef Cow Operations and Inventory**

Herd Size (No. Cows)	Operations		Inventory	
	Number (thou)	Percent of Total	No. Cows (mil)	Percent of Total
1–49	727.7	80.1	11.0	31.0
50–99	105.5	11.6	6.8	19.2
100–499	70.2	7.7	12.6	35.6
500+	5.8	0.6	5.0	14.2
	Total 909.2		**Total** 35.3	

Source: USDA (*Agricultural Statistics*, 1995–96.)

excessively finished. Larger-framed, later-maturing cattle usually are more efficient and profitable if they go directly to the feedlot after weaning.

Feedlot Cattle Production.   Feedlot cattle are fed in small pens or fenced areas, where harvested feed is brought to them. Some cattle are finished for market on pasture, but they represent only 10–15% of the slaughter steers and heifers. They are sometimes referred to as *nonfed cattle* because they are fed little, if any, grain or concentrate feeds.

The cattle-feeding areas in the United States (Fig. 2.7) correspond to the primary feed-producing areas where cultivated grains and roughage are grown.

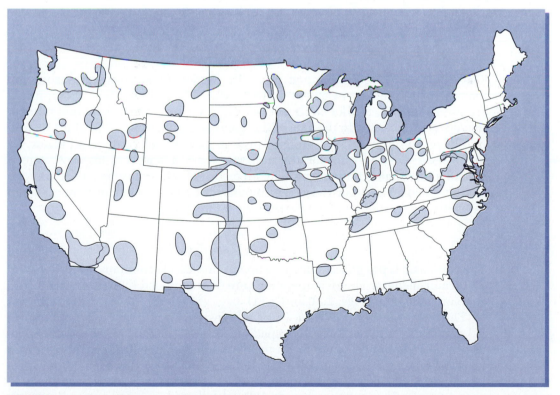

FIGURE 2.7   Cattle-feeding areas in the United States. The areas (in blue) represent location but not volume of cattle fed or backgrounded. Courtesy of the USDA.

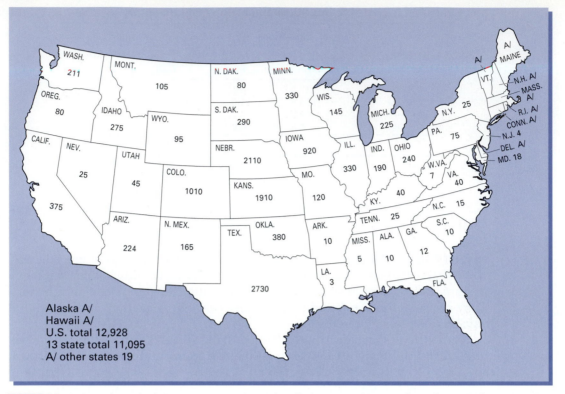

FIGURE 2.8 U.S. cattle on feed, January 1, 1994 (1,000 head). Courtesy of Livestock Marketing Information Center, Lakewood, Colo.

These locations are determined primarily by soil type, growing season, and amount of rainfall or irrigation water. Figure 2.8 shows where the approximately 26 million feedlot cattle are fed in the various states. The total number of fed cattle marketed in the 13 leading states is shown in Fig. 2.9. The 23.8 million head represent 90% of the 26.4 million head of fed cattle in all states. By contrasting Figs. 2.8 and 2.9, the number of cattle marketed for each state is considerably higher than the number on feed. This is because most commercial feedlots feed more than twice the number of cattle of their one-time feedlot capacity.

Cattle Feeding

The two basic types of cattle-feeding operations are (1) **commercial feeders** and (2) **farmer-feeders.** The two operations are distinguished by type of ownership and size of feedlot (Figs. 2.10 and 2.11).

The farmer-feeder operation is usually owned and operated by an individual or a family and has a feedlot capacity of under 1,000 head. The commercial feedlot is sometimes owned by an individual or partnership, but more often a corporation owns it, especially as feedlot size increases. It has a feedlot capacity of over 1,000 head. Approximately 87% of fed cattle are fed in feedlots with over 1,000-head capacity, whereas 13% of fed cattle are fed in feedlots with under 1,000-head capacity. There continues to be an increased marketing of fed

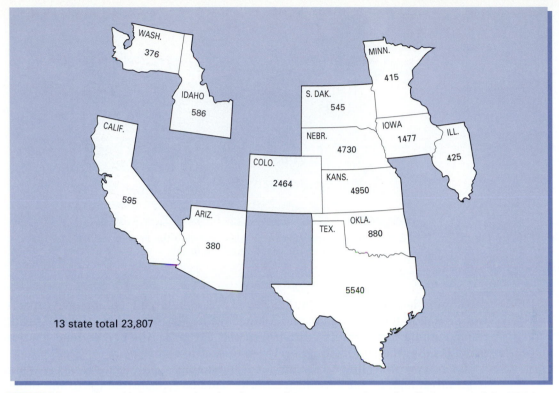

WASH.
376

MINN.
415

IDAHO
586

S. DAK.
545

IOWA
1477

CALIF.

NEBR.
4730

ILL.
425

COLO.
2464

KANS.
4950

595

ARIZ.
380

OKLA.
880

TEX.

5540

13 state total 23,807

FIGURE 2.9 Number of fed cattle marketed in the United States in 1995 (1,000 head). Courtesy of the USDA.

FIGURE 2.10 A large commercial feedlot where thousands of cattle can be easily fed in fenceline bunks using feed trucks. Courtesy of *BEEF*.

FIGURE 2.11   A farmer feedlot. In this operation the feed is fed in troughs located within the pen. Courtesy of *BEEF*.

cattle from larger feedlots. A number of U.S. commercial feedlots have capacities of 40,000 head or higher, and a few have capacities of over 100,000 head. Some commercial feedlots *custom-feed* cattle; that is, the commercial feedlot provides the feed and feeding service to the cattle owner.

Each type of feeding operation has its advantages and disadvantages. What is an advantage to one type of feedlot is usually a disadvantage to the other type. The large commercial feedlot usually enjoys economic advantages associated with its size as well as professional expertise in nutrition, health, marketing, and financing. The farmer-feeder has the advantages of distributing labor over several enterprises by using high-roughage feeds effectively, creating a market of homegrown feeds through cattle, and more easily closing down the feeding operation during times of unprofitable returns.

Chapters 22 and 23 examine the breeding, feeding, and management of beef cattle in more detail.

THE DAIRY CATTLE INDUSTRY

Global Perspective

Milk and milk products are produced and consumed by most countries of the world. Although buffalo, goat, and other types of milk are important in some areas, our focus here is on cows' milk and its products.

Table 2.9 shows world dairy cattle numbers, fluid milk production, and per-capita consumption of milk. Of the 225 million dairy cows in the world, India leads all other countries with its 30 million. The United States has the highest total fluid milk production. The average annual world production per cow is 4,700 lb, with production in Israel the highest at nearly 19,000 lb. Countries with high milk-production levels per cow have excellent breeding, feeding, and health-management programs.

World butter production is 14 billion lb. The former Soviet Union, the United States, and Germany are the leading countries in total butter production, while New Zealand, France, and Germany have the highest per-capita consumption of butter.

The leading countries in cheese production are the United States, France, and the former Soviet Union, with all countries producing 24 billion lb of cheese. Greece, France, and Italy have the highest per-capita consumption of cheese.

Milk is produced worldwide but dairy trade is dominated by manufactured dairy products. Dairy products traded internationally account for about 5% of the global milk production on a milk-equivalent basis. Trade restrictions keep trade volume and prices lower than if trade was more liberalized. Most developed countries, including the United States, extensively regulate their dairy industries by subsidizing production and, often, exportation.

Table 2.10 shows the leading countries and total world trade for cheese and butter. Nonfat dry milk is another milk product produced in large quantities—world production is 7 billion lb. France, Germany, and the former Soviet Union produce the most nonfat dry milk.

United States

The U.S. dairy industry has changed dramatically since the days of the family milk cow. Today it is a highly specialized industry that includes the production, processing, and distribution of milk. A large investment is required in cows, machinery, barns, and milking parlors where cows are milked. Dairy operators who produce their own feed need additional money for land on which to grow the feed. They also require machinery to produce, harvest, and process the crops.

Although the size of a dairy operation can vary from less than 30 milking cows to more than 5,000 milking cows (Fig. 2.12), the average U.S. dairy has approximately 100 milking cows, 30 dry cows, 30 heifers, and 25 calves. Average dairy producers farm 200–300 acres of land, raise much of the forage, and market the milk through cooperatives, of which they are members. The producers sell about 4,100 lb of milk daily, or about 1.5 million lb annually, valued at about $200,000. Their average total capital investment may exceed $500,000. The average dairy producer has a partnership (with a family member or another person) to make the management of time and resources easier.

Nearly half of the U.S. dairy herd was concentrated in large dairy farms (with 100 or more milk cows) in 1993. These large dairies represented just

TABLE 2.9   World Dairy Cattle Numbers, Milk Production, and Consumption

Country	No. Dairy Cattle (mil head)	Country	Fluid Milk Production (bil lb)	Country	Per-Capita Fluid Milk Consumption (lb)
1. India	30	1. United States	153	1. Ireland	421
2. Brazil	20	2. Russian Fed.	97	2. Sweden	362
3. Russian Fed.	20	3. India	66	3. Poland	341
4. United States	10	4. Germany	62	4. Finland	310
5. Ukraine	8	5. France	55	5. Austria	310
World total	225	**World total**	1,009	**World average**	186

Sources: USDA, FAS (*World Livestock Situation*); 1994 *FAO Production Yearbook*.

TABLE 2.10   World Dairy Trade

Exports			
Cheese		Butter	
Country	(mil lb)	Country	(bil lb)
1. Netherlands	1.1	1. New Zealand	0.5
2. France	0.7	2. Netherlands	0.4
3. Germany	0.6	3. Belgium	0.3
4. Denmark	0.5	4. Ireland	0.3
5. New Zealand	0.2	5. France	0.2
World total	4.4	**World total**	2.6

Imports			
Cheese		Butter	
Country	(mil lb)	Country	(bil lb)
1. Germany	0.8	1. USSR (former)	0.6
2. Italy	0.6	2. Germany	0.3
3. United Kingdom	0.4	3. Belgium	0.2
4. Japan	0.3	4. Netherlands	0.2
5. France	0.2	5. United Kingdom	0.2
World total	3.9	**World total**	2.0

Sources: USDA, ERS (*World Livestock Situation*).

13.6% of all U.S. farms with milk cows, but they were responsible for about 50% of total milk production. New technologies have required extensive capital investment that is most feasible for large dairy operations. Since 1977, farms with fewer than 30 milk cows have declined continuously as a share of all farms with milk cows. The share of farms with 30–49 milk cows gradually increased until 1990, but then began a slow decline. The share of farms with 50 or more milk cows increased in recent years, with farms having 100 or more milk cows increasing most in both number and share of all farms with milk cows. The largest farms are increasing most in the West and Southwest. The traditional milk-producing states of the Northeast and Lake States have seen their share of milk production become stable and then decline in recent years.

One way to stress the importance of dairying in the United States is to examine the number of cows in the states (Fig. 2.13). The five leading states in thousands of dairy cows are, respectively, Wisconsin (1,490), California (1,254), New York (703), Pennsylvania (642), and Minnesota (599). The five leading states in milk production per cow are, respectively, California (19,425 lb), Washington (19,377 lb), New Mexico (19,272 lb), Arizona (18,402 lb), and Colorado (18,175 lb). And the five leading states in total pounds of milk produced (given in million pounds) are, respectively, Wisconsin (23.0), California (22.9), New York (11.4), Pennsylvania (10.2), and Minnesota (9.7).

As Fig. 2.14 shows, today's 9.5 million dairy cows are approximately one-third the number of cows 50 years ago, yet total milk production continues to increase due to increased production per cow. This marked improvement is the result of effective breeding, feeding, health, and management programs.

Factors affecting dairy cattle productivity and profitability are covered in Chapters 24 and 25.

FIGURE 2.12 Dairies in the United States continue to increase in herd size. Many dairy cow herds are intensively managed in excellent facilities. Courtesy of Colorado State University.

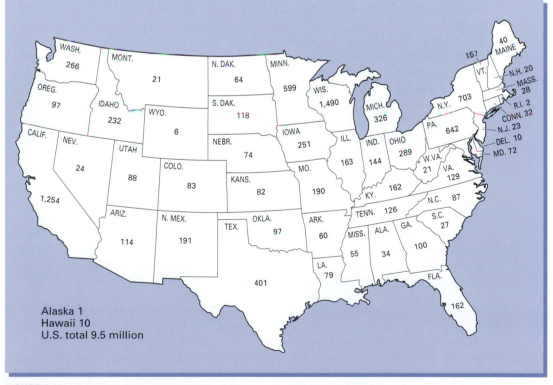

FIGURE 2.13 U.S. milk cow numbers, 1995 (1,000 head). Source: USDA.

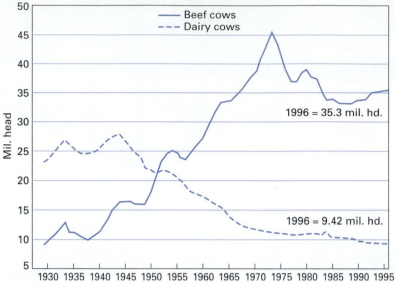

FIGURE 2.14 U.S. beef cow and dairy cow numbers. Courtesy of the USDA.

THE HORSE INDUSTRY

Global Perspective

The horse was domesticated about 5,000 years ago. It was one of the last farm animals to be domesticated. Horses were first used as food, then for war and sports, and also for draft purposes. They were used for transporting people swiftly and for moving heavy loads. In addition, horses became important in farming, mining, and forestry.

The donkey, descended from the wild ass of Africa, was domesticated in Egypt prior to the domestication of the horse. Donkeys and zebras are in the same genus (*Equus*), but are different species from horses (see Fig. 14.9, Chapter 14). Horses mate with donkeys and zebras, but the offspring produced are sterile.

Table 2.11 shows the world numbers of horses (58.2 million), donkeys (43.8 million), and mules (15.0 million). The five leading countries are given for each of the three species, with China having the most of all three. Brazil, Mexico, and the United States each have approximately 3–6 million head of horses.

Horses have been companions for people since their domestication. Once important in wars, mail delivery, farming, forest harvesting, and mining, the horse today is used in shows, in racing, in the handling of livestock, and for companionship, recreation, and exercise.

United States

There were no horses on the North American continent when Columbus arrived. However, there is fossil evidence that the early ancestor of the horse was here some 50–60 million years ago. The eohippus (or dawn horse), a four-toed animal less than a foot high, is believed to be the oldest relative of the

TABLE 2.11   World Horse, Donkey, and Mule Numbers

Horses		Donkeys		Mules	
Country	(mil head)	Country	(mil head)	Country	(mil head)
1. China	9.9	1. China	10.9	1. China	5.5
2. Mexico	6.2	2. Ethiopia	5.2	2. Mexico	3.2
3. Brazil	5.8	3. Pakistan	3.9	3. Brazil	2.1
4. United States	3.9	4. Mexico	3.2	4. Columbia	0.6
5. Argentina	3.4	5. Iran	1.9	5. Ethiopia	0.6
World total	58.2	**World total**	43.8	**World total**	15.0

Source: 1994 *FAO Production Yearbook.*

horse. Through evolutionary changes, the mesohippus (about the size of a collie dog) was believed to have foraged on the prairies of the Great Plains.

The early ancestors of the horse disappeared in pre-Columbian times, supposedly by crossing from Alaska into Siberia. It is from these animals that horses may have evolved in Asia and Europe. The draft horses and Shetland ponies developed in Europe, whereas the lighter, more agile horses developed in Asia and the Middle East. The Spaniards and colonists introduced modern horses to the Americas.

In the early 1900s, there were approximately 25 million horses and mules in the United States. Shortly after World War I, however, horse numbers began a rapid decline. The war stimulated the development and use of motor-powered equipment, such as automobiles, trucks, tractors, and bulldozers. Railroads were heavily used for transporting people and for moving freight long distances. By the early 1960s, horses and mules had declined to a mere 3 million in the United States.

With the shorter work week and greater affluence of the working class, there has been more time and money available for recreation. An estimated 1 million American horse owners are involved in the industry. Pleasure riding is the main contribution made by horses, though many people think of horse racing as more important. Attendance at U.S. racetracks exceeds 69 million people and annual wagering exceeds $13 billion.

The U.S. horse industry employs thousands of people and gives pleasure to millions. It also generates more than $15 billion annually, including income for salaries and services; taxes on pari-mutuel betting; income from the breeding, showing, and selling of horses; expenditures for feeds and medicines; and investments in the animals, land, and facilities. Horse shows generate $223 million per year and rodeos $104 million; foreign sales are $200 million annually. On a state level, California's horse industry generates the most dollars with a GNP of $2 billion annually, followed by New York's $1.3 billion, and Texas's $1 billion.

The American Horse Council estimates that there were 5.25 million economically productive horses in the United States in 1987, with Texas (478,000) and California (389,000) the leading states in horse numbers. The American Veterinary Medical Association estimates some 6.6 million equine in the United States in 1988. Approximately 85% of the horses are used for pleasure, not for racing or work. Horses used for working cattle are estimated at 400,000–500,000 head.

Horse numbers in the United States increased rapidly from 1960–76, with slower growth after that time period. Horse numbers have declined since the mid-1980s when applicable tax laws were passed by Congress. That removed some of the advantages of horse ownership. Horse numbers are stabilized at the present time. Even though horse numbers are lower than in previous decades, there are more individual horse owners who demand more expensive horses, and they spend more to take care of the horses.

Horses that have served their productive usefulness are usually slaughtered. Their meat is consumed by both humans and pets. Approximately 200,000 horses are slaughtered annually in the United States with some of the horse-meat being exported (Table 2.3).

Horse breeding, feeding, and management practices are presented in Chapters 30 and 31.

THE POULTRY INDUSTRY

Global Perspective

The term *poultry* applies to chickens, turkeys, geese, ducks, pigeons, peafowls, and guineas. Chickens, ducks, and turkeys dominate the world poultry industry. In parts of Asia, ducks are commercially more important than broilers (young chickens), and in areas of Europe, there are more geese than other poultry because they are more economically important.

Chickens originated in Southeast Asia and were kept in China as early as 1400 B.C. Charles Darwin concluded in 1868 that domestic chickens originated from the Red Junglefowl, although three other Junglefowl species were known to exist.

Table 2.12 shows the leading countries in world poultry numbers, production, and consumption. China has the highest poultry numbers, while the United States leads in poultry production and consumption. World hen egg production is approximately 79.4 billion pounds with the leading countries the same as poultry numbers in Table 2.12. Eggs are used for both hatching and human consumption.

France and the United States export the largest amounts of poultry, while Germany, Japan, and the former Soviet Union are the leading importers (Table 2.13).

World governments influence world poultry trade by controlling production and pricing and by placing tariffs on incoming goods—all barriers to free international trade. Trade liberalization by countries with industrial market economies would likely increase the trade of and decrease the price of poultry meat. Countries with efficient producers (such as the United States, Brazil, and Thailand), combined with consumers from countries with considerable trade protection (such as Japan, Canada, and 15 countries in the European Union), would benefit most from liberalized trade.

Poultry is the fastest-growing source of meat for people. The industrialized countries produce and export more than 50% of the poultry meat in the world.

United States

The annual U.S. income from broilers, turkeys, and eggs exceeds $17 billion (broilers, $11 billion; eggs, $3.8 billion, and turkeys, $2.7 billion). Figure 2.15

TABLE 2.12   World Poultry Numbers, Production, and Consumption

Country	No. Poultry (mil head)	Country	Production (bil lb)[a]	Country	Per-Capita Consumption (lb)[a]
1. China	3,136	1. United States	29.4	1. United States	96
2. United States	1,624	2. China	14.2	2. Hong Kong	89
3. Indonesia	667	3. Brazil	7.7	3. Israel	83
4. Brazil	696	4. France	4.2	4. Singapore	78
5. Russian Fed.	623	5. Russian Fed.	3.0	5. Saudi Arabia	70
World total	12,927[b]	**World total**	108.1	**World average**[c,d]	19

[a] Ready-to-cook equivalent.
[b] Includes 12.0 million chickens, 0.7 million ducks, and 0.3 million turkeys.
[c] Not all countries included.
[d] Estimated by dividing world production by world population.
Source: USDA, FAS (*World Poultry Situation*); 1994 *FAO Yearbook.*

TABLE 2.13   World Poultry Meat Trade

Exports		Imports	
Country	(mil lb)[a]	Country	(mil lb)[a]
1. United States	1,648	1. Germany	1,100
2. France	1,375	2. Japan	895
3. Netherlands	979	3. Hong Kong	686
4. Brazil	825	4. Saudi Arabia	550
5. Thailand	363	5. Netherlands	374
World total[b]	7,185	**World total**[b]	6,019

[a] Ready-to-cook equivalent.
[b] Not all countries included.
Source: USDA, FAS (*World Poultry Situation*).

shows the leading states in broiler, turkey, and egg production. The U.S. poultry industry is concentrated primarily in the southeastern area of the country.

From 1900–40, the primary concerns of the U.S. poultry industry were egg production by chickens and meat production by turkeys and waterfowl. Meat production by chickens was largely a by-product of the egg-producing enterprises. The broiler industry, as it is known today, was not yet established. Egg production was well established near large population centers, but the quality of eggs was often low because of seasonal production, poor storage, and the absence of laws to control grading standards.

Before 1940, large numbers of small farm flocks existed in the United States, but the management practices applied today were practically unknown. The modern mechanized poultry industry of the United States emerged during the late 1950s as the number of poultry farms and hatcheries decreased and the number of birds per installation dramatically increased. Larger cage-type layer operations appeared, and egg-production units grew, with production geared to provide consumers with eggs of uniform size and high quality. Huge broiler farms (Fig. 2.16) that provided consumers with fresh meat throughout the year were established, and large dressing plants capable of dressing 50,000 or more broilers a day were built. The U.S. poultry industry was thus revolutionized.

One of the most striking achievements of the poultry industry is its increased production of eggs and meat per hour of labor. Poultry operations

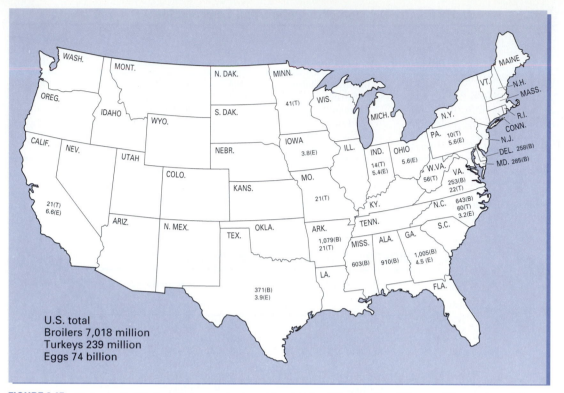

WASH.

MONT.

N. DAK.

MINN.

MAINE

OREG.

IDAHO

WYO.

S. DAK.

WIS.
41(T)

VT.

N.H.

MASS.

N.Y.

MICH.

CALIF.

NEV.

UTAH

NEBR.

IOWA
3.8(E)

ILL.

IND.
14(T)
5.4(E)

OHIO
5.6(E)

PA.
10(T)
5.6(E)

R.I.

CONN.

N.J.

DEL. 258(B)

MD. 285(B)

COLO.

MO.
21(T)

W.VA.
56(T)

VA.
253(B)
22(T)

21(T)
6.6(E)

KANS.

KY.

N.C.
643(B)
60(T)
3.2(E)

ARIZ.

N. MEX.

OKLA.

ARK.
1,079(B)
21(T)

TENN.

S.C.

TEX.

LA.

MISS.
603(B)

ALA.
910(B)

GA.
1,005(B)
4.5(E)

371(B)
3.9(E)

FLA.

U.S. total
Broilers 7,018 million
Turkeys 239 million
Eggs 74 billion

FIGURE 2.15   States in the United States producing at least 250 million broilers (B), 10 million turkeys (T), or 3 billion eggs (E) in 1994. Source: USDA.

FIGURE 2.16   Modern broiler houses are ventilated facilities where broilers are raised from chicks to maturity. Each house holds thousands of broilers. Courtesy of Grant Heilman Photography, Inc.

with 1 million birds at one location are not uncommon. Labor reduction has been accomplished by automatic feeding, watering, egg collecting, egg packing, and manure removal.

Dramatic changes came between 1955 and 1975 with the introduction of integration. It began in the broiler industry and is applied to a lesser degree in egg and turkey operations. *Integration* brings all phases of an enterprise under the control of one head, frequently the corporate ownership of breeding flocks, hatcheries, feed mills, raising, dressing plants, services, and marketing and distribution of products (Fig. 2.17). After 1975, poultry entered a new period of consolidation, in which vertically integrated companies purchased, acquired, or merged with each other, creating a small number of superintegrated companies.

The actual raising of broilers is sometimes accomplished on a contract basis between the person who owns the houses and equipment and who furnishes the necessary labor, and the corporation that furnishes the birds, feed, field service, dressing, and marketing. Payment for raising birds is generally based on a certain price for each bird reared to market age. Bonuses are usually paid to those who do a commendable job of raising birds. An important advantage to the integration system is that all phases are synchronized to ensure the utmost efficiency.

Broiler production is concentrated in the states of Arkansas, Georgia, Alabama, North Carolina, Mississippi, Texas, Maryland, Delaware, California, and Virginia. California is the only state with large production located outside the

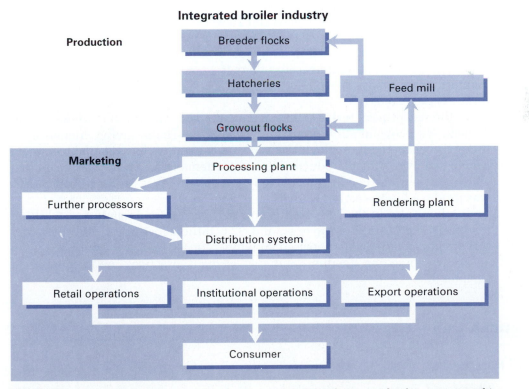

FIGURE 2.17 Most of the activities in producing and marketing broilers are under the same ownership and management. This integrated system produces broilers very efficiently. Courtesy of CSU Graphics.

"broiler belt." These 10 states produce nearly 85% of the total number of broilers in the United States.

Broiler production has increased tremendously over the past three decades, from 3.7 billion lb of ready-to-cook broilers in 1960 to 24 billion lb in 1995. The U.S. broiler industry is concentrating its growout operations into fewer but larger farms.

Although the number of laying hens in the United States has declined over the years, egg production is higher because of improved performance of the individual hen. In 1880, for example, the average laying hen produced 100 eggs per year; in 1950, the average was 175; in 1986, it increased to 250; and in 1993 it was 254 eggs.

The USDA estimates that 85% of the market eggs produced in the United States are from large commercial producers (those that maintain 1–11 million birds). The 45 largest egg-producing companies in the United States have more than 97 million layers, which is 35.5% of the nation's total.

Important changes have also occurred in U.S. turkey production. Fifty years ago, turkeys were raised in small numbers on many farms; today, they are raised in larger numbers on fewer farms. The average turkey farmer is producing well over 50,000 turkeys per year. The leading turkey-producing states are North Carolina, Minnesota, California, Arkansas, and Missouri.

Most of the turkeys produced today are the heavy or large, broad-breasted white type. At one time, the fryer-roaster-type turkey (5–9 lb) was popular; it bridged the gap between the turkey and the broiler chicken. Today, the broiler industry has taken over the fryer-roaster market because it can produce fryer-roaster chickens at a much lower price.

It takes 16 weeks for turkey hens and 19 weeks for turkey toms to reach market size. Because there is a heavy demand for turkey meat in November and December, turkey eggs are set in large numbers in April, May, and June. However, eggs are set year-round because there is a constant, though lower, demand for fresh turkey meat throughout the year. The normal incubation period for turkey eggs is 28 days.

In the recent past, turkey was considered a seasonal product by most consumers, with consumption occurring primarily at Thanksgiving. Increasingly, turkey is consumed throughout the year, and consumers have available to them a large variety of turkey products. The greater demand for turkey has meant that approximately 1 billion lb more of ready-to-cook turkey is produced today than was produced in the early 1980s.

Product sales in the U.S. poultry industry are approximately $11.4 billion for broilers, $3.8 billion for eggs, and $2.6 billion for turkeys. Broilers and turkeys are experiencing increased growth while egg sales are declining.

Poultry breeding, feeding, and management practices are presented in Chapter 32.

THE SHEEP AND GOAT INDUSTRY

Global Perspective

Sheep and goats are closely related, with both originating in Europe and the cooler regions of Asia. Sheep are distinguished from goats by the absence of a beard, less odor (males only), and glands in all four feet. The horns are spiral in different directions; goat horns spiral to the left, while sheep horns spiral to the right (like a corkscrew).

Sheep and goats are important ruminants in temperate and tropical agriculture. They provide fibers, milk, hides, and meat, making them versatile and efficient, especially for developing countries. Sheep and goats are better adapted than cattle to arid tropics, probably because of their superior water and nitrogen economy. Cattle, sheep, and goats can be grazed together because they utilize different plants. Goats graze **browse** (shrubs) and some **forbs** (broadleafed plants), cattle graze tall grasses and some forbs, and sheep graze short grasses and some forbs. Many of the forbs in grazing areas are broadleaf weeds.

More than 60% of all sheep are in temperate zones and fewer than 40% are in tropical zones. Goats, though, are mostly (80%) in the tropical or subtropical zones (0–40°N). Temperature and type of vegetation are the primary factors encouraging sheep production in temperate zones and goat production in tropical zones.

Sheep originated in the dry, alternately hot-and-cold climate of Southwest Asia. To succeed in tropical areas, then, sheep had to adapt the abilities to lose body heat, resist diseases, and survive in an adverse nutritional environment. For sheep to lose heat easily, they require a large body surface-to-mass ratio. Such a sheep is a small, long-legged animal. In addition, a hairy coat allows ventilation and protects the skin from the sun and abrasions. In temperate areas, sheep with large, compact bodies, a heavy fleece covering, and storage of subcutaneous fat have an advantage. Sheep can also constrict or relax blood vessels to the face, legs, and ears for control of heat loss.

The productivity of sheep is much greater in temperate areas than it is in tropical environments. This difference is the result not only of a more favorable environment (temperature and feed supply) but also of a greater selection emphasis on growth rate, milk production, lambing percentage, and fleece weight. Sheep from temperate environments do not adapt well to tropical environments; therefore, it may be more effective to select sheep in the production environment, rather than introduce sheep from different environments.

World sheep numbers were 1.1 billion head in 1994. Australia, China, New Zealand, and Iran are the leading sheep countries (Table 2.14). New Zealand, Australia, and Saudi Arabia have the highest per-capita consumption of mutton, lamb, and goat meat.

Countries highly involved in the export and import of lamb and mutton are shown in Table 2.15. New Zealand and Australia are the leading exporters, while France and the United Kingdom are the top importers.

The 526 million head of goats in the world are concentrated primarily in India and China, with Pakistan, Nigeria, and Bangladesh also having large goat populations (Table 2.16). In many countries, goats are important for both milk and meat production. Table 2.16 shows the leading countries in goat meat and milk production.

United States

The number of sheep in the United States reached a high of 56 million in 1942, thereafter declining to approximately 9 million in 1996 (Fig. 2.18). United States sheep are located on about 111,000 different operations.

Although the price received for lambs increased from $0.15–$0.70 per pound between 1961 and 1985, this increase has been largely offset by

TABLE 2.14 **World Sheep Numbers, Production, and Consumption**

Country	No. Sheep (mil head)	Country	Production (bil lb)[a]	Country	Per-Capita Consumption (lb)[a]
1. Australia	133	1. China	1.8	1. New Zealand	55
2. China	112	2. Australia	1.4	2. Australia	40
3. New Zealand	50	3. New Zealand	1.0	3. Saudi Arabia	34
4. Iran	45	4. Russian Fed.	0.8	4. Greece	32
5. India	44	5. Turkey	0.7	5. Ireland	18
World total	1,087	**World total**	15.1	**World average**[b]	4

[a] Carcass weight of lamb and mutton.
[b] Estimated by dividing world production by world population.
Sources: USDA, ERS; 1994 *FAO Production Yearbook.*

TABLE 2.15 **World Lamb, Mutton, and Goat Meat Trade**

Exports		Imports	
Country	(mil lb)[a]	Country	(mil lb)[a]
1. New Zealand	979	1. France	341
2. Australia	594	2. United Kingdom	275
3. United Kingdom	196	3. Japan	246
4. Ireland	147	4. Germany	90
5. Korea	44	5. USSR (former)	70
World total	2,105	**World total**	1,419

[a] Carcass weight.
Source: USDA, FAS, *World Livestock Situation.*

TABLE 2.16 **World Goat Numbers, Meat Production, and Goat Milk Production**

Country	No. Goats (mil head)	Country	Goat Meat Production (bil lb)[a]	Country	Milk Production (bil lb)
1. India	118	1. China	1.6	1. India	5.2
2. China	106	2. India	1.0	2. Iran	2.0
3. Pakistan	41	3. Pakistan	0.9	3. Pakistan	1.4
4. Bangladesh	28	4. Nigeria	0.3	4. Sudan	1.3
5. Nigeria	26	5. Iran	0.2	5. Greece	1.0
World total	609	**World total**	6.7	**World total**	23.1

[a] Carcass weight.
Sources: USDA, FAS (*World Livestock Situation*); 1994 *FAO Production Yearbook.*

increased production costs, scarcity of good sheep herders, and heavy losses from disease, predators, and adverse weather.

The number of ewes on U.S. farms in 1996 is shown in Fig. 2.19. The five leading states in ewe numbers are, respectively, Texas (1.08 million), Wyoming (0.46 million), California (0.39 million), South Dakota (0.33 million), and Utah (0.31 million). The Midwest and West Coast areas are important in farm-flock sheep production, whereas other western states are primarily range-flock areas. The midwestern states generally raise more lambs per ewe than the western states do, though Washington, Oregon, and Idaho resemble the midwestern states in this regard.

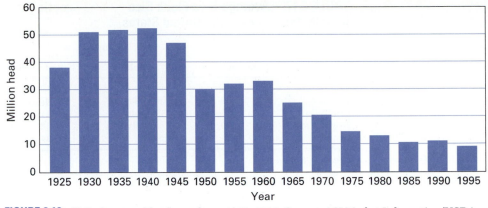

FIGURE 2.18 U.S. sheep and lamb numbers, 1925–1995. Source: ASI Market Information/USDA.

Table 2.17 shows average ewe flock sizes in selected states. Most of the decline in U.S. sheep numbers during the past 50 years has occurred in the West. However, this region, with its extensive public and private rangelands, still produces 80% of the sheep raised in the country today. Most U.S. sheep growers have small flocks (50 or fewer sheep) and raise sheep as a secondary enterprise. About 40% of the West's sheep producers maintain flocks of more than 50 sheep; these flocks contain about 93% of the sheep in that region. Only

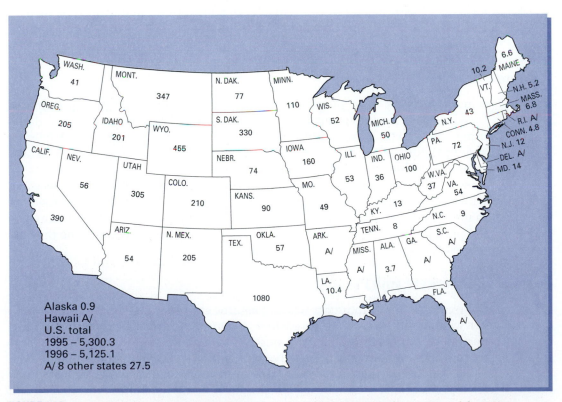

FIGURE 2.19 Number of ewes in the United States, January 1, 1996 (1,000 head). Courtesy of the USDA.

TABLE 2.17   Average Ewe Flock Size in Selected States in the United States

Range Flocks		Farm Flocks	
State	No. of Ewes	State	No. of Ewes
Wyoming	563	Minnesota	48
Arizona	389	Nebraska	43
Colorado	362	Michigan	41
New Mexico	309	Pennsylvania	38
Nevada	259	Iowa	34

Source: USDA.

about one-third of the operators in the West who have flocks of 50 or more sheep specialize in sheep; the other two-thirds have diversified livestock operations.

About 23% of all sheep born in the western United States and 13% of those born in the North-Central region are lost before they are marketed. In the West, predators, especially coyotes, and weather are the most frequent causes of death. In the North-Central region, the primary causes of death include weather and disease before docking and disease and attacks by dogs after docking.

The range operator produces some slaughter lambs if good mountain range is available, but most range-produced lambs are *feeders* (e.g., lambs that must have additional feed before they are slaughtered). Feeder lambs are fed by producers, sold to feedlot operators, or fed on contract by feedlot operators. Some feedlots have a capacity for 20,000 or more lambs on feed at one time, and many feedlots that have a capacity of 5,000 are in operation. The leading lamb-feeding states are Colorado, California, Texas, Wyoming, South Dakota, and Kansas.

Purebred Breeder

Some producers raise purebred sheep and sell **rams** for breeding. Purebred breeders have an important responsibility to the sheep industry: They determine the genetic productivity of commercial sheep. The purebred breeder should rigidly select breeding animals and then sell only those rams that will improve productivity and profitability for the commercial producer.

Records are essential to indicate which ram is bred to each of the ewes and which ewe is the mother of each lamb. All ram lambs are usually kept together to identify those that are desirable for sale. The ram lambs are often retained by the purebred breeder until they are 7–12 months of age, at which time they are offered for sale.

Commercial Market Lamb Producers

Commercial lamb producers whose pastures are productive can produce market wether and ewe lambs on pasture, whereas those whose pastures are less desirable produce feeder lambs. Lambs that are neither sufficiently fat nor large for market at weaning time usually go to feedlot operators, where they are fed to market weight and condition. However, most feedlot lambs are obtained from producers of range sheep. The producer of market lambs, whose feed and pasture conditions are favorable, strives to raise lambs that finish at 90–120 days of age, weighing approximately 120–130 lb each.

Commercial Feedlot Operator

Lambs are sent to the feedlot if they need additional weight prior to slaughter. The feedlot operator hopes to profit by increasing the value of the lambs per unit of weight and by increasing their weight. Lambs on feed that gain about 0.5–0.8 lb per head per day are considered satisfactory gainers. Feeder buyers prefer feeder lambs weighing 70–80 lb over larger lambs because of the need to put 20–30 lb of additional weight on the lambs to finish them. A feeding period of 40–60 days is usually sufficient for finishing healthy lambs.

Chapters 28 and 29 discuss in detail sheep breeding, feeding, and management practices. Chapter 33 covers goat production.

THE SWINE INDUSTRY

Global Perspective

Swine are widely distributed throughout the world, even though 40% of the world's pigs are located in China (Table 2.18). China far exceeds all other countries in number of pigs and total pork production. However, when pork production (carcass weight) is evaluated on a per-capita basis, then the United States exceeds China. Other countries where swine numbers and production are important include Germany, the Russian Federation, and Brazil (Table 2.18).

Table 2.19 shows the countries most highly involved in the export and import markets for pork. Several European countries, Japan, and Canada lead the world pork trade.

United States

Swine production in the United States is concentrated heavily in the nation's midsection known as the Corn Belt. This area produces most of the nation's

TABLE 2.18   World Swine Numbers, Production, and Consumption

Country	No. Swine (mil head)	Country	Production (bil lb)[a]	Country	Per-Capita Consumption (lb)[a]
1. China	403	1. China	74	1. Denmark	145
2. United States	58	2. United States	18	2. Sweden	121
3. Brazil	30	3. Germany	8	3. Poland	116
4. Russian Fed.	29	4. Russian Fed.	5	4. Czechoslovakia	113
5. Germany	26	5. Spain	5	5. Austria	112
World total	875	**World total**	174	**World average**[b]	31

[a] Carcass weight.
[b] Estimated by dividing world production by world population.
Sources: USDA, FAS (*World Livestock Situation*); 1994 *FAO Production Yearbook*.

TABLE 2.19   World Swine Trade

Exports		Imports	
Country	(bil lb)[a]	Country	(bil lb)[a]
1. Denmark	2.3	1. Germany	2.0
2. Netherlands	2.2	2. Japan	1.5
3. Belgium, Luxembourg	1.0	3. Italy	1.3
4. France	0.7	4. United Kingdom	1.0
5. Canada	0.7	5. France	1.0
World total	10.1	**World total**	9.9

[a] Carcass weight.
Source: USDA, FAS (*World Livestock Situation*).

corn, the principal feed used for swine. Many swine producers farm several hundred acres of land and have other livestock enterprises. The Corn Belt states of Iowa, Missouri, Illinois, Indiana, Ohio, northwest Kentucky, southern Wisconsin, southern Minnesota, eastern South Dakota, and eastern Nebraska produce nearly 75% of the nation's swine, with Iowa alone accounting for approximately 25% (Fig. 2.20). Since 1990, North Carolina has increased swine numbers faster than most other states and currently there are more than 8 million pigs in the state.

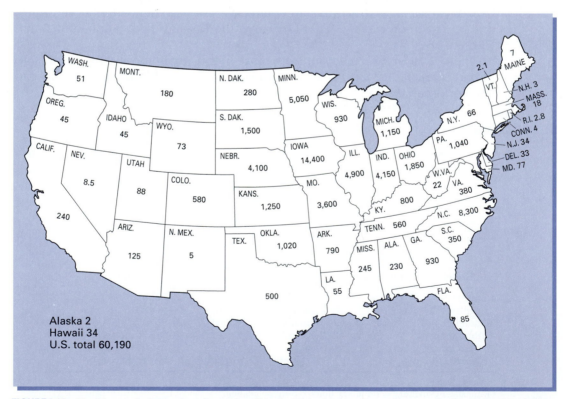

FIGURE 2.20   Total hog and pig numbers (1,000 head) in the United States, December 1, 1995. Courtesy of the USDA.

Figure 2.21 shows that during the past 20 years the number of hog farms has significantly decreased. However, during this same period of time, pork production has continued to increase. This implies that the number of hogs raised per farm has increased. Production costs, based on per hundred weight of hogs produced, decrease as number of hogs raised per farm increases (Fig. 2.22).

The cash receipts from hogs (Table 2.2) was nearly $10 billion. This amount received by producers expands into large amounts when economic activities beyond the farm gate are considered. For example, total hog-related activity in the U.S. economy is estimated at $66 billion of output, $23 billion of personal income, $26 billion of value-added, and 764,000 jobs are directly and indirectly linked to the pork industry in the United States.

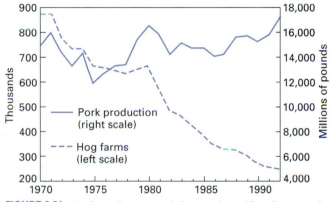

FIGURE 2.21   Pork production and the number of hog farms in the United States. Source: USDA.

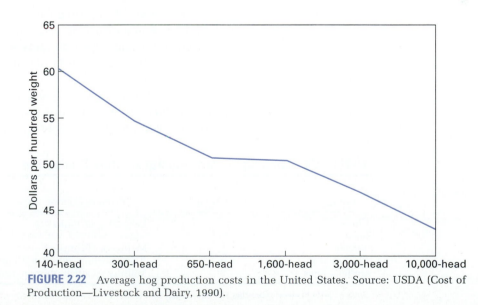

FIGURE 2.22   Average hog production costs in the United States. Source: USDA (Cost of Production—Livestock and Dairy, 1990).

Types of Swine Operations

There are four primary types of swine operations. In **feeder-pig production,** the producer maintains a breeding herd and produces feeder pigs for sale at an average weight of 40 lb. In **feeder-pig finishing,** the feeder pigs are purchased and then fed to slaughter weight. The **farrow-to-finish** operation maintains a breeding herd as pigs are produced and fed to market weight on the same farm. Finally, **seedstock** operations are similar to farrow-to-finish, except the salable product is primarily breeding boars and gilts.

The various swine operations and the management factors affecting them are discussed in Chapters 26 and 27.

OTHER ANIMAL INDUSTRIES

Aquaculture

Aquaculture is the farming or rearing of aquatic plants and animals. In a sense, it is underwater agriculture. Farming implies some form of intervention in the rearing process to enhance production such as stocking, feeding, and protection from predators. The most common aquatic animals raised in the United States include finfish (catfish, trout, and salmon), crustaceans (shrimp and crawfish), and mollusks (oysters and clams).

Global animal production from aquaculture approaches 15 million tons valued at approximately $30 billion U.S. dollars, and growth continues. Table 2.20 shows the leading countries.

Most of the world's wild-catch fisheries are already exploited at sustainable yield levels. Health-conscious consumers will most likely increase the demand for fish worldwide. Production increases to meet this new demand will most likely come from aquaculture because wild catches are expected to decline over the next decade.

Aquaculture in the United States exceeds 1.2 billion pounds with a value in excess of $600 million. Production has increased significantly over the past several years. Catfish farming is the largest aquaculture activity in the United States with annual sales of $350 million involving approximately 450 million pounds. Mississippi produces about 70% of the U.S. total. Other major catfish producing states are Alabama, Arkansas, and Louisiana.

Trout production is the second largest sector in the U.S. aquaculture industry, with Idaho accounting for approximately three-fourths of the $75 million annual sales. North Carolina has had increasing trout production sales.

TABLE 2.20   Leading Countries in Aquaculture Production

Country	% of World Total	Primary Commodities
China	35	Mollusks, shrimp, carp
Japan	20	Algae, mollusks, yellow tail
Taiwan	5	Mollusks, shrimp, eels
Philippines	3	Algae, shrimp, milkfish, tilapia
United States	3	Mollusks, catfish, salmon, trout

Source: FAO Fisheries Circular, No. 886, 1995.

The third largest sector in the U.S. aquaculture industry is crawfish production. Concentrated mostly in Louisiana, the industry produces around 25 to 50 million pounds a year.

There has been increased interest in producing other aquatic species, although most of them are in the experimental stages. Two species that have garnered publicity are hybrid striped bass and tilapia. Tilapia is a warm-water fish, native to Africa and the Middle East, but its production is expanding throughout the world. U.S. production in 1994 was 15 million pounds, up 25% from 1993.

One of the most attractive aspects of aquatic farming is its low feed conversion ratio. Because fish are suspended in water, they don't expend many calories to move about, and because they are cold-blooded, they are more efficient to produce than any of the major land-raised meats. Today, fish farmers are achieving average feed conversion ratios as low as 1.3.

U.S. aquaculture producers have a number of competitive advantages. First, the United States is one of the world's largest seafood markets, and the domestic live market is one that pays premium prices. Second, although most aquaculture firms are in rural areas, they still are close to good transportation facilities. Third, U.S. producers have access to a growing network of researchers and companies working on aquaculture. Fourth, the United States is a major producer of grains and has large supplies of livestock by-products for use in fish feeds, which can account for up to 50% of variable costs. Fifth, the United States has a wide variety of climates and long coastlines for marine aquaculture, in addition to large freshwater resources.

On the other hand, U.S. growers also have a number of disadvantages. First, many of the biggest farm-raised species are tropical and, with the exception of Hawaii, the United States does not have any tropical areas. Second, land costs are generally higher in the United States than in less developed countries, especially the coastal properties needed for marine aquaculture. Third, in most cases, labor costs are considerably higher in the United States than in competing countries. Fourth, many countries have fewer regulations controlling aquaculture production and processing practices.

Detailed aquaculture production practices can be obtained by reading the reference materials noted at the end of the chapter.

Ostrich and Emu Farming and Ranching

Ostrich and emu, native to Africa and Australia, respectively, belong to the ratite family of flightless birds. The ostrich is the largest living bird. A general description of emu and ostrich is provided in Table 2.21.

Emu farming in the United States is relatively new. In 1989, a national association was started in Texas by emu farmers.

TABLE 2.21　Productivity of the Emu and Ostrich

Name	Adult Size (male)		No. Eggs Per Year	Egg Weight (lb)	Products
	Height (ft)	Weight (lb)			
Emu	5–6	150	30–40	1–1½	Meat, leather, oil, feathers, carved eggs
Ostrich	up to 8	300	30–50	3–3½	Meat, hides, feathers, egg shells

INDUSTRY ISSUES AND CHALLENGES

There are several significant issues and challenges that the animal industries have encountered and will continue to face. These issues are discussed in Chapter 35.

CHAPTER SUMMARY

■ The animal industries are described by numbers, cash receipts, per-capita product consumption, and their contribution to world trade.

■ U.S. farm cash receipts in 1995 were $185.7 billion with livestock contributing $98.8 billion. Cash receipts for the major livestock species was cattle ($33.9 bil), dairy products ($19.9 bil), broilers ($11.8 bil), hogs ($10.1 bil), eggs ($4.0 bil), and turkeys ($2.8 bil).

■ The major animal industries are becoming more integrated and concentrated. The poultry industry is highly integrated and swine production units continue to become larger. In the beef cattle industry, four major packers slaughter most of the cattle; however, most of the cattle are produced in herd sizes of less than 100 cows.

■ Sheep numbers have shown a significant decline in the past 50 years. Dairy cow numbers have decreased; however, milk production has remained about the same. Milk production per cow continues to increase.

■ Aquaculture and other types of animal farming (e.g. ostrich and emu) are increasing in the United States.

REVIEW QUESTIONS

1. *True or False:* Agricultural products contribute to the trade deficit in the United States.
2. *True or False:* In some countries, the most important product of cattle is draft power and milk rather than meat.
3. What are the two reasons why world cattle numbers have continued to increase during the last 30 years?
4. What country has the greatest number of cattle?
5. What country has the greatest production of beef?
6. What are the three phases of beef cattle production?
7. From what segment of cow–calf production do most of the potential slaughter steers and heifers come from?
8. Seedstock cow–calf producers primarily produce what two products?
9. The primary basis of stocker–yearling operations is to market _____ _____.
10. What two segments comprise the feedlot beef industry?
11. *True or False:* Commercial feeders have greater production capacity (head of cattle) than farmer-feeders.

12. What country produces the most fluid milk?

13. What are the three dairy products produced throughout the world in the greatest quantity after fluid milk?

14. What country has the greatest number of horses, donkeys, and mules?

15. Why did the numbers of horses, donkeys, and mules in the United States decline after World War I?

16. Today, horses in the United States are used primarily for what purposes?

17. What is the leading country in number of poultry?

18. What is the leading country in production of poultry meat?

19. *True or False:* Most sheep are located in temperate zones, whereas goats occur overwhelmingly in tropical and subtropical zones.

20. What is the leading country for number of sheep?

21. What is the leading country for number of goats?

22. What is the leading country for goat meat production?

23. What is the leading country for number of swine and production of pork?

34. What are the four segments of the swine industry?

SELECTED REFERENCES

Publications

American Poultry History. 1996. Mount Morris, IL: Watt Publishing Co.

Directions: A 21st Century Vision. National Cattlemen, July, 1996. Englewood, CO: National Cattlemen's Association.

Economic Impact of the U.S. Pork Industry, 1994. Des Moines, IA: National Pork Producers' Council.

FAO Production Yearbook. Vol. 46, 1994. Rome: FAO.

Helming, B. 1990. Meat's decade of danger. *Meat and Poultry* 36:30.

Horse Industry Directory. 1996. Washington, D.C.: American Horse Council, Inc.

Majeskie, J. L. and Eastwood, B. R. 1990. Status of United States Dairy Cattle. National Cooperative Dairy Herd Improvement Program Handbook, Fact Sheet K-7.

Rankings of States and Commodities by Cash Receipts, 1995. Washington, DC: USDA.

Stillman, R., Crawford, T., and Aldrich, L. 1990. *The U.S. Sheep Industry.* Washington, D.C.: USDA, ERS.

Swinker, A. M. and Heird, J. C., 1994. *Horse Industry: Trends, Opportunities and Issues. Encyclopedia of Agricultural Science.* San Diego: Academic Press, Inc.

The Changing U.S. Pork Industry, 1993. *Economic Review* (2nd qtr.).

USDA. 1992. *Economic Indicators of the Farm Sector: Costs of Production— Livestock and Dairy.* Washington, D.C.: GPO.

USDA. 1996. *Foreign Agricultural Trade of the United States.* Washington, D.C.: GPO.

USDA. 1996. *Livestock and Poultry Situation and Outlook Report.* LPS39-41.

USDA. 1996. *World Livestock Situation.*

USDA, FAS, 1996. *World Dairy Situation.* FD1-90. Washington, D.C.: GPO.

U.S. Sheep Industry Market Situation Report. 1995–96. American Sheep Industry Association, Englewood, CO.

Watt Poultry Yearbook. 1996. Mount Morris, IL: Watt Publishing Co.

Visuals

An Introduction to the Poultry Industry (77 slides and audiotape; 12 min). Poultry Science Department, Ohio State University, Columbus, OH.

CHAPTER 3

Red Meat Products

Red meat products come primarily from cattle, swine, sheep, goats, and, to a lesser extent, horses and other animals. Poultry meat, sometimes called *white meat,* is discussed in Chapter 6. Pork has been advertised as "the other white meat"; however, it is considered a red meat.

Red meats are named according to their source: **beef** is typically from cattle over a year of age; **veal** is from calves 5 months of age or younger (veal carcasses are distinguished from beef by their grayish-pink color of the lean); **pork** is from swine; **mutton** is from mature sheep; **lamb** is from young sheep; **chevon** is from goats, but it is more commonly called **goat meat.**

PRODUCTION

World red meat supply is approximately 320 billion pounds, with China, the United States, and Brazil leading all other countries (Table 3.1). Buffalo meat comes from the true buffalo; it should not be confused with bison in the United States. The true buffalo has no hump; thus the name *buffalo* belongs only to water buffalo in Asia and the African buffalo.

Red meat production in the United States is shown in Fig. 3.1. Beef and pork comprise most of the annual production of nearly 40 billion lb. The total production has been relatively stable over the past several years. Some red meat has to be imported into the United States (Chapter 2, Table 2.3) to meet consumer demand and global trade agreements. Most of the imported meat is beef which is needed to fill the U.S. deficit in processing beef.

Figures 3.2, 3.3, and 3.4 show where cattle, hogs, and sheep, respectively, are slaughtered in the United States. Kansas, Nebraska, Texas, Colorado, and

TABLE 3.1   World Red Meat Supply

Product	Bil lb[a]	Leading Countries (bil lb)
Pork	174	China (74); United States (18); Germany (8); Russian Fed. (5)
Beef and veal	111	United States (25); Russian Fed. (7); Brazil (7); Argentina (6)
Mutton and lamb	15	China (1.8); Australia (1.4); New Zealand (1.0); Russian Fed. (0.8)
Goat meat	7	China (1.6); India (1.0); Pakistan (0.9); Nigeria (0.3)
Buffalo meat	6	India (2.6); Pakistan (1.0); China (0.6); Egypt (0.4)
Horsemeat	1	Mexico (0.18); China (0.13); Italy (0.13); Argentina (0.13)
World total	320	China (84); United States (44); Brazil (14); Russian Fed. (13)

[a] Carcass weight.
Source: 1994 *FAO Production Yearbook.*

Iowa are the leading states in cattle slaughter. Iowa slaughtered more than 30% of the nation's hogs in 1995, and Illinois, Minnesota, Nebraska, South Dakota, North Carolina, Michigan, Ohio, Indiana, Missouri, and Kentucky together slaughtered an additional 46%. The leading lamb-slaughtering states are Colorado, California, Texas, Iowa, Kansas, and Minnesota.

The United States has an annual supply of horsemeat of approximately 80 million lb. While most of the horsemeat goes into pet food, several thousand tons are shipped to Europe for human consumption. Several U.S. packing plants specialize in processing horses, shipping the meat to France, Belgium, the Netherlands, Switzerland, and Italy. Most of the U.S. horsemeat comes from horses that are no longer functional in their typical activities.

Meat animals are processed in packing plants located near areas of live animal production. This reduces the transportation costs because carcasses, wholesale cuts, and retail cuts can be transported at less cost than live animals. The top 10 meat packers in the United States are identified in Table 3.2.

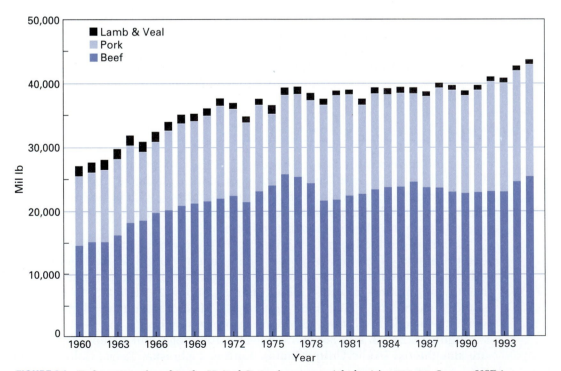

FIGURE 3.1   Red meat produced in the United States (carcass weight basis), 1960–95. Source: USDA.

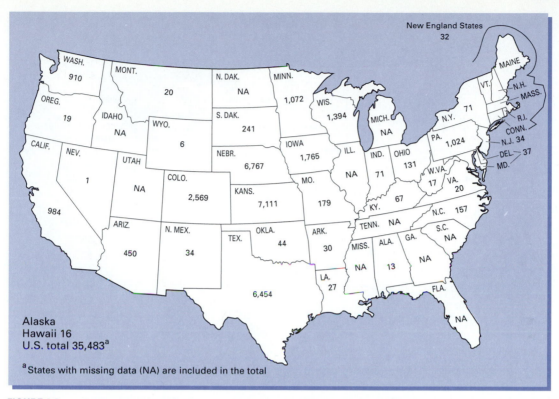

FIGURE 3.2   U.S. commercial cattle slaughter (1,000 head), 1995. Courtesy of USDA.

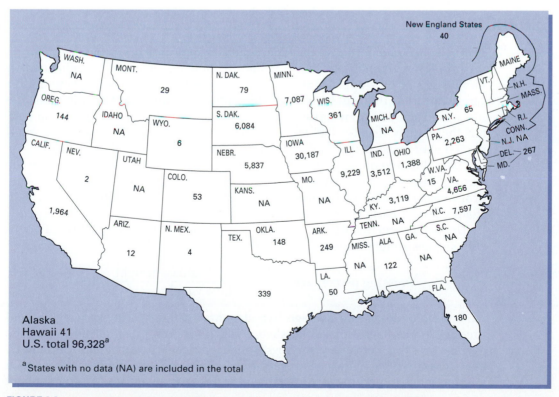

FIGURE 3.3   U.S. commercial hog slaughter (1,000 head), 1995. Courtesy of the USDA.

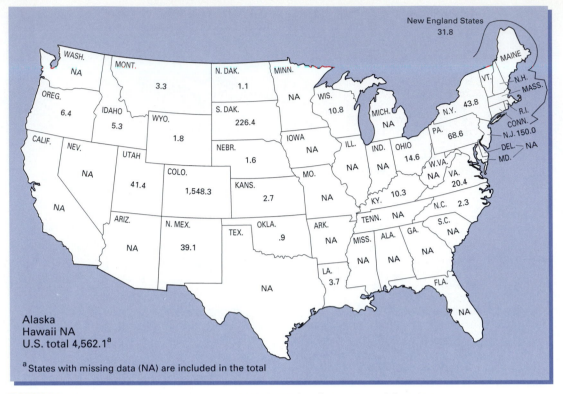

WASH.
NA

MONT.
3.3

N. DAK.
1.1

MINN.
NA

WIS.
10.8

MICH.
NA

New England States
31.8

MAINE

VT.

N.H.

MASS.

N.Y.
43.8

R.I.

OREG.
6.4

IDAHO
5.3

WYO.
1.8

S. DAK.
226.4

IOWA
NA

ILL.
NA

IND.
NA

OHIO
14.6

PA.
68.6

CONN.

N.J. 150.0

DEL. NA

MD.

CALIF.
NA

NEV.
NA

UTAH
41.4

COLO.
1,548.3

NEBR.
1.6

KANS.
2.7

MO.
NA

KY.
10.3

W.VA.
NA

VA.
20.4

N.C. 2.3

ARIZ.
NA

N. MEX.
39.1

OKLA.
.9

TEX.

ARK.
NA

TENN. NA

MISS.
NA

ALA.
NA

GA.
NA

S.C.
NA

LA.
3.7

FLA.
NA

NA

Alaska
Hawaii NA
U.S. total 4,562.1[a]

[a] States with missing data (NA) are included in the total

FIGURE 3.4   U.S. commercial sheep slaughter, 1995 (1,000 head). Courtesy of the USDA.

PRODUCTS

Animals are transported to packing plants, where they are processed. During the initial processing stage, the animals are made unconscious by using carbon dioxide gas or by stunning (electrical or mechanical). The jugular vein and/or carotid artery is then cut to drain the blood from the animal. After bleeding, the hides are removed from cattle and sheep. Hogs are scalded to remove the hair, but the skin is usually left on the carcass. A few packers (especially small plants) skin hogs, as it is more energy efficient than leaving the skin on.

After the hide or hair is removed, the internal organs are separated from the carcass. Those parts removed from the carcass are sometimes referred to as the **drop, viscera, offal,** or **by-products.** Typically these are the head, hide, hair, shanks (lower parts of legs and feet), and internal organs.

Dressing percentage (sometimes referred to as **yield**) is the relation of hot or cold carcass weight to live weight. It is calculated as follows:

$$\text{Dressing percentage} = \frac{\text{Hot or cold carcass weight}}{\text{Live weight}} \times 100$$

Table 3.3 shows the average live slaughter weights, carcass weights, and dressing percentages for cattle, sheep, and hogs. For practical purposes, the average dressing percentage for hogs is 72%; cattle, 60%; and sheep, 50%; however, dressing percentage can vary several percentage points within each species. The primary factors affecting dressing percentage are *fill* (contents of

TABLE 3.2   Leading Red Meat Packing and Processing Companies

Rank	Company (location)	1995 Sales ($ bil)	No. Plants	No. Employees	Products[a]
1	IBP (Dakota City, Nebr.)	$12.7	24	34,000	B-P
2	Cargill (Wichita, Kans.)	9.3	13	15,000	B-P
3	ConAgra (Omaha, Nebr.)	8.7	80	90,870	B-P-L
4	Monfort[b] (Greeley, Colo.)	7.8	35	90,000	B-P
5	Hardees (Rocky Mount, N.C.)	5.0	3	50,000	B-P
6	Sara Lee Corp. (Chicago, Ill.)	3.2	29	15,000	P
7	Hormel (Austin, Minn.)	3.1	12	10,600	P
8	Oscar Mayer (Madison, Wis.)	2.2	11	10,000	P
9	National Beef (Liberal, Kans.)	1.8	2	3,730	B
10	Smithfield Foods (Smithfield, Va.)	1.5	14	8,000	P

[a] B = beef; P = pork; L = lamb.
The top four companies, for each species, harvest 81% of the steers and heifers, 46% of the hogs, and 72% of the lambs.
[b] Owned by ConAgra.
Source: Meat Marketing and Technology, May 1996.

digestive tract), fatness, muscling, pregnancy status, weight of hide, and, in sheep, weight of wool.

Beef and pork carcasses are split down the center of the backbone (sheep carcasses are not split), giving two sides approximately equal in weight. After the carcasses are washed, they are put in a cooler to chill. The cooler temperature of approximately 28–32°F removes the heat from the carcass and keeps the meat from spoiling. Carcasses can be stored in coolers at 32–35° for several weeks; however, this is uncommon in large packing plants as profitability is dependent on processing live animals one day and shipping carcasses or products out the next day or two. Smaller packers may keep carcasses in the coolers for as long as 2 weeks. This allows the carcasses to age. This **aging** process increases meat tenderness through an enzyme activity that decreases the tensile strength of muscle fibers. Recent research has shown that $CaCl_2$ (calcium chloride) injection into beef carcasses improves tenderness.

TABLE 3.3   U.S. Livestock Slaughter and Meat Production, 1995

Species/Class	Meat Products	No. Head (mil)	Average Liveweight (lb)	Total Carcass Weight (bil lb)	Average Carcass Weight (lb)	Average Dressing Percentage
Cattle	Beef	35.6	1,172	25.1	705	60%
Calves[a]	Veal	1.4	363	0.3	215	60
Hogs	Pork	96.3	256	17.8	185	72
Sheep/lamb	Mutton/lamb	4.6	125	0.3	62	50

[a] Part of these are classified as calf carcasses, which are intermediate in age and size between veal and beef carcasses.
Source: USDA. AMI, *Meat and Poultry Facts.*

In the past, most red meat was shipped in carcass form, but today most packing plants process the carcasses into wholesale or primal cuts and even into retail cuts (Table 3.4; Color Plates A to D). The latter process involves shipping meat in boxes; thus the terms **boxed beef, boxed pork,** and **boxed lamb.**

Most red meats are marketed as fresh products. Vitamin E is fed to beef cattle to extend the shelf life of the fresh product. Vitamin E extends the length of time the beef will hold its cherry red color. Some meats are cured, smoked, processed, or canned. Table 3.5 identifies the forms in which major meat products are prepared for sale.

KOSHER AND MUSLIM MEATS

Both the Jewish and Muslim faiths have specific requirements for the slaughter of religiously acceptable animals. The primary difference from the general practices in the United States is that the animals are not stunned prior to slaughter, although a mild mechanical stunning has been approved by some Muslim authorities in certain countries.

All **kosher meat** comes from animals that have split hooves, that chew their cud, and that have been slaughtered in a manner prescribed by the Torah (Orthodox Jewish law). Meat from undesirable animals or from animals not properly slaughtered or with imperfections are called **nonkosher** (trefah). Red meats, poultry, and all other foods are classified as kosher or nonkosher. Foods that are kosher bear the endorsement symbolⓊ.

Kosher slaughter, inspection, and supervision are performed only by trained individuals approved by rabbinic authority. The trachea and esophagus of the animal must be cut in swift continuous strokes with a perfectly smooth, sharp knife causing instant death and minimum pain to the animal. Various internal organs are inspected for defects and adhesions.

Only the forequarters are identified as kosher, as the hindquarters contain the forbidden sciatic nerve, which is embedded deep in the muscle tissue and is difficult to remove. Hindquarters are sold as nonkosher meats.

The Torah forbids the eating of blood, including blood of arteries, veins, and muscle tissue. For this reason, certain arteries and veins must be removed from cattle; neck veins of poultry must be severed to allow the free flow of blood. Blood is extracted from meat by broiling and salting. There are specific steps required in proper broiling and salting to endorse the meat as kosher.

TABLE 3.4   Wholesale Cuts of Beef, Veal, Pork, and Lamb

Beef	Veal	Pork	Lamb
Round	Leg/round	Leg/ham	Leg
Sirloin	Sirloin	Loin	Loin
Short loin	Loin	Blade shoulder	Rib
Rib	Rib	Jowl	Shoulder
Chuck	Shoulder	Arm shoulder	Neck
Foreshank	Foreshank	Spareribs	Foreshank
Brisket	Breast	Side	Breast
Short plate			
Flank			

Source: The National Live Stock and Meat Board.

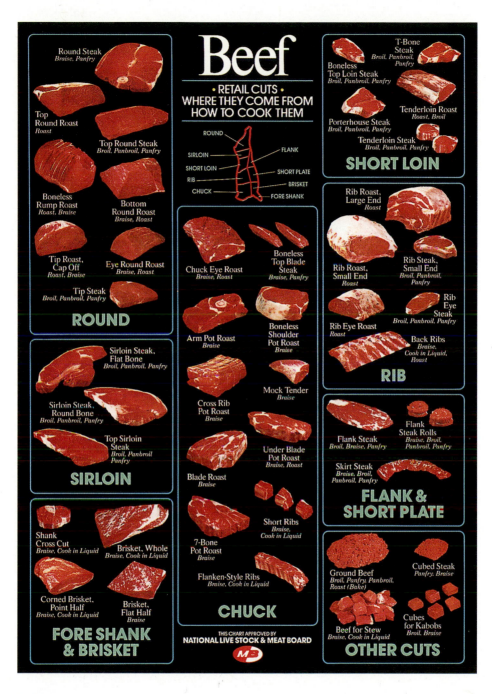

PLATE A.

Wholesale and retail cuts of beef with recommended types of cooking. Courtesy of the National Cattlemen's Beef Association © 1986.

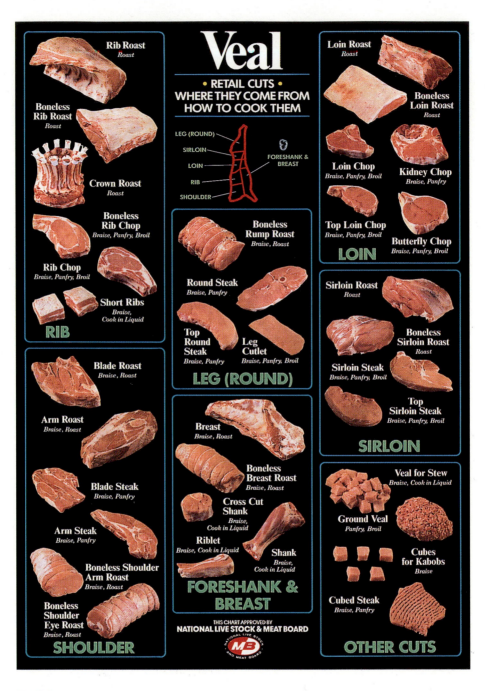

PLATE B.

Wholesale and retail cuts of veal with recommended types of cooking. Courtesy of the National Cattlemen's Beef Association © 1986.

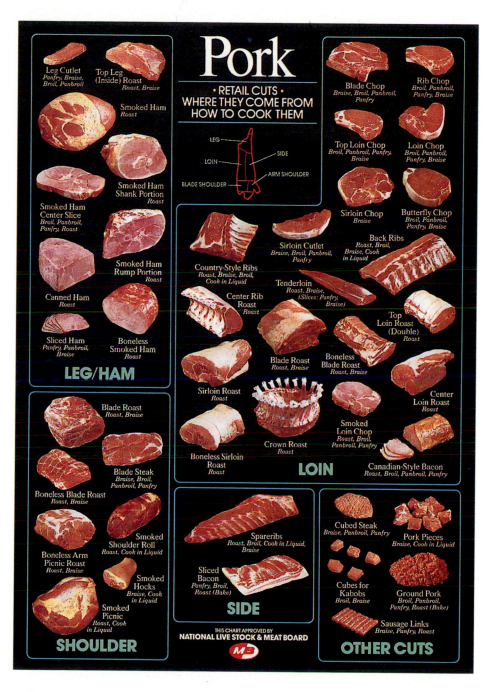

PLATE C.

Wholesale and retail cuts of pork with recommended types of cooking. Courtesy of the National Cattlemen's Beef Association and the National Pork Producers Council © 1986.

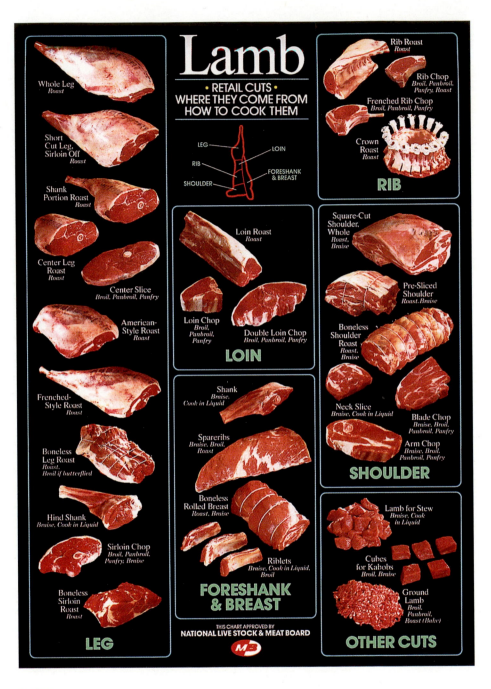

PLATE D.

Wholesale and retail cuts of lamb with recommended types of cooking. Courtesy of the National Cattlemen's Beef Association © 1986.

TABLE 3.5   Major Red Meat Products (fresh, frozen, and processed) under Federal Inspection

Meat Product	Mil lb
Cured	
Beef briskets	215
Other beef	160
Pork	3,651
Other meats	22
Smoked, Dried, or Cooked	
Hams	
Bone-in	99
Bone-in, water added	314
Semiboneless	9
Semiboneless, water added	63
Boneless	104
Boneless, water added	637
Sectioned and formed	58
Sectioned and formed, water added	348
Hams, dry cured	104
Total hams	1,735
Pork, regular	199
Pork, water added	246
Total other pork	445
Bacon	1,534
Total pork	3,714
Beef, cooked	481
Beef, dried	32
Other meats	350
Fresh/Frozen Products	
Beef cuts	9,944
Beef boning	8,857
Pork cuts	9,679
Pork boning	3,172
Other cuts	486
Other boning	230
Mechanically processed beef	4
Mechanically processed pork	33
Mechanically processed other	4
Steaks, chops, roasts	1,887
Steaks, chops (chopped and formed)	267
Meat patties	483
Hamburger/ground beef	3,659
Other fresh/frozen	962
Convenience Foods	
Pizza	773
Pies	210
Dinners	214
Entrées	425
Other	393

Sources: USDA: AMI (*Meat Facts*).

There are approximately 2 million Orthodox Jews in the United States who keep kosher. Estimates on quantities of kosher meat consumed are not available.

The Muslim code is found in the Quran. Any Muslim may slaughter an animal while invoking the name of Allah. In cases where Muslims cannot kill their own animals, they may use meat killed by a "person of the book," e.g., a Christian or a Jew. Usually this "Halal" slaughter is done by regular plant workers while Muslim religious leaders are present and reciting the appropriate prayers.

COMPOSITION

Meat composition can be defined in either physical or chemical terms. Physical composition is observed visually and with objective measurements; chemical composition is determined by chemical analysis.

Physical Composition

The major physical components of meat are lean (muscle), fat, bone (Fig. 3.5), and connective tissue. The proportions of fat, lean, and bone change from

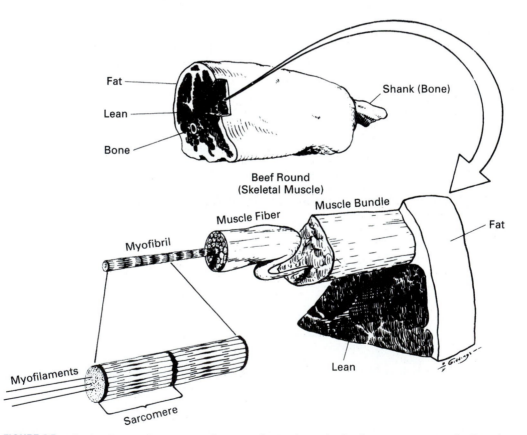

FIGURE 3.5 The fundamental structure of meat and muscle in the beef carcass. Drawing by Dennis Giddings.

birth to slaughter time. Chapter 18 discusses these compositional changes in more detail. Connective tissue, which to a large extent determines meat tenderness, exists in several different forms and locations. For example, tendons are composed of connective tissue (collagen), which attaches muscle to bone. Other collagenous connective tissues hold muscle bundles together and provide the covering to each muscle fiber. Figure 3.5 shows the physical structure of muscle: **myofibrils** are component parts of muscle fibers; muscle fibers combined together comprise a muscle bundle; and groups of muscle bundles are components of whole muscles or muscle systems. Within the myofibrils are two types of myofilaments—namely, thick (*myosin*) and thin (*actin*) filaments.

Chemical Composition

Since muscle or lean meat is the primary carcass component consumed, only its chemical composition is discussed here. Chemical composition is important because it largely determines the nutritive value of meat.

Muscle consists of approximately 65–75% water, 15–20% protein, 2–12% fat, and 1% minerals (ash). As the animal increases in weight, water and protein percentages decrease and fat percentage increases.

Fat-soluble vitamins (A, D, E, and K) are contained in the fat component of meat. Most B vitamins (water-soluble) are abundant in muscle.

The major protein in muscle is actomyosin, a globulin consisting of the two proteins actin and myosin. Most of the other nitrogenous extracts in meats are relatively unimportant nutritionally. However, these other extracts provide aroma and flavor in meat, which stimulate the flow of gastric juices.

Simple carbohydrates in muscle are less than 1%. Glucose and glycogen are concentrated in the liver. They are not too important nutritionally, but they do have an important effect on meat quality, particularly muscle color and water-holding capacity.

TABLE 3.6   Nutrient Contents of Red Meats (3 oz of cooked, lean portions) and Their Contribution to RDAs (Recommended Daily Allowances)[a]

Nutrient	Beef Amount	% RDA	Pork Amount	% RDA	Lamb Amount	% RDA
Calories	192	10%	198	10%	176	9%
Protein	25 g	56	23 g	51	24 g	54
Fat[b]	9.4 g	14	11.1 g	17	8.1 g	12
Cholesterol[b]	73 mg	24	79 mg	26	78 mg	26
Sodium	57 mg	2	59 mg	2	71 mg	2
Iron	2.6 mg	14	1.1 mg	6	1.7 mg	10
Zinc	5.9 mg	39	3.0 mg	20	4.5 mg	30
Thiamin	0.08 mg	6	0.59 mg	39	0.09 mg	6
Niacin	3.4 mg	17	4.3 mg	22	5.3 mg	27
B₁₂	2.4 mg	79	0.7 mg	24	2.2 mg	37

[a] Based on a 2,000-calorie diet and the RDA for women aged 23–51.
[b] The American Heart Association recommends that fat contribute not more than 30% of total daily calories and that cholesterol be limited to 300 mg/day.
Source: USDA.

NUTRITIONAL CONSIDERATIONS

Nutritive Value

Heightened health awareness in the United States has made today's consumers more astute food buyers than ever before. Consumers look for foods that meet their requirements for a nutritious, balanced diet as well as for convenience, versatility, economy, and ease of preparation. Processed meat products have long been among American's favorite foods. Consumers today choose from more than 200 styles or forms of meat, including over 50 cold cuts, dozens of sausages, and a wide variety of hams and bacon. Nutrient-rich red meats—beef, pork, lamb, and veal—are primary ingredients in these products. Meat products in all delicious and unique varieties contain the same nutrients as the fresh meats from which they are made (Table 3.6). More than 40 nutrients are needed by the body for good health. These must be obtained from a variety of foods; no single food contains all nutrients in the desired proportions.

Meat is nutrient-dense and is among the most valuable sources of important vitamins and minerals in a balanced diet. **Nutrient density,** a measurement of food value, compares the amount of essential nutrients to the number of calories in food.

Red meats are abundant sources of iron, a mineral necessary to build and maintain blood hemoglobin, which carries oxygen to body cells. Heme iron, one form of iron found in red meat, is used by the body most effectively, and meat enhances the absorption of nonheme iron from meat and other food sources. Red meat is also a rich source of zinc, an essential trace element that contributes to tissue development, growth, and wound healing.

Red meats contain a rich supply of the B vitamins. Pork is a particularly rich source of thiamine. Thiamine is essential to convert carbohydrates into energy. Vitamin B_{12} occurs only in animal products, such as meat, fish, poultry, and milk, and it is essential to protect nerve cells and for the formation of blood cells in bone marrow. Niacin, riboflavin, and vitamin B_6 are also found in significant amounts in red meats.

Finally, red meats are an excellent source of protein, a fundamental component of all living cells. Fresh and processed meats contain high-quality or complete protein with all essential amino acids in good quantity.

Issues related to red meat consumption and human diet and health are presented in Chapter 35.

CONSUMPTION

Meat is consumed for both nutrient content and eating satisfaction (Figs. 3.6–3.8). There are food products other than meat that cost less per unit of protein. However, consumers have a preference for how the protein is packaged, particularly if it is more palatable in tenderness, flavor, and juiciness. Meat has these preferred palatability characteristics.

Most foods are required to have a label identifying their nutrient content. Meat has been exempted from this requirement; however, most processed meats and some fresh meats show the Nutrition Facts label (Fig. 3.9).

Consumer acceptance of red meat and poultry is shown in Fig. 3.10. Poultry is included here to demonstrate that over the past 20 years the total annual

FIGURE 3.6 A thick, tender, and juicy steak is the most highly preferred beef product. Courtesy of the National Live Stock and Meat Board.

FIGURE 3.7 A favorite pork dish is smoked, boneless ham. Courtesy of the National Live Stock and Meat Board.

FIGURE 3.8 Processed beef and pork are popular meat items for sandwich lovers. Courtesy of the National Live Stock and Meat Board.

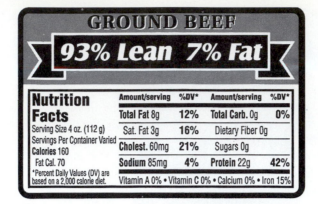

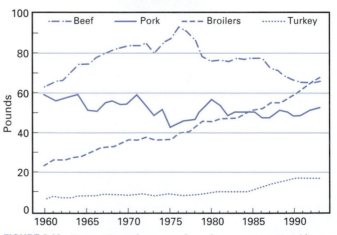

FIGURE 3.9 Examples of Nutrition Facts labels for ground beef and ham.

FIGURE 3.10 Per-capita red meat and poultry consumption (disappearance), boneless weight basis. Lamb and mutton consumption is approximately 1 lb per-capita. Courtesy of the USDA.

per-capita meat consumption has been increasing, primarily with greater amounts of poultry and less red meat.

Tables 3.7 and 3.8 show the per-capita consumption (disappearance) of red meat, poultry, and fish on a carcass weight, retail-weight, and boneless basis. Even the boneless data do not reflect the actual amount consumed, as there is fat trim, bone, cooking losses, and plate waste. Consumption data correcting most of these losses is shown in Fig. 3.11. The daily per-capita red meat consumption of 3.4 ounces is consistent with the dietary guidelines recommended in the Food Guide Pyramid (Chapter 35).

During the late 1980s, consumers continued to demand less external fat on retail cuts. Surveys have shown that most retail cuts of beef, pork, and lamb have 0.1 in. or less of external fat covering the lean muscle tissue.

Data on the consumption and disappearance of meat and other food items can be misleading. What people say they eat and what they purchase does not determine the amount actually consumed. In one study, Dr. Bill Rathje (University of Arizona) evaluated garbage in the Tucson area in an attempt to determine what people eat by what they throw away. Over 10 years of his research, data show (1) that people throw away approximately 15% of all the solid food they purchase; and (2) when questioned, consumers in middle- and low-income brackets consistently overestimate their meat consumption, while those in higher-income brackets underestimate their meat consumption.

TABLE 3.7   Per-Capita Consumption (disappearance) of Beef, Pork, and Poultry[a]

Year	Beef (lb) Carcass	Retail	Boneless	Pork (lb) Carcass	Retail	Boneless	Poultry (lb) Carcass	Retail	Boneless
1960	85.7	63.4	—	78.3	59.1	—	—	34.1	—
1965	101.0	74.7	—	68.1	51.8	—	—	41.2	—
1970	114.4	84.6	79.6	73.1	56.0	48.0	—	48.7	33.8
1975	119.2	88.2	83.0	55.9	43.0	38.7	—	47.7	32.9
1980	103.4	76.6	72.1	74.0	57.3	52.1	—	59.1	40.6
1985	106.7	78.9	74.6	66.3	51.7	47.7	—	65.7	45.2
1990	96.1	67.8	63.9	64.1	49.8	46.4	88.7	80.6	55.3
1995	96.6	67.1	63.2	67.5	52.4	48.8	99.0	89.7	61.6

[a] Carcass weight is shown to compare consumption data from other countries (see Chapter 2).
Source: USDA.

TABLE 3.8   Per-Capita Consumption (disappearance) of Red Meat, Poultry, and Fish

Year	Red Meat (lb) Retail	Boneless[a]	Poultry (lb) Retail	Boneless	Fish (lb) Retail	Total Meat (lb) Retail	Boneless
1960	131.2	—	34.1	—	10.3	176.1	—
1965	134.1	—	41.2	—	10.8	186.1	—
1970	145.9	131.7	48.7	33.8	11.8	206.4	177.3
1975	136.3	125.8	47.7	32.9	12.2	196.2	170.9
1980	136.8	126.5	59.1	40.6	12.5	208.4	179.4
1985	133.8	124.9	65.7	45.2	15.1	214.6	185.1
1990	120.1	112.3	80.6	55.3	15.0	215.7	182.7
1995	122.1	114.2[a]	89.7	61.6	15.0	226.8	198.7

[a] Boneless lb for red meat species are: beef (63.2); pork (52.4); veal (0.9); lamb (1.1).
Source: USDA.

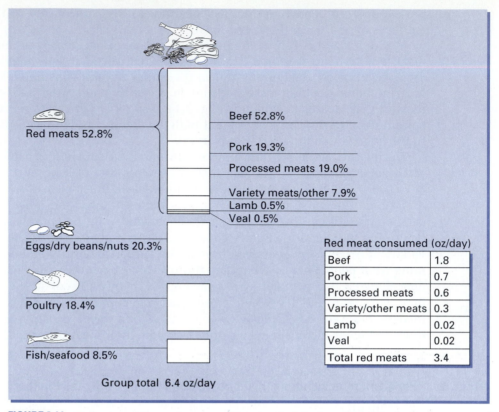

Red meats 52.8%

Beef 52.8%

Pork 19.3%

Processed meats 19.0%

Variety meats/other 7.9%

Lamb 0.5%

Veal 0.5%

Eggs/dry beans/nuts 20.3%

Poultry 18.4%

Fish/seafood 8.5%

Red meat consumed (oz/day)	
Beef	1.8
Pork	0.7
Processed meats	0.6
Variety/other meats	0.3
Lamb	0.02
Veal	0.02
Total red meats	3.4

Group total 6.4 oz/day

FIGURE 3.11 Daily per-capita meat consumption as part of meat, poultry, fish, dry beans, eggs, and nuts group (Food Guide Pyramid). Source: National Live Stock and Meat Board.

Per-capita disposable income does affect the total amount of money spent on meat and the amount of product consumed (Table 3.9). However, consumers are spending a smaller percentage of their disposable income on meat and food, indicating that these items are relatively inexpensive compared to other purchases.

Table 3.9 shows that per-capita disposable income has risen more rapidly than the per-capita red meat expenditure. This change reflects either a decreased preference for red meat or a preference for other goods and services as income has increased above a certain level. Lower poultry prices relative to

TABLE 3.9   Disposable Per-Capita Income and Red Meat Expenditures

			Expenditure for Red Meat as Percentage of	
Year	Per-Capita Income	Red Meat Expenditures	Per-Capita Income (%)	Total Food Expenditures (%)
1970	$3,489	$140.38	4.0%	24.5%
1975	5,291	214.31	4.1	24.6
1980	8,421	286.02	3.4	21.2
1985	11,861	294.67	2.5	17.3
1990	16,236	378.81	2.3	17.1
1995	19,703	413.55	2.2	16.7

Source: U.S. Departments of Agriculture and Commerce; AMI, *Meat and Poultry Facts.*

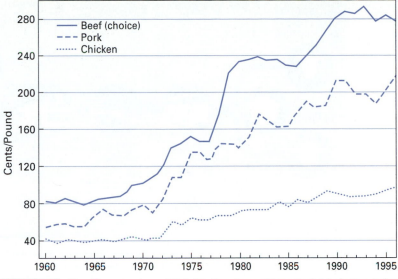

FIGURE 3.12 Retail meat prices, 1960–96. Courtesy of Livestock Marketing Information Center, Lakewood, Colo. or Denver, Colo.

red meat prices has influenced some producers to shift their preference from red meat to poultry (Fig. 3.12).

Table 3.10 shows the consumption patterns for hamburger, processed beef, and beef cuts (steaks and roasts). Comparative prices for hamburger and the average price of retail cuts from choice beef are also shown.

MARKETING

Figure 3.13 shows channels of meat production and processing prior to reaching the retail meat outlets and, eventually, the consumers. On hog farms, most pigs are raised to slaughter weight. Cattle and sheep are sold by producers to feedlots for added weight gain prior to slaughter.

Packers and processors purchase the animals through several different marketing channels (Table 3.11). Direct purchases can occur at the packing plant,

TABLE 3.10   Consumption and Prices of Various Beef Products, 1970–95

| | Per-Capita Consumption (lb retail) | | | | Price ($/lb) | |
Year	Hamburger	Processed Beef	Beef Cuts	Total Beef[a]	Hamburger	Choice Beef
1970	21.9	10.8	51.9	84.6	0.66	1.00
1975	30.1	11.7	46.4	88.2	0.88	1.52
1980	24.7	7.4	44.5	76.6	1.36	2.34
1985	27.2	6.0	45.7	78.9	1.24	2.29
1990	28.5	5.1	34.2	67.8	1.59	2.81
1995	26.6	4.7	36.2	67.5	1.48	2.83

[a] Retail-weight basis.
Sources: AMI, *Meat Facts;* USDA.

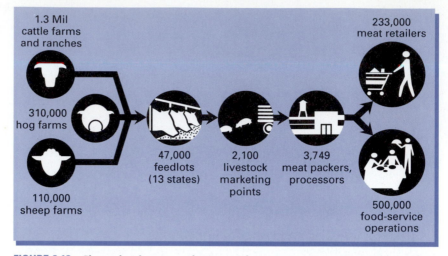

FIGURE 3.13   Channels of meat production and processing in reaching the retail segment. Courtesy of the National Live Stock and Meat Board.

the feedlot, or a buying station near where the animals are produced. Terminal markets are large livestock collection centers where an independent organization serves as a selling agent for the livestock owner. Oklahoma City, Omaha, and Sioux City, Iowa, are examples of some of the larger terminal markets, most of which are located in the midsection of the United States.

Approximately 2,000 auctions, sometimes called *sale barns,* are located throughout the United States. The auction company has pens, a sale ring, and an auctioneer. Auctioneers sell the livestock to the highest bidder among the buyers attending the auction. Producers pay a marketing fee to the auction company for the sale service.

Most cattle, sheep, and swine are purchased on a live-weight basis, with the buyer estimating the value of the carcass and other products. Some animals are purchased on the basis of their carcass weight and the desirability of the carcass produced. For cattle and swine there has been an increased number of purchases on a carcass basis during the past several years (Table 3.12). In this method of marketing, sometimes called **grade and yield,** the seller is paid on a carcass-merit basis.

TABLE 3.11   **Market Outlets Used by U.S. Packers in Purchasing Red Meat Animals for Slaughter**

Animals	Total (mil head)	Direct Purchases[a] (%)	Public Markets[b] (%)
Cattle	27.9	86	14
Calves	0.8	82	18
Hogs	83.5	94	6
Sheep and lambs	3.8	84	16

[a] Direct purchases at packing plants, buying stations, country points, and feedlots.
[b] Public markets consist of terminal markets and auction markets.
Source: USDA, *Packers and Stockyards' Statistical Resume,* 1994.

TABLE 3.12   Livestock Purchased by Packers on Carcass Grade and Weight Bases

Year	Cattle and Calves		Hogs		Sheep and Lambs	
	No. Head (mil)	Percent of Total Purchases (%)	No. Head (mil)	Percent of Total Purchases (%)	No. Head (mil)	Percent of Total Purchases (%)
1975	8.6	23%	6.0	9%	0.82	10%
1980	8.4	27	10.2	11	1.54	28
1985	10.2	30	13.0	16	2.19	37
1990	12.1	36	9.0	11	2.01	42
1995	14.5	44	27.7	31	2.13	47

Source: USDA, *Packers and Stockyards' Statistical Resume.*

The Packers and Stockyards Act, passed in 1921 and continually updated, is enforced by the USDA. The act provides for uniform and fair marketing practices. Federal supervision is given at market outlets, including packing plants, for fee and yardage charges, testing of scales, commission charges, and other marketing transactions. Carcass grade and weight sales are covered as well as live animal sales. Marketing protection is provided to both the buyer and the seller.

Consumer attitudes and preferences for meat changed during the 1980s. These attitudes were surveyed, and some are identified in Table 3.13. This information shows that the market is segmented and that different products are needed for each segment. Other market surveys vividly demonstrate that many consumers have a preference for leanness, which in the minds of these consumers means a low level of trimmable fat. Consumers are demanding more convenient (easier to prepare) meat products, as the numbers of two-income families and single parents have increased dramatically since 1975. There were more (50% versus 33%) active, health-oriented consumers in 1985 than there were only 2–3 years previously.

In the early 1990s, the convenience-oriented or active lifestyle consumer group increased more rapidly than the other consumer segments. The new traditionalist consumer group, which combines the meat lovers and creative cooks, continues to be a strong second.

Advertising dollars spent by the livestock industry have been low in recent years. Even though there has been an increase in media advertising, the meat industry's total advertising has been small compared to that of other industries.

Those consumers who are price-driven (Table 3.13) are continually comparing the prices of different meats. Figure 3.12 shows how retail meat prices for beef, pork, turkey, and chicken have changed over the past several years.

TABLE 3.13   Consumer Attitudes Affecting Meat Consumption, 1983–87

Consumer Segments	1983	1985	1987
Meat lovers	22%	10%	7%
Creative cooks	20	17	21
Price-driven	25	23	22
Active lifestyle	16	26	22
Health-oriented	17	24	27

Figure 3.14 shows consumer expenditures for red meat and poultry. Currently, consumers spend approximately $100 more compared to dollars spent in 1980, with most of this increase going to broilers. Percent of total expenditures for pork has stayed about the same since 1980, while beef has been losing consumer market share (Fig. 3.15). The primary reason for beef's loss, according to the 1995 National Beef Quality audit is beef's inconsistency and lack of uniformity in palatability (tenderness, juiciness, and flavor). Other reasons were that carcasses and beef cuts were too large and that beef was too high priced for the value received. Obviously beef has serious challenges to stabilizing or increasing consumer market share.

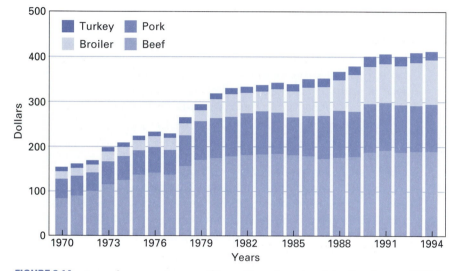

FIGURE 3.14   Annual, per-capita expenditures for red meat and poultry. Source: USDA.

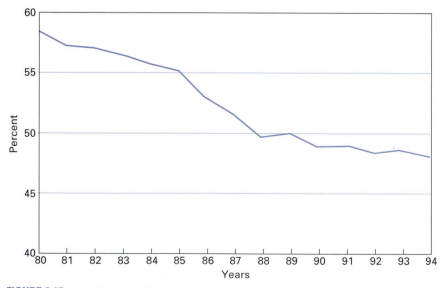

FIGURE 3.15   Beef's share of consumer meat expenditures. Source: National Live Stock and Meat Board.

TABLE 3.14 Consumer Expenditures for Meat Products, Poultry, and Fish in Grocery Stores

Food Item	$Bil 1993	$Bil 1983	Percent of Food and Beverages 1993	1983	Percent of Total 1993	1983
Fresh meat	$26.7	$26.7	9.6%	14.4%	7.2%	11.1%
Beef	26.8	21.1	7.8	11.4	5.9	8.7
Veal	0.4	0.6	0.1	0.4	0.1	0.3
Lamb	1.0	1.1	0.3	0.6	0.3	0.4
Pork	3.6	3.9	1.3	2.1	1.0	1.6
Packaged meats	14.3	13.2	5.1	7.1	3.9	5.5
Cured hams and picnics	2.0	2.3	0.7	1.2	0.5	0.9
Packaged bacon	1.8	2.3	0.6	1.2	0.5	1.0
Frankfurters	2.0	1.5	0.7	0.8	0.5	0.6
Sausage products	3.7	4.0	1.3	2.1	1.0	1.7
Cold cuts and other	4.8	3.1	1.7	1.7	1.3	1.3
Frozen meat	0.2	0.2	0.1	0.1	0.1	0.1
Canned meat	1.1	1.5	0.4	0.8	0.3	0.6
Total meat	42.3	41.6	15.2	22.5	11.4	17.3
Poultry	13.2	7.3	4.7	3.9	3.6	3.0
Fish and seafood	6.2	4.6	2.2	2.5	1.7	1.9
Total meat, poultry, and seafood	61.7	53.5	22.1	28.9	16.6	22.2
Total grocery store sales	370.7	241.0	—	—	—	—
Food and beverages	279.4	185.2	—	—	75.2	76.8
Nonfood and general merchandise	91.3	55.8	—	—	24.8	23.2

Sources: Supermarket Business magazine; AMI, Meat Facts.

Table 3.14 shows the dollars generated by red meat products through grocery stores and large supermarket chains. Total red meat represents 17% of total retail sales and 13% of food sales. Note the changes in red meat and poultry sales during the past decade.

CHAPTER SUMMARY

- World red meat production is 320 billion pounds with China (84 bil lb), the U.S. (44 bil lb), and Brazil (14 bil lb) the leading countries.
- Most of the 44 billion pounds of red meat production in the United States is beef (25 bil lb) and pork (18 bil lb).
- The major contribution of red meats to human nutrition is protein, iron, and vitamin B_{12}.
- Per capita red meat consumption in the U.S. is 122 pounds of retail weight with beef (67.1 lb) and pork (52.4 lb) being the primary contributors.

REVIEW QUESTIONS

1. What does dressing percentage represent?
2. How is dressing percentage calculated?
3. What is the rule of thumb or approximate dressing percentages of lamb, beef, and hogs?

4. What are the primary factors affecting dressing percentage?

5. What are the four major physical components of meat?

6. _____ are components of muscle fibers and are comprised of two types of myofilaments, myosin and actin.

7. What is the amount of essential nutrients to the number of calories in food referred to as?

8. *True or False:* Meat is nutrient dense, meaning that it provides a large proportion of certain nutrients such as essential amino acids, B-vitamins, and iron.

9. *True or False:* Most retail cuts of beef, pork, and lamb have greater than 0.1 inch of external fat covering lean muscle.

10. *True or False:* Even though advertising dollars spent by the livestock industry have increased in recent years, it is far below that of other industries such as the automotive, tobacco, fast-food, communications, and retail sales industries.

PUBLICATIONS

Boggs, D. L., and Merkel, R. A. 1993. *Live Animal, Carcass Evaluation and Selection Manual.* Dubuque, IA: Kendall-Hunt.

Cassens, R. G. 1994. *Meat Processing. Encyclopedia of Agricultural Science.* San Diego: Academic Press, Inc.

Eating in America Today. 2e. 1994. Chicago, IL: National Live Stock and Meat Board.

Facts About Beef; Facts About Pork; Facts About Lamb. Chicago: The National Live Stock and Meat Board.

Hedrick, H. B., Aberle, E. D., Forrest, J. C., Judge, M. D., and Merkel, R. A., 1994. *Principles of Meat Science.* Dubuque, IA: Kendall-Hunt.

Helming, B. 1990. Meat's decade of danger. *Meat and Poultry* 36:30.

Horsemeat. 1980. *Livestock* (Sept. 1980).

Kashruth. Handbook for Home and School. 1972. New York: Union of Orthodox Jewish Congregations of America.

Meat and Poultry Facts. 1995. Washington, D.C.: American Meat Institute.

Regenstein, J. M. and T. Grandin. 1992. Religious slaughter and animal welfare—an introduction for animal scientists. Reciprocal Meat Conference Proceedings 45:155.

Romans, J. R., Jones, K. W., Costello, W. J., Carlson, C. W., and Ziegler, P. T. 1985. *The Meat We Eat.* 12th ed. Danville, IL: Interstate Printers and Publishers.

The National Beef Quality Audit. 1995. Englewood, CO: National Cattlemen's Beef Association.

Visuals

Excel Beef Plant: Slaughter. 1997 (video; 45 min.) and *Excel Beef Plant: Fabrication.* 1997. (video: 45 min.) CEV, P. O. Box 65265, Lubbock, TX 79464.

Lamb Carcass Evaluation Multi-Media Kit. Vocational Education Productions, California Polytechnic State University, San Luis Obispo, CA 93407.

Livestock Slaughter, Carcass Fabrication; Packer to Consumer; Beef Carcass Grading; and *Retail Cut Identification* (videotapes). CEV, P. O. Box 65265, Lubbock, TX 79464-5265.

Meat Cut Identification (slide set). Vocational Education Productions, California Polytechnic State University, San Luis Obispo, CA 93407.

Meat Identification (slide set), *Meat Evaluation Handbook,* and *The Meat Board Guide to Identifying Meat Cuts* (17-203). The National Live Stock and Meat Board, 444 N. Michigan Ave., Chicago, IL 60611.

Packer to Consumer (video; 20 min). AAVIM, 220 Smithonia Road, Winterville, GA 30683.

CHAPTER 4

By-Products of Meat Animals

By-products are products of considerably less value than the major product. However, the value of by-products can account for as much as 8–10% of the total value of a fed steer. In the United States, meat animals produce meat as the major product; hides, fat, bones, and internal organs are considered by-products. In other countries, primary products may be draft (work), milk, hides, and skins, with meat considered a by-product when old, less useful animals are slaughtered.

There are numerous by-products resulting from animal slaughter and the processing of meat, milk, and eggs. Animals dying during their productive life are also processed into several by-products; however, these by-products are not used in human consumption.

By-products are commonly classified into two categories based on human consumption: edible and inedible (Fig. 4.1; Table 4.1). Some by-products that are not considered edible by humans are edible to other animals. By-products that are processed into animal feeds are good examples.

Figure 4.2 shows the major by-products obtained from a steer in addition to the retail beef products. Average total value of cattle by-products ranges between $0.05 and $0.09 per pound of liveweight, with the hide comprising the largest part of the total value. Because hides have cyclic prices, as is often the case with commodity products, their value, as percent of the total by-product value, has ranged from 45% to 65% during the past several years.

EDIBLE BY-PRODUCTS

Variety meats are edible products originating from organs and body parts other than the carcass. Liver, heart, tongue, tripe, and sweetbread are among the more

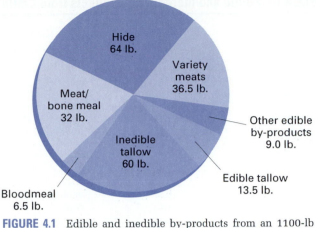

FIGURE 4.1 Edible and inedible by-products from an 1100-lb steer. Courtesy of the USDA.

typical variety meats. Tripe comes from the lining of the stomach; sweetbread is the thymus gland.

An average 1,100-lb slaughter steer produces approximately 36 lb of variety meats. Since the U.S. per-capita disappearance is 9 lb, surplus variety meats are exported into countries that have a preference for them.

Other edible products are fats used to produce lard and tallow. These products are eventually used in shortenings, margarine, pastries, candy, and other

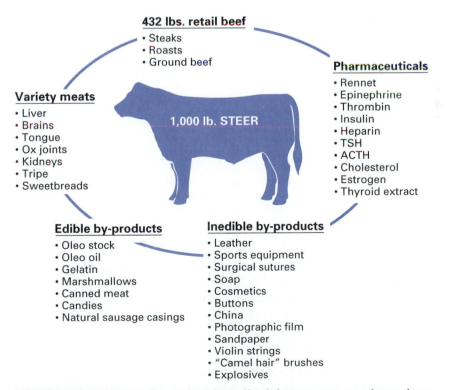

FIGURE 4.2 In addition to the retail product of beef, there are numerous by-products that come from beef. Courtesy of the National Live Stock and Meat Board.

TABLE 4.1   Edible and Inedible By-Products from Cattle and Swine

By-Product	1,100-lb Steer		230-lb Hog	
	Expected Yield Per Head (lb)	Market Value Per Cwt	Expected Yield Per Head (lb)	Market Value Per Cwt
Edible				
Select liver	16.20	0.40	1.30	0.10
Heart	3.80	0.30	0.30	0.20
Cheek meat	4.41	0.59	0.30	0.55
Head meat	1.80	0.39	0.40	—
Salivary glands	1.25	—	0.10	0.15
Feet	—	—	0.80	0.17
Lips	1.68	0.76	0.25	—
Oxtail	2.95	1.14	—	—
Spleen	1.97	0.06	—	—
Tongue	3.45	1.65	0.20	0.58
Weasand meat[a]	0.47	0.61	0.12	—
Tripe (honeycomb)	1.90	0.95	—	—
Stomach	—	—	0.40	0.42
Sweetbread	0.62	0.49	—	—
Ears	—	—	0.26	0.65
Kidneys	1.74	0.12	0.20	—
Snouts	—	—	0.65	0.27
Tallow/lard	—	0.21	9.00	0.20
Inedible				
Lung	5.30	0.07	—	—
Gullet	1.00	0.05	—	—
Rendered fat	60.00	0.19	—	—
Meat and bone meal	32.00	0.09	—	—
Dried blood	6.50	0.19	—	—
Hide (native steer)	64.00	0.78	—	—
Pharmaceutical				
Fetal blood[b]		64.00		—
Fresh bile		0.11		
Pancreas	0.69	0.69	—	0.62
Adrenal glands	—	2.80	—	—
Pituitary glands	—	19.50	—	180.00
Thyroid glands	—	2.00	—	4.25
Pepsin linings	—	—	—	0.88
Ovaries	—	8.75	—	—
Spleens	—	—	—	—
Lungs (lobe only)	—	0.22	—	—
Trachea	0.99	0.20	—	—

[a] *Weasand* is the muscular lining that surrounds the esophagus from larynx to first stomach.
[b] Per liter.
Sources: Monfort of Colorado; *National Provisioner* (1996).

food items. The United States produces approximately 1.5 billion lb of edible tallow and 6.4 billion lb of inedible tallow. Nearly 50% of the inedible tallow is exported with nearly 60% of edible tallow and lard exported.

INEDIBLE BY-PRODUCTS

Tallow, hides (skins), and inedible organs are the higher-valued inedible by-products. Not all skins are inedible, as some pork skins are processed into consumable food items. Life-saving and life-supporting pharmaceuticals originate from animal by-products, even though many of the same pharmaceuticals are made synthetically. Table 4.2 shows many of these pharmaceuticals and their animal by-product source.

Table 4.3 lists other inedible by-products and their uses. Animals contribute a large number of useful products. The many uses of inedible fats are discussed later in the chapter.

Hides and skins are valuable as by-products or as major products on a worldwide basis. Countries that are most important in hide and skin production are shown in Table 4.4.

Cattle and buffalo hides comprise 80% of the farm animal hides and skins produced in the world. The United States has a major contribution in the world hide and skin market, as most of the 1,972 million lb of cattle hides produced are exported to other countries (Chapter 2). The United States produces very

TABLE 4.2 Selected Pharmaceuticals from Red Meat Animals—Their Source and Utilization

Pharmaceutical	Source	Utilization
Amfetin	Amniotic fluid	Reduces postoperative pain and nausea and enhances intestinal peristalsis
Cholesterol	Nervous system	Male sex hormone synthesis
Chymotrypsin	Pancreas	Removes dead tissue; for treatment of localized inflammation and swelling
Corticosteroids	Adrenal gland	Treatment for shock, Addison's disease
Corticotrophin (ACTH)	Pituitary gland	Diagnostic assessment of adrenal gland function; treatment of psoriasis, allergies, mononucleosis, and leukemia
Cortisone	Adrenal gland	Treatment for shock, arthritis, and asthma
Desoxycholic acid	Bile	Used in synthesis of cortisone for asthma and arthritis
Epinephrine	Adrenal gland	Relief of hay fever, asthma, and other allergies; heart stimulation
Fibrinolysin	Blood	Dead tissue removal; wound-cleansing agent; healing of skin from ulcers or burns
Glucagon	Pancreas	Counteracts insulin shock; treatment of some psychiatric disorders
Heparin	Intestines	Natural anticoagulant used to thin blood, retards clotting, especially during organ implants
Heparin	Lungs	Anticoagulant; prevention of gangrene
Hyaluranidase	Testes	Enzyme that aids drug penetration into cells
Insulin	Pancreas	Treatment of diabetes for 10 million Americans (primarily synthetic insulin is used)
Liver extracts	Liver	Treatment of anemia
Mucin	Stomach	Treat ulcers
Norepinephrine	Adrenal gland	Shrinks blood vessels, reducing blood flow and slowing heart rate
Ovarian hormone	Ovaries	To treat painful menstruation and prevent abortion
Ox bile extract	Liver	Treatment of indigestion, constipation, and bile tract disorders
Parathyroid hormone	Parathyroid gland	Treatment of human parathyroid deficiency
Plasmin	Blood	Digests fibrin in blood clots, used to treat patients with heart attacks
Rennet	Stomach	Assists infants in digesting milk; cheese making
Thrombin	Blood	Assists in blood coagulation; treatment of wounds; skin grafting
Thyroid extract	Thyroid gland	Treatment of cretinism
Thyrotropin (TSH)	Pituitary gland	Stimulates functions of thyroid gland
Thyroxin	Thyroid	Treatment of thyroid deficiency
Vasopressin	Pituitary gland	Control of renal function

TABLE 4.3   Other Inedible By-Products and Their Uses

By-Product	Use
Hog heart valves	They replace injured or weakened human heart valves; since the first operation in 1971, more than 35,000 hog heart valves have been implanted in humans.
Pig skins	Used in treating massive human burns; these skins help prepare the patient for permanent skin grafting.
Gelatin (from skin)	Coatings for pills and capsules.
Brains	Cholesterol for an emulsifier in cosmetics.
Blood	Sticking agent for insecticides; a leather finish; plywood adhesive; fabric printing and dyeing.
Hides and skins	Many leather goods from coats, handbags, and shoes to sporting goods.
Bones	Animal feed, glue, buttons, china, and novelties.
Gallstones	Shipped to Orient for use as ornaments in necklaces and pendants.
Hair	Paint and other brushes; insulation; padding in upholstery, carpet padding, filters.
Meat scraps and blood	Animal feed.
Inedible fats	Animal feed, fatty acids.
Wool (pulled from pelts)	Clothing, blankets, lanolin.
Poultry feathers	Animal feed, arrows, decorations, bedding, brushes.

few goat skins but ranks 17th in sheepskins, with an annual production of approximately 37 million lb. Pork skin usually remains on the carcass, although some pigs are skinned.

The U.S. hide export market yields approximately $1.3 billion on a yearly basis. Hides, skins, and pelts are made into useful leather products through the tanning process (Table 4.5). One cow hide can yield approximately 144 baseballs, 20 footballs, 18 volleyballs or soccer balls, 12 baseball gloves, or 12 basketballs. The surplus hides are exported primarily to Japan, Korea, Mexico, Taiwan, and Romania. Leather utilization in the United States is categorized as 40% for upholstery, 50% for shoes and shoe leather, and 10% for other uses.

The general term *hide* refers to a beef hide weighing more than 30 lb. Those weighing less than 30 lb are called *skins*. Skins come from smaller animals, such as pigs, sheep, goats, and small wild animals. Those skins from sheep with the wool left on are usually called *sheep pelts*. Hides, skins, and pelts are

TABLE 4.4   Leading Countries in Fresh Hide and Skin Production

Cattle and Buffalo Hides		Sheepskins		Goatskins	
Country	(mil lb)	Country	(mil lb)	Country	(mil lb)
1. India	2,012	1. China	411	1. China	292
2. United States	1,896	2. Australia	287	2. India	248
3. Brazil	950	3. New Zealand	220	3. Pakistan	237
4. China	845	4. Turkey	142	4. Bangladesh	68
5. Argentina	792	5. Pakistan	126	5. Nigeria	45
World total	13,822	**World total**	2,746	**World total**	1,147

Source: 1994 FAO *Production Yearbook.*

TABLE 4.5   Leather Uses Related to Types of Hides and Skins

Skin Origin	Use
Cow and steer	Shoe and boot uppers, soles, insoles, linings; patent leather; garments; work gloves; waist belts; luggage and cases; upholstery; transmission belting; sporting goods; packings
Calf	Shoe uppers; slippers; handbags and billfolds; hat sweatbands; bookbindings
Sheep and lamb	Grain and suede garments; shoe linings; slippers; dress and work gloves; hat sweatbands; bookbindings; novelties
Goat and kid	Shoe uppers, linings; dress gloves; garments; handbags
Pig	Shoe suede uppers; dress and work gloves; billfolds; fancy leather goods
Horse	Shoe uppers; straps; sporting goods

Source: New England Tanners Club, *Leather Facts.*

classified according to (1) species, (2) weight, (3) size and placement of brand, and (4) type of packer producing them.

The value of hides may be reduced by branding, by nicking the hide while skinning, by warbles (larvae of heel flies), mange, lice, biting and sucking insects, grubs, water, mud, and urine damage. Warbles emerge from the back region of cattle in the spring, making holes in the hide. Sheep hides may be damaged by outgrowths of grasses in the production of seeds (called *beards*), which can penetrate the skin, as well as by external parasites called keds.

Leather tanning and finishing is a $1.5 billion industry in the United States, employing over 20,000 people in 330 establishments and accounting for $250 million in exports. About 160 plants process raw hides or skins directly into tanned leather. Some tanneries are relatively small concerns specializing in the manufacture of a particular kind of leather, whereas others employ several hundred people and produce a variety of leathers. Tanning activities are concentrated in the Northeast, Midwest, Middle Atlantic states, and California. There is increased interest in further processing cattle hides close to their origin. For example, there are more than 20 million head of cattle slaughtered in Texas, Kansas, Nebraska, and eastern Colorado. In 1992, IBP opened a 60,000-square-foot hide tannery in Amarillo, Texas, making the company the biggest hide tanner in the world. Other areas in the southern Great Plains are considering building tanneries because of value added in processing the hides.

A fed steer produces a 65–75-pound hide that is worth nearly $1 per pound. By converting the hide into the *blue* stage (which means it is treated so it will not deteriorate during shipping), the hide loses about 15 pounds. The value, however, has increased to $80–$90. A 60-pound blue hide will produce 40 square feet of shoe leather. Shoe leather sells for $2.50 a square foot so the hide is now worth approximately $100.

Every year, U.S. tanneries convert millions of raw hides and skins into leather. The tanners' value added by manufacture constitutes over $500 million annually. Their product (leather) serves in turn as a raw material for the shoe and leather goods industries that provide jobs for over 200,000 people. Table 4.6 shows the quantity of exported hides and skins.

Table 4.6 shows that less than half of U.S. cattle hides are converted to leather domestically, most being exported for tanning. Some sheepskins are

TABLE 4.6   U.S. Hide and Skin Exports

	Per 1000 Pieces		
Item	1992	1993	1994
Cattle and buffalo hides	19,028	18,226	17,911
Sheep and lamb skins	4,825	3,618	3,686
Calf and pig hides	2,895	2,622	4,566

Source: USDA, 1995.

imported, as the domestic supply does not meet the demand. Pigskins are used primarily as food, but interest in leather production is increasing.

Just as meat is perishable, so too are hides and skins. If not cleaned and treated, they begin to decompose and lose leather-making substance within hours after removal from the carcass. Hides are commonly treated (cured) by adding salt as the principal curing agent. The salt solution penetrates the hide in about 12 hours, then the hides are bundled for shipment (Fig. 4.3).

THE RENDERING INDUSTRY

A focal point of by-products is the rendering industry, which recycles offal, fat, bone, meat scraps, and entire animal carcasses. The sources of these raw materials are packing and processing plants, butcher shops, restaurants, supermarkets, farmers, and ranchers. Some large packing and processing plants have their own rendering plants integrated with other operations.

Renders have a regular pickup service that amounts to more than 70 million lb of animal material daily. This amount has been declining in recent years because less fat is shipped out of packing plants. This pickup service is essential to public health, as it reduces a major garbage-disposal problem. Also, since animal by-products have value, consumers can buy meat at a lower price than would otherwise be possible.

FIGURE 4.3 Beef hide room where hides are cured, tied, and stored in preparation for shipping. Courtesy of Iowa Beef Processors, Inc.

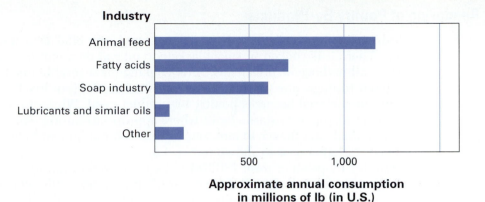

FIGURE 4.4   The major uses of animal fat. Courtesy of the National Renderers Association.

Rendering of Red Meat Animal By-Products

Animal fat and animal protein are the primary products of the renders' art. Originally, animal fats went almost entirely into soap and candles. Today, from the same basic material, renders produce many grades of tallow and semiliquid fat. The major uses of rendered fat are for animal feeds, fatty-acid production, and soap manufacture (Fig. 4.4).

Fatty acids are used in the manufacture of many products, such as plastic consumer items, cosmetics, lubricants, paints, deodorants, polishes, cleaners, caulking compounds, asphalt tile, printing inks, and others. The fatty-acid industry has had a tremendous growth since 1950.

The rendered animal proteins are processed into several high-protein (more than 50% protein) feed supplements among which are meat and bone meals and blood meals. The supplements are more commonly fed to young monogastric animals—swine, poultry, and pet animals. These animals require a high-quality protein, particularly the amino acid lysine, which is typical of animal protein supplements.

Blood not processed into blood meal is used to produce products used in the pet food industry. Such products are blood protein (fresh, frozen whole blood) and blood cell protein (frozen, fresh, dewatered blood).

By-products enjoy a relatively strong export market as indicated by Table 4.7.

TABLE 4.7   Value of Export ($1,000)

	1990	1992	1994
Fats, oils, greases	459,157	497,662	514,931
Lard, rendered pig fat	22,447	31,117	29,020
Tallow, inedible	344,079	321,658	335,375
Other	91,631	144,847	150,536
Offals, edible, variety meat	370,674	420,004	524,715
Offals, inedible	251,873	270,122	283,562
Hides and skins	1,793,785	1,335,562	1,438,607

Rendering of Poultry By-Products

Within a few hours after poultry is slaughtered, the offal is at the rendering plant. There it is cooked in steam-jacketed tanks at temperatures sufficient to destroy all pathogenic organisms. After cooking for several hours, the material is passed through presses that remove most of the fat (poultry fat), while the remaining material becomes poultry by-product meal. While it is still hot, the poultry fat is piped to sterile tanks where appropriate antioxidants or *stabilizers* are added, and thence to tank cars or drums for shipment to feed manufacturers or for other uses as shown in Fig. 4.4.

Poultry by-product meal (PBPM) is high in protein (approximately 58%) and contains 12–14% fat. PBPM consists of ground dry-rendered clean parts of the carcass of slaughtered poultry, such as heads, feet, undeveloped eggs, and intestines, exclusive of feathers, except in such trace amounts as might occur unavoidably in good factory practice. It should not contain more than 16% ash and not more than 4% acid ash.

EXPORT MARKET

By-products from the livestock and poultry industries are an extremely important part of the export market of livestock products. Hides and skins, fats and oils, and variety meats comprise more than one-third of the export market.

CHAPTER SUMMARY

- Hides, fat, bones, and internal organs are the primary by-products, with hides usually the highest value.
- Variety meats (liver, heart, tongue, tripe, etc.) are examples of edible by-products, while hide and other internal organs are inedible.
- Numerous inedible by-products produce useful human products—for example, pig heart valves (human heart valves), pig skin (human graft skin), hides and skins (leather, bone buttons, china, etc.), and many pharmaceuticals (e.g., corticosteroids, epinephrine, heparin, insulin, thyrotropin, etc.).
- Hides, fats, and variety meats are by-products that comprise most of the total value of U.S. livestock exports.

REVIEW QUESTIONS

1. What is the most valuable by-product obtained from the slaughter of livestock?
2. What important consumer product are hides used to produce?
3. What is the recycling of nonedible animal by-products and carcasses referred to as?

SELECTED REFERENCES

Publications

Field, T. G. 1996. Quantification of the utilization of edible and inedible beef by-products. Colorado State University and The National Cattlemen's Beef Association, Englewood, CO.

Hog Is Man's Best Friend; Meat; By-products. Chicago: The National Live Stock and Meat Board.

Kinsman, D. M. 1994. Animal By-Products From Slaughter. *Encyclopedia of Agricultural Science.* San Diego: Academic Press, Inc.

Lawrence, J. D. 1994. An economic assessment of trends in pork by-product values. Iowa State University Swine Research Report.

Renderer Recycling for a Better Tomorrow. 1981. National Renderers Association, 2250 East Devon Ave., Des Plaines, IL 60018.

Romans, J. R., Jones, K. W., Costello, W. S., Carlson, C. W., and Ziegler, P. T. 1985. Packing house by-products. Chap. 10 in *The Meat We Eat.* Danville, IL: Interstate Printers and Publishers.

Visuals

Leather Production (VHS Video). VEP, Calif. State Polytechnic Univ., San Luis Obispo, CA 93407.

CHAPTER 5

Milk and Milk Products

The world's population obtains most of its milk and milk products from cows, water buffalo, goats, and sheep (Fig. 5.1). Horses, donkeys, reindeer, yaks, camels, and sows contribute a smaller amount to the total human milk supply. Milk, with its well-balanced assortment of nutrients, is sometimes called "nature's most nearly perfect food." While milk is an excellent food product in many ways, it is not perfect. Nor is any other food. Milk and numerous milk products (e.g., cheese, butter, ice cream, and cottage cheese) are major components of the human diet in many countries.

Changes in dietary preferences and milk marketing are some of the challenges facing the dairy industry. Health considerations and new products are changing the consumption patterns of milk and milk products. Milk surpluses in the United States and the world result in an abundance of consumer products, but the surpluses represent economic challenges to milk producers.

This chapter focuses primarily on milk as human food, but the importance of milk in nourishing suckling farm mammals should not be overlooked. When the term *milk* is used, reference is to milk from dairy cows unless otherwise specified.

MILK PRODUCTION

Table 5.1 shows the most important sources of milk for humans in the world and the United States. Total world milk production has increased during the past 25 years, but not at the same rate as world human population (49% versus 53%). Dairy cows produce over 90% of the world fluid milk supply (Fig. 5.2). During the past 30 years, milk yield from buffalo and sheep has been increasing, while

FIGURE 5.1 (A) Milk plays an important role in human nutrition throughout the world. Courtesy of Heifer Project International. (B) Milking native sheep in the desert of Iran. Courtesy of R. E. McDowell, Cornell University. (C) Goats being milked in Mexico. Courtesy of Winrock International. (D) Milking dairy cows in a modern U.S. dairy. Courtesy of Colorado State University.

goat milk production has been decreasing. Some milk from dairy goats in the United States is marketed through supermarkets and other outlets similar to the marketing of cow's milk (Fig. 5.3). However, most milk from goats is produced and consumed by individuals or families who raise a few goats.

Leading countries for milk production are also shown in Table 5.1. The United States produces more milk than the Russian Federation (153 billion lb versus 97 billion lb). It is interesting to note the difference in cow numbers and productivity per cow. The 42 million cows in the former Soviet Union have an average annual production per cow of approximately 5,000 lb, whereas the 9.5 million cows in the United States have an average yearly production of more than 16,400 lb per cow.

TABLE 5.1   Major Sources of Milk Production

Species	Total Production (bil lb)	Leading Countries (bil lb)
Cows	1,009	United States (153), Russian Fed. (97), India (66), France (62)
Buffalo	106	India (69), Pakistan (26), China (4), Egypt (3)
Goats	23	India (5.2), Iran (2.0), Pakistan (1.4), Sudan (1.3)
Sheep	18	Turkey (2.3), Iran (1.8), China (1.5), Italy (1.4)
All others	52	
World total	1,208	

Source: 1994 *FAO Production Yearbook.*

The national dairy herd in the United States has decreased from over 20 million cows in 1956 to approximately 9.5 million in 1995. Yet total milk production has increased. Reduction in cow numbers has been offset by an increased production per cow—from 5,800 lb in 1956 to more than 16,400 lb in 1995. The 16,400 lb per cow comes to 1,903 gal (1 gal = 8.62 lb), which would supply the annual per-capita, fluid-milk consumption for approximately 78 people in the United States. However, since part of the milk is made into other dairy

FIGURE 5.2 Dairy cows have the unique ability to convert feedstuffs into milk. Milk is an important source of nutrients to humans throughout the world. Courtesy of Dairy Council, Inc.

FIGURE 5.3 Milk products from dairy goats. Courtesy of George F. W. Haenlein, University of Delaware.

products (Fig. 5.4), the milk from one cow provides milk and manufactured products (cheese, frozen dairy products, and others) for more than 20 people.

Fluid milk sales are important to the U.S. economy. The wholesale and retail values of fluid milk are approximately $19 billion and $26 billion, respectively.

MILK COMPOSITION

Milk is a colloidal suspension of solids in liquid. Fluid whole milk is approximately 88% water, 8.6% **solids-not-fat (SNF),** and 3–4% milk fat. The SNF is the total solids minus the milk fat. It contains protein, lactose, and minerals.

FIGURE 5.4 Milk, cheese, and yogurt are among the most highly preferred milk products. Courtesy of the American Dairy Association.

Even though milk is a liquid, its 12% total solids is similar to the solids content of many solid foods. Differences in milk composition for several species of animals are given in Chapter 19.

The first milk a female produces after the young is born is called **colostrum.** True colostrum is milk obtained during the first milking. For the next several milkings, the milk is called *transitional milk* and is not legally salable until the 11th milking. Colostrum differs greatly in composition from milk. Colostrum is higher in protein, minerals, and milk fat, but it contains less lactose than milk. The most remarkable difference between colostrum and milk produced later is the extremely high immunoglobulin content of colostrum. These protein compounds accumulate in the mammary gland, and they are the antibodies that are transferred through milk to the suckling young. People seldom use colostrum from animals because of its unpleasant appearance and odor, although a few may use it for pudding. Additionally, the high globulin content causes a precipitation of proteins when colostrum is pasteurized.

Milk contains a high nutrient density—more than 100 milk components have been identified. **Nutrient density** refers to the concentration of major nutrients in relation to the caloric value of the food. In other words, milk contains important amounts of several nutrients while being relatively low in calories. The major components of milk solids can be grouped into nutrient categories including proteins, fats, carbohydrates, minerals, and vitamins.

Milk Fat

In whole milk, the approximate 3–4% milk fat is a mixture of lipids existing as microscopic globules suspended in the milk. The fat contributes about 48% of the total calories in whole milk. Fat-soluble vitamins (A, D, E, and K) are normal components in milk fat. Milk fat contains most of the flavor components of milk, so when milk fat is decreased, there may be a concurrent reduction in flavor.

Approximately 500 different fatty acids and fatty-acid derivatives from 2–26 carbon atoms in length have been identified in milk. Milk fat from ruminant animals contains both short-chain and long-chain fatty acids.

Carbohydrates

Lactose, the predominant carbohydrate in milk, is synthesized in the mammary gland. Approximately 4.8% of cow's milk is lactose. It accounts for approximately 54% of the SNF content in milk. Lactose is only about one-sixth as sweet as sucrose, and it is less soluble in water than other sugars. It contributes about 30% of the total calories in milk. Milk is the only natural source of lactose.

Proteins

Milk contains approximately 3.3% protein. Protein accounts for about 38% of total SNF and about 22% of the calories of whole milk. The proteins of milk are of high quality. They contain varying amounts of all amino acids required by humans. A surplus of the amino acid lysine offsets the low lysine content of vegetable proteins and particularly cereals.

Casein, a protein found only in milk, is approximately 82% of the total milk protein. Whey proteins, primarily lactalbumin and lactoglobulin, constitute the remaining 18%. Immunoglobulins, the antibody proteins of colostral milk, were discussed earlier in the chapter.

Vitamins

All vitamins essential in human nutrition are found in milk. Fat-soluble vitamins are in the milk fat portion of milk, and water-soluble vitamins are in the nonfat portion. Milk is usually fortified with vitamin D during processing.

Milk fat from Jersey and Guernsey cows has a rich, yellow color due to carotene (a precursor of vitamin A), which is yellow. Milk fat from Holstein cows has a pale, yellow color, and goat milk fat is white. Carotene can be split into two molecules of vitamin A. Milk fat of Holstein cows is pale yellow because most of the carotene has been split, and the milk fat of goats is white because all of the carotene has been changed to colorless vitamin A.

Water-soluble vitamins (C and B) are relatively constant in milk and are not greatly influenced by vitamin content of the cow's ration. The B vitamins are produced by rumen microorganisms, and vitamin C is formed by healthy epithelial tissue in most animals (excluding humans, other primates, and guinea pigs).

Minerals

Milk is a rich source of calcium for the human diet and a reasonably good source of phosphorus and zinc. Calcium and vitamin D are needed in combination to contribute to bone growth in young humans and to prevent **osteoporosis** in adults, particularly women. Milk is not a good source of iron, and the iodine content of milk varies with the iodine content of the animal's feed.

Table 5.2 and Fig. 5.5 show how the milk supply is used to produce various milk products. More than 80% of the milk produced is marketed as fluid milk, cream, cheese, and butter.

The amount of milk needed to produce each product depends primarily on the milk fat content of the milk. Table 5.3 gives the approximate amount of milk used to produce each product.

TABLE 5.2   Milk and Milk Products Resulting from the 1994 Milk Supply

Whole-Milk Product	Total Production (bil lb)	Percent of Total
Fluid milk and cream	54.7	34.8%
Cheese	51.1	33.2
Butter	24.5	16.0
Frozen dairy products	12.5	8.2
Evaporated and condensed milk	1.2	1.1
Uses on farms where produced	1.8	1.1
Other uses	5.5	4.9
Total	151.3	

[a] Equivalent to 26 billion quarts.
Source: Milk Industry Foundation, 1995 *Milk Facts.*

FIGURE 5.5 The modern dairy tree showing the many products and by-products of milk. Courtesy of *J. Dairy Sci.* 64:1005.

MILK PRODUCTS IN THE UNITED STATES

Fluid Milk

Approximately 92% of the 155 billion lb of milk produced in the United States is grade A quality milk. Grade A milk in excess of fluid milk demand is processed into other milk products. How fluid milk typically moves from farm to consumer is shown in Fig. 5.6.

TABLE 5.3 **Approximate Amount of Milk Used to Produce Several Selected Dairy Products**

Dairy Product (lb)	Whole Milk (lb)
Butter	21.2
Whole-milk cheese	10.0
Evaporated milk	2.1
Condensed milk	7.4
Ice cream (1 gal)	12.0
Cottage cheese	7.2[a]
Nonfat dry milk	11.0[a]

[a] Skim milk, not whole milk.
Sources: USDA; Milk Industry Foundation, 1995 *Milk Facts.*

Depending on the milk fat content and the processing at the plant, the fresh fluid product is labeled *whole milk, low-fat milk,* or *skim milk.* In further discussion in this chapter, *milk* refers to whole milk. Whole milk is defined as a lacteal secretion, and when it is packaged for beverage use it must contain not less than 3.25% milk fat and not less than 8.25% milk SNF. Low-fat milk usually has had some or most milk fat removed to have one of the following milk fat levels: 0.5, 1.0, 1.5, or 2.0% with not less than 8.25% milk SNF. Skim milk has had most of the milk fat removed (it contains less than 0.5% milk fat). It must contain at least 8.25% SNF and may be fortified with nonfat solids to 10.25%.

Most fluid milk consumed in the United States is **homogenized** to prevent milk fat from separating from the liquid portion and rising to the top. Homogenization is a physical process resulting in a stable emulsion of milk fat. A *cream line* does not appear in homogenized milk, nor does it form butter if churned.

Fat globules in raw milk average about 6 μm in diameter. To visualize how small the fat globules in homogenized milk are, note that 25,000 μm equal approximately 1 in. Homogenized milk will likely deteriorate by becoming rancid more rapidly than nonhomogenized milk. Rancidity of homogenized milk is forestalled by pasteurizing milk prior to or immediately following homogenization, thus destroying the action of lipolytic enzymes.

Evaporated and Condensed Milk

Evaporated milk is produced by preheating to stabilize proteins and removing about 60% of the water. It is sealed in the container and then heat-treated to sterilize its contents. Milk fat and SNF of evaporated milk must be at least 7.5% and 25%, respectively. Evaporated skim milk must have at least 20% SNF and not more than 0.5% milk fat. Evaporated milk requires no refrigeration until opened. Once the can is opened, refrigeration is necessary to avoid spoilage.

Concentrated, or condensed, milk has milk fat (7.5%) and SNF (25.5%) requirements similar to those for evaporated milk. Concentrated milk also has water removed but, unlike evaporated milk, it is not subjected to further heat treatment to prevent spoilage. Most concentrated milk is sold bulk for industry use. A common form of concentrated milk on grocery shelves is sweetened (sugar added) condensed milk; it is used for candy and other confections.

Milk from farm to family

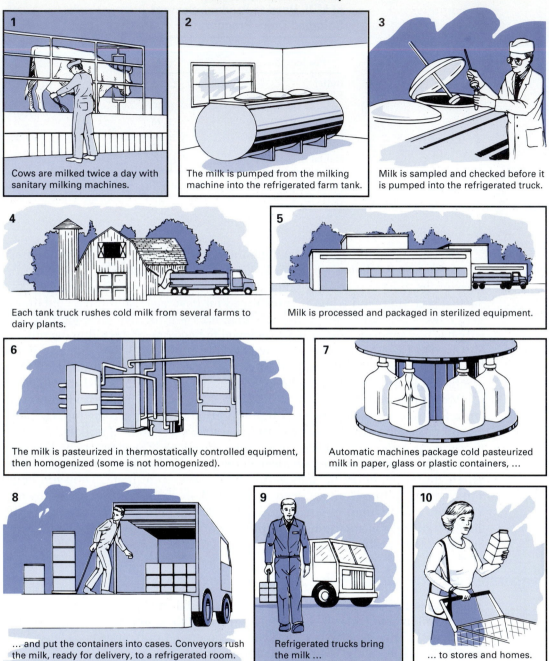

1 Cows are milked twice a day with sanitary milking machines.

2 The milk is pumped from the milking machine into the refrigerated farm tank.

3 Milk is sampled and checked before it is pumped into the refrigerated truck.

4 Each tank truck rushes cold milk from several farms to dairy plants.

5 Milk is processed and packaged in sterilized equipment.

6 The milk is pasteurized in thermostatically controlled equipment, then homogenized (some is not homogenized).

7 Automatic machines package cold pasteurized milk in paper, glass or plastic containers, ...

8 ... and put the containers into cases. Conveyors rush the milk, ready for delivery, to a refrigerated room.

9 Refrigerated trucks bring the milk ...

10 ... to stores and homes.

FIGURE 5.6 The processes involved in making milk available to consumers.

Dry Milk

Dry milk is prepared by removing water from milk, low-fat milk, or skim milk. All of these products are to contain no more than 5% moisture by weight. Dry milk can be stored for long periods of time if it is sealed in an atmosphere of nonoxidizing gas, such as nitrogen or carbon dioxide. The spray-dried or foam-dried product can be reconstituted easily in warm or cold water with agitation.

Fermented Dairy Products

Buttermilk, yogurt, sour cream, and other cultured milk products are produced under rigidly controlled conditions of sanitation, inoculation, incubation, acidification, and/or temperature. The word *cultured* appearing in a product name indicates the addition of appropriate bacteria cultures to the fluid dairy product and subsequent fermentation. Selected cultures of bacteria convert lactose into lactic acid, which produces a tart flavor. The specific bacterial cultures used and other controls in the fermentation process determine whether buttermilk, sour cream, yogurt, or cheese is the end product.

Today buttermilk is a cultured product rather than the by-product of churning cream into butter as was the procedure in the past. At the correct stage of acid and flavor development, the buttermilk is stirred gently to break the curd that has formed. It is then cooled to stop the fermentation process.

The word *acidified* in the product name or production process indicates that the food was produced by souring milk or cream with or without the addition of microbial organisms.

Cottage cheese dry curd is soft, unripened cheese; it is produced by culturing or direct acidification. Finished dry curd contains less than 0.5% milk fat and not more than 80% moisture. Cottage cheese is prepared for market by mixing cottage cheese dry curd with a pasteurized creaming mixture called *dressing*. The finished product contains not less than 4% milk fat and not over 80% moisture. Low-fat cottage cheese is made by the same process; however, the milk fat range is 0.5–2%.

Sour cream is generally a lactic acid fermentation, although rennet extract is often added in small quantities to produce a thicker-bodied product. Federal standards provide that cultured sour cream contain not less than 18% milk fat. When stored for more than 3–4 weeks, sour cream may develop a bitter flavor as a result of continued bacterial proteolytic enzyme activity.

Yogurt, as a liquid or gel, can be manufactured from fresh whole, low-fat, or skim milk that is heated before fermentation. Federal standards specify that yogurt contain not less than 3.25% milk fat, low-fat yogurt between 0.5–2% milk fat, and nonfat yogurt not more than 0.5% milk fat before any bulky flavors are added.

Today three main types of yogurt are produced: (1) flavored containing no fruit; (2) flavored containing fruit (fruit may be at the bottom or blended in); and (3) unflavored.

Cream

Cream is a liquid milk product, high in fat that has been separated from milk. Federal standards require that cream contain not less than 18% milk fat.

Several cream products are marketed (Fig. 5.7). Half-and-half is a mixture of milk and cream containing not less than 10.5% but less than 18% milk fat. Light cream (coffee or table cream) contains not less than 18% but less than 30% milk fat. Light whipping cream or whipping cream contains not less than 30% but less than 36% milk fat. Heavy cream or heavy whipping cream

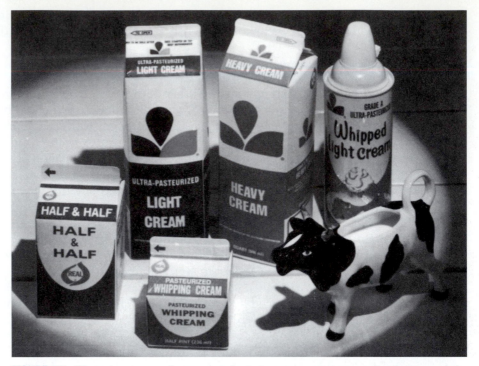

FIGURE 5.7   There is a range of cream products to meet consumer needs. Courtesy of the American Dairy Association.

contains not less than 36% milk fat. Dry cream is produced by removal of water only from pasteurized milk and/or cream. It contains not less than 40% but less than 75% milk fat and not more than 5% moisture.

Butter

It is known from food remnants in vessels found in early tombs that the Egyptians cooked with butter and cheese. For many centuries, butter making and cheese making were the only means of preserving milk and cream.

Butter churns are the oldest dairy equipment, dating from prehistoric days when nomads carried milk in a type of pouch made from an animal's stomach. Slung on the back of a horse or camel, the pouch bounced as the animal moved, churning the cream or milk into butter.

Wooden or crockery dasher-type churns used at the turn of the century are today's attic treasures. By jouncing a long-handled, wooden dasher up and down in the deep churn, the cream was sloshed around until butter particles separated from the remaining liquid, called *buttermilk.*

Butter is the dairy product made exclusively from milk or cream, or both, which contains not less than 80% by weight of milk fat. Federal standards establish U.S. grades for butter based on flavor, color, and salt characteristics.

In modern butter making, fresh sweet milk is weighed, tested for milk fat content, and checked for quality. The cream is then separated by centrifugation to contain 30–35% milk fat for batch-type churning or 40–45% milk fat for continuous churning.

Continuous butter-making operations, which can produce 1,800–11,000 lb an hour, are an industry trend. New continuous butter-making processes employ several different scientific principles to form butter.

Cheese

There are more than 400 different kinds of cheese that can be made (Fig. 5.8). The cheeses have more than 2,000 names, since the same cheese may have two or more different names. For example, cheddar cheese is one of the American-type cheeses in the United States, along with Colby, washed or stirred curd, and Monterey.

Most cheeses and cheese products are classified into one of four main groups: (1) soft, (2) semisoft, (3) hard, and (4) very hard. Classification is based on moisture content in the cheese. Thus the body and texture of cheeses range from soft, unripened cheese (such as cottage cheese with 80% moisture) to very hard, grated, shaker cheeses such as Parmesan and Romano. These latter, ripened cheeses have 32–34% moisture.

Cheese flavor varies from bland cottage cheese to pungent Roquefort and Limburger. Modern microbiology makes it possible to add specific species and strains of microorganisms required to produce the desired product. For example, a fungus (*Penicillium roqueforti*) is used to produce Roquefort cheese.

The most dramatic increase in cheese production in the United States has occurred with Italian varieties. The phenomenal growth of the pizza industry has caused an increased production and importation of mozzarella cheese.

Cheese making involves a biochemical process called *coagulation,* or *curdling.* First the milk is heated, then a liquid starter culture is added. Bacteria

FIGURE 5.8   There are numerous kinds of cheeses to satisfy varying consumer preferences. Courtesy of the American Dairy Association.

from the culture form acids and turn the milk sour. At a later time, rennet is added to thicken the milk. *Rennet,* obtained from the stomachs of young calves, contains the enzyme rennin. Other enzymes of microbial origin have been used successfully to replace a declining supply of rennin. After stirring this mixture, a custardlike substance called *curd* is formed. Then the liquid part (*whey*) is removed. The cheese-making process reduces 100 lb of milk to 8–16 lb of cheese.

In cheese plants, disposal of whey is a problem. It is currently estimated that 35 billion lb of whey containing more than 4 billion lb of solids is produced. Slightly more than half the whey is used to produce human food or animal feeds. The remaining whey poses waste-disposal problems. There continues to be an increasing industrial use of whey for production of food, feed, fertilizer, alcohol, and insulation.

Table 5.4 identifies the leading countries in yearly cheese production and the most important cheese-producing states in the United States.

Ice Cream

Although they come in many variations, ice cream and a group of similarly frozen foods are made in a similar way and have many of the same ingredients. Ice cream, frozen custard, French ice cream, and French custard ice cream are frozen dairy products highest in milk fat and milk solids. Ice cream may contain egg yolk solids. If egg yolks are in excess of 1.4% by weight, the product is called frozen custard, French ice cream, or French custard ice cream.

Ice milk contains less milk fat, protein, and total solids than ice cream. Ice milk usually has more sugar than does ice cream. Soft ice milk and soft ice cream are soft and ready to eat when drawn from the freezer. About three-fourths of the soft-serve products are ice milks. Some of the soft-serve frozen dairy foods contain vegetable fats that have replaced all or most of the milk fat.

Frozen yogurt has become a popular dairy product. Frozen yogurt has less milk fat and higher acidity than ice cream and less sugar than sherbet.

Sherbet is low in both milk fat and milk solids. It has more sugar than ice cream. The tartness of fruit sherbet comes from the added fruit and fruit acid. Nonfruit sherbet is flavored with such ingredients as spices, coffee, or chocolate.

Water ices are nondairy frozen foods. They contain neither milk ingredients nor egg yolk. Ices are made similarly to sherbet.

Mellorine differs from ice cream in that milk fat may be replaced partially by a vegetable fat, another animal fat, or both.

TABLE 5.4   Leading Cheese-Producing Countries and States

Country	Total Production (bil lb)	State	Total Production[a] (mil lb)
United States	7	Wisconsin	2,018
Germany	3	California	926
France	3	Minnesota	658
Italy	2	New York	560
Russian Federation	1	Pennsylvania	348
World total	33	**U.S. total**	6,730

[a] Does not include cottage cheese.
Sources: USDA; 1994 *FAO Production Yearbook.*

Eggnog

Eggnog contains milk products, egg yolk, egg white, and a nutritive carbohydrate sweetener. In addition, eggnog may contain salt, flavoring, and color additives. Federal standards specify that eggnog shall contain not less than 6% milk fat and 8.25% SNF.

Imitation Dairy Products

The FDA established regulations in 1973 to differentiate between imitation and substitute products. It defined an *imitation product* as one that looks like, tastes like, and is intended to replace the traditional counterpart but is nutritionally inferior to it. A *substitute product* resembles the traditional food but also meets the FDA's definition of nutritional equivalency.

Imitation milks usually contain ingredients such as water, corn syrup solids, sugar, vegetable fats, and protein from soybean, sodium caseinate, or other sources. Although imitation fluid milk may not contain dairy products per se, it may contain derivatives of milk such as whey, lactose, casein, salts of casein, and milk proteins other than casein.

The dairy industry has developed a program for identification of real dairy foods. A "REAL" seal (as seen in Fig. 5.7) on a carton or package identifies milk or other dairy foods made from U.S.-produced milk that meets federal and/or state standards. This seal assures consumers that the food is not an imitation or substitute.

HEALTH CONSIDERATIONS

Nutritive Value of Milk

Milk and other dairy products make a significant contribution to the nation's supply of dietary nutrients. Particularly noteworthy are the relatively large percentages of calcium, phosphorus, protein, and B vitamins (Table 5.5).

Human milk is regarded as the best source of nourishment for infants. Cow's milk for infant feeding is modified to meet the nutrient and physical requirements of infants. Cow's milk is heated, homogenized, or acidified so the nutrients can be utilized efficiently by infants. Sugar is usually added to cow's milk for infant feeding to make milk more nearly like human milk.

Milk is low in iron; therefore, young animals consuming nothing but milk may develop anemia. Baby pigs produced in confinement need a supplemental source of iron, usually given as an injection. Babies typically receive supplemental iron, and young children who consume large amounts of milk at the expense of meat should be given supplemental iron.

Milk and milk products are excellent sources of nutrients to meet the dietary requirements of young children, adolescents, and adults. With aging, there is an increased prevalence of osteoporosis, a problem of loss of bone mass. Since osteoporosis involves the loss of bone matrix and minerals, a diet generous in protein, calcium, vitamin D, and fluoride is recommended to overcome the problem. Thus, milk, owing to its content of calcium and other nutrients, is an important food for this segment of the population. Furthermore, for the elderly as well as for the infant and young child, milk is an efficient source of nutrients readily tolerated by a sometimes weakened digestive system.

TABLE 5.5   Percentage Contribution of Dairy Foods (excluding butter) to Nutrients Consumed in the United States

Nutrients	1970	1980	1990
Energy	10	10	9
Protein	20	20	20
Fat	12	12	12
Carbohydrate	6	6	5
Cholesterol	14	13	15
Minerals			
Calcium	74	75	75
Phosphorus	35	34	34
Magnesium	20	19	18
Iron	2	2	2
Zinc	18	18	19
Vitamins			
Ascorbic acid	4	3	3
Thiamin	9	7	7
Riboflavin	34	31	31
Vitamin B_6	11	11	10
Vitamin B_{12}	17	18	20
Vitamin A	17	17	18

Source: USDA.

Wholesomeness

Milk is among the most perishable of all foods owing to its excellent nutritional composition and its fluid form. As it comes from the cow, milk provides an ideal medium for bacterial growth. Properly processed milk can be kept for 10–14 days under refrigeration. With ultra-high-temperature (UHT) processing, milk can be kept for several weeks at room temperature.

Protecting the quality of milk is a responsibility shared by public health officials, the dairy industry, and consumers. The U.S. Food and Drug Administration describes milk as "one of the best controlled, inspected and monitored of all food commodities."

The most important safeguards from a health standpoint are requirements that bacterial counts be low and that milk be pasteurized (Fig. 5.9). Milk sold through commercial outlets is certified to be from herds that are tested and found to be free from brucellosis and tuberculosis. The consumer who buys milk and milk products can be assured of obtaining a safe, desirable, wholesome food. There are definite health risks in consuming milk from a cow or goat where the milk has not been pasteurized. This concern is important where families or individuals have purchased the animal with an unknown health background or purchased milk that has not been processed.

Milk Processing

Milk is taken from the cow by a sanitized milking machine, then transported through sanitized pipes into holding tanks. When withdrawn from the cow, milk is at the cow's body temperature of about 100°F (38°C). It flows into a refrigerated tank, where it is rapidly cooled to 40–42°F. Cold temperature maintains the high quality of the milk while it is held for delivery.

Tank truck drivers, who pick up milk, inspect it to see if it is cold and has an acceptable aroma. They take a milk sample for testing later at the processing

FIGURE 5.9   Milk samples are checked for undesirable bacteria. Courtesy of Dairy
Council, Inc.

plant. Then the cold milk is pumped from the refrigerated farm tank through a
sanitized hose into the insulated tank on the truck (Fig. 5.10).

After the milk sample passes several tests at the processing plant, it is
pumped through sanitized pipes into the processing plant's refrigerated or
insulated holding tanks.

Milk is pasteurized at the processing plant. *Pasteurization* is a process of
exposing milk to a temperature that destroys all pathogenic bacteria but neither

FIGURE 5.10   Large milk trucks move milk from dairy farms to processing plants.
The tanks are washed and sanitized after each delivery. Courtesy of Dairy Council,
Inc.

reduces the nutritional value of milk nor causes it to curdle. Milk is most commonly pasteurized at 161°F (71.5°C) for 15 sec.

Ultrapasteurized milk and UHT-processed milk are heated to 280°F (138°C) for at least 2 sec; this sterilizes milk and increases shelf life.

As it flows out of the pasteurizer, most milk is homogenized by being pumped through a series of valves under pressure. The milk fat is broken up into particles too small to coalesce. As a result, they remain suspended throughout the milk rather than separating to the top as a layer of cream.

The pasteurized, ultrapasteurized, or UHT-processed milk is cooled rapidly to below 45°F (7°C). UHT milk is packaged into presterilized containers and aseptically sealed. Since bacteria cannot enter the UHT milk, it can be kept unrefrigerated for at least 3 months. However, once the container is opened, the UHT milk picks up organisms from the air. Then the UHT milk must be handled and stored like any other fluid milk product.

Milk Intolerance

Milk is the main dietary source of a carbohydrate lactose. In the 1960s, the potential problem of lactose intolerance was emphasized when reports revealed low levels of the enzyme lactase in digestive tracts of 70% of black and 10% of the white persons in the United States. Symptoms include bloating, abdominal cramps, nausea, and diarrhea. Worldwide, lactose intolerance is relatively high among nonwhite populations. However, results of many studies have disclosed the difference between lactose intolerance and milk intolerance. While a large segment of certain populations may be diagnosed as lactose-intolerant, most of these individuals, once adapted, can tolerate the amount of lactose contained in typical servings of dairy foods.

For the few individuals who are truly milk-intolerant, suitable alternatives are available. These include consumption of recommended amounts of milk in smaller but more frequent servings throughout the day, most cheeses, many fermented and culture-containing dairy products (e.g., yogurt), lactose-hydrolyzed milk, and some dairy products containing up to 75% less lactose.

Milk protein allergy may be considered as another form of milk intolerance. An allergic reaction to the protein component of cow's milk may occur in a few infants who experience allergies early in life. The incidence is probably 1% or less in the infant and child population in industrialized countries, although a range of 0.3–7.5% has been reported. The condition is usually outgrown by 2 years of age, beyond which time true allergic reactions to milk in the general population are rare.

CONSUMPTION

Population growth by the year 2000 will no doubt cause an increase in consumption of dairy products; however, per-capita consumption is expected to decrease (*J. Dairy Sci.* 64:959).

Table 5.6 shows the per-capita consumption of dairy products in several selected countries. Ireland and Sweden are highest in fluid milk consumption, and New Zealand leads in butter consumption. Greece, France, and Sweden are the countries with the highest per-capita cheese consumption.

TABLE 5.6   Yearly Per-Capita Consumption of Selected Dairy Products in Selected Countries

Leading Countries and the United States[a]	Per-Capita Consumption (lb)		
	Milk	Butter	Cheese
Austria	310	11.4	—
Belgium	—	13.1	28.3
Denmark	—	11.1	34.0
Finland	310	16.2	23.7
France	—	19.8	49.4
Germany	—	15.1	25.7
Greece	—	—	51.6
Ireland	421	—	—
Italy	—	—	39.9
Netherlands	267	—	32.5
New Zealand	286	32.5	—
Poland	341	—	—
Sweden	362	12.2	37.4
Switzerland	—	14.1	34.8
United States	211	4.2	26.9

[a] U.S. data shown for comparison to leading countries.
Sources: USDA; Milk Industry Foundation, 1995 *Milk Facts.*

Table 5.7 shows the per-capita sales of milk and milk products in the United States in 1975, 1985, 1990, and 1995. There is a noticeable consumer preference toward low-fat products. Figures 5.11 and 5.12 reflect the changes in consumer preferences during the past several years, with ice milk, skim milk, yogurt, and cheese showing the most significant increases.

TABLE 5.7   Yearly Per-Capita Sales of Dairy Products in the United States (lb)

Product	1975	1985	1990	1995
Plain whole milk	165.0	116.7	85.5	72.1
Lowfat milk	53.2	83.3	98.2	94.6
Skim milk	11.5	12.6	22.6	31.2
Flavored milk and drinks	9.6	9.7	9.4	10.0
Buttermilk	4.7	4.4	3.5	2.8
Yogurt	2.1	4.1	4.1	4.8
Eggnog	0.4	0.5	0.5	0.4
Half-and-half	2.4	3.0	3.0	3.1
Light cream	0.4	0.4	0.3	0.3
Heavy cream	0.6	1.0	1.3	1.4
Sour cream and dips	1.6	2.3	2.5	2.7
Butter	4.4	3.9	3.7	4.2
American cheese	7.9	9.6	11.1	11.7
Other cheese	6.1	10.4	13.5	15.4
Cottage cheese	4.6	4.1	3.4	2.8
Dry milk	3.2	2.2	3.4	3.9
Evaporated milk	8.7	7.4	7.9	8.0
Ice cream	18.6	18.1	15.8	16.1
Ice milk	7.6	6.9	7.7	7.7

Sources: USDA; Milk Industry Foundation, 1995 *Milk Facts.*

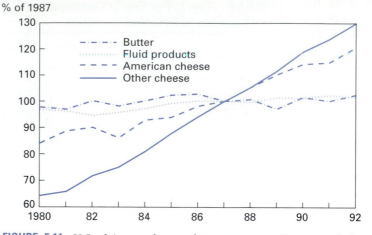

FIGURE 5.11 U.S. dairy products sales, 1980–1992. Courtesy of the USDA.

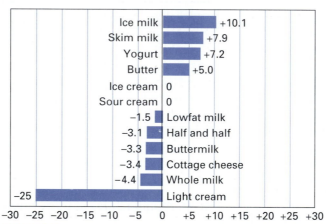

FIGURE 5.12 Percent change in per-capita sales of selected dairy products, 1993–94. Source: *Milk Facts,* 1995, Milk Industry Foundation.

MARKETING

World

Only a small percentage of the world's production of milk products enters world trade: butter (4%), cheese (4%), skim milk powder (18%), and casein (80%). The European Union (EU) and New Zealand account for about 80% of total world exports of dairy products.

The EU, Canada, and the United States have surpluses of skim milk powder in part because of price stabilization programs and because production costs exceed world market prices. Exports of skim milk powder go principally to countries with a milk deficit and are used in a variety of recombined products. Much of New Zealand and Australian production of casein is exported to the United States.

Nutritionally, the world may need milk, but there is no real market for its present surpluses of dairy products. Surpluses continue to be a major marketing challenge both in the United States and the world.

United States

Most milk produced on U.S. dairy farms goes to plants and dealers for processing. More than 50% of the fluid milk is marketed by supermarkets, primarily in plastic gallon containers.

Prices

Cooperative milk marketing associations, located near large population centers, give producers more bargaining power, as they control approximately 90% of the milk produced. The cooperatives hire professionals to market the milk to prospective purchasers.

There are classes and grades of milk that determine price. Grade A (fluid or market milk) and grade B (manufacturing milk) are determined by the sanitary and microbial quality of the milk. Class I, the highest-price class, is grade A milk for fluid use. Classes II and III are for milk used in manufactured products. Class II includes milk for cottage cheese, cream, and frozen desserts; class III is milk used for butter and cheese. Class II is grade A milk, and class III is surplus grade A or grade B milk. Producers receive a *blend* price based on the proportion of milk used in each price class.

Forty-four federal *milk-marketing orders* and 14 states establish the prices that processors must pay dairy farmers for about 95% of the fluid milk and fluid milk products consumed in the United States. In some states, milk control commissions not only determine what farmers are to be paid but the price that stores can charge customers. Federal milk orders are to promote orderly marketing conditions for dairy producers and ensure consumers an adequate supply of milk.

Milk prices in recent years are shown in Chapter 2, Fig. 2.2. Lower milk prices in 1990 and 1991 reflect the surplus of dairy products. Total farm receipts from the sale of milk in 1995 were $19.9 billion, while the retail value of the fluid milk industry was $26.0 billion.

CHAPTER SUMMARY

■ Most of the 1,208 billion pounds of world milk production is contributed by cows (1,009 bil lb), water buffalo (106 bil lb), goats (23 bil lb), and sheep (18 bil lb).

■ The United States has the highest milk production (153 bil lb) from cows with the Russian Federation (97 bil lb) and India (66 bil lb) the next leading countries.

■ In addition to fluid milk for drinking and cooking, cheese, butter, fermented dairy products (e.g., cottage cheese, sour cream, and yogurt), and frozen dairy products (e.g., ice cream, yogurt, and sherbet) are important consumer dairy products.

- Dairy products contribute significant amounts of protein, calcium, phosphorus, and riboflavin to the human diet.
- The leading contribution to the 290 lb annual per capita consumption of milk products in the U.S. are lowfat milk (95 lb), whole milk (72 lb), skim milk (31 lb), cheese (31 lb), and ice cream (24 lb).
- Ice milk, skim milk, and yogurt have increased sales in recent years while cream, whole milk, and cottage cheese have experienced decreased per capita sales.

REVIEW QUESTIONS

1. What is the term used to describe all of the milk components exclusive of water and milk fat?
2. The solids non-fat component of milk consists of which three constituents?
3. What is the first milk produced by a female after giving birth called?
4. What is the most remarkable difference between milk and colostrum?
5. Milk is a rich source of which mineral required in the diet of humans?
6. What is homogenization?
7. What is the process used to destroy all pathogens in milk during milk processing?
8. *True or False:* Pasteurization reduces the nutritional value of milk.

SELECTED REFERENCES

Publications

Dairy Producer Highlights. National Milk Producers Federation, 1840 Wilson Blvd., Arlington, VA 22201.

Hedrick, T. I., Harmon, L. G., Chandan, R. C., and Seiberling, D. 1981. Dairy products industry in 2006. *J. Dairy Sci.* 64:959.

Lowenstein, M., Speck, S. J., Barnhart, H. M., and Frank, J. F. 1980. Research on goat milk products: A review. *J. Dairy Sci.* 63:1629.

Milk Facts. 1995. Milk Industry Foundation, 888 Sixteenth St. NW, Washington, DC 20006.

Miller, G. D. et al. 1995. *Handbook of Dairy Foods and Nutrition.* Rosemont, IL: National Dairy Council.

Newer Knowledge of Cheese and Other Cheese Products. 1983. National Dairy Council, 6300 N. River Road, Rosemont, IL 60018.

Newer Knowledge of Milk and Other Fluid Dairy Products. 1983. National Dairy Council, 6300 N. River Road, Rosemont, IL 60018.

Product Information Sheets (Butter and Cream; Milk; and Ice Cream). 1983. National Dairy Council, 6300 N. River Road, Rosemont, IL 60018.

Quality of U.S. Agricultural Products (milk and dairy products, p. 247) 1996. Ames, IA: Council for Agricultural Science and Technology.

Sellars, R. L. 1981. Fermented dairy foods. *J. Dairy Sci.* 64:1070.

Speckmann, E. W. L., Brink, M. F., and McBean, L. D. 1981. Dairy foods in nutrition and health. *J. Dairy Sci.* 64:1008.

Tobias, J., and Muck, G. A. 1981. Ice cream and frozen desserts. *J. Dairy Sci.* 64:1077.

Tong, P. 1994. Dairy Processing and Products. *Encyclopedia of Agricultural Science.* San Diego: Academic Press, Inc.

Visuals

The Dairy Industry (VHS video). VEP, California Polytechnic State University, San Luis Obispo, CA 93407.

The Dairy Plant (VHS video; 28 min). Creative Educational Video, Inc., 5147-A 69th St., Lubbock, TX 79424.

CHAPTER 6

Poultry and Egg Products

Poultry meat and eggs are nutritious and relatively inexpensive animal products used by humans throughout the world. Feathers, down, livers, and other **offal** are additional useful products and by-products obtained from poultry.

Application of genetics, nutrition, and disease control, along with sound business practices, has advanced the commercial poultry industry to where eggs and poultry meat can be produced very efficiently. The industry has developed two specialized and different types of chickens—one for meat production and one for egg production. Broiler lines are superior in efficient, economical meat production but have a lower egg-producing ability than the egg-production lines. However, the egg-type birds produce large numbers of eggs very efficiently. They have been bred to mature at light weights and they are therefore slower-growing and inefficient meat producers.

POULTRY MEAT AND EGG PRODUCTION

Broiler chickens provide most of the world's production and consumption of poultry meat. Turkeys and roaster chickens, mature laying hens (fowl), ducks, geese, pigeons, and guinea hens are consumed in smaller quantities, although some of these are important food sources in some areas of the world. World poultry and egg production is presented in Chapter 2. The more common kinds of poultry used for meat production in the United States are shown in Table 6.1.

Poultry are slaughtered and processed in large plants owned by integrated poultry companies. Two such companies are Tyson Foods and Perdue Farms with 1995 sales of $5.5 billion and $1.6 billion, respectively. Some packing and

TABLE 6.1   Age and Weights of Various Kinds of Poultry Used for Meat Production

Classification	Typical Age (mo)	Average Slaughter Wt (lb)	Average Dressed Wt (lb)	Dressing (%)
Young chickens				
Broiler (fryer)	1.5–2	4–5	3–4	78
Rock Cornish	1.0	2	1.5	75
Roaster	3–5	6–8	5	75
Fowl (mature for stewing)	19	5	3.8	75
Turkeys	4–6	10–25	8–20	83
Ducks	<2	6–7	4–5	70
Geese	6	11	8–9	75

Sources: USDA and various others.

processing companies that slaughter and process poultry also slaughter and process red meat animals. Con Agra, with $24.1 billion sales in 1995, is the leading meat and poultry processor.

COMPOSITION

Meat

Table 6.1 shows chickens yield approximately 78% of their liveweight as carcass weight whereas the dressing percent for turkeys is 83%. The edible raw meat from carcasses of chickens and turkeys is approximately 67% and 78%, respectively.

The gross chemical composition of chickens and turkeys is shown in Table 6.2. Dark meat is higher in calories, lower in protein, and higher in cholesterol than light meat for both chicken and turkey. If the skin remains on poultry, it usually adds cholesterol and total fat.

TABLE 6.2   Composition of Chicken and Turkey per 100 g, Edible Portion

	Water (g)	Food Energy (calories)	Protein (g)	Total Fat (g)	Cholesterol (mg)	Ash (g)
Young chicken						
Light meat (no skin)	74.9	114	23.2	1.6	58	0.98
Dark meat (no skin)	76.0	125	20.1	4.3	80	0.94
Turkey (hen)						
Light meat (no skin)	73.6	116	23.6	1.7	58	1.0
Dark meat (no skin)	74.0	130	20.1	4.9	62	0.95

Source: USDA, *Agricultural Handbook* no. 8-5.

Eggs

An egg has a spherical shape with one end being rather blunt and larger than the other smaller, more pointed end. Figure 6.1 shows a longitudinal section of a hen's egg. The mineralized shell surrounds the contents of the egg. Immediately inside the shell are two membranes; one is attached to the shell itself, and the other tightly encloses the content of the egg. The air cell usually forms between the membranes in the blunt end of an egg shortly after it is laid, as the contents of the egg cool and contract.

Seven thousand to 17,000 tiny pores are distributed over the shell surface, a greater number at the larger end. As the egg ages, these tiny holes permit moisture and carbon dioxide to move out and air to move in to form the air cell. The shell is covered with a protective covering called the *cuticle* or *bloom.* By blocking the pores, the cuticle helps to preserve freshness and prevent microbial contamination of the contents.

The egg white, or *albumen,* exists in four distinct layers. One of these layers surrounds the yolk and keeps the yolk in the center of the egg. The spiral movement of the developing egg in the magnum causes the mucin fibers of the albumen to draw together into strands that form the chalaziferous layer and chalazae. The twisting and drawing together of these mucin strands tend to squeeze out the thin albumen to form an inner thin albumen layer.

The yellowish-colored yolk is in the center of the egg, its contents surrounded by a thin, transparent membrane called the *vitelline membrane.*

Egg weight varies from 1.5–2.5 oz, with the average egg weighing approximately 2 oz. The component parts of an egg are shell and membranes (11%), albumen (58%), and yolk (31%). The gross composition of an egg is shown in Table 6.3. The mineral content of the shell is approximately 94% calcium carbonate. The yolk has a relatively high fat and protein percent, although there is more total protein in the albumen.

The weight classes and grades of eggs are discussed in Chapter 8. Egg formation is presented in Chapter 10.

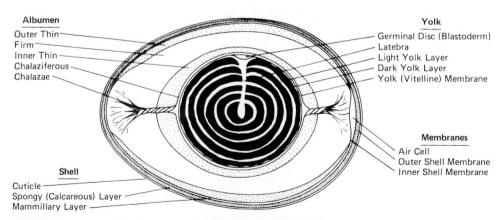

Albumen
Outer Thin
Firm
Inner Thin
Chalaziferous
Chalazae

Yolk
Germinal Disc (Blastoderm)
Latebra
Light Yolk Layer
Dark Yolk Layer
Yolk (Vitelline) Membrane

Membranes
Air Cell
Outer Shell Membrane
Inner Shell Membrane

Shell
Cuticle
Spongy (Calcareous) Layer
Mammillary Layer

FIGURE 6.1   Longitudinal section of a hen's egg. Courtesy of the USDA.

TABLE 6.3 **Chemical Composition of the Egg, Including the Shell**

	Percent of Total	Water (%)	Protein (%)	Fat (%)	Ash (%)
Whole egg	100	65.5	11.8	11.0	11.7
White	58	88.0	11.0	0.2	0.8
Yolk	31	48.0	17.5	32.5	2.0
	Percent of Total	Calcium Carbonate (%)	Magnesium Carbonate (%)	Calcium Phosphate (%)	Organic Matter (%)
Shell and shell membranes	11	94.0	1.0	1.0	4.0

Source: USDA.

POULTRY PRODUCTS

Meat

More than one-half of all broilers leave the processing plant as cut-up chicken (Table 6.4) or selected parts rather than whole birds (compared to less than 30% a decade ago).

Approximately 60% of the broilers, fowl (mature chickens), and turkeys are further processed beyond the ready-to-cook carcass or parts stage. The meat is separated from the bone and formed into products (e.g., turkey roll fingers and nuggets), cut and diced (e.g., for preparation of salads and casseroles), canned, or ground. Some of the cured and smoked ground products are frankfurters, bologna, pastrami, turkey ham, and salami. More than 1 out of 8 lb of all broilers is processed into such products as chicken patties, nuggets, battered and cooked chicken, and hot dogs. Even more of the broilers are expected to be processed in future years.

Eggs

Eggs are unique, prepackaged food products that are ready to cook in their natural state. Eggs find their way into the human diet in numerous ways. They are

TABLE 6.4 **The Retail Parts and Cooked Edible Meat of a Broiler-Fryer Chicken**

Part	(% of carcass)	Edible[a]
Breast	28	63
Thighs	18	60
Drumsticks	17	53
Wings	14	30
Back and neck	18	27
Giblets	5	100
Total	100	53

[a] Net average yield of meat is 67% without neck and giblets.
Sources: USDA; National Broiler Council.

most popular as a breakfast entrée; however, they are commonly used in sandwiches, salads, beverages, desserts, and several main dishes.

Shell eggs can be further processed into pasteurized liquid eggs and dried eggs and other egg products. Pasteurized liquid egg can be blast-frozen and sold to bakeries and other food-processing plants as an ingredient in other fabricated food products or transported to users as a liquid product in refrigerated tank trucks. The dried egg products are sold to food manufacturers to produce products such as cake mixes, candy, and pasta.

The initial step in making egg products is breaking the shell and separating the yolks, whites, and shells. This is done in most egg-breaking plants by equipment completely automated to remove eggs from egg-filler flats, wash and sanitize the shells, break the eggs for individual inspection of the contents by an operator, and separate the yolks from the whites. Processing equipment can also be used to produce various white and yolk products or mixtures of them. Present automatic systems are capable of handling 18,000–50,000 eggs per hour (Fig. 6.2).

Some breeds of chickens lay white-shelled eggs and other breeds produce brown-shelled eggs. The brown shell color comes from a reddish-brown pigment (ooporphyrin) derived from hemoglobin in the blood. Although there are shell color preferences by consumers in certain parts of the United States, there are no significant nutritional differences related to egg shell color.

Feathers and Down

In some areas of the world, ducks and geese are raised primarily for feathers and down. *Down* is a small, soft feather found beneath the outer feathers of ducks and geese. Feathers and down provide stuffing for pillows, quilts, and upholstery. They are also used in the manufacture of hats, clothing (outdoor wear), sleeping bags, fishing flies, and brushes.

The duck and goose industry in the United States is too small to meet the manufacturing demand for down. Thus, raw feathers and down are primarily imported (approximately 90% of the product used) from China, France, Switzerland, Poland, and several other European and Asian countries.

Breeder geese are often used for down production, as the older birds produce the best down. Four ducks or three geese will produce 1 lb of feather-and-down mixture, of which 15–25% is down. Down and feathers are separated by machine, then washed and dried.

Other Products and By-Products

Goose livers are considered a gourmet product in European countries. These enlarged fatty livers, weighing as much as 2 lb each, are produced by force-feeding corn to geese three times a day for 4–8 weeks. The liver is used to make a flavored paste called *pâté de foie gras* (for hors d'oeuvres or sandwiches).

Poultry by-products include hydrolyzed feather meal and poultry meat and bone scraps. These by-products are occasionally included in livestock and poultry rations as protein sources (refer to Chapter 4 for more discussion on these by-products).

A

B

C

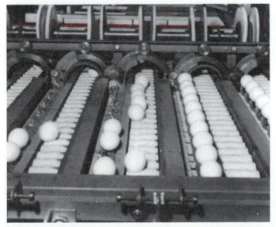

D

FIGURE 6.2 Automated egg-handling equipment. (A) Overall view. (B) Automatic egg washer. (C) Eggs on the flash-candling area of the in-feed conveyor. (D) Inline scales. Eggs of different sizes are weighed and ejected at different points on the line. Courtesy of the USDA, *Egg Grading Manual.*

NUTRITION CONSIDERATIONS

Nutritive Value of Poultry Meat

The white meat from fowl is approximately 33% protein, and dark meat is about 28% protein. This protein is easily digested and is of high quality, containing all the essential amino acids. The fat content of poultry meat is lower than that found in many other meats. Poultry meat is also an excellent source of vitamin A, thiamin, riboflavin, and niacin. The nutritive value of poultry and eggs to the U.S. population is shown in Table 6.5.

Nutritive Value of Eggs

Since the egg (Fig. 6.3) contains many essential nutrients, it is recognized as one of the important foods in the major food groups recommended in *Dietary Guidelines for Americans.* These guidelines were developed by the U.S. Department of Agriculture to use in planning diets contributing to good health.

Eggs have a high nutrient density in that they provide excellent protein and a wide range of vitamins and minerals and have a low calorie count. The abundance of readily digestible protein, containing large amounts of essential amino acids, makes eggs a least-cost source of high-quality protein.

Table 6.6 lists the amounts of selected nutrients found in eggs. When the values of these nutrients are compared with the dietary allowances recommended by the Food and Nutrition Board of the National Research Council (1980), it is found that one large egg supplies a relatively high percentage of essential nutrients (Table 6.7). Eggs are an especially rich source of high-quality protein, vitamins (A, D, E, folic acid, riboflavin, B_{12}, and pantothenic acid), and

TABLE 6.5   Contribution of Poultry and Eggs to Nutrient Supplies Available to U.S. Civilian Consumption

Nutrients	1970	1980	1990	1970	1980	1990
	Meat, Poultry,[a] and Fish (%)			Eggs (%)		
Food energy	22	20	17	2	2	1
Protein	44	44	41	5	2	2
Fat	37	35	30	3	2	2
Cholesterol	44	46	47	37	36	33
Minerals						
Calcium	4	4	3	2	2	2
Phosphorus	30	29	27	5	5	4
Iron	26	23	18	4	3	2
Magnesium	15	15	14	1	4	1
Vitamins						
Vitamin A	29	26	23	6	5	5
Thiamin	30	25	21	1	1	1
Riboflavin	25	22	20	9	8	6
Niacin	48	42	40	—	—	—
Vitamin B_6	42	41	39	3	3	2
Vitamin B_{12}	77	76	75	4	4	4
Ascorbic Acid	2	2	2	0	0	0

[a] Poultry comprises approximately 35% of the per-capita retail weight consumption of meat, poultry, and fish.
Source: USDA.

FIGURE 6.3 "The incredible, edible egg." Courtesy of the American Egg Board.

minerals (phosphorus, iodine, iron, and zinc). Some consumers believe that the dark yellow yolk is of higher nutrient value than the paler yolks, but there is no evidence to support this belief.

Eggs are relatively high in cholesterol with most of the cholesterol and fat concentrated in the yolk (Table 6.6). Individuals who desire to reduce fat and cholesterol in their diet and still enjoy eggs can: (1) limit egg yolks to one serving for scrambled eggs and omelets (use more whites for larger servings) and (2) substitute two egg whites for one whole egg when baking.

Table 6.7 lists the U.S. recommended daily allowances (RDAs) supplied by one large egg.

CONSUMPTION

Meat

Table 6.8 shows the poultry and egg consumption for the leading countries of the world. The United States has the highest per-capita consumption of poultry meat, while per-capita egg consumption is highest in Israel. The annual

TABLE 6.6   Estimated Nutrient Values for a Large Egg (based on 60.9-g shell weight with 55.1-g total liquid whole egg, 38.4-g white, and 16.7-g yolk)

Nutrients and Units (proximate)	Whole	White	Yolk
Solids (g)	13.47	4.6	8.81
Calories (kcal)[a]	84	19	64
Protein (N × 6.25) (g)	6.60	3.88	2.74
Total lipids (g)	6.00	—	5.80
Cholesterol (mg)	213	—	213
Ash (g)	0.55	0.26	0.29

[a] A medium-sized egg has 66 calories, while a jumbo egg has 94 calories.
Source: Poultry Sci. 58:131; USDA.

TABLE 6.7   U.S. Recommended Daily Allowance (RDAs) in Relation to Nutrient Content of One Large Egg

Nutrient	RDA	Percent RDA, One Egg[a]
Protein	45.0 g	15%
Vitamins		
A	5,000.0 IU	6
D	400.0 IU	6
E	30.0 IU	2
C	60.0 mg	
Folic acid	0.4 mg	6
Thiamin	1.5 mg	2
Riboflavin	1.7 mg	15
Niacin	20.0 mg	
B_6	2.0 mg	4
B_{12}	6.0 µg	8
Biotin	0.3 mg	4
Pantothenic acid	10.0 mg	6
Minerals		
Calcium	1.0 g	2
Phosphorus	1.0 g	8
Iodine	150.0 µg	15
Iron	18.0 mg	4
Magnesium	400.0 mg	2
Copper	2.0 mg	2
Zinc	15.0 mg	4

[a] One large egg will also contain approximately 80 calories and 5 grams of fat.
Source: American Egg Board (Eggcyclopedia).

per-capita consumption of broilers in the United States is approximately 70 lb. Broiler consumption and production continue to expand during the 1990s (Fig. 6.4). Poultry consumption is approximately one-half of the total per-capita consumption of red meat and poultry (Fig. 6.5). In recent years, poultry consumption has increased at the expense of red meat consumption. This is due, in part, to the price differential existing between the various meats (Fig. 6.6).

The dollar amount expended for broilers and turkey is shown in Table 6.9. Both total dollars and percent of per-capita income have increased for both of these poultry products during the past 10–15 years.

An important reason for the marked increase in chicken consumption has been the speed of preparing and serving chicken. The Kentucky Fried Chicken Corporation (KFC) led in preparation and serving chicken in a short time, and now chicken has become a major item in fast-food service. The KFC restaurants

TABLE 6.8   World Per-Capita Poultry Meat and Egg Consumption

Country	Poultry Meat (lb)[a]	Country	Hen Eggs (no.)
1. United States	96	1. Israel	353
2. Israel	79	2. Czechoslovakia	328
3. Hong Kong	78	3. Hungary	281
4. Singapore	68	4. Hong Kong	271
5. Saudi Arabia	69	5. Japan	260

[a] Ready-to-cook equivalent.

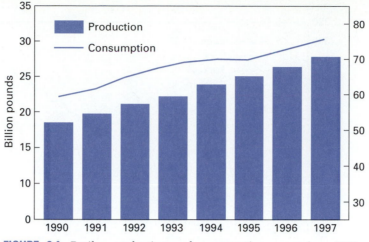

FIGURE 6.4 Broiler production and consumption expansion, 1990–1997. Source: USDA.

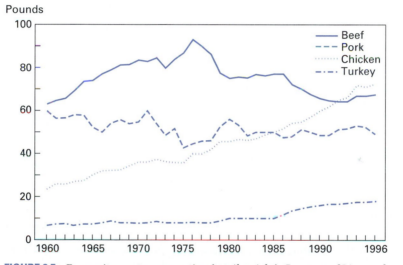

FIGURE 6.5 Per-capita meat consumption (retail weight). Courtesy of Livestock Marketing Information Center, Lakewood, Colo.

are found in 56 countries and they sell more than 3 billion pieces of chicken a year. In recent years, other fast-food restaurants have featured chicken or added more chicken items to their menus.

Heavy hens, the type used to produce hatching eggs for broiler production, are sometimes available in retail stores as *stewing hens* or *baking hens.* These account for about 10% of the poultry meat consumed by Americans each year.

Consumption of turkey meat per person has increased from 1.7 lb in 1935 and 6.1 lb in 1960 to 8.0 lb in 1970 and 18 lb in 1995. Most of this increase in turkey consumption was stimulated by the development of processed products such as turkey rolls, roasts, pot pies, and frozen dinners. The selling of prepackaged turkey parts in small packages has also increased consumption.

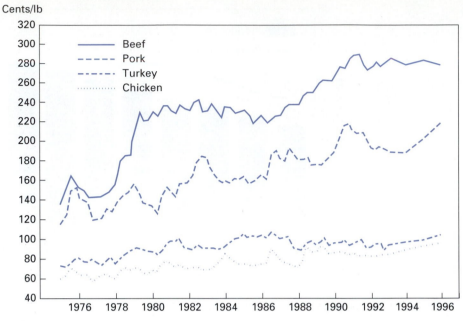

FIGURE 6.6   Retail prices for chicken and turkey with comparative beef and pork prices. Courtesy of Livestock Marketing Information Center, Lakewood, Colo.

Eggs

Per-capita egg consumption in the United States is approximately 235 eggs per year which is a decrease from 310 eggs in the early 1970s and 265 eggs in the early 1980s. Most of these are consumed as shell eggs (85%) rather than processed eggs (Fig. 6.7). The per-capita consumption is nearly the number of eggs (254) a hen lays per year.

MARKETING

In the United States, large amounts of poultry meat are processed in commercial facilities and then transported under refrigeration to local and world markets. Broilers are the primary poultry meat exported, with more than 900 million pounds exported each year. Approximately 100 million dozen eggs are

TABLE 6.9   Expenditures Per Person for Poultry

	Broilers		Turkey		Total Poultry	
Year	Dollars	Percent[a]	Dollars	Percent[a]	Dollars	Percent[a]
1970	$15.22	0.15%	$4.54	0.05%	$19.77	0.20%
1980	45.08	0.38	9.93	0.08	55.01	0.46
1985	59.89	0.45	12.66	0.10	75.55	0.55
1990	88.90	0.63	17.35	0.12	106.25	0.76
1995	93.89	0.67	17.40	0.12	111.29	0.79

[a] Percent of per-capita disposable income.
Source: USDA, *Livestock and Poultry Situation and Outlook.*

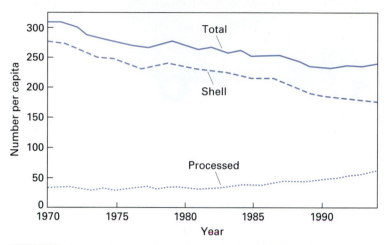

FIGURE 6.7   Per-capita egg consumption in the United States. Source: USDA.

exported annually from the United States. The export value of broiler meat is shown in Fig. 6.8.

Chicken meat is usually marketed fresh in the United States, whereas turkeys are marketed frozen. However, some turkeys are sold fresh year-round in some markets. Freshly frozen poultry meat can be marketed on a year-round basis through the use of frozen storage and environmentally controlled units.

Most of the broilers are sold to consumers through retail stores (Fig. 6.9). Fresh chicken can reach the retail market counter the day following slaughter. "Sell-by" date on each package is usually 7 days from processing and is the last date recommended for sale of chicken. However, with proper refrigeration (28–32°F), shelf life can be extended up to 3 days longer. Some processors guarantee 14 days of shelf life for their broilers.

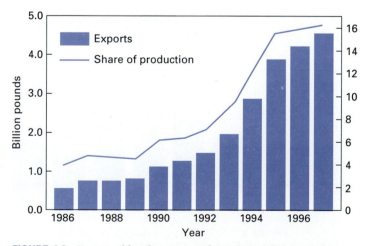

FIGURE 6.8   Export of broiler meat and its share of U.S. production. Source: USDA.

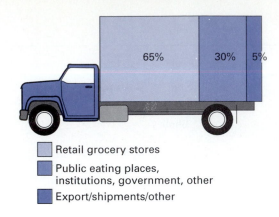

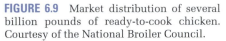

■ Retail grocery stores

■ Public eating places,
  institutions, government, other

■ Export/shipments/other

FIGURE 6.9 Market distribution of several
billion pounds of ready-to-cook chicken.
Courtesy of the National Broiler Council.

Almost all commercially available chicken and turkey meat is inspected for
wholesomeness by federal or state government employees. About two-thirds of
the chickens are graded by the USDA for quality.

Over 40% of the broilers are marketed to consumers under the producer's
brand names. In certain markets, such as New York, over 90% of the broilers are
sold with brand names. Poultry brands identify the single firm that raised, processed, and marketed the bird, not just the firm that processed the retail product, as is the case of red meats. The four largest vertically integrated poultry
firms represented more than 40% of the broiler market in 1992, compared with
only 17% in 1973.

Before and during the 1950s, the marketing of turkey was highly seasonal,
as 90% of the turkeys were consumed during the last quarter of the year for
the Thanksgiving and Christmas holidays. Today less than 40% of turkey
sales are made during that period. Current trends indicate that turkey is
being consumed more on a year-round basis and has become a meat staple
(Fig. 6.10).

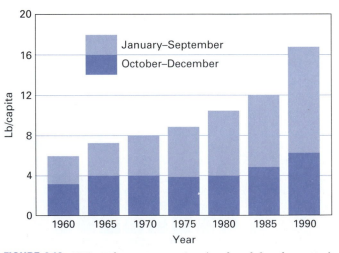

FIGURE 6.10 U.S. turkey consumption (total and fourth quarter),
1960–90. Courtesy of the USDA.

CHAPTER SUMMARY

■ Poultry meat and egg products are nutritious and relatively inexpensive animal products used by humans throughout the world.

■ Poultry meat and eggs are excellent sources of protein, vitamin A, and several B vitamins for human nutrition.

■ Annual per capita consumption of broilers is 70 lb and has been increasing rapidly over the past several years. Per capita turkey consumption is 19 lb and increasing. Egg consumption has been decreasing and currently is 235 eggs per year.

■ Broiler and turkey retail prices are much lower than prices of beef and pork.

REVIEW QUESTIONS

1. What major products come from poultry?
2. What essential nutrient is poultry meat very high in?
3. What three important by-products are produced by geese?
4. What is the primary poultry meat exported by the United States?

SELECTED REFERENCES

Publications

Boiler Industry (periodical). Watt Publishing Co., 122 S. Wesley Ave., Mt. Morris, IL 61054-1497.

Egg Industry (periodical). Watt Publishing Co., 122 S. Wesley Ave., Mt. Morris, IL 61054-1497.

Eggcyclopedia. 1989. American Egg Board, 1460 Renaissance Dr., Park Ridge, IL 60068.

Moreng, R. E., and Avens, J. S. 1985. *Poultry Science and Production.* Reston, VA: Reston Publishing.

Poultry Digest (periodical). Watt Publishing Co., 122 S. Wesley Ave., Mt. Morris, IL 61054-1497.

Quality of U.S. Agricultural Products (poultry, p. 195). 1996. Ames, IA: Council for Agricultural Science and Technology.

Stadelman, W. J., and Cotterill, O. J. 1995. *Egg Science and Technology.* Westport, CT: Avi.

USDA. *Egg Grading Manual.* Apr. 1983. USDA. Agricultural Handbook no. 75. Revised.

USDA. 1992–94. Livestock and poultry Situation and Outlook Report. Washington, D.C.

Scanes, A. R. 1994. Poultry Processing and Products. *Encyclopedia of Agricultural Science.* San Diego: Academic Press, Inc.

Stadelman, W. J. 1994. Egg Production, Processing, and Products. *Encyclopedia of Agricultural Science.* San Diego: Academic Press, Inc.

Watt Poultry Yearbook. 1996. Mount Morris, IL: Watt Publishing Co.

Visuals

Egg quality film (loan). Information Division, AMS, USDA, Washington, D.C. 20250.

Egg quality slides. Photography Division, GPA, USDA, Washington, D.C. 20250.

The Incredible Edible Egg on Video. Northwest Egg Producers Coop. Assoc., P.O. Box 1038, Olympia, WA 98507.

Processing Chicken Broilers (80 slides and tape; 13 min); and *Egg Breaking Operations* (80 slides and tape; 12 min). Poultry Science Dept., Ohio State University, 674 W. Lane Ave., Columbus, OH 43210.

Turkey product slides. National Turkey Federation, 11319 Sunset Hills Rd., Reston, VA 22090.

Wool, Mohair, and Other Fibers

Skins of common mammals have a covering to which various terms are applied depending on the nature of the growth. For example, cattle, pigs, horses, and dairy goats have hair; sheep have wool; mink and non-Angora rabbits have fur; Angora rabbits have angora; Angora goats have mohair; and cashmere goats produce cashmere hair and cashmere down (fibers <30 microns). Fibers that grow from the skin of animals give protection from abrasions to the skin and help keep animals warm.

Hair from most mammals has little commercial value (it is used mostly in padding and cushions). Fur of mink and non-Angora rabbits is either naturally beautiful or can be dyed to give attractive colors; therefore, it has considerable value as fur. Fur is composed of fine, short fibers and relatively long, coarse guard hairs, in contrast to the skin covering of cattle, horses, and pigs, which is composed entirely of guard hairs.

Furs made from rabbit, mink, fox, and bison (American buffalo) may be classed as status clothing because they are usually costly and attractive in appearance. White rabbit furs can be dyed any color (including the pastels), but colored furs cannot. Because colored furs cannot be dyed, color **mutations** have been important in the mink business because they provide a wide array of fur colors. The mink industry in the United States has suffered economically because of competition from furs made in other countries and high costs of producing mink.

Hides from young market lambs are often marketed with wool left on to be processed into heavy winter lambskin coats. Lambs from the **Karakul** breed of fat-tail sheep are used to produce Persian lambskins. The black, tight curls of fiber are more like fur than wool. Persian lambskins have been produced in the United States, but larger numbers are currently produced in Afghanistan, Iran, and the former Soviet Union.

Hides from sheep with longer wool on the pelts also may be marketed intact for processing to produce ornamental rug pieces. The long wool may be dyed. The long wool may also be loosened and removed from the hides after slaughter and sold separately from the hides.

GROWTH OF HAIR, WOOL, AND MOHAIR

Hair or wool fiber grows from a **follicle** located in the outer layers of the skin. Growth occurs at the base of the follicle, where there is a supply of blood, and cells produced are pushed outward. The cells die after they are removed from the blood supply because they can no longer obtain nutrients or eliminate wastes. A schematic drawing of a wool follicle is presented in Fig. 7.1.

The cuticle causes the fibers to cling together. The intermingling of wool fibers is known as **felting.** The felting of wool is advantageous in that wool fibers can be entangled to make **woolens,** but it is also responsible for the shrinkage that occurs when wool becomes wet.

All wool and hair fibers have a similar gross structure, consisting of an outer thin layer (**cuticle**) and a cortex that surrounds an inner core (**medulla**) in medullated fibers (Fig. 7.2). Only medium and coarse wools have a medulla. It is absent in fine wools.

Wool fibers have waves called **crimp** (Fig. 7.3). Crimp is caused by the presence of hard and soft cellular material in the cortex. The soft cortex is more elastic and is on the outer side of the crimp.

Hair does not exhibit any crimp because all the cells in the cortex are hard. Also, the inner core of hair is not solid, unlike the inner core of most wool fibers. Some fibers in wool of a low quality are large, lack crimp, and do not have a solid inner core. These fibers are called *kemp,* and they reduce the value of the fleece.

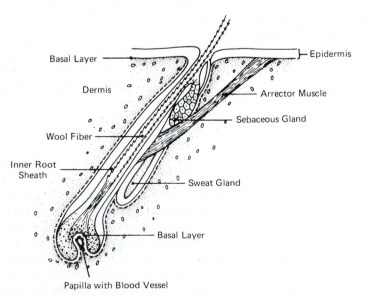

FIGURE 7.1 Schematic drawing of a wool follicle. Courtesy of R. W. Henderson.

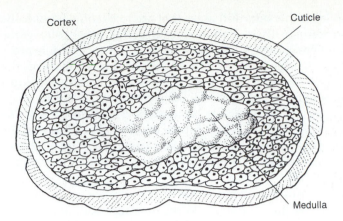

FIGURE 7.2 Cross section of a medullated wool fiber. Adapted from *The Sheep Production Handbook.*

Cortex

Cuticle

Medulla

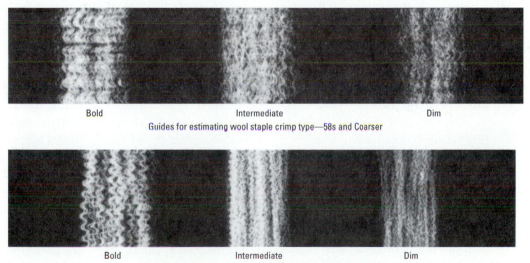

Bold   Intermediate   Dim

Guides for estimating wool staple crimp type—58s and Coarser

Bold   Intermediate   Dim

Guides for estimating wool staple crimp type—60s and Finer

FIGURE 7.3 Several expressions of crimp in coarser and finer wools. Courtesy of the USDA.

Mohair fiber follicles develop in groups consisting of three primary follicles each. Secondary follicles develop later. Mohair fibers have very little crimp (less than 1 crimp for each inch of length). The mohair cortex is composed largely of so-called *ortho* cells, in contrast to the cortex of wool, which is composed of both *ortho* and *para* cells. Mohair has a different scale structure from wool, but it does have scales and will felt. Kemp and **medullated fibers** reduce the value of wool and mohair fleeces because such fibers do not dye well and because they show in apparel made from fleeces containing such fibers.

Skin of young lambs is made up of two layers: the surface skin (outer skin) is the *epidermis,* and the underlying skin is the *dermis* (Fig. 7.1). Between these two layers is the *basal layer.* Development of wool follicles occurs in skin of the fetus. The basal layer thickens and areas that are to become wool follicles push down into the dermis. Two glands develop: the *sebaceous gland,* which secretes *sebum* (a greasy secretion) and the *sweat gland,* which produces and secretes sweat. The downward growth of the basal layer rests on the papilla, which has a supply of blood. The wool fiber grows toward the skin's surface, and as it becomes removed from the blood supply, the cells die.

Two major types of follicles are produced. The first type to develop is known as the *primary follicles.* They appear on the poll and face of the fetus by 35–50 days' gestation and over all other parts of the body by 60 days. They are arranged in groups of three. All primary follicles are fully developed and are producing fiber at birth.

A second type of follicle, associated with the primary follicles, develops later and is known as the *secondary follicles.* These follicles have an incomplete set of accessary structures—usually the sweat gland and arrector muscle are absent. The primary and associated secondary follicles are grouped into follicle bundles.

Since all the primary follicles are formed prior to birth of the lamb, the density of primary follicles per square inch of skin declines rapidly during the first 4 months of postnatal life and then declines gradually until the lamb approaches maturity.

Secondary follicles show little activity during the first week of a lamb's life, after which there is a burst of activity of secondary follicles. Activity is maximum between 1 and 3 weeks of age with marked reduction after the third week.

Adverse prenatal environmental conditions can affect the number of secondary follicles initiated in development. Also, early postnatal influence could affect secondary follicle fiber production.

Growth of the wool fiber takes place in the *root bulb,* which is located in the follicle. Permanent dimensions of the fiber are determined in the area of the bulb, and there are no changes in growth characteristics of the fiber after it is formed. Defective portions of a fiber are caused by a reduction in the size of the fiber that creates a weakened area when sheep are under stressed conditions, such as poor nutrition and high body temperatures. Such fibers are likely to break under pressure.

Two types of undesirable fibers are kemp and medullated fibers. Both are hollow and brittle, and both are produced by primary follicles rather than secondary follicles, as is the case with true wool.

FACTORS AFFECTING THE VALUE OF WOOL

The two most important factors under the control of the producer that affect the amount of wool produced by sheep are nutrition and breeding. The amount of feed available to the sheep (energy intake) and the percentage of protein in the diet influence wool production. Wool production is decreased when sheep are fed diets having less than 8% protein. When the diet contains more than 8% protein, the amount of energy consumed is the determining factor in wool production.

Improving wool production per animal through breeding involves selection for increased clean fleece weight, **staple length, fineness** and uniformity of length, and fineness throughout the fleece. Much progress can be made by performance-testing rams for quantity and quality of wool production, but for overall improvement of the flock, both lamb and wool production must be included (Fig. 7.4).

Sheep and Angora goat producers are interested in maximizing net income from their sheep and goat operations. Increasing net income from wool and mohair can be accomplished by selecting to improve wool and mohair production and grade. The values of wool and mohair that are produced can be enhanced by giving attention to the following items:

1. The sheep or goats should be shorn when the wool or mohair is dry.

2. The portions of clips that are loaded with dung (called *tags*) should be sorted from fleeces and sacked separately.

3. Wool or mohair should be sacked by wool grades so that when **core samples** are taken from a sack of otherwise good wool, a few fleeces of low-grade wool (or mohair) will not cause the entire sack of wool or mohair to be placed in a low grade.

4. Wool should be shorn without making many double clips.

5. Fleeces should be tied with paper twine after they are properly folded with the clipped side out.

6. A lanolin-based paint should be used for branding animals. This type of paint is scourable.

7. Fleeces from black-faced sheep should be packed separately from other wool. Also, black fleeces should be packed separately. Black fibers do not take on light-colored dyes and consequently stand out in a garment made from black-and-white fibers dyed a light pastel color.

8. All tags, sweepings, and wool or mohair from dead animals should be packed separately.

9. Hay that has been baled with twine should be carefully fed to sheep; otherwise, pieces of twine will get into the wool or mohair and markedly lower its value.

FIGURE 7.4 The ewe has been bred primarily for wool production with her lamb being a secondary product. Her heavy fleece will soon be shorn from her body. Courtesy of Woolknit Associates, Inc.

10. Environmental stresses should be avoided. Being without feed or water for several days or having a high fever can cause a break or weak zone in the wool fibers.

11. Wool containing coarse fibers, or kemp, should be avoided, or kempy fleeces should be packed separately. Cloth and wool containing these coarse fibers are highly objectionable and thus are low in price.

Figure 7.5 identifies several types of undesirable fleeces. These fleeces have a lower value than fleeces containing white fibers and minimal foreign material. Some of the fleeces in Fig. 7.5 not previously discussed include the following:

Burry—contains vegetable matter, such as grass seeds and prickly seeds, which adheres tenaciously to wool

Chaffy—contains vegetable matter such as hay, straw, and other plant material

Cotted—wool fibers are matted or entangled

Dead—wool pulled from sheep that have died but that have not been slaughtered

Murrain—wool obtained from decomposed sheep

CLASSES AND GRADES OF WOOL

The value of wool is determined by its class and grade. Staple length determines the class, and fineness of fibers determines the grade. The grades of wool are given by any one of three methods, all of which are concerned with a value that indicates the fineness (diameter) of the wool fibers. The three methods of reporting wool grades are these:

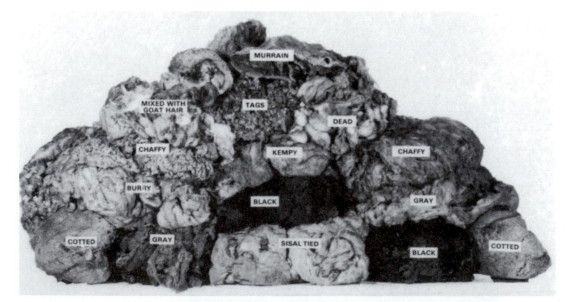

FIGURE 7.5 Several types of undesirable fleeces are identified. Each of these should be sorted and packed separately from the more desirable fleeces. Courtesy of the USDA.

1. The **American grade,** or blood (outdated terminology), method is based on the theoretical amount of fine wool breeding (Merino and Rambouillet) represented in the sheep producing the wool (Fig. 7.6). The terms *fine, ½ blood, ⅜, ¼ blood,* and *low-¼ blood* are used to describe typical fiber diameter. Wool classified as *fine* or *⅝ blood* has small fiber diameters, whereas *¼ blood* wool is coarse.

2. The **spinning count** system refers to the number of *hanks* of yarn, each 560 yards long, which can be spun from 1 lb of **wool top.** These grades range from 80 (fine) to 36 (coarse). Thus, a grade of 50 by the spinning count method means that 28,000 yards (560 yd × 50 = 28,000 yd) of yarn can be spun from 1 lb of wool top.

3. The **micron diameter** method is based on the average of actual measurements of several wool fibers. This is the most accurate method of determining the grade of wool. A micron is ½₅,₄₀₀ of an inch.

Wool samples from each of the major grades are shown in Figs. 7.6 and 7.7. The systems of grading, along with the breeds from which the grades of wool come, are presented in Table 7.1. The Merino and Rambouillet breeds are considered fine wool breeds with the Cotswold, Lincoln, and Romney identified as coarse wool breeds. In Table 7.1, breeds listed between the fine and coarse wool breeds are called *medium wool breeds.*

Classes of wool are staple, French combing, and clothing; however, fibers of finer grades that do not meet the length requirement may go into a higher class than they would if they were not fine. Much of the staple length wool is combed into **worsted-**type yarns for making garments of a hard finish that hold a press well.

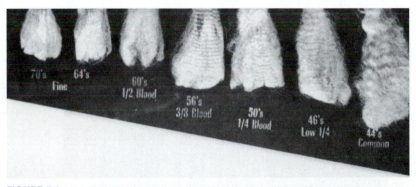

FIGURE 7.6   Wool samples of the major grades of wool based on spinning count and the blood system. Courtesy of J. H. Landers.

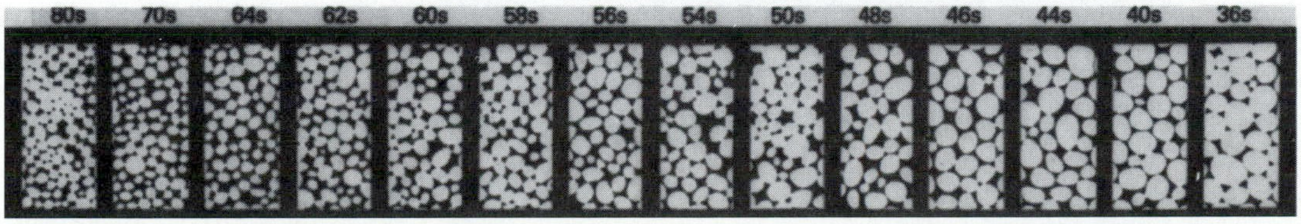

FIGURE 7.7   Cross section of magnified wool fibers demonstrating the wool grades based on fiber diameter (micron diameter). Courtesy of the USDA.

TABLE 7.1 Market Grades of Wool and Breeds of Sheep That Produce These Grades of Wool

USDA Grades Based on Spinning Count	USDA Grades Based on American (Blood) System	Grades Based on Micron Diameter	Breeds According to Average Grades
80s	Fine	17.70–19.14	Merino
70s	Fine	19.15–20.59	Rambouillet
64s	Fine	20.60–22.04	Targhee, Southdown
62s	½ blood	22.05–23.49	Corriedale, Columbia
50s to 60s	¼ blood to ½ blood	23.50–30.99	Panama, Romeldale
48s to 56s	Low-¼ blood to ½ blood	26.40–32.69	Shropshire, Hampshire, Suffolk, Oxford, Dorset, Cheviot
46s to 48s	Common to low ¼ blood	32.69–38.09	Romney
36s to 40s	Common and braid	36.20–40.20	Cotswold
36s to 46s	Common and braid	32.70–40.20	Lincoln

The fineness of wool fibers from a sheep depends on the area of the body from which the fibers came. The body of a sheep can be divided into seven areas. Area 1 is the shoulder and area 2 is the neck, where wool grows longer and finer than average wool from the same sheep. Area 3 is the back. It has long, average fineness, though it has been exposed to the most weathering. Area 4 is the side. It has long, average fineness and constitutes most of the wool. Area 5 is the tags. It is long, coarse, dirty, and stained wool. Area 6 is the britch. It is long, coarser-than-average wool that tends to be medullated. Area 7 is the belly. It has short, fine, matted wool that may be very heavy in vegetable matter. Preference is given to a fleece that has uniformity of length and fineness in all areas of the sheep's body.

PRODUCTION OF WOOL AND MOHAIR

Production of wool can be expressed as **greasy wool** or **scoured** (clean) **wool.** Greasy wool is wool or the fleece shorn once each year from sheep (Fig. 7.8).

FIGURE 7.8 Sheep being shorn for their yearly production of wool. Courtesy of the National Wool Growers (USDA photo).

TABLE 7.2 World Production of Greasy and Scoured Wool

Country	Greasy Wool (mil lb)	Scoured Wool (mil lb)
Australia	1,617	1,280
New Zealand	614	450
China	572	292
Russia Fed.	319	191
Kazakhstan	220	132
United States[a]	68	36
World total	5,865	3,770

[a] U.S. production shown for comparison; other countries are higher than the United States.
Source: 1994 FAO Production Yearbook.

Scoured wool originates in the initial wool-processing stage. Scouring is the washing and rinsing of wool to remove grease, dirt, and other impurities.

Table 7.2 shows the production of greasy and scoured wool for several countries of the world. The world grease wool production is approximately 6 billion lb, with the United States producing approximately 69 million lb. The U.S. wool production and total yearly supply is shown in Table 7.3. U.S. wool production by states is shown in Fig. 7.9. Scoured or clean wool production represents 50–60% of the greasy wool produced.

Growth of wool varies greatly among sheep breeds. Fine-wool sheep grow 4–6 inches of wool per year, ¼ to ½ blood breeds of sheep grow 5–10 inches of wool per year, and common to braid wool breeds grow 9–18 inches of wool per year. Mohair growth is approximately 12 inches per year.

Weights of **fleeces** also vary greatly among the breeds of sheep. Even though fine-wool breeds have fleeces of short length, weight of their fleeces varies from 9–12 lb; ¼ to ½ blood breeds of sheep produce fleeces weighing 7–15 lb; and coarse wool (common and braid) breeds produce fleeces weighing 9–16 lb.

Annual world mohair production is approximately 59 million lb with the leading countries of the United States and South Africa producing approximately half the world total. Texas produces about 97% of the mohair in the United States (Fig. 7.10). The pounds and value of mohair production in Texas is shown in Table 7.4. Mohair growth is about 1–2 inches per month; therefore, goats are usually shorn twice per year, with length of clip averaging 6–12 inches.

TABLE 7.3 U.S. Shorn Wool Production and Supply

Year	No. Head Shorn (mil)	Ave. Fleece Wt. (lb)	Price (cents/lb)[a]	Domestic Production (mil lb)	Total Supply (mil lb)[b]
1985	11.2	7.9	$0.63	88.1	124.6
1990	11.2	7.8	0.80	88.0	115.5
1994	8.9	7.7	0.78	68.6	125.1

[a] Prior to incentive payment.
[b] Includes imports, primarily from Australia and New Zealand.
Source: USDA (Agricultural Statistics, 1995–96).

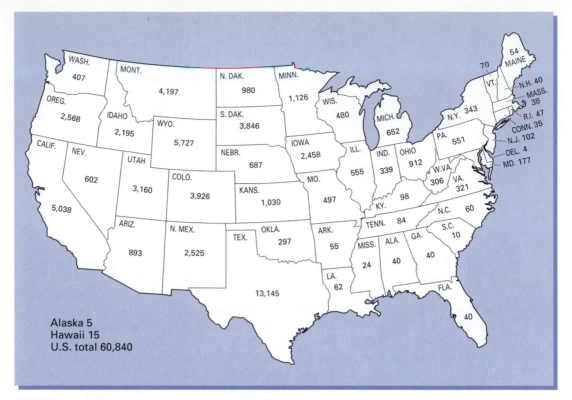

FIGURE 7.9 1996 U.S. wool production (1,000 lb). Source: USDA.

FIGURE 7.10 Angora goats with a full growth of mohair. Courtesy of Woolknit Associates, Inc.

TABLE 7.4   Mohair Production in Texas

Year	No. Goats Clipped (mil)	Ave. Wt. Per Clip (lb)	Total Production (mil lb)	Price per lb	Total Value ($ mil)
1985	1.7	7.7	13.3	$3.45	$45.9
1990	1.9	7.8	14.5	0.95	13.8
1994	1.6	7.3	11.7	2.62	30.6

Source: USDA (Agricultural Statistics, 1995–96).

USES OF WOOL AND MOHAIR

Fibers from sheep, Angora rabbits, and Angora goats (Fig. 7.11) are used in making cloth and carpets. The consumption of wool in the United States is small when compared to cotton and numerous manufactured fibers (Fig. 7.12). Figure 7.13 shows the U.S. production and consumption of wool.

FIGURE 7.11 Mohair, kid-mohair, and cashmere, strengthened by different proportions of synthetic fibers, are popular yarns for making attractive garments that "breathe" and reduce perspiration. Courtesy of George F. W. Haenlein, University of Delaware.

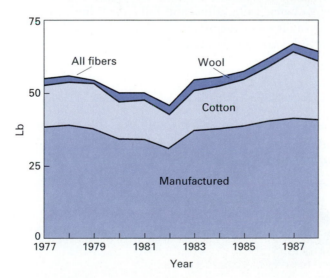

FIGURE 7.12 U.S. per-capita consumption of fibers, 1977–87. All fibers do not include flax and silk. Manufactured fibers include synthetic fibers such as nylon, dacron, and orlon. Courtesy of the USDA.

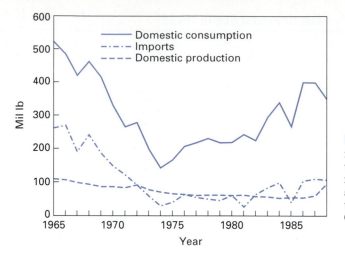

FIGURE 7.13 U.S. production, imports, and consumption of wool, 1965–88. Imports are on a raw-wool basis. Domestic consumption includes textiles containing wool that are imported. Courtesy of the USDA.

Cloth made from wool has both highly desirable and undesirable qualities. Wool has a pleasant, warming quality and can absorb considerable quantities of moisture while still providing warmth. It is also highly resistant to fire. However, wool has a tendency to shrink when it becomes wet, and cloth from wool causes some people to itch.

Researchers have contributed greatly to the alteration of wool so that fabrics or garments will not shrink when washed. A process called the WURLAN treatment makes woolen garments machine-washable. In this treatment, the wool fibers are coated with a very thin layer of resin, which adds only 1% to the weight of the wool. The resin does not alter wool in any significant way except to prevent the fibers from absorbing water. There is an increasing trend to blend wool with other fibers. This improves aesthetics and dyability and provides materials with more durability and comfort.

Only fabrics that have passed the quality tests demanded by the International Wool Secretariat and its U.S. branch, the Wool Bureau, Inc., are eligible to carry the Woolmark or Woolblend label (Fig. 7.14).

Woolen garments are as old as the Stone Age, yet as new as today. Wool takes many forms: wovens, knits, piles, and felts, each the result of a different process. It is a long way from a bale of raw wool to a beautiful bolt of fabric and, ultimately, to beautiful, versatile garments (Fig. 7.15).

CHAPTER SUMMARY

- Wool, mohair, and hair from animals provide useful human products.
- Wool value is determined by fleece weight, cleanliness, staple length, and fineness (fiber diameter).

FIGURE 7.14 The sewn-in Woolmark label (pure wool) and Woolblend label (wool blend) give assurance of quality-tested fabrics. Courtesy of the Wool Bureau, Inc.

FIGURE 7.15 Wool sweaters, wool-knit suits, and wool dresses offer comfort and fashion to the most discerning customers. Courtesy of Woolknit Associates, Inc. (A) Wool sweaters in landscape patterns by Panache and Alps. Knitted in America. (B) The wool suit by Castleberry is the well-dressed delight. Knitted in America. (C) A wool knit dress is designed in a swingy trapeze shape by Geoffrey Beene.

■ Mohair production from Angora goats is concentrated primarily in Texas.

■ Cotton, synthetic fibers, and wool imports have provided stiff competition for domestic wool production.

REVIEW QUESTIONS

1. From what structure located in the outer layer of skin does wool and hair grow?

2. What is "felting" of wool? What is responsible for it and why is it important?

3. What are the three basic layers in a wool fiber?

4. What are waves in the wool fiber and what creates them?

5. What are hairlike fibers in wool called and why are they important?

6. What is staple length?

7. What are tags?

8. What three systems are used to grade wool?

9. *True or False:* The finer the wool fiber, the more valuable it is.

10. Which two breeds of sheep produce the finest grade of wool?

11. What is the process of cleaning wool called?

SELECTED REFERENCES

Publications

Botkin, M. P., Field, R. A., and Johnson, C. L. 1988. *Sheep and Wool: Science, Production, and Management.* Englewood Cliffs, NJ: Prentice-Hall.

Hyde, N. 1988. Fabric of history: Wool. *National Geographic,* 173:552.

Lupton, C. J. 1994. Wool and Mohair Production and Processing. *Encyclopedia of Agricultural Science.* San Diego: Academic Press, Inc.

Lupton, C. J. 1996. Prospects for expanded mohair and cashmere production and processing in the United States of America. *J. Anim. Sci.* 74:1164.

Quality of U.S. Agricultural Products (Wool, p. 174). 1996. Ames, IA: Council for Agricultural Science and Technology.

Rogers, G. E., et al. 1989. *The Biology of Wool and Hair.* New York: Chapman and Hall.

Shelton, M., and Lupton, C. 1986. Where there's wool, there's a way. *Science of Food and Agriculture* (Nov. 6).

Terrill, C. E. 1971. Mohair. In *Encyclopedia of Science and Technology.* 3d ed. New York: McGraw-Hill.

Visuals

From Fleece to Fashion. Pendleton Woolen Mills, P.O. Box 1691, Portland, OR 97207.

Mohair presentation (video, 1989, 6 min). Mohair Council of America, P.O. Box 5337, San Angelo, TX 76902.

CHAPTER 8

Market Classes and Grades of Livestock, Poultry, and Eggs

The production and movement of more than 80 billion lb of highly perishable livestock and poultry products to U.S. consumers is accomplished through a vast and complex marketing system. Marketing is the transformation and pricing of goods and services through which buyers and sellers move livestock and livestock products from the point of production to the point of consumption. Producers need to understand marketing if they are to produce products preferred by consumers, decide intelligently among various marketing alternatives, understand how animals and products are priced, and eventually raise productive animals profitably.

Market classes and grades have been established to segregate animals, carcasses, and products into uniform groups based on preferences of buyers and sellers. The USDA has established extensive classes and grades to make the marketing process simpler and more easily communicated. Use of USDA grades is voluntary. Some packers have their own private grades, though these are often used in combination with USDA grades. An understanding of market classes and grades helps producers recognize quantity and quality of products they are supplying to consumers.

MARKET CLASSES AND GRADES OF RED MEAT ANIMALS

Slaughter Cattle

Slaughter cattle are separated into classes based primarily on age and sex. Age of the animal has a significant effect on tenderness, with younger animals typically producing more tender meat than older animals. Age classifications for meat from cattle are veal, calf, and beef. Veal is from young calves, 1–3 months of age, with carcasses weighing less than 150 lb. Calf is from animals ranging in age from 3–10 months with carcass weights between 150–300 lb. Beef comes from mature cattle (over 12 months of age) having carcass weights higher than 300 lb. Classes and grades established by the USDA for cattle are based on sex, quality grade, and yield grade, all of which are used in the classification of both live cattle and their carcasses (Table 8.1).

The sex classes for cattle are **heifer, cow, steer, bull,** and **bullock.** Occasionally, the sex class of **stag** is used by the livestock industry to refer to males that have been castrated after their secondary sex characteristics have developed. Sex classes separate cattle and carcasses into more uniform carcass weights, tenderness groups, and processing methods. **Quality grades** are intended to measure certain consumer palatability characteristics, and **yield grades** measure amounts of fat, lean, and bone in the carcass. Slaughter steers representing some of the eight quality grades and five yield grades are shown in Figs. 8.1 and 8.2.

Quality grades are based primarily on two factors: (1) **maturity** (physiological age) of the carcass, and (2) amount of **marbling.** Maturity is determined primarily by observing bone and cartilage structures. For example, soft, porous, red bones with a maximum amount of pearly white cartilage characterize A maturity, whereas very little cartilage and hard, flinty bones characterize C, D, and E maturities. As maturity in carcasses increases from A to E, the meat becomes less tender.

Marbling is intramuscular fat or flecks of fat within the lean, and it is evaluated at the exposed rib-eye muscle between the 12th and 13th ribs. Ten degrees of marbling, ranging from abundant to devoid, are designated in the USDA marbling standards. Various combinations of marbling and maturity that identify the carcass quality grades are shown in combinations of marbling and maturity in Fig. 8.3. Figure 8.4 shows rib-eye sections representing several different quality grades and several different degrees of marbling. Note that C maturity starts at 42 months of age. Slaughter cows more than 42 months of age will grade commercial, utility, cutter, or canner regardless of amount of marbling.

Yield grades, sometimes referred to as *cutability grades,* measure the quantity of *boneless, closely trimmed retail cuts* (BCTRC) from the major wholesale cuts of beef (round, loin, rib, and chuck). BCTRC should not be confused with the total percentage of retail cuts (including hamburger) from a beef carcass. A

TABLE 8.1 Official USDA Grade Standards for Live Slaughter Cattle and Their Carcasses

Class of Kind	Quality Grades (highest to lowest)	Yield Grades (highest to lowest)
Beef		
Steer and heifer	Prime, choice, select, standard, commercial, utility, cutter, canner	1, 2, 3, 4, 5
Cow	Choice, select, standard, commercial, utility, cutter, canner	1, 2, 3, 4, 5
Bullock	Prime, choice, select, standard, utility	1, 2, 3, 4, 5
Bull	(No designated quality grades)	1, 2, 3, 4, 5
Veal	Prime, choice, select, standard, utility	NA
Calf	Prime, choice, select, standard, utility	NA

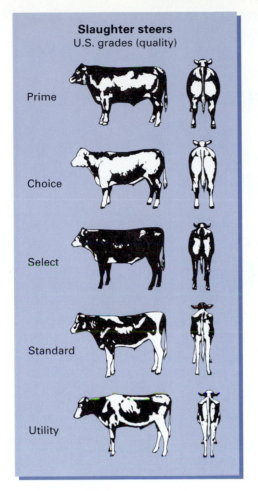

Slaughter steers
U.S. grades (quality)

Prime

Choice

Select

Standard

Utility

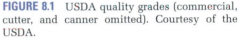

FIGURE 8.1 USDA quality grades (commercial, cutter, and canner omitted). Courtesy of the USDA.

numerical scale of 1–5 is used to rate yield grade, with 1 denoting the highest percentage of BCTRC. Although marketing communications generally quote yield grades in whole numbers, these yield grades are often shown in tenths in research data and information in carcass contests.

Packers can elect to have carcasses quality-graded. Table 8.2 shows the yield grades and their respective percentage of BCTRC. As an example, a carcass with a yield grade of 3.0 has a BCTRC of 50%; a 700-lb carcass with this yield grade would produce 350 lb of boneless, closely trimmed retail cuts from the round, loin, rib, and chuck.

Yield grades are determined from the following four carcass characteristics: (1) amount of fat measured in tenths of an inch over the rib-eye muscle, also known as the *longissimus dorsi* (Fig. 8.5); (2) kidney, pelvic, and heart fat (usually estimated as percent of carcass weight); (3) area of the rib-eye muscle, measured in square inches (Fig. 8.6); and (4) hot carcass weight. The last measurement reflects amount of intermuscular fat. Generally, as the carcass increases in weight, amount of fat between the muscles increases as well.

Measures of fatness in beef carcasses have the greatest effect in determining yield grade. A preliminary yield grade (PYG) can be determined by estimating

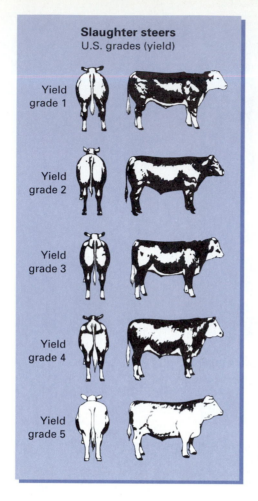

FIGURE 8.2 USDA yield grades for slaughter cattle. Courtesy of the USDA.

or measuring the outside fat of the rib-eye muscle. Figure 8.7 shows the five yield grades with varying amounts of fat over the rib-eye muscle and the area of the rib eye.

Quality grading and yield grading of beef carcasses by packers are voluntary. Approximately 84% of the 25 billion lb of federally inspected carcass beef produced each year is quality-graded and 82% is yield-graded. Most of the graded beef is slaughter steers and heifers from feedlots. The distribution of quality grades and yield grades is shown in Fig. 8.8. The percent of choice of the 18 million lb that was quality-graded in 1995 was 62%. Of the beef quality-graded, the Select grade has increased the most in recent years—from 4.6% in 1988 to 35.4% in 1995.

Feeder Cattle

The revised 1979 USDA feeder grades for cattle are intended to predict feedlot weight gain and the slaughter weight end point of cattle fed to a desirable fat-to-lean composition. The two criteria used to determine feeder grade are frame size and thickness. The three measures of frame size and thickness are shown

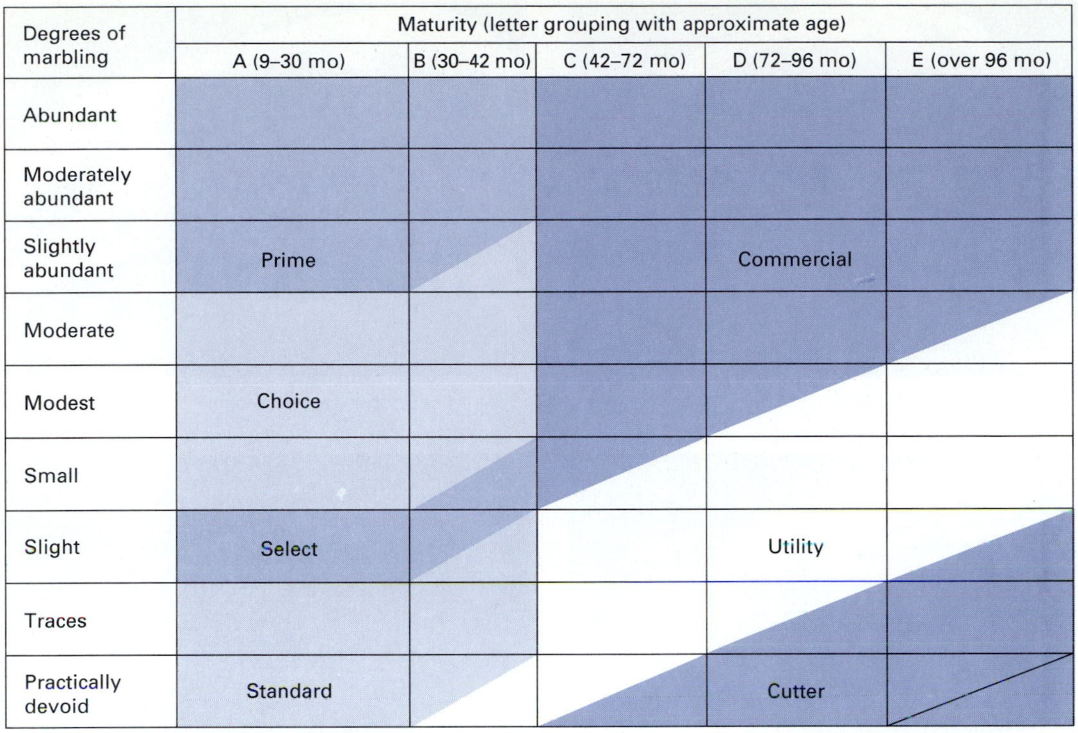

Degrees of marbling	Maturity (letter grouping with approximate age)				
	A (9–30 mo)	B (30–42 mo)	C (42–72 mo)	D (72–96 mo)	E (over 96 mo)
Abundant					
Moderately abundant					
Slightly abundant	Prime			Commercial	
Moderate					
Modest	Choice				
Small					
Slight	Select			Utility	
Traces					
Practically devoid	Standard			Cutter	

FIGURE 8.3 Marbling and the maturity of the carcass are the major factors that determine the quality grade of the beef carcass. Courtesy of the USDA.

in Figs. 8.9 and 8.10. Some examples of feeder cattle grade terminology are "large no. 1," "medium no. 2," and "small no. 2." Feeder cattle are given a USDA grade of "inferior" if the cattle are unhealthy or double-muscled. These cattle would not gain satisfactorily in the feedlot.

Although frame size and ability to gain weight in the feedlot are apparently related in the sense that large-framed cattle usually gain fastest, frame size appears to predict more accurately carcass composition or yield grade at different slaughter weights than it does gaining ability. The USDA feeder grade specifications identify the different liveweights from the three frame sizes when they reach the choice grade (Table 8.3).

Slaughter Swine

Sex classes of swine are **barrow, gilt, sow, boar,** and **stag.** Boars and sows are older breeding animals, whereas gilts are younger females that have not produced any young. The barrow is the male pig castrated early in life, and the stag is the male pig castrated after it has developed boar flavor in the meat. Because of relationships between sex and sex condition and the acceptability of prepared meats to the consumer, separate grade standards have been developed for barrow and gilt carcasses and for sow carcasses. There are no official grade standards for boar and stag carcasses.

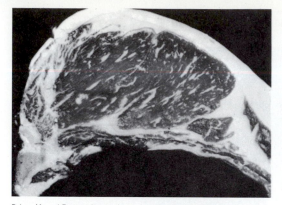

Prime (A and B maturity; moderately abundant marbling)

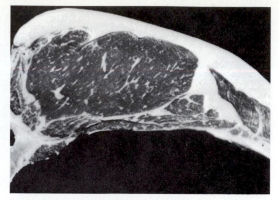

Choice (A and B maturity; modest marbling)

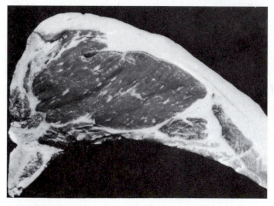

Select (A and B maturity; slight marbling)

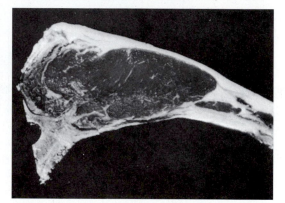

Standard (A and B maturity; traces of marbling)

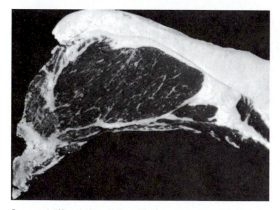

Commercial (C maturity; small marbling)

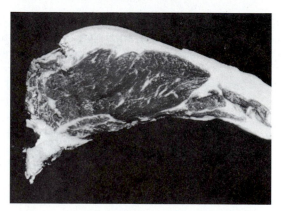

Utility (C maturity; slight marbling)

FIGURE 8.4 Exposed rib-eye muscles (between twelfth and thirteenth ribs) showing various degrees of marbling of several beef carcass quality grades. Courtesy of the National Live Stock and Meat Board.

TABLE 8.2   Beef Carcass Yield Grades and the Yield of BCTRC (boneless, closely trimmed retail cuts) from the Round, Loin, Rib, and Chuck

Yield Grade	BCTRC (%)	Yield Grade	BCTRC (%)	Yield Grade	BCTRC (%)
1.0	54.6	2.8	50.5	4.6	46.4
1.2	54.2	3.0	50.0	4.8	45.9
1.4	53.7	3.2	49.6	5.0	45.4
1.6	53.3	3.4	49.1	5.2	45.0
1.8	52.8	3.6	48.7	5.4	44.5
2.0	52.3	3.8	48.2	5.6	44.1
2.2	51.9	4.0	47.7	5.8	43.6
2.4	51.4	4.2	47.3	—	—
2.6	51.0	4.4	46.8	—	—

Source: USDA.

The grades for barrow and gilt carcasses are based on two general criteria: (1) quality characteristics of the lean, and (2) expected combined yields of the four lean cuts (ham, loin, blade (Boston) shoulder, and picnic shoulder). There are only two quality grades for lean in a pork carcass: acceptable and unacceptable. Quality of the lean is assessed by observing the exposed surface of a cut muscle, usually between the 10th and 11th ribs. Acceptable lean is gray-pink in color, has fine muscle fibers, and has fine marbling. Carcasses that have unacceptable lean quality (too dark or too pale, soft, or watery) or bellies too thin for suitable bacon production are graded U.S. utility, as they are soft and oily carcasses. Carcasses with acceptable lean quality are graded U.S. no. 1, U.S. no. 2, U.S. no. 3, or U.S. no. 4. These grades are based on expected yields of the four lean cuts, as shown in Table 8.4.

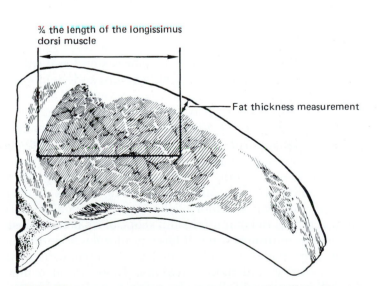

FIGURE 8.5   Location of the fat measurement over the rib-eye (*longissimus dorsi*) muscle. Courtesy of Dennis Giddings.

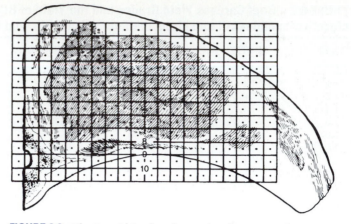

FIGURE 8.6 Plastic grid is placed over the rib-eye muscle to measure the area. Each square represents 0.1 square inch. Courtesy of Dennis Giddings.

Carcasses differ in their yields of the four lean cuts primarily because of differences in amount of fatness and muscling in relation to skeletal size. Backfat thickness and an evaluation of muscling are the two factors used to determine the numerical yield grade (Table 8.5).

Figure 8.11 shows the muscling scores used to determine grades.

Feeder Pig Grades

The USDA grades of feeder pigs are U.S. nos. 1, 2, 3, 4, and utility (Fig. 8.12). These grades correspond to USDA grades for market swine when slaughtered at 220 lb. Utility grade is for unthrifty or unhealthy feeder pigs. Thus, feeder pig grades combine an evaluation for thriftiness and slaughter potential.

Slaughter Sheep

Slaughter sheep are classified by their sex and maturity. Live sheep and their carcasses are also graded for quality grades and yield grades (Table 8.6).

Lamb carcasses, ranging from approximately 2–14 months of age, always have the characteristic break joint on one of their shanks following the removal of their front legs (Fig. 8.13). Mutton carcasses are distinguished from lamb carcasses by the appearance of the spool joint instead of the break joint (Fig. 8.13). The break joint ossifies as the sheep matures. Yearling mutton carcasses ranging from 12–15 months of age usually have the spool joint present but may occasionally have a break joint. Yearling mutton is also distinguished from lamb and mutton by color of the lean (intermediate between pinkish red of lamb and dark red of mutton) and shape of rib bones. Most U.S. consumers prefer lamb to mutton because it has a milder flavor and is more tender.

Quality grades are determined from a composite evaluation of conformation, maturity, and **flank streaking.** There are also minimum standards for **flank firmness and fullness;** however, most lambs meet these standards. Conformation is an assessment of overall thickness of muscling in the lamb carcass.

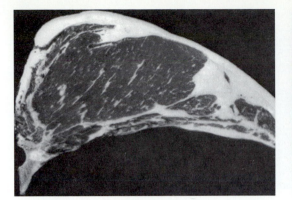

Yield Grade 1
(Fat 0.2 in. ribeye area 13.9 sq in.)

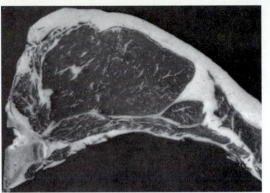

Yield Grade 2
(Fat 0.4 in. ribeye area 12.3 sq in.)

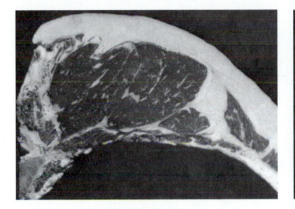

Yield Grade 3
(Fat 0.6 in., ribeye area 11.8 sq in.)

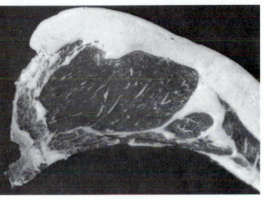

Yield Grade 4
(Fat 0.9 in., ribeye area 10.5 sq in.)

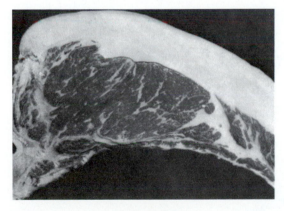

Yield Grade 5
(Fat 1.1 in., ribeye area 10.9 sq in.)

FIGURE 8.7   The five yield grades of beef shown at twelfth and thirteenth ribs. Courtesy of the National Live Stock and Meat Board.

FIGURE 8.8 Quality grades and yield grades of beef, 1995. Note: 16 and 18% of total carcasses are not quality or yield graded, respectively. Quality and yield grade percentages do not equal 100% due to rounding error. Source: USDA.

a. Large frame

b. Medium frame

c. Small frame

FIGURE 8.9 The three frame sizes of the USDA feeder grades for cattle. Courtesy of the USDA.

No. 1

No. 2

No. 3

FIGURE 8.10   The three thickness standards of the USDA feeder grades for cattle. Courtesy of the USDA.

TABLE 8.3   Slaughter Weights of Large-, Medium-, and Small-Frame Slaughter Cattle When They Reach the Choice Grade

	Slaughter Weight	
Frame Size	Steers (lb)	Heifers (lb)
Large	>1,200	>1,000
Medium	1,000–1,200	850–1,000
Small	<1,000	<850

TABLE 8.4 Expected Yields of the Four Lean Cuts, Based on Percent of Chilled Carcass Weight

Grade	Four Lean Cuts (%)
U.S. no. 1	53.0 and higher
U.S. no. 2	50.0–52.9
U.S. no. 3	47.0–49.0
U.S. no. 4	<47.0

TABLE 8.5 Preliminary Grade Based on Back Fat Thickness over the Last Rib (assumes average muscle thickness)

Preliminary Grade[a]	Backfat Thickness (in.)
U.S. no. 1	<1.00
U.S. no. 2	1.00–1.24
U.S. no. 3	1.25–1.49
U.S. no. 4	1.50 and over[b]

[a] Swine with thick muscling qualify for next higher grade; those with thin muscling are downgraded to next lower grade.
[b] Animals with an estimated last-rib backfat thickness of 1.75 inches or over have to be U.S. no. 4 and cannot be graded U.S. no. 3, even with thick muscling.
Source: USDA.

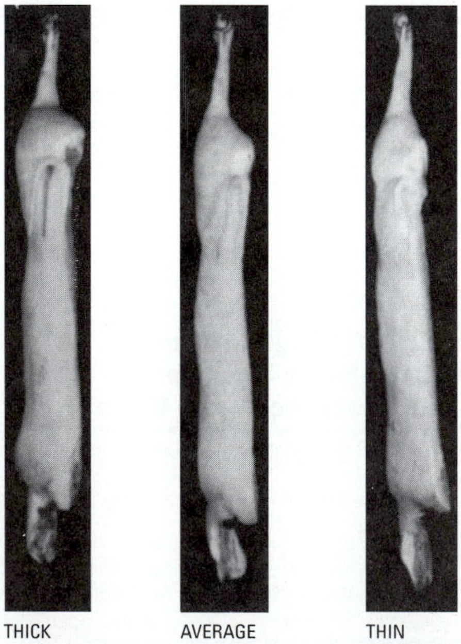

THICK   AVERAGE   THIN

FIGURE 8.11 Three degrees of muscling in pork carcasses. Thick and average, as shown, represent the minimum accepted for each of these two degrees. Courtesy of the USDA.

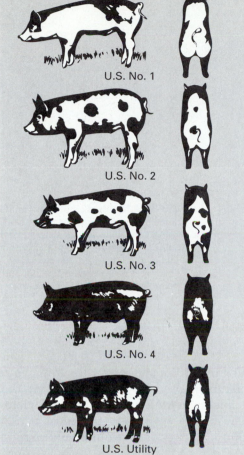

FIGURE 8.12 The USDA grades for feeder pigs. Courtesy of the USDA.

Maturity of lamb carcasses is determined by bone color and shape and muscle color. Flank streaking (streaks of fat within the flank muscle) predicts marbling since lamb carcasses are not usually ribbed to expose marbling in the rib-eye muscle. Flank firmness and fullness are determined by taking hold of the flank muscle with the hand. Most lamb carcasses grade either prime or choice, with few carcasses grading in the lower grades.

Lamb carcasses must be both quality-graded and yield-graded. Yield grades estimate boneless, closely trimmed retail cuts from the leg, loin, rack, and shoulder. The approximate percentage of retail cuts for selected yield grades are shown in Table 8.7.

TABLE 8.6 **USDA Maturity Groups, Sex Classes, and Grades of Slaughter Sheep**

Maturity Group	Sex Class	Quality Grade (highest to lowest)	Yield Grade (highest to lowest)
Lamb	Ewe, wether, or ram	Prime, choice, good, utility	1, 2, 3, 4, 5
Yearling mutton	Ewe, wether, or ram	Prime, choice, good, utility	1, 2, 3, 4, 5
Mutton	Ewe, wether, or ram	Choice, good, utility, cull	1, 2, 3, 4, 5

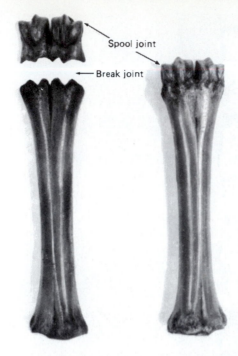

Spool joint

Break joint

FIGURE 8.13 Break and spool joints. The cannon bone on the left exhibits the typical break joint in which the foot and pastern are removed at the cartilaginous junction. In the cannon bone on the right, the cartilaginous junction has ossified, making it necessary for the foot and pastern to be removed at the spool joint. Courtesy of the USDA.

Yield grades are determined by fat thickness over the loin eye muscle. Figure 8.14 shows cross sections of lamb carcasses representing the five yield grades and amount of fat thickness over the loin eye for each yield grade. The yield grade can be estimated by multiplying the fat thickness by 10 and adding 0.4. For example, a lamb with 0.12 inches of fat would have an estimated yield grade of 1.6 (e.g., $(0.12 \times 10) + 0.4 = 1.6$).

Feeder Lamb Grades

Choice and prime slaughter lambs are produced in relatively large numbers, fed on grass and their mother's milk. Lambs weighing less than 100 lb at weaning are considered feeder lambs. They require additional feeding to produce a more desirable carcass.

TABLE 8.7   Lamb Carcass Yield Grades and the Percentage of Retail Cuts

Yield Grade	Percent Retail Cuts	Yield Grade	Percent Retail Cuts
1.0	49.0%	3.5	44.6%
1.5	48.2	4.0	43.6
2.0	47.2	4.5	42.8
2.5	46.3	5.0	41.8
3.0	45.4	5.5	41.0

Fat .05 in.
Yield Grade 1.7

Fat .10 in.
Yield Grade 2.4

Fat .25 in.
Yield Grade 3.6

Fat .35 in.
Yield Grade 4.6

Fat .45 in.
Yield Grade 5.5

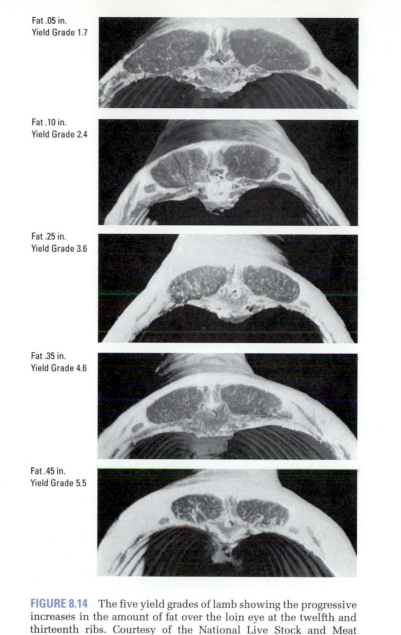

FIGURE 8.14 The five yield grades of lamb showing the progressive increases in the amount of fat over the loin eye at the twelfth and thirteenth ribs. Courtesy of the National Live Stock and Meat Board.

There are no official USDA grades for feeder lambs. Some research has been done to evaluate some possible feeder lamb grades. The research looks promising. These grades, based on frame size, project the slaughter weight of the lambs when they would reach 0.25 inch of fat thickness over the loin eye muscle. The frame sizes and slaughter weights are: *large* (over 120 lb), *medium* (100–120 lb), and *small* (less than 100 lb). There are also body-thickness categories based on conformation of slaughter lambs. They are no. 1 (prime conformation), no. 2 (choice conformation), and no. 3 (less than choice conformation).

MARKET CLASSES AND GRADES OF POULTRY PRODUCTS

Poultry Meat

The class of poultry must be displayed on the package label or a tag on the wing of the bird. The age class indicates tenderness, as meat from younger birds is more tender than that from older birds. Table 8.8 shows the age-classification labels for poultry meat.

Meat from the young age classes is usually prepared by broiling, frying, roasting, or barbecuing. The mature, less tender poultry meat is best prepared by baking, stewing, or fabricating, or by including it in other prepared dishes.

The grades are U.S. grade A, U.S. grade B, and U.S. grade C for each of the classes, with A being the highest grade (Fig. 8.15). Carcasses of A quality are free of deformities that detract from their appearance or that affect normal distribution of flesh. They have a well-developed covering of flesh and a well-developed layer of fat in the skin. They are free of pinfeathers and diminutive feathers, exposed flesh on the breast and legs, and broken bones. They have no more than one disjointed bone, and they are practically free of discolorations of the skin and flesh and defects resulting from handling, freezing, or storage. Carcasses of B quality may have moderate deformities. They have a moderate covering of flesh, sufficient fat in the skin to prevent a distinct appearance of the flesh through the skin, and no more than an occasional pinfeather or diminutive feather. They may have moderate areas of exposed flesh and discoloration of the skin and flesh. They may have disjointed parts but no broken bones, and they may have moderate defects resulting from handling, freezing, or storage.

Eggs

The grading of shell eggs involves classifying individual eggs according to established standards. Eggs are graded by sorting them into groups, each group having similar weight and quality characteristics. Table 8.9 shows how eggs are classified according to a size-and-weight relationship.

TABLE 8.8   Labels Used to Identify Poultry Age Classes

Type	Young	Mature
Chicken	Young chicken	Mature chicken
	Rock Cornish game hen	Old chicken
	Broiler	Hen
	Fryer	Stewing chicken
	Roaster	Fowl
	Capon	
Turkey	Young turkey	Mature turkey
	Fryer-roaster	Yearling turkey
	Young hen	Old turkey
	Young tom	
Duck	Duckling	Mature
	Young duckling	Old
	Broiler duckling	
	Fryer duckling	
	Roaster duckling	

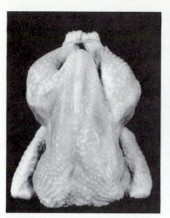
Broiler or fryer, A quality

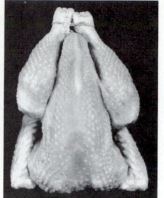
Broiler or fryer, B quality

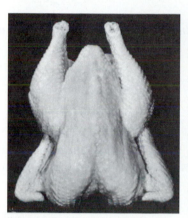
Hen or stewing chicken, A quality

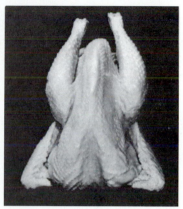
Hen or stewing chicken, B quality

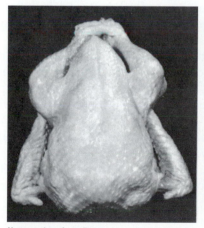
Young turkey, A quality

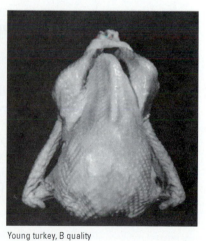
Young turkey, B quality

FIGURE 8.15 Classes and grades of ready-to-cook poultry. Courtesy of the USDA.

TABLE 8.9   Consumer Weight Classes of Eggs, Minimum Net Weight Per Dozen

Size	Ounces (per doz)
Jumbo	30
Extra large	27
Large	24
Medium	21
Small	18
Peewee	15

Source: USDA.

The USDA quality standards used to grade individual shell eggs are as follows:

Exterior Quality Factors

Cleanliness of shell

Soundness of shell (cracks, checks, and texture)

Shape

Interior Quality Factors

Albumen thickness

Condition of yolk

Size and condition of air cell

Abnormalities (e.g., blood spots, meat spots)

Exterior quality factors are apparent from external observation; interior quality factors involve an assessment of egg content. The latter is accomplished through a process called **candling** (visually appraising the eggs while light is shown through them).

Although shell color is not a factor in the U.S. standards and grades, eggs are usually sorted by color and sold as either "whites" or "browns." Eggs sell better when sorted by color and packed separately. Contrary to popular opinion, there are no differences between similar quality brown- and white-shelled eggs other than color.

The U.S. standards for quality of individual shell eggs are applicable to eggs from domestic chickens only; these standards are summarized in Table 8.10. Consumer egg grades are U.S. grade AA, U.S. grade A, and U.S. grade B. Figure 8.16 shows selected shields for communicating the quality grade.

FIGURE 8.16   U.S. Department of Agriculture grade shields showing two egg quality grades. Courtesy of the USDA.

Color Plate Section

CARCASS COMPOSITION DISPLAY

The proper ratio of fat to lean in an animal carcass has been a controversial issue among producers, feeders, and packers since the industry began. In the 1940s, fat-type animals, such as the lard-type hog, baby beef, and fat lamb, were the preferred type. Today, however, vegetable oils have largely replaced lard and other animal fats; lean beef has replaced the fatted calf; and the lamb has been made leaner through modern production techniques. Today's consumer demands lean meat, and today's grading system reflects those demands.

By visually appraising carcasses, a person can quickly identify the lean-to-fat compositional differences. Appraisal of similar differences in live animals, however, can be a more difficult and discriminating challenge. The following set of photographs graphically demonstrates a lean-fat ratio both on individual animals and as a comparison between different yield grades. The pictures are courtesy of Iowa State University and Colorado State University, where carcasses were frozen in a standing position. This process allowed technicians to remove different layers of tissue, such as the hide, fat, and muscling, and to cross-section entire carcasses for easy comparison. The photographs are organized by color plates, with the caption for each photo presented here by plate and photo number:

Plate E compares body conformation of different yield grades in both live animals and their carcasses.

1. Thickness of muscling in two medium-framed feeder steers is shown using cutaway sections of the round. The steer on the left has more thickness of lean meat in the round than the steer on the right.

2 & 3. Pounds of fat trimmed from one-half of the body of a yield grade 4 steer (92 lb) and a yield grade 2 steer (47 lb). Black stripes are hide (with fat underneath) that remain at key locations on the body.

4. Rear view showing how the fat increases in thickness from the middle of the back to the edge of the loin.

5. Two yearling Hereford bulls of approximately the same weight. Breeding cattle can be visually appraised for fat and lean compositional differences. The bull on the left would sire slaughter steers having yield grade 4 or 5 carcasses, while the bull on the right would sire yield grade 1 or 2 steers. This assumes that the bulls would be bred to cows similar in frame-size to themselves, and that the steers would be slaughtered at approximately 1,150 pounds. Compare to plates F and G.

Plates F and **G** compare fat-to-lean composition on a yield grade 2 steer (plate F) and a yield grade 5 steer (plate G).

1. Live animals—side view

2. Black ribbons identify where cross sections were made (3, 4, 5, and 6— cross sections from rump, hip, mid-carcass, and shoulder).

3. The percentages of fat, lean, and bone found in this carcass.

Plates H and **I** compare the fat-to-lean composition of lamb carcasses on yield grade 1 lamb (plate H) and yield grade 5 lamb (plate I).

1. Side view.

2. Rear view.

3. Black ribbons identify where cross sections were made (4, 5, and 6— cross sections from shoulder, mid-carcass, and rump).

Plates J and **K** compare fat-to-lean composition of a U.S. no. 1 pig (plate J) with a U.S. no. 4 pig (plate K).

1. Live animals—side view.

2. Rear view.

3. Black ribbons identify where cross sections were made (4, 5, and 6— cross sections from shoulder, mid-carcass, and rump).

Plate L shows the composition and appearance of a steer carcass at various stages of removal of hide, fat, and muscle.

1. Live steer with hair clipped from one side.

2. Frozen steer with hide removed and fat exposed.

3. Fat removed from one-half of steer's body.

4. Rear view with fat removed from the left side. Note cod and twist fat on the right side.

5. Skeleton of the beef animal after all the muscle has been removed.

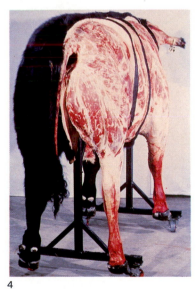

1

2

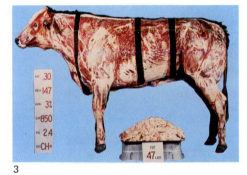

3

4

5

PLATE E.
Courtesy of Colorado State University.

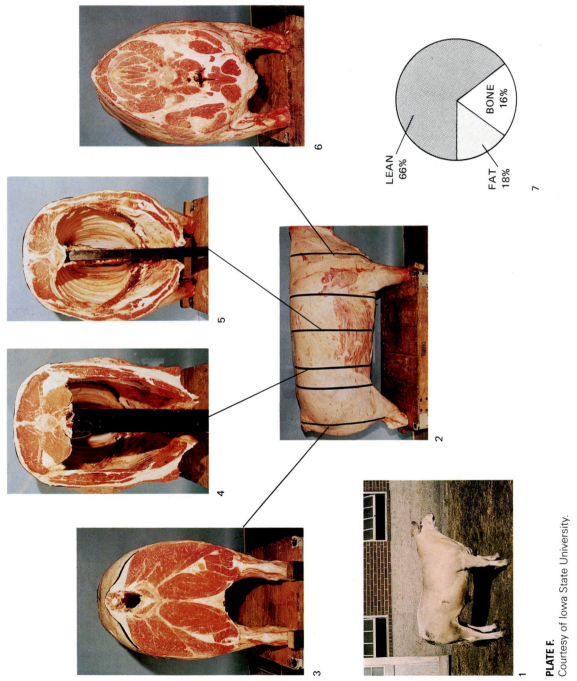

LEAN
66%

FAT
18%

BONE
16%

7

6

5

4

3

2

1

PLATE F.
Courtesy of Iowa State University.

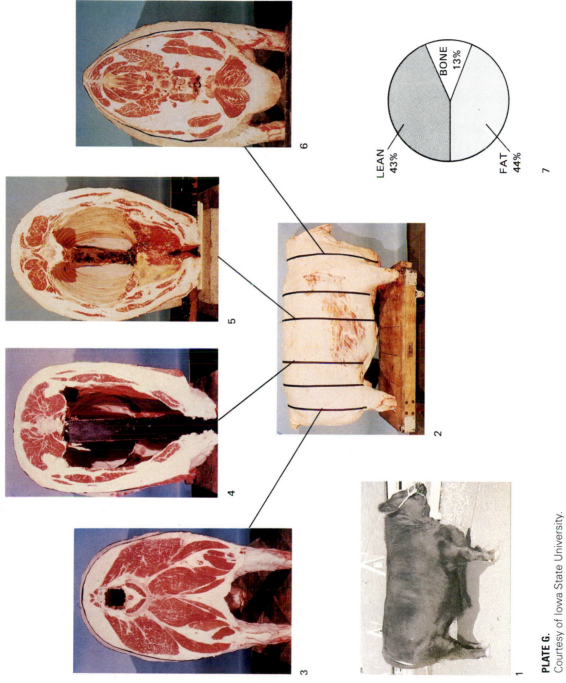

LEAN
43%

BONE
13%

FAT
44%

7

6

5

4

3

2

1

PLATE G.
Courtesy of Iowa State University.

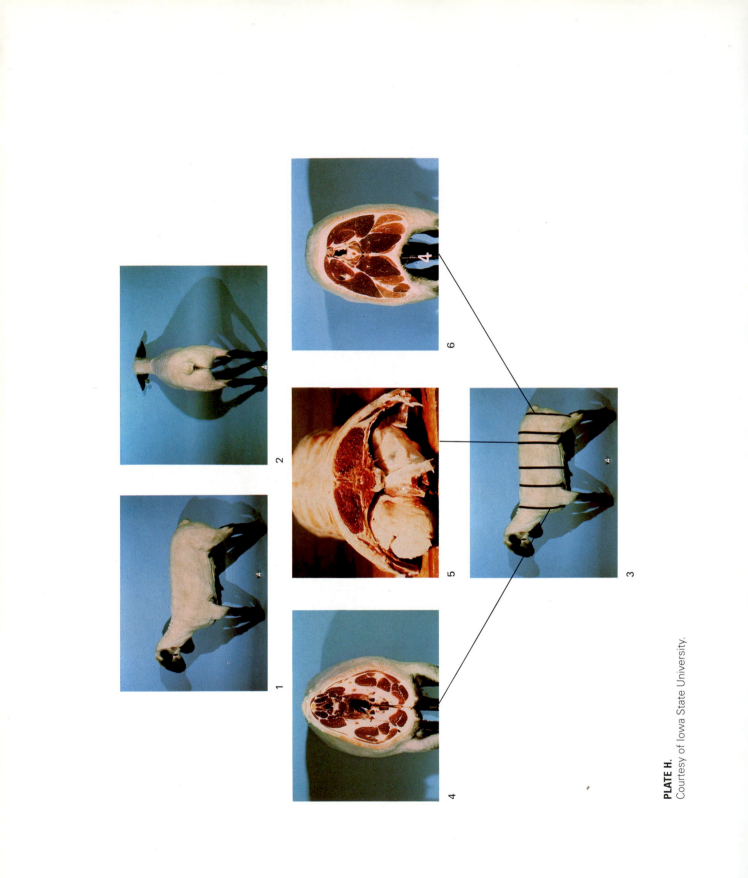

PLATE H.
Courtesy of Iowa State University.

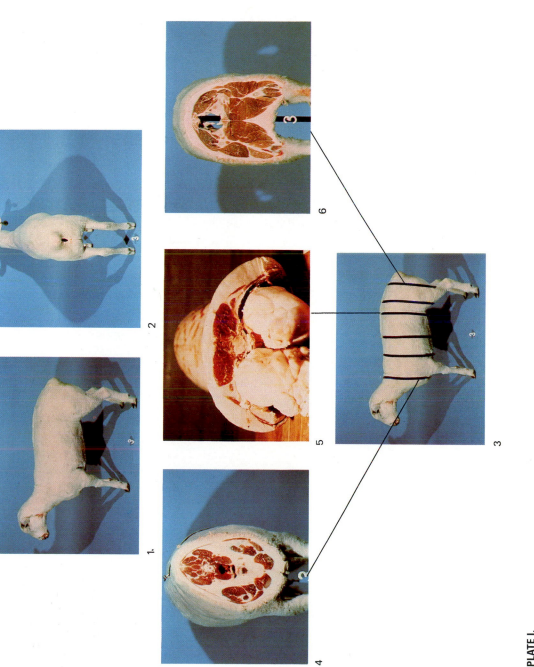

PLATE I.
Courtesy of Iowa State University.

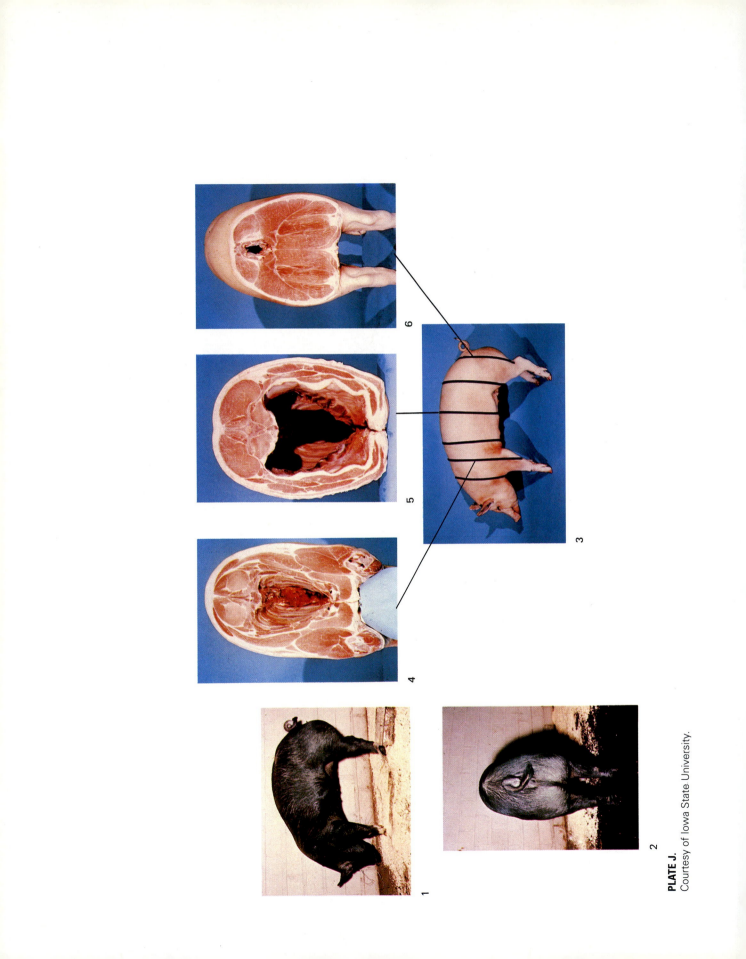

PLATE J.
Courtesy of Iowa State University.

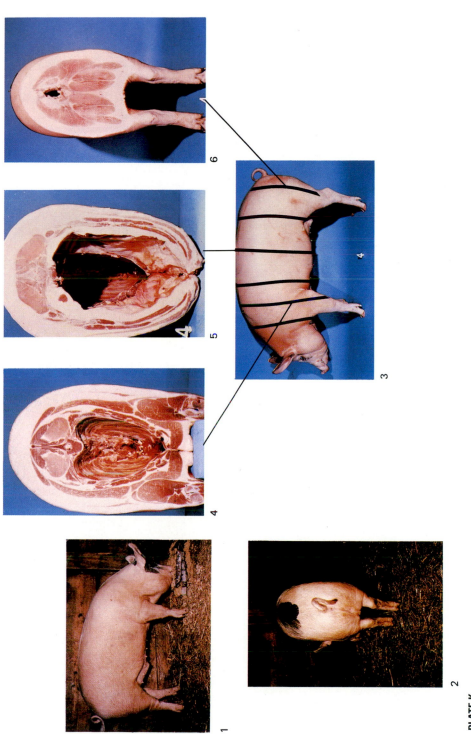

PLATE K.
Courtesy of Iowa State University.

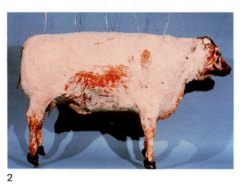

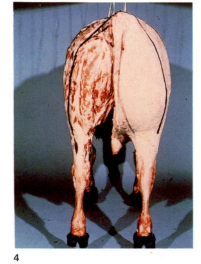

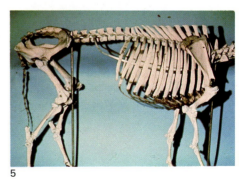

PLATE L.
Courtesy of Iowa State University.

TABLE 8.10   Summary of U.S. Standards for Quality of Individual Shell Eggs

	Specifications for Each Quality Factor				
Quality Factor	AA Quality	A Quality	B Quality	Dirty	Check
Shell	Clean, unbroken; practically normal	Clean, unbroken; practically normal	Clean to slightly stained,[a] unbroken; abnormal	Unbroken; adhering dirt or foreign material, prominent stains moderate stained areas in excess of B quality.	Broken or cracked shell but membranes intact, not leaking[c]
Air cell	$\frac{1}{8}$ in. or less in depth; unlimited movement and free or bubbly	$\frac{3}{16}$ in. or less in depth; unlimited movement and free or bubbly	Over $\frac{3}{16}$ in. in depth; unlimited movement and free or bubbly		
White	Clear, firm	Clear, reasonably firm	Weak and watery; small blood and meat spots present[b]		
Yolk	Outline slightly defined; practically free from defects	Outline fairly well defined; practically free from defects	Outline plainly visible; enlarged and flattened; clearly visible germ development, but no blood; other serious defects		

[a] Moderately stained areas permitted ($\frac{1}{32}$ of surface if localized, or $\frac{1}{16}$ if scattered).
[b] If they are small (aggregating not more than $\frac{1}{8}$ in. in diameter).
[c] Leaker has broken or cracked shell and membranes, and contents are leaking or free to leak.
Source: USDA, *Egg Grading Manual.*

CHAPTER SUMMARY

- Market classes and grades have been established to segregate animals, carcasses, and products into uniform groups based on preferences of buyers and sellers.

- The USDA (United States Department of Agriculture) classes and grades makes the marketing process simpler and more easily communicated.

- Red meat animals are separated into classes based primarily on age and sex. Age of the animal significantly affects tenderness. Sex class separates animals and carcasses into more uniform carcass weights, tenderness groups, and processing methods.

- Feeder and slaughter grades exist for most red meat animals. For slaughter animals, quality grades are used to measure consumer palatability characteristics while yield grades measure the amount of fat, lean, and bone in the carcass.

- Poultry meat classes are based on age (affects tenderness) and the grades evaluate thickness and fullness of meat in the breast and thigh.

- Eggs are graded for size (weight), shell condition, and interior quality (yolk, white, and air cell).

REVIEW QUESTIONS

1. Slaughter cattle are separated into market classes based upon what two characteristics?

2. Quality grades for carcasses, which estimate eating quality of meat, are based on what two characteristics?

3. What do beef carcass yield grades estimate?

4. What do U.S. numerical grades of pork carcasses estimate?

5. What do lamb carcass yield grades estimate?

6. What process is used to appraise the interior quality of table eggs from chickens?

SELECTED REFERENCES

Publications

Boggs, D. L., and Merkel, R. A. 1993. *Live Animal Carcass Evaluation and Selection Manual.* 3d ed. Dubuque, IA: Kendall-Hunt.

Council for Agricultural Science and Technology. Mar. 1980. *Foods from Animals: Quantity, Quality, and Safety.* CAST Report no. 82.

McCoy, J. H. 1988. *Livestock and Marketing.* Westport, CT: AVI.

Meat Evaluation Handbook. 2d ed., 1977. Chicago: National Live Stock and Meat Board.

Quality of U.S. Agricultural Products (Beef, Pork, and Lamb, p. 203; poultry, p. 195). 1996. Ames, IA: Council for Agricultural Science and Technology.

USDA. Apr. 1983. *Egg Grading Manual.* USDA Agriculture Handbook no. 75.

USDA. April 1980. *Facts About U.S. Standards for Grades of Feeder Cattle.* Agricultural Marketing Service AMS-586.

USDA. *Official United States Standards for Grades of Feeder Pigs* (Apr. 1, 1969) and *Grades of Slaughter Swine* (Jan. 14, 1985); *Grades of Pork Carcasses* (Jan. 14, 1985); *Grades of Veal and Calf Carcasses* (Oct. 6, 1980); *Grades of Lamb, Yearling Mutton and Mutton Carcasses* (Oct. 17, 1982); *Grades of Carcass Beef and Slaughter Cattle* (Apr. 9, 1989). Agricultural Marketing Service.

Visuals

Beef Quality Grading 1990. (Video; 24 min.) and *Beef Yield Grading.* 1990. (Video; 21 min.). CEV, P.O. Box 65265, Lubbock, TX 79464.

Grading Eggs for Quality (sound filmstrip). Vocational Education Productions, California Polytechnic State University, San Luis Obispo, CA 93407.

Egg Grades . . . A Matter of Quality (16mm film, 12 min). National Audio Visual Center, Order Section, 8700 Edgeworth Dr., Capitol Heights, MD 20743-3701.

Egg Grading (videotape). CEV, P.O. Box 65265, Lubbock, TX 79464-5265.

Feeder Cattle Evaluation. 1991 (Video; 28 min.) and *Feeder Pig Evaluation.* 1991 (Video; 20 min.) CEV, P.O. Box 65265, Lubbock, TX 79464.

Market Animal Grading—Feeder Cattle, Slaughter Cattle, Slaughter Lambs, and Slaughter Hogs (four videotapes). CEV, P.O. Box 65265, Lubbock, TX 79464-5265.

United States Standards of Poultry and Poultry Parts (three-part slide series: part 1—whole fryer; part 2—fryer parts; part 3—turkey). 1986. Learning Resources Center, VPI, Blacksburg, VA 24061.

U.S. Standards for Quality of Individual Shell Eggs (photos illustrating interior and exterior grade factors; chart, 15 × 22 in.). 1984. Superintendent of Documents, U.S. Government Printing Office, Washington, DC 20402.

CHAPTER 9

Visual Evaluation of Slaughter Red Meat Animals

Most red meat animals, ready for slaughter, are purchased at market time based on visual appraisal of their apparent carcass merit. Livestock producers attempt to combine high productivity and visual acceptability in these slaughter animals in anticipation of high market prices and positive net returns.

Productivity of breeding and slaughter meat animals is best identified by combining meaningful performance records and effective visual appraisal. Opinions regarding the relationship of form and function in red meat animals (cattle, sheep, swine) differ widely among producers. They continually discuss the value of visual appraisal, weights, and body measurements in breeding programs, in determining market grades, and in defining the so-called ideal types. It is important to separate true relationships from opinion in the area of animal form and function.

Type is defined as an ideal or standard of perfection combining all the characteristics that contribute to an animal's usefulness for a specific purpose. *Conformation* implies the same general meaning as type and refers to form and shape of the animal. Both type and conformation describe an animal according to its external form and shape that can be evaluated visually or measured more objectively with a tape or some other device.

Red meat animals have three productive stages: (1) breeding (reproduction), (2) feeder (growth), and (3) slaughter (carcass or product). It has been well demonstrated that performance records are much more effective than visual appraisal in improving reproduction and growth stages. Visual appraisal does have importance in these stages, primarily in identifying reproductive soundness, skeletal soundness, and health status. Visual appraisal has some importance in complementing performance records (such as ultrasonic measurements for fatness), as visual appraisal is important in evaluating slaughter animals.

Visual appraisal of slaughter red meat animals can be used effectively to predict carcass composition (fat and muscling) primarily when relatively large differences exist.

EXTERNAL BODY PARTS

Effective communication in many phases of the livestock industry requires a knowledge of the external body parts of the animal. The locations of the major body parts for swine, cattle, and sheep are shown in Figs. 9.1, 9.2, and 9.3, respectively.

LOCATION OF THE WHOLESALE CUTS IN THE LIVE ANIMAL

The next step for effective visual appraisal, after becoming familiar with the external parts of the animal, is to understand where major meat cuts are located in the live animal. Wholesale and retail cuts of the carcasses of beef, sheep, and swine are identified in Chapter 3. A carcass is evaluated after the animal has been slaughtered, eviscerated, and, in the case of beef and swine, split into two halves. The lamb carcass remains as a whole carcass. Furthermore, when it is evaluated, the carcass is hanging by the hind leg from the rail in the packing plant. This makes it difficult to perceive how the carcass would appear as part of the live animal standing on all four legs. Figure 9.4, which shows the location of the wholesale cuts on the live animal, assists in correlating carcass evaluation to live-animal evaluation.

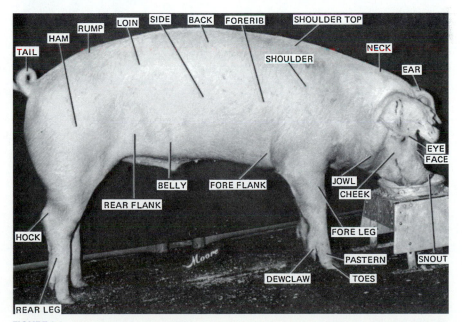

FIGURE 9.1   The external parts of swine. Courtesy of the Chester White Record Association.

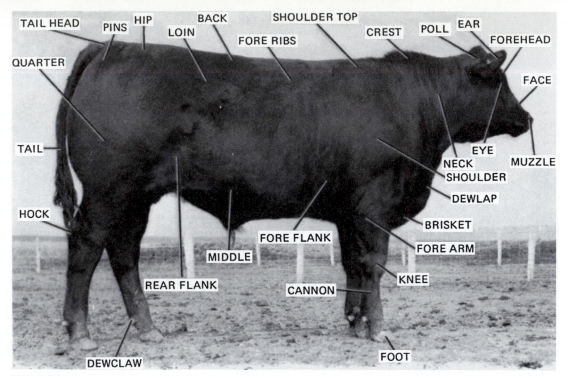

FIGURE 9.2   The external parts of beef cattle. Courtesy of the American Angus Association.

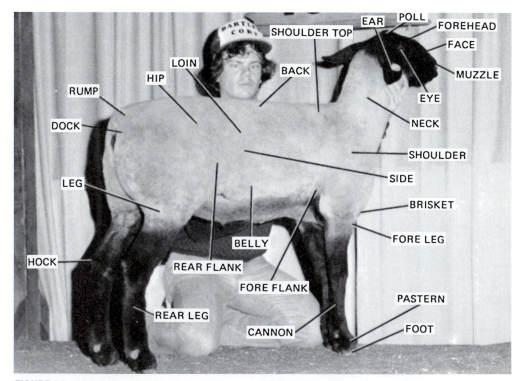

FIGURE 9.3   The external parts of sheep. Courtesy of *Sheep Breeder and Sheepman* magazine.

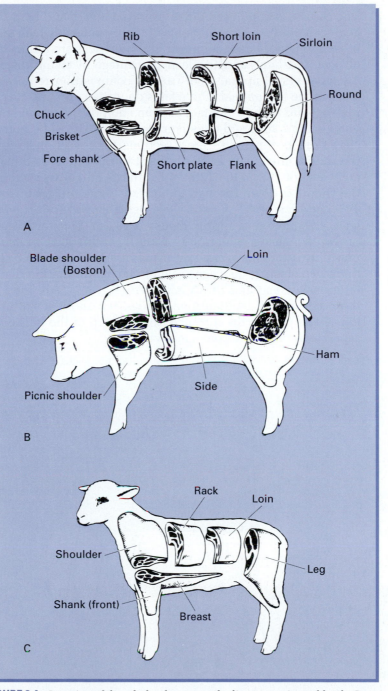

FIGURE 9.4 Location of the wholesale cuts on the live steer, pig, and lamb. Courtesy of Dennis Giddings.

VISUAL PERSPECTIVE OF CARCASS COMPOSITION OF THE LIVE ANIMAL

The carcass is composed of fat, lean (red meat), and bone. The meat industry's goal is to produce large amounts of highly palatable lean and minimal amounts of fat and bone. These composition differences are reflected in the yield grades of beef and lamb and the lean percentage of swine discussed in Chapter 8. Effective visual appraisal of carcass composition requires knowing the body areas of the live animal where fat is deposited and muscle growth occurs. Color plates F and G show the fat and lean composition at several cross sections of a yield grade 2 steer and a yield grade 5 steer. The ribbon on the frozen carcasses shows the location of the cross sections. Cross section 3 removes the bulge from the round; cross section 4 is in front of the hip bone down through the flank, cross section 5 is at the 12th and 13th rib, and cross section 6 is at the point of the shoulder down through the brisket. Note the contrasts in fat (18% versus 44%) and lean (66% versus 43%) percentages. Both steers graded choice and were slaughtered at approximately 1,100 lb. The conformation characteristics of the two steers, which are primarily influenced by fat deposits and some muscling differences, are contrasted in Figs. 9.5 and 9.6. The conformation of these two steers is also contrasted to the conformation of a slaughter steer that is underfinished and thinly muscled (Fig. 9.7). All body parts referred to in these figures can be identified in Figs. 9.2 and 9.4, with the exception of the twist. The *twist* is the distance from the top of the tail to where the hind legs separate as observed from a rear view.

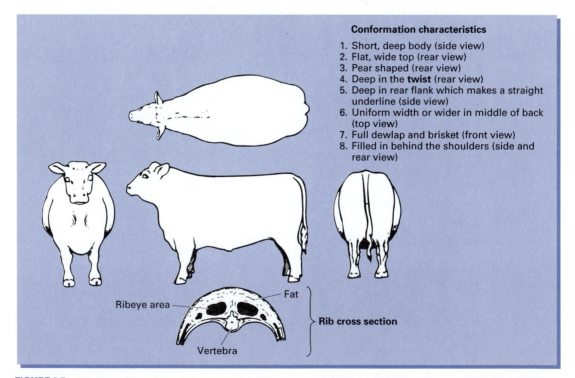

Conformation characteristics

1. Short, deep body (side view)
2. Flat, wide top (rear view)
3. Pear shaped (rear view)
4. Deep in the **twist** (rear view)
5. Deep in rear flank which makes a straight underline (side view)
6. Uniform width or wider in middle of back (top view)
7. Full dewlap and brisket (front view)
8. Filled in behind the shoulders (side and rear view)

Ribeye area

Fat

Vertebra

Rib cross section

FIGURE 9.5 Conformation characteristics of slaughter steer typical of yield grade 5. Compare with Plate G. Courtesy of Dennis Giddings.

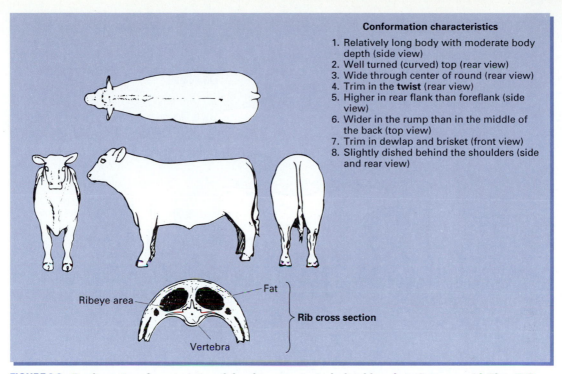

Conformation characteristics

1. Relatively long body with moderate body depth (side view)
2. Well turned (curved) top (rear view)
3. Wide through center of round (rear view)
4. Trim in the **twist** (rear view)
5. Higher in rear flank than foreflank (side view)
6. Wider in the rump than in the middle of the back (top view)
7. Trim in dewlap and brisket (front view)
8. Slightly dished behind the shoulders (side and rear view)

Ribeye area

Fat

Rib cross section

Vertebra

FIGURE 9.6 Conformation characteristics of slaughter steer typical of yield grade 2. Compare with Plate F. Courtesy of Dennis Giddings.

Color plates H and I show the cross sections of a yield grade 1 lamb and a yield grade 5 lamb. Cross section 4 is cut through the shoulder area, cross section 5 is at the back, and cross section 6 is through the leg area. Note that many fat deposits and muscle areas are similar to those described in Fig. 9.5.

Color plates J and K show the cross section of a U.S. no. 1 pig and a U.S. no. 4 pig. Percent lean cuts is used in pork carcass evaluation instead of yield grades. Although the terminology is different, they measure fat-to-lean composition. Percent lean cuts are measured by obtaining the total weight of the trimmed Boston shoulder, picnic shoulder, loin, and ham (Fig. 9.4), and then dividing the total by carcass weight. The lean cuts percentage for the U.S. no. 1 would be 53% or more, as contrasted with less than 47% for the U.S. no. 4. The ribbon on the frozen carcass marks the location of the cross sections. Cross section 4 is through the shoulder area, 5 is through the middle of the back, and 6 is through the ham area.

Even though the size and shape of slaughter cattle, sheep, and swine are different, these three species are remarkably similar in muscle structure and fat-deposit areas. Therefore, what a person learns from visual evaluation of one species can be applied to another species. Regardless of species, an animal that shows a square appearance over the top of its back and appears blocky and deep from a side view usually has a large accumulation of fat. Fat accumulates first in flank areas, brisket, dewlap, and throat (jowl); between the hind legs; and over the edge of the loin. Fat also fills in behind the shoulders and gives the animal a smooth appearance. Movement of shoulder blade can be observed when lean cattle and swine walk. Slaughter red meat animals having an oval

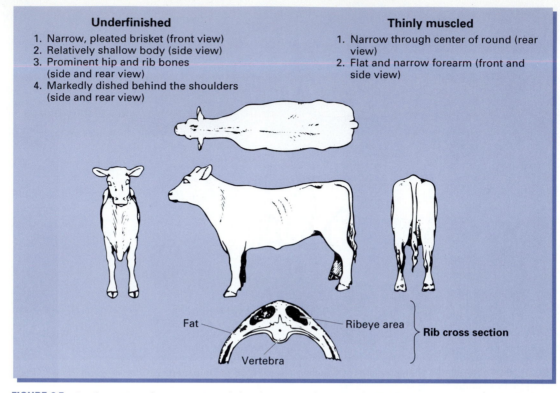

Underfinished
1. Narrow, pleated brisket (front view)
2. Relatively shallow body (side view)
3. Prominent hip and rib bones (side and rear view)
4. Markedly dished behind the shoulders (side and rear view)

Thinly muscled
1. Narrow through center of round (rear view)
2. Flat and narrow forearm (front and side view)

Fat

Ribeye area

Vertebra

Rib cross section

FIGURE 9.7 Conformation characteristics of slaughter steer that is underfinished (contrast with Fig. 9.5) and thinly muscled (contrast with Fig. 9.6). Courtesy of Dennis Giddings.

turn to the top of their back, while having thickness through the center part of their hind legs (as viewed from the rear), have a high proportion of lean to fat. It is important that slaughter animals have an adequate amount of fat because thin animals typically do not produce a highly palatable consumer product.

The wool covering of sheep can easily camouflage the fat covering. The amount of fat in sheep can be determined by pressing the fingers of the closed hand lightly over the last two ribs and over the spinal processes of the vertebrae. Sheep that have a thick padding of fat is these areas will produce poor yield-grading carcasses.

Accuracy in visually appraising slaughter red meat animals is obtained by making visual estimates of yield grades and percent lean cuts and their component parts, and then comparing the visual estimates with the carcass measurements. Accurate visual appraisal can be used as one tool in producing red meat animals with a more desirable carcass composition of lean to fat.

During the late 1980s and early 1990s the amount of fat on retail cuts has been significantly reduced—e.g., down to ¼ inch, ⅛ inch, or no fat on many cuts. This fat reduction has occurred primarily by trimming excess fat at the packing plant. Certain breeding and feeding practices could reduce the trimmable fat more economically; however, marketing practices do not encourage the implementation of these practices (see value-based marketing in Chapter 35).

CHAPTER SUMMARY

■ Numerous red meat slaughter animals are evaluated and priced on the visual appraisal of the animal's carcass merit.

■ Carcass composition (fat, lean, and bone) can be appraised quite accurately by visually appraising live animals.

■ Understanding the anatomical structure, especially the location of major muscle and patterns of fat deposition, is needed for accurate visual appraisal.

■ Thickness through the center of the round (cattle), ham (pig), and leg (lamb) gives a visual estimate of muscling. Squareness over the animal's top and rectangular appearance from a side view usually gives a visual assessment that the animal is excessively fat.

REVIEW QUESTIONS

1. What are the three productive stages of red meat-producing animals?

2. How do production records compare to visual appraisal for improving reproductive and growth stages of livestock?

3. What is being estimated during visual evaluation of livestock for carcass merit?

4. Carcasses are primarily composed of what three types of tissues?

SELECTED REFERENCES

Publications

Boggs, D. L., and Merkel, R. A. 1993. *Live Animal Carcass Evaluation and Selection Manual.* 3d ed., Dubuque, IA: Kendall-Hunt.

Crouse, J. E., Dikeman, M. E., and Allen, D. M. 1974. Prediction of beef carcass composition and quality by live-animal traits. *J. Anim. Sci.* 38:264.

Kauffman, R. G., Grummer, R. H., Smith, R. E., Long, R. A., and Shook, G. 1973. Does live-animal and carcass shape influence gross composition? *J. Anim. Sci.* 37:1112.

Visuals

Slaughter Cattle Evaluation, Slaughter Hog Evaluation, and Slaughter Lamb Evaluation. 1991 (three videotapes). CEV, P.O. Box 65265, Lubbock, TX 79464-5265

Swine Evaluation (VHS video). VEP, California Polytechnic State University, San Luis Obispo, CA 93407.

CHAPTER 10

Reproduction

Reproductive efficiency in farm animals, as measured by number of calves or lambs per 100 breeding females or number of pigs per litter, is a trait of great economic importance in farm animal production. It is essential to understand the reproductive process in the creation of new animal life because it is a focal point of overall animal productivity. Producers who manage animals for high reproductive rates must understand the production of viable sex cells, estrous cycles, mating, pregnancy, and birth. (Some aspects of reproductive behavior are presented in Chapter 34.)

FEMALE ORGANS OF REPRODUCTION AND THEIR FUNCTIONS

Figures 10.1 and 10.2 show the reproductive organs of the cow and sow. The female reproductive anatomy of the various farm animal species is similar, although there are a few obvious differences.

The organs of reproduction of the typical female farm mammal include a pair of ovaries, which are suspended by ligaments just back of the kidneys, and a pair of open-ended tubes, the oviducts (also called the *Fallopian tubes*), which lead directly into the uterus (womb). The uterus itself has two horns, or branches, that in farm mammals merge together at the lower part into one structure so that the lower opening, or exit, from the uterus is a canal. This canal is called the *cervix*. Its surface is fairly smooth in the mare and the sow, but is folded in the cow and ewe. The cervix opens into the vagina, a relatively large canal or passageway that leads posteriorly to the external parts, which are the vulva and clitoris. The urinary bladder empties into the vagina through the urethral opening.

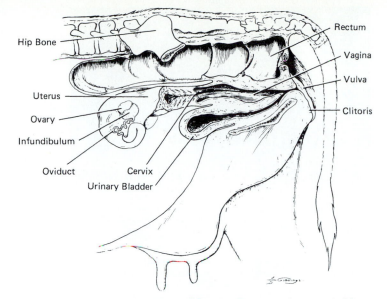

FIGURE 10.1 Reproductive organs of the cow. Courtesy of Dennis Giddings.

Ovaries

Ovaries produce *ova* (female sex cells, also called *eggs*) and the female sex hormones estrogen and progesterone. Each ovum (Fig. 10.3) is the largest single cell in the body and it develops inside a recently formed follicle within the ovary (Fig. 10.4). Some tiny follicles develop and ultimately attain maximum

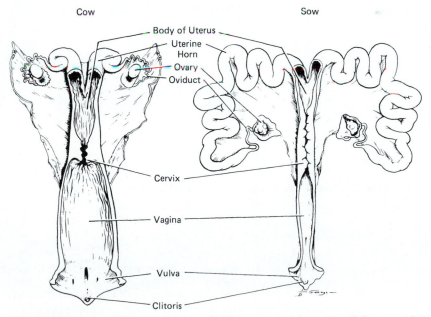

FIGURE 10.2 A dorsal view of the reproductive organs of the cow and the sow. The most noticeable difference is the longer uterine horns of the sow compared to the cow. Courtesy of Dennis Giddings.

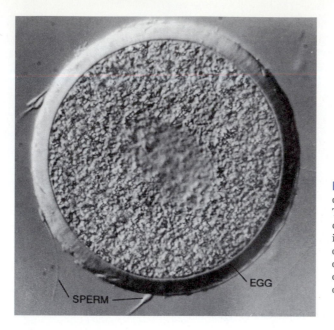

FIGURE 10.3 Bull sperm and cow egg, each magnified 300×. The ovum is about ½₀₀ in. in diameter, while the sperm is ⅙₀₀₀ in. in diameter. Each is a single cell and contains one-half the chromosome number typical of other body cells. Courtesy of Colorado State University.

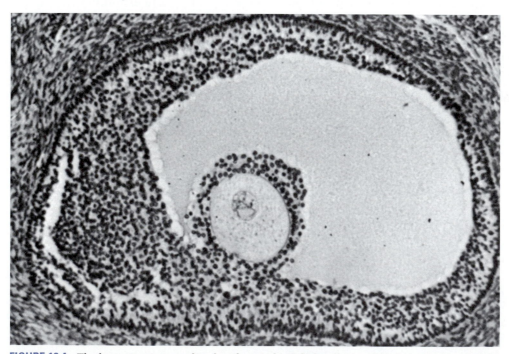

FIGURE 10.4 The large structure, outlined with a circle of dark cells, is a follicle located on a cow's ovary (magnified 265×). The smaller circle, near the center, is the egg. The large light-gray area is the fluid that fills the follicle. When the follicle ruptures, the egg will move into the oviduct by anatomical action of the infundibulum. Courtesy of Colorado State University.

size (about 0.8–1.5 in. in diameter) after having migrated from deep in the ovary to the surface of the ovary. These growing follicles produce estrogens. These mature (Graafian) follicles rupture, thus freeing the ovum (ovulation). Many of the tiny follicles grow to various stages, cease growth, deteriorate, and are absorbed (Fig. 10.5).

After the ovum escapes from the mature follicle, cells of the follicle change into a corpus luteum, or *yellow body*. The corpus luteum produces progesterone, which becomes a vitally important hormone for maintaining pregnancy.

The Oviducts

Immediately after ovulation, the ova are caught by the infundibulum of the oviduct. The ova are tiny (approximately ½₀₀ in. in diameter), which is approximately the size of a dot made by a sharp pencil. Sperm are transported through the uterus into the oviduct after the female is inseminated (naturally or artificially). Therefore, the oviducts are the sites where ova and sperm meet and

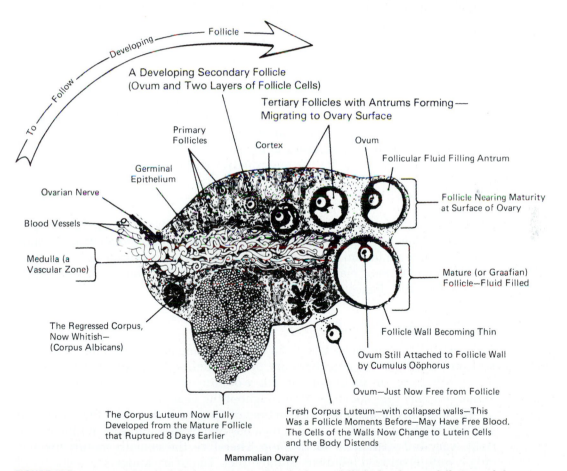

Mammalian Ovary

FIGURE 10.5 A cross section of the bovine ovary showing how a follicle develops to full size and then ruptures, allowing the egg to escape. The follicle then becomes a "yellow body" (*corpus luteum*), which is actually orange-colored in cattle. The corpus luteum degenerates in time and disappears. Of course, many follicles cease development, stop growing, and disappear without ever reaching the mature stage. From J. F. Bone, *Animal Anatomy and Physiology*, 4th ed. (Corvallis: Oregon State University Book Stores, © 1975).

where fertilization takes place. After fertilization, 3–5 days are required in cows and ewes (and probably about the same amount of time in other farm animals) for the ova to travel down the remaining two-thirds of the oviduct. From the oviduct, the newly developing embryos pass to the uterus and soon attach to it.

The Uterus

The uterus varies in shape from the type that has long, slender left and right horns, as in the sow, to the type that is primarily a fused body with short horns, as in the mare. In the sow, the embryos develop in the uterine horn; in the mare, the embryo develops in the body of the uterus. Each surviving embryo develops into a fetus and remains in the uterus until parturition (birth).

The lower outlet of the uterus is the cervix, an organ composed primarily of connective tissue that constitutes a formidable gateway between the uterus and the vagina. Like the rest of the reproductive tract, the cervix is lined with mucosal cells. These cells make significant changes as the animal goes from one estrous cycle to another and during pregnancy. The cervical passage changes from one that is tightly closed or sealed in pregnancy to a relatively open, very moist canal at the height of estrus.

The Vagina

The vagina serves as the female organ of copulation at mating and as the birth canal at parturition. Its mucosal surface changes during the estrous cycle from very moist, when the animal is ready for mating, to almost dry, even sticky, between periods of heat. The tract from the urinary bladder joins the posterior ventral vagina; from this juncture to the exterior vulva, the vagina serves the dual role of a passageway for the reproductive and urinary systems.

The Clitoris

A highly sensitive organ, the clitoris is located ventrally and at the lower tip of the vagina. The clitoris is the homologue of the penis in the male (e.g., it came from the same embryonic source as the penis). Some research indicates that clitoral stimulation or massage, following AI (artificial insemination) in cattle, will increase the chance of conception.

Reproduction in Poultry Females

The hen differs from farm animals in that the young are not suckled, the egg is laid outside the body, and there are no well-defined estrous cycles or pregnancy. Since eggs are an important source of human food, hens are selected and managed to lay eggs consistently throughout the year.

The anatomy of the reproductive tract of the hen is shown in Figs. 10.6 and 10.7. At hatching time, the female chick has two ovaries and two oviducts. The right ovary and oviduct do not develop. Therefore, the sexually mature hen has only a well-developed left ovary and oviduct. The ovary appears as a cluster of tiny gray eggs or yolks in front of the left kidney and attached to the back of the hen. The ovary is fully formed, although very small, when the chick is hatched. It contains approximately 3,600–4,000 miniature ova. As the hen reaches sexual maturity, some of the ova develop into mature yolks (yellow part of the laid

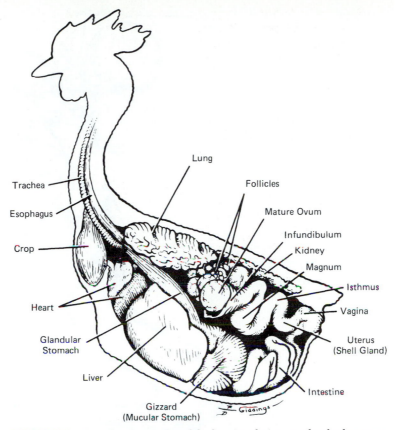

FIGURE 10.6 Reproductive organs of the hen in relation to other body organs. The single ovary and oviduct are on the hen's left side; an underdeveloped ovary and oviduct are sometimes found on the right side, having degenerated in the developing embryo. Courtesy of Dennis Giddings.

egg). The remaining ova are variable in size from the nearly mature to those of microscopic size.

The oviduct is a long, glandular tube leading from the ovary to the *cloaca* (common opening for reproductive and digestive tracts). The oviduct is divided into five parts: (1) the infundibulum (3–4 in. long), which receives the yolk; (2) the magnum (approximately 15 in. long), which secretes the thick albumen, or white, of the egg; (3) the isthmus (about 4 in. long), which adds the shell membranes; (4) the uterus (approximately 4 in. long), or shell gland, which secretes the thin white albumen, the shell, and the shell pigment; and (5) the vagina (about 2 in. long).

Ovulation is the release of a mature yolk (ovum) from the ovary. When ovulation occurs, the infundibulum engulfs the yolk and starts it on its way through the 25–27-in. oviduct. The yolk moves by peristaltic action through the infundibulum into the magnum area in about 15 minutes.

During the 3-hour passage through the magnum, more than 50% of the albumen is added to the yolk. The developing egg passes through the isthmus in about 1¼ hours. Here, water and mineral salts and the two shell membranes are added. In the 21-hour stay in the uterus, the remainder of the albumen is added,

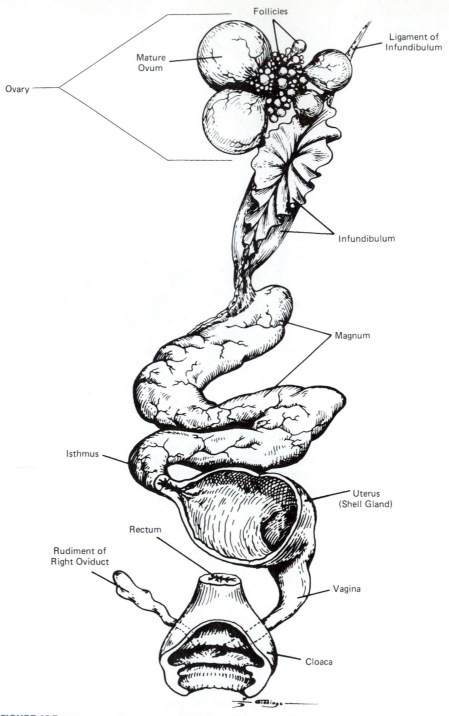

FIGURE 10.7 Reproductive organs of the hen. Sections of the uterus and cloaca are cut away to better view internal structure. Courtesy of Dennis Giddings.

followed by the addition of shell and shell pigment. Moving finally into the vagina, the fully formed egg enters the cloaca and is laid. The entire time from ovulation to laying is usually slightly more than 24 hours. About 30 minutes after a hen has laid an egg, she releases another yolk into the infundibulum, and it will likewise travel the length of the oviduct.

After the fertilized egg is incubated for 21 days, the chick is hatched. The egg is biologically structured to support the growth and life processes of the developing chick embryo during incubation and for 3–4 days after the chick is hatched.

There are several egg abnormalities that occur because of factors affecting ovulation and the developmental process. Double-yolked eggs result when two yolks are released about the same time or when one yolk is lost into the body cavity for a day and is picked up by the funnel when the next day's yolk is released. Yolkless eggs are usually formed about a bit of tissue that is sloughed off the ovary or oviduct. The tissue stimulates the secreting glands of the oviduct and a yolkless egg results. The abnormality of an egg within an egg is due to reversal of direction of an egg by the wall of the oviduct. One day's egg is added to the next day's egg, and shell is formed around both. Soft-shelled eggs generally occur when an egg is laid prematurely and insufficient time in the uterus prevents the deposit of the shell. Thin-shelled eggs may be caused by dietary deficiencies, heredity, or disease. Glassy- and chalky-shelled eggs are caused by malfunctions of the uterus of the laying bird. Glassy eggs are less porous and will not hatch, but may retain their quality.

MALE ORGANS OF REPRODUCTION AND THEIR FUNCTIONS

Figures 10.8 and 10.9 show the reproductive organs of the bull and boar. The organs of reproduction of a typical male farm mammal include two testicles, which are held in the scrotum. Male sex cells (called *sperm* or *spermatozoa*) are formed in the tiny seminiferous tubules of the testicles. The sperm from each testicle then pass through very small tubes into the *epididymis,* which is a highly coiled tube that is held in a covering on the exterior of the testicle. Each epididymal tube leads to a larger tube, the *vas deferens* (also called the *ductus deferens*). The two vasa deferentia converge at the upper end of the urethral canal, where the urinary bladder opens into the urethra. In some species, the wall of the upper end of the vas deferens is thickened and forms a secretory gland called the *ampulla.* The urethra is the large canal that leads through the penis to the outside of the body. The penis has a triple role: It serves as a passageway for semen and urine and it is the male organ of copulation.

The left and right parts of the seminal vesicles, which lie against the urinary bladder, consist of glandular tissue that secretes a substance into the urethra, which supplies nutrients for the sperm. The prostate gland contains 12 or more tubes, each of which empties into the urethra. Another gland, the bulbourethral (Cowper's) gland, which also empties its secretions into the urethral canal, is posterior to (behind) the prostate.

Testicles

The testicles produce (1) sperm cells that fertilize the ova of the female, and (2) a hormone called *testosterone* that conditions the male so that his appearance

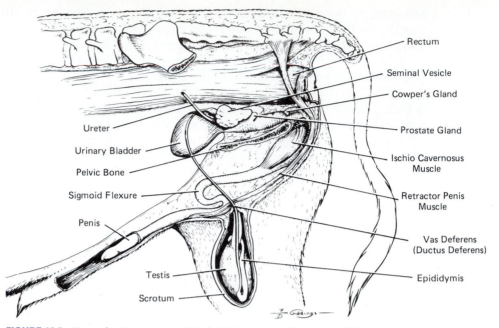

FIGURE 10.8   Reproductive organs of the bull. Courtesy of Dennis Giddings.

and behavior are masculine. Details of the structure of the spermatozoa of the bull are shown in Figs. 10.10–10.12. The diameter of sperm is approximately $\frac{1}{6000}$ in.

If both testicles are removed (as is done in castration), the individual loses his sperm factory and is sterile. Also, without testosterone his masculine appearance is not apparent, and he approaches the status of a neuter—an

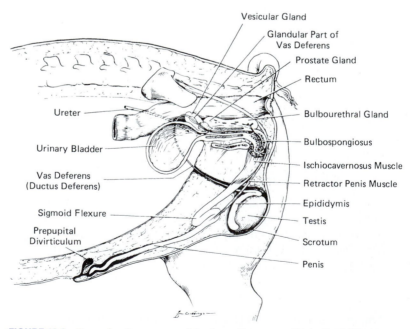

FIGURE 10.9   Reproductive organs of the boar. Courtesy of Dennis Giddings.

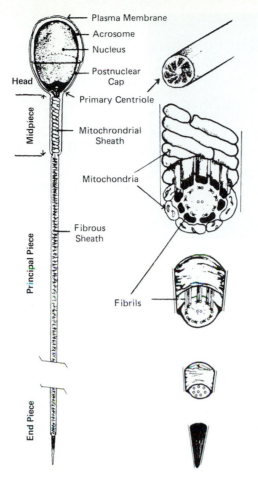

FIGURE 10.10 A diagrammatic sketch of the structure of bull sperm. Courtesy of Dr. Arthur S. H. Wu, Oregon State University.

FIGURE 10.11 A spermatozoa (sperm cell) from a bull (7000×) viewed through the electron microscope after treatment with 0.15 N NaOH at 25°C for 16 h. Note how the covering membrane of the neck region of the sperm has been removed, exposing the nine fibrils of the axial filament. Three filaments are larger than the rest. From A. S. H. Wu and F. F. McKenzie, "Microstructure of Spermatozoa After Denudation as Revealed by Electron Microscope," *Journal of Animal Science* 14(4):1151–66, 1955.

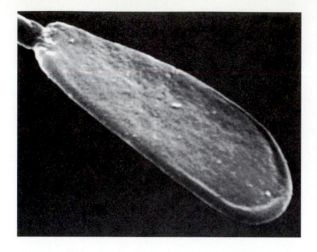

FIGURE 10.12 Bull sperm (12,000×) viewed with the scanning electron microscope showing the depth of the sperm head covered by the raised acrosome. Courtesy of Dr. Arthur S. H. Wu, Oregon State University.

individual whose appearance is somewhere between that of a male and that of a female. A steer does not have the crest, or powerful neck, of the bull. The bull has heavier, more muscular shoulders and a deeper voice than a counterpart steer. If a bull calf is castrated, the reproductive organs, such as the vas deferens, seminal vesicles, and prostate and bulbourethral glands, all but cease further development. If castration is done in a mature bull, the remaining genital organs shrink in size and in function.

Within each testicle, sperm cells are generated in the seminiferous tubules, and testosterone is produced in the cells between the tubules, called *Leydig cells* or *interstitial cells* (Fig. 10.13).

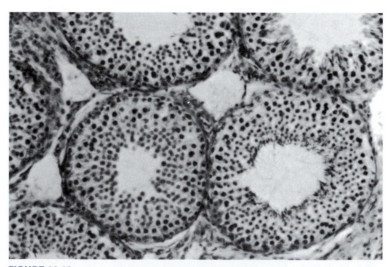

FIGURE 10.13 A cross section through the seminiferous tubules of the testis of the bull (magnified 240×). The tubule in the lower right-hand corner demonstrates the more advanced stages of spermatogenesis as the spermatids are formed near the lumen (opening) of the tubule. Courtesy of Colorado State University.

The Epididymis

The epididymis is the storage site for sperm cells, which enter it from the testicle to mature. In passing through this very long tube (95–115 ft in the bull, longer in the boar and stallion), the sperm acquire the potential to fertilize ova. Sperm taken from the part of the epididymis nearest the testicle and inseminated into females are not likely to be able to fertilize ova, whereas sperm taken near the vas deferens have the potential to fertilize.

In the sexually mature male animal, sperm reside in the epididymis in large numbers. In time, the sperm mature, then degenerate, and are absorbed in the part of the epididymis farthest from the testicle, unless they have been moved on into the vas deferens to be ejaculated.

The Scrotum

The scrotum is a two-lobed sac that contains and protects the two testicles. It also regulates temperature of the testicles, maintaining them at a temperature lower than body temperature (3–7°F lower in the bull and 9–13°F lower in the ram and goat). When the environmental temperature is low, the tunica dartos muscle of the scrotum contracts, pulling the testicles toward the body and its warmth; when the environmental temperature is high, this muscle relaxes, permitting the testicles to drop away from the body and its warmth. This heat-regulating mechanism of the scrotum begins at about the time of puberty. Hormone function precedes puberty by 40–60 days.

When the environmental temperature is elevated such that the testicles cannot cool sufficiently, the formation of sperm is impeded, and a temporary condition of lowered fertility results. Providing shade, keeping the males in the shade during the heat of the day, even providing air conditioning, are ways to manage and prevent this temporary sterility.

The Vas Deferens

The vas deferens is essentially a transportation tube that carries the sperm-containing fluid from each epididymis to the urethra. The vasa deferentia join the urethra near its origin as the urethra leaves the urinary bladder. In the mature bull, the vas deferens is about 0.1 in. in diameter except in its upper end, where it widens into a reservoir, or ampulla, about 4–7 in. long and 0.4 in. wide.

Under the excitement of anticipated mating, the secretion loaded with spermatozoa from each epididymis is propelled into each vas deferens and accumulates in the ampulla of the deferent duct. This brief accumulation of semen in the ampulla is an essential part of sexual arousal. The sperm reside briefly in the ampulla until the moment of ejaculation, when the contents of each ampulla are pressed out into the urethra, and then through the urethra and the penis en route to their deposition in the female tract.

The ampulla is found in the bull, stallion, goat, and ram—species that ejaculate rapidly. It is not present in the boar or dog, animals in which ejaculation normally takes several minutes (8–12 minutes is typical in swine). In such animals, large numbers of sperm travel all the way from the epididymis through the entire length of the vas deferens and the urethra. On close observation of the boar at the time of mating, one can see the muscles over the scrotum quivering rhythmically as some of the contents of each epididymis are propelled into the vasa deferentia and on into the urethra. This slow ejaculation of the boar contrasts to the sudden expulsion of the contents of the ampulla of the vas deferens at the height of the mating reaction, or orgasm, in the bull, stallion, ram, and goat.

The Urethra

The urethra is a large, muscular canal extending from the urinary bladder. The urethra runs posteriorly through the pelvic girdle and curves downward and forward through the full length of the penis. Very near the junction of the bladder and urethra, tubes from the seminal vesicles and tubes from the prostate gland join this large canal. The bulbourethral gland joins the urethra at the posterior floor of the penis.

Accessory Sex Glands

The ampullae, seminal vesicles, prostate, and bulbourethral glands are known as the *accessory sex glands.* Their primary functions are to add volume and nutrition to the sperm-rich fluid coming from the epididymis. Semen consists of two components: the sperm and the fluids secreted by the accessory sex glands. The semen characteristics of some farm animals are shown in Table 10.1.

The Penis

The penis is the organ of copulation. It provides a passageway for semen and urine. It is an organ characterized especially by its spongy, erectile tissue that fills with blood under considerable pressure during periods of sexual arousal, making the penis rigid and erect.

The penis of the bull is about 3 ft in length and 1 in. in diameter, tapering to the free end, or glans penis. In the bull, boar, and ram, the penis is S-shaped when relaxed. This S curve, or sigmoid flexure, becomes straight when the penis is erect. The sigmoid flexure is restored after copulation, when the relaxing penis is drawn back into its sheath by a pair of retractor penis muscles. The stallion penis has no sigmoid flexure; it is enlarged by engorgement of blood in the erectile tissues.

TABLE 10.1   Semen Characteristics of Several Male Animals

Animal	Volume per Ejaculate (ml)	Composition of Ejaculate	Sperm Concentration per ml × 10^9	Total Sperm per Ejaculate × 10^9
Bull (cattle)	3–10	Single fraction	0.8–1.2	4–18
Ram (sheep)	0.5–2.0	Single fraction	2–3	1–4
Boar (swine)	150–250	Fractionated	0.2–0.3	30–60
Stallion (horse)	40–100	Fractionated	0.15–0.40	8–50
Buck (goat)	0.5–2.5	—	2.0–3.5	1–8
Dog (dog)	1.0–5.0	—	2–7	4–14
Buck (rabbit)	0.5–6.5	—	0.3–1.0	1.5–6.5
Tom (turkey)	0.1–0.7	—	8–30	1–20
Cock (chicken)	0.1–1.5	—	0.4–1.5	0.05–2.0

The free end of the penis is termed the *glans penis.* The opening in the ram penis is at the end of a hairlike appendage that extends about 0.8–1.2 in. beyond the larger penis proper. This appendage also becomes erect and, during ejaculation, whirls in a circular fashion, depositing semen in the anterior vagina. It does not regularly penetrate the ewe's cervix, as some investigators claim it does. Only a small portion of the penis of the bull, boar, ram, and goat extends beyond its sheath during erection. The full extension awaits the thrust after entry into the vagina has been made. The stallion and ass usually extend the penis completely before entry into the vagina.

All these accessory male sex organs depend on testosterone for their tone and normal function. This dependence is especially apparent when the testicles are removed (castration); the usefulness of the accessory sex organs is then diminished or even terminated.

Reproduction in Male Poultry

The reproductive tract of male poultry is shown in Fig. 10.14. There are several differences when compared to the reproductive tracts of the farm mammals previously described. The testes of male poultry are contained in the body cavity. Each vas deferens opens into small papillae, which are located in the cloacal wall. The male fowl has no penis but does have a rudimentary organ of copulation. The sperm are transferred from the papillae to the rudimentary copulatory organ, which transfers the sperm to the oviduct of the hen during the mating process. The sperm are stored in primary sperm-host glands located in the oviduct. These sperm are then released on a daily basis and transported to secondary storage glands in the infundibulum. Fertilization occurs in the infundibulum. Sperm stored in the oviduct are capable of fertilizing the eggs for 30 days in turkeys and 10 days in chickens.

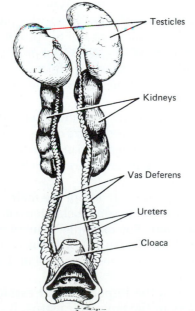

Testicles

Kidneys

Vas Deferens

Ureters

Cloaca

FIGURE 10.14 Male poultry reproductive tract (ventral view). Courtesy of Dennis Giddings.

WHAT MAKES TESTICLES
AND OVARIES FUNCTION

Testicular Function

Testicles produce their hormones under stimuli coming to them from the anterior pituitary (AP) gland situated at the base of the brain. The AP produces and secretes two hormones important to male reproductive performance. Luteinizing hormone (LH) and follicle-stimulating hormone (FSH) are known as *gonadotropic hormones* because they stimulate the gonads (ovary and testicle). LH produces its effect on the interstitial tissue (Leydig cells) of the testicle, causing the tissue to produce the male hormone, testosterone. FSH stimulates cells in the seminiferous tubules to nourish the developing spermatozoa.

Some species that respond to change in length of daylight exhibit more seasonal fluctuation than others in reproductive activities. These influences of daylight are on the neurophysiological mechanism in the brain. The hypothalamus is the floor and part of the wall of the third ventricle of the brain and secretes releasing factors through blood vessels that affect the AP and its production of FSH and LH (see Figs. 18.11 and 19.3 in Chapters 18 and 19, respectively).

Ovarian Function

Ovarian hormones in the sexually mature female owe the rhythmicity on their production to hormones that originate in the AP and to the interplay between gonad-stimulating hormones produced there and ovarian hormones whose level and potency vary as the estrous cycle progresses. FSH (the same hormone as FSH in the male) circulates through the bloodstream and affects the responsive follicle cells of the ovary, which respond by secreting the so-called estrogens or estrus-producing hormones, estradiol and estrone. When the amount of estradiol and estrone in the blood reaches a sufficient level, the pituitary is caused to reduce its production of FSH. With this drop in FSH production, the production of estradiol and estrone subsides. When the estrus-producing hormones have been lost from the body and their depressing effect on the AP has been spent, the AP again steps up its production of FSH and the cycle is repeated.

Luteal cells, the successors of the follicle cells, secrete a hormone called *progesterone,* which plays a role in the female reproductive cycle. Progesterone helps cause the occurrence and recurrence of the desire to mate, which is called **estrus** or **heat.** Fig. 10.15 shows the major events and hormone changes in the estrous cycle of the cow.

Estrus is the period of time when the female will accept the male for breeding purposes. The female of each species exhibits some behavior patterns that demonstrate she is in heat (see Chapter 34). For example, a mare in estrus, when "teased" with the presence of a stallion (Fig. 10.16), will not avoid or kick him. The mare in heat will stand solidly, sometimes squatting and urinating when approached by the stallion.

Synchronized with the estrous cycle is the important and essential phenomenon of ovulation. Ovulation occurs in the cow after estrus. It occurs in

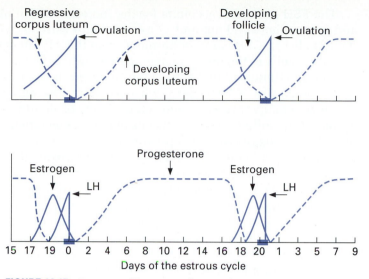

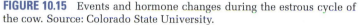

FIGURE 10.15 Events and hormone changes during the estrous cycle of the cow. Source: Colorado State University.

the sow, ewe, goat, and mare toward the latter part of, but nevertheless during, estrus. These species all ovulate spontaneously; that is, ovulation takes place whether copulation occurs or not. By contrast, copulation (or some such stimulation) is necessary to trigger ovulation in such animals as the rabbit, cat, ferret, and mink, which are considered *induced ovulators.* Ovulation in these animals takes place at a fairly consistent time after mating. Ovulation is controlled by hormones. The follicle of the ovary grows, matures, fills with fluid, and softens a few hours before rupturing. The follicle ruptures owing to a sudden release of LH rather than bursting as a result of pressure inside.

FIGURE 10.16 Estrus is determined in the mare by "teasing" her with the presence of a stallion. Courtesy of Colorado State University.

The FSH of the AP accounts for the increase in size of the ovarian follicle and for the increased amount of estradiol and estrone, which are products of the follicle cells. LH, another hormone from the AP, alters the follicle cells and granulosa cells of the ovary, changing them into luteal cells. These luteal cells are in turn stimulated by LH from the AP to produce progesterone.

In the course of 7–10 days, what was formerly an egg-containing follicle develops into the corpus luteum, a luteal body of about the same size and shape as the mature follicle. The luteal cells of the corpus luteum produce a sufficient quantity of progesterone to depress FSH secretion from the AP until the luteal cells reach maximum development (in nonpregnant animals), cease their development, and (in 3 weeks) lose their potency and disappear.

If pregnancy occurs, the corpus luteum continues to function, persisting in its progesterone production and preventing further estrous cycles. Thus, no more heat occurs until after pregnancy has terminated. The process of follicle development, ovulation, and corpus luteum development and regression is shown in Fig. 10.5. Table 10.2 shows the length of estrus, estrous cycles, and time of ovulation for the different farm animals.

The changing length of daylight is a potent factor that influences the estrous cycle, the onset of pregnancy, and the seasonal fluctuations in male fertility. The amount of daylight acts both directly and indirectly on the animal. It acts directly on the central nervous system by influencing the secretion of hormones and indirectly by affecting plant growth, thus altering the level and quality of nutrition available. In cattle, increasing length of day is associated with increased reproductive activity in both males and females. In sheep, the breeding season reaches its height in the autumn, as the hours of daylight shorten. Of course, individuals of both sexes vary in their intensity of sex drive and level of fertility. When selection has resulted in improvements in the traits associated with reproduction, individuals exhibit higher levels of fertility (e.g., more intense expression of estrus, occurrence of estrus over more months of the year, or occurrence of spermatogenesis at a high level over more months of the year) than unselected individuals.

Table 10.3 summarizes the major hormones affecting the reproductive process. These hormones are discussed throughout this chapter.

TABLE 10.2   Duration and Frequency of Heat and Time of Ovulation

Animal	Duration of Heat Average	Duration of Heat Range	Length of Cycle (days) Average	Length of Cycle (days) Range	Approximate Time of Ovulation
Heifer, cow (cattle)	12 hours	6–27 hours	21	19–23	30 hours after beginning of heat
Ewe (sheep)	30 hours	20–24 hours	17	14–19	26 hours after beginning of heat
Mare (horse)	6 days	1–37 days	21	10–37	1 day before the end of heat
Gilt, sow (swine)	44 days	1½–4 days	21	19–23	30–38 hours after beginning of heat
Doe (goat)	39 hours	20–80 hours	20	12–27	On second day of heat
Doe (rabbit)[a]		Constant estrus			8–10 hours after mating
Queen (cat)	5 days	4–7 days	—	14–21	24 hours after mating
Bitch (dog)[a,b]	9 days	4–13 days	—	120–240	24–48 hours after heat begins
Jill (mink)	2 days	Seasonal breeding	(March)		40–50 hours

[a] The dog and rabbit may exhibit a pseudo or false pregnancy after mating has occurred.
[b] The dog has no cycle. There are generally two heats per year—in the fall and spring.

TABLE 10.3   The Major Reproductive Hormones, Their Source, Target Organs, and Functions

Hormone	Source	Target	Function(s)
Cortisol (fetal)	Fetal adrenal gland	Dam's uterus	Initiation of parturition causes $PGF_2\alpha$ release causes decreased progesterone
Estrogen	Ovary	Uterus, brain, mammary system	Female mating behavior Secondary sex characteristics Decreased progesterone
Follicle stimulating hormone (FSH)	Anterior pituitary	Ovary, testes	Sperm production Follicular growth
Gonadotropin releasing hormone (GnRH)	Hypothalamus	Anterior pituitary	Release of LH and FSH
Luteinizing hormone	Anterior pituitary	Ovary, testes	Testosterone release Ovulation Corpus luteum formation and maintenance
Melatonin	Pineal gland	Hypothalamus	Ovulatory initiation in sheep and horses
Oxytocin	Posterior pituitary	Uterus, oviduct	Helps sperm and ova transport Stimulates uterine contractions during parturition
Progesterone	Corpus luteum	Uterus, mammary tissue	Maintenance of pregnancy Mammary development
Prostaglandin ($PGF_2\alpha$)	Uterus	Ovary	Regression of the corpus luteum
Relaxin	Ovary	Cervix, pelvis	Dilation of cervix Pelvic expansion
Testosterone (androgens)	Testis	Male tract, brain, seminiferous tubules	Male mating behavior Sperm production Secondary male sex characteristics

PREGNANCY

When the sperm and the egg unite (fertilization), conception occurs, which is the beginning of the gestation period. The fertilized egg begins a series of cell divisions (Fig. 10.17). About every 20 hours, embryonic cells duplicate their genes and divide, progressing through the 2-, 4-, 8-, and 16-cell stages, and so on. In farm mammals, the embryo migrates through the oviduct to the uterus in 3–4 days. By then it has developed to the 16- or 32-cell stage. The chorionic and amniotic membranes develop around this new embryo, and the chorion attaches to the uterus. The embryo (and later the fetus) obtains nutrients and discharges wastes through these membranes. This period of attachment (20–30 days in cattle and 14–21 days in swine) is critical. Unless the environment is sufficiently favorable, the embryo dies. Embryonic mortality causes a significant economic loss in farm animals, especially swine, where multiple ovulations and embryos are typical of the species. Good management will help protect the female and embryos early in pregnancy. High temperatures will cause embryonic death. The female should enter the breeding season in a thrifty, weight-gaining condition.

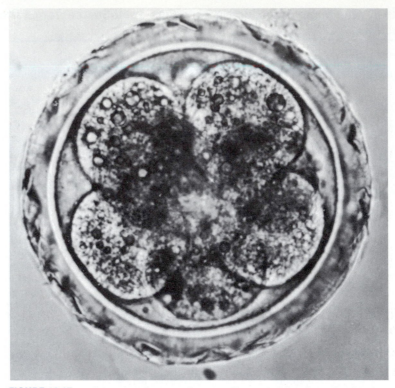

FIGURE 10.17 A bovine embryo in the six-cell stage of development (magnified 620×). Courtesy of Colorado State University.

The *embryonic* stage is defined as that period when body parts differentiate and essential organs are formed. This period lasts 45 days in cattle.

When the embryonic stage is completed, the young organism is called a *fetus*. The *fetal* period, which lasts until birth, is mainly a time of growth. The duration of pregnancy for several species of animals is included in Table 10.4. Length of pregnancy varies chiefly with the breed and age of the mother.

TABLE 10.4 **Gestation Length and Number of Offspring Born**

Animal	Gestation Length (days)	Usual Number of Offspring Born
Cow (cattle)	285	1
Ewe (sheep)	147	1–3
Mare (horse)	336	1
Sow (swine)	114	6–14
Doe (goat)	150	2–3
Doe (rabbit)	31	4–8
Jill (mink)	50	4
Queen (cat)	63	4
Bitch (dog)	65	7

PARTURITION

Parturition (birth) marks the termination of pregnancy (Fig. 10.18). Extraembryonic membranes that had been formed around the embryo in early pregnancy are shed at this time and are known as *afterbirth.* These membranes attach to the uterus during pregnancy and become known as the *placenta,* which is responsible for the transfer of nutrients and wastes between mother and fetus. The placenta produces hormones, especially estrogens and progesterone, in some farm animal species.

The parturition process is initiated by release of the hormone cortisol from the fetal adrenal cortex. Progesterone levels decline, whereas estrogen, prostaglandin $F_{2\alpha}$, and oxytocin levels rise, resulting in uterine contractions.

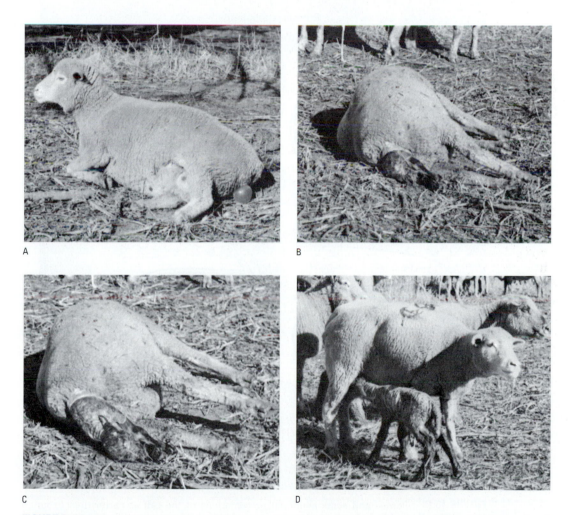

A

B

C

D

FIGURE 10.18 Parturition in the ewe. (A) The water bag (a fluid-filled bag) becomes visible. (B) The head and front legs of the lamb appear. (C) The lamb is forced out by the uterine contractions of the ewe. (D) In a few minutes after birth, the lamb is nursing the ewe. From H. H. Cole and W. N. Garrett (eds.), *Animal Agriculture,* 2nd ed. (New York: W. H. Freeman, © 1980).

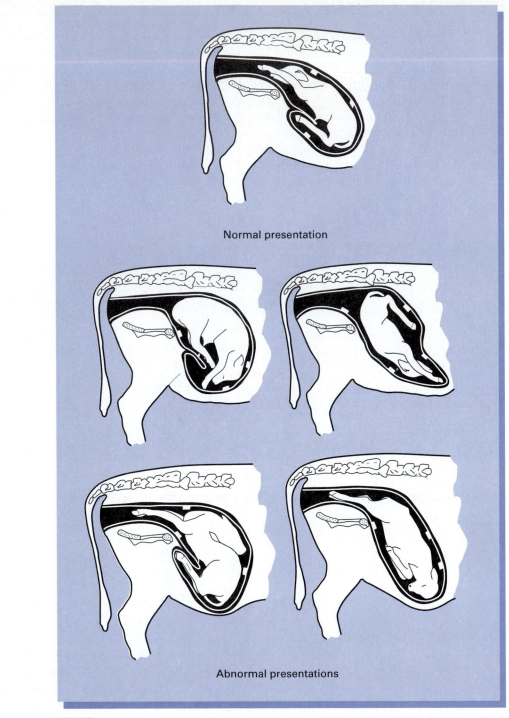

Normal presentation

Abnormal presentations

FIGURE 10.19 Normal and some abnormal presentations of the calf at parturition. From R. A. Battaglia and V. B. Mayrose, *Handbook of Livestock Management Techniques* (New York: Macmillan, 1981), pp. 131, 134, 135.

Parturition is a synchronized process. The cervix, until now tightly closed, relaxes. Relaxation of the cervix, along with pressure generated by uterine muscles on the contents of the uterus, permits the passage of the fetus into the vagina and on to the exterior. Another hormone, relaxin, is thought to aid in parturition. Relaxin, which originates in the corpus luteum or placenta, helps to relax cartilage and ligaments in the pelvic region.

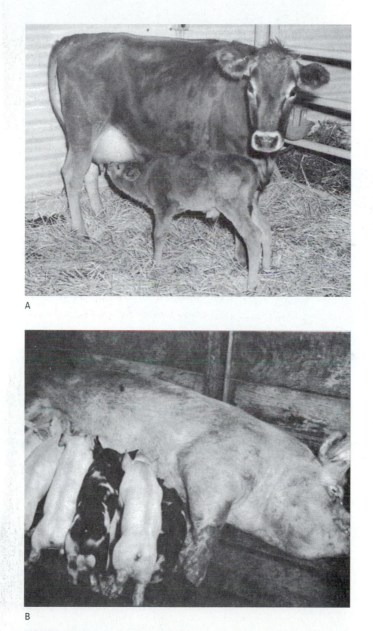

FIGURE 10.20 Live offspring are born to each breeding female in the herd or flock when the intricate mechanisms of reproduction function properly. (A) Dairy cow and calf. Courtesy of Colorado State University. (B) Sow and litter of pigs. Courtesy of D. C. England, Oregon State University.

C

D

E

F

FIGURE 10.20 (*continued*).   (C) Mare and foal. Courtesy of *The Western Horseman.* (D) Ewe and lambs. Courtesy of Colorado State University. (E) Chicks in the process of hatching. Courtesy of *Poultry Digest.* (F) Beef cow and calf. Courtesy of the American Polled Hereford Association.

At the beginning of parturition, the offspring typically assumes a position that will offer the least resistance as it passes into the pelvic area and through the birth canal. Fetuses of the cow, mare, and ewe assume similar positions in which the front feet are extended with the head between them (Fig. 10.19). Fetal piglets do not orient themselves in any one direction, which does not appear to affect the ease of birth. Calves, lambs, or foals may occasionally present themselves in a number of abnormal positions (Fig. 10.19). In many of these situations, assistance needs to be given at parturition. Otherwise, the offspring may die or be born dead, and in some instances, the mother may die as well. An abnormally small pelvic opening or an abnormally large fetus can cause some mild to severe parturition problems.

Producers can manage their herds and flocks for high reproductive rates. Management decisions are most critical that cause males and females, selected for breeding, to reach puberty at early ages, have high conception rates, and have minimum difficulty at parturition. Keeping animals healthy, providing adequate levels of nutrition, selecting genetically superior animals, and producer attention to parturition are some of the more critical management inputs to minimize reproductive loss. Live offspring born to each breeding female is the key end point to successful farm animal reproduction (Fig. 10.20).

CHAPTER SUMMARY

- Reproductive efficiency, as measured by number of offspring born in a herd or flock of breeding females, is a trait of high economic importance.

- A knowledge of the anatomy and physiology of the male and female reproduction organs is essential in understanding the production of ova and sperm, estrous cycles, mating, pregnancy, and parturition.

- Follicle-stimulating hormone, luteinizing hormone, estrogen, progesterone, and testosterone are important hormones in the reproductive process.

- Reproduction in poultry is unique in that ova (eggs) are produced outside the body from only one ovary and there are no well-defined estrous cycles or pregnancy.

REVIEW QUESTIONS

1. What are the two functions of the female gonads or ovaries?
2. What structure on the ovary produces the ovum?
3. Where does fertilization take place?
4. In what structure does the fetus develop during pregnancy in mammals?
5. In what three features does reproduction in poultry differ from that in mammals?
6. How does the reproductive tract of female poultry differ from that of female mammals?
7. What are the two functions of the male gonads or testicles?
8. Where are the spermatozoa produced within the testicles?

9. What cells within the testicles produce testosterone?

10. What are two functions of the epididymis?

11. The scrotum of males helps to maintain the testicles at temperatures several degrees cooler than body temperature. Why is this knowledge important to livestock producers?

12. What are the functions of the accessory sex glands in male mammals?

13. What are the organs of copulation in male and female mammals? What is a second function of these organs?

14. What two hormones from the anterior pituitary gland are responsible for stimulating the gonads?

15. What is that period when the female is sexually receptive to the male called? Why is this knowledge important to livestock producers?

16. What hormone is responsible for the occurrence of estrus or heat?

17. What hormone is responsible for the male sex drive or libido?

18. What is the termination of pregnancy, resulting in birth, called?

19. Why is the knowledge of the gestation length for individual species of livestock important to producers?

SELECTED REFERENCES

Publications

Battaglia, R. A., and Mayrose, V. B. 1981. *Handbook of Livestock Management Techniques.* New York: Macmillan.

Bearden, H. J., and Fuquay, J. W., 3d ed., 1997. *Applied Animal Reproduction.* Upper Saddle River, NJ: Prentice-Hall.

Bone, J. F. 2d ed., 1988. *Animal Anatomy and Physiology.* Reston, VA: Reston Publishing.

Frandson, R. D. 2d ed., 1986. *Anatomy and Physiology of Farm Animals.* Philadelphia: Lea and Febiger.

Hafez, E. S. E. (ed.). 1993. *Reproduction in Farm Animals.* 6th ed. Philadelphia: Lea and Febiger.

King, G. J. (ed.). 1993. *Reproduction in Domesticated Animals.* Amsterdam, The Netherlands: Elsevier Science Publishers.

Murdoch, W. J. 1994. Animal Reproduction. *Encyclopedia of Agricultural Science.* San Diego: Academic Press, Inc.

Pickett, B. W., Voss, J. L., Squires, E. L., and Amann, R. P. 1981. *Management of the Stallion for Maximum Reproductive Efficiency.* Fort Collins: Colorado State University Press, Animal Reprod. Lab. Gen. Series 1005.

Visuals

Beef Cattle Reproduction. 1995. (three videotapes). CEV, P.O. Box 65265, Lubbock, TX 79464-5265.

Embryo Development of the Chick (silent filmstrip). Vocational Education Productions, California Polytechnic State University, San Luis Obispo, CA 93407.

Equine Reproduction. 1995. (video; 27 min.). CEV, P.O. Box 65265, Lubbock, TX 79464.

Foaling. 1997. (video; 28 min) and *Equine Reproduction.* 1995. (video; 27 min.). CEV, P.O. Box 65265, Lubbock, TX 79464.

Fundamental Livestock Parturition. 1997. (video; 30 min.). CEV, P.O. Box 65265, Lubbock, TX 79464.

Heat Detection in Dairy Cows; Artificial Insemination; and *The Calving Process* (videotapes). Agricultural Products and Services, 2001 Killebrew Dr., Suite 333, Bloomington, MN 55420.

Poultry Reproduction: Male. 1996. (video; 18 min.) and *Poultry Reproduction: Female.* 1996. (video; 41 min). CEV, P.O. Box 65265, Lubbock, TX 79464.

Swine Reproduction. 1994. (two videotapes). CEV, P.O. Box 65265, Lubbock, TX 79464-5265.

CHAPTER 11

Artificial Insemination, Estrous Synchronization, and Embryo Transfer

In the process of artificial insemination (AI), semen is deposited in the female reproductive tract by artificial techniques rather than by natural mating. AI was first successfully accomplished in the dog in 1780 and in horses and cattle in the early 1900s. AI techniques are also available for use in sheep, goats, swine, poultry, laboratory animals, and bees.

The primary advantage of AI is that it permits extensive use of outstanding sires to optimize genetic improvement. For example, a bull may sire 30–50 calves naturally per year over a productive lifetime of 3–8 years. In an AI program, a bull can produce 200–400 units of semen per ejaculate, with four ejaculates typically collected per week. If the semen is frozen and stored for later use, hundreds of thousands of calves can be produced by a single sire (one calf per 1.5 units of semen), and many of these offspring can be produced long after the sire is dead. AI also can be used to control reproductive diseases, and sires can be used that have been injured or are dangerous when used naturally.

SEMEN COLLECTING AND PROCESSING

There are several different methods of collecting semen. The most common method is the artificial vagina (Fig. 11.1), which is constructed to be similar to an actual vagina. The artificial vagina is commonly used to collect semen from bulls, stallions, rams, buck goats, and rabbits. The semen is collected by having the male mount an estrous female or training him to mount another animal or object (Figs. 11.2, 11.3). When the male mounts, his penis is directed into the artificial vagina by the person collecting the semen, and the semen accumulates in the collection tube. Semen from the boar and dog is not typically collected

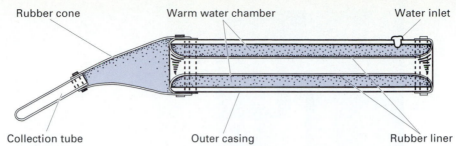

Rubber cone   Warm water chamber   Water inlet

Collection tube   Outer casing   Rubber liner

FIGURE 11.1 Longitudinal section of an artificial vagina. Courtesy of Dennis Giddings.

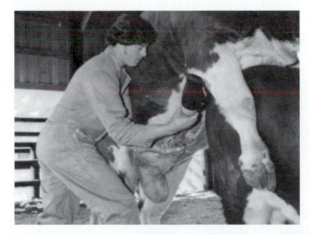

FIGURE 11.2 Using the artificial vagina to collect semen from the stallion. The handler of the stallion is holding the front leg to prevent the horse from striking the collector. Courtesy of Colorado State University.

FIGURE 11.3 Collecting semen from a bull using the artificial vagina. Courtesy of Colorado State University.

with an artificial vagina. It is collected by applying pressure with a gloved hand, after grasping the extended penis, as the animal mounts another animal or object.

Semen can also be collected by using an electroejaculator, in which a probe is inserted into the rectum and an electrical stimulation causes ejaculation. This is used most commonly in bulls and rams that are not easily trained to use the artificial vagina and from which semen is collected infrequently.

FIGURE 11.4 Approximately 4 ml of bull semen in the collection tube attached to the artificial vagina. Note that the collection tube is immersed in water to control the temperature. The livability of the sperm is decreased when they are subjected to sudden temperature changes. Courtesy of Colorado State University.

If semen is collected too frequently, the number of sperm per ejaculate decreases. Semen from a bull is typically collected twice a day for 2 days a week (Fig. 11.4). Semen from rams can be collected several times a day for several weeks, but bucks (goats) must be ejaculated less frequently. Boars and stallions give large numbers of sperm per ejaculate, so semen from them is usually collected every other day at most. Semen is collected from tom turkeys 2–3 times per week. The simplest and most common technique for collecting sperm from toms is the abdominal massage. This technique usually requires two persons: the "collector" helps hold the bird and operates the semen collection apparatus, while the "milker" stimulates a flow of semen by massaging the male's abdomen (Fig. 11.5).

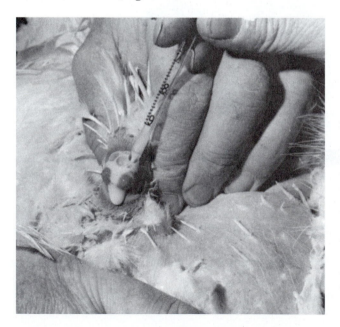

FIGURE 11.5 Collection of semen from a turkey. Two technicians are usually involved in the collection process. One holds the turkey while the other manually stimulates semen into the cloaca while drawing semen into the collection tube. Courtesy of P. E. Lake, and J. M. Stewart, *Artificial Insemination in Poultry,* 1978, Ministry of Agriculture, Fisheries, and Food. No. 213, London: Her Majesty's Stationery Office, British Crown Copyright.

After semen is collected, it is evaluated for volume, sperm concentration, motility of the sperm, and sperm abnormalities (Fig. 11.6). The semen is usually mixed with an extender that dilutes the ejaculate to a greater volume. This greater volume allows a single ejaculate to be processed into several units of semen, where one unit of semen is used for each female inseminated. The extender is usually composed of nutrients such as milk and egg yolk, a citrate buffer, antibiotics, and glycerol. The amount of extender used is based on the projected number of viable sperm available in each unit of extended semen. For example, each unit of semen for insemination in cattle should contain 10 million motile, normal spermatozoa.

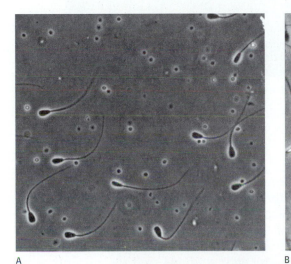

A

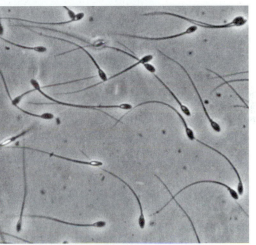

B

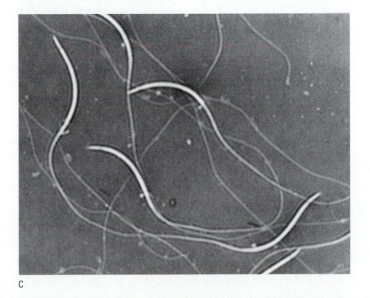

C

FIGURE 11.6   (A) Normal bull semen. (B) Normal stallion semen. (C) Normal semen from domestic fowl as viewed under the microscope. Courtesy of Colorado State University (A and B); Dr. P. E. Lake and British Crown Copyright (C).

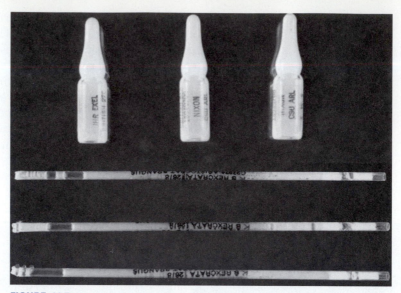

FIGURE 11.7 Semen is typically stored and frozen in ampules or straws. Three ampules appear at the top of the figure, with three straws of bull semen below. The animal's name, registration number, and location of collection are printed on the ampules and straws. Courtesy of Colorado State University.

Some semen is used fresh; however, it can be stored this way only for a day or two. Most semen is frozen in liquid nitrogen and stored in ampules (glass vials), plastic straws (Fig. 11.7), or pellets. Bull semen can be stored in this manner for an indefinite period of time and retain its fertilization capacity. Most cattle are inseminated with frozen semen that has been thawed. The fertilizing capacity of frozen semen from boars, stallions, and rams is only modestly satisfactory; however, improvement has been noted in recent years. Turkey semen cannot be frozen satisfactorily. For maximum fertilization capacity, it should be used within 30 minutes of collection.

INSEMINATION OF THE FEMALE

Prior to insemination, the frozen semen is thawed. Thawed semen cannot be refrozen and used again because refreezing will kill the sperm cells.

High conception rates using AI depend on the female's cycling and ovulating; detecting estrus; using semen that has been properly collected, extended, and frozen; thawing and handling the semen satisfactorily at the time of insemination; insemination techniques; and avoiding extremes in stress and excitement to the animal being inseminated.

Detecting Estrus

Estrus must be detected accurately because it signals time of ovulation and determines proper timing of insemination. The best indication of estrus is the condition called *standing heat,* in which the female stands still when mounted by a male or another female.

Cows are typically checked for estrus twice daily, in the morning and evening. They are usually observed for 30 minutes to detect standing heat. Other observable signs are restlessness, attempting to mount other cows, and a clear, mucous discharge from the vagina. Some producers use sterilized bulls or hormone-treated cows as *heat checkers* in the herd. These animals are sometimes equipped with a chin marker that greases or paints marks on the back of the cow when she is mounted.

Estrus in sheep or goats is checked using sterilized males equipped with a brisket-marking harness. Gilts and sows in heat assume a rigid stance with ears erect when hands are placed firmly in their backs. The vulva is usually red and swollen. The presence of a boar and the resulting sounds and odors can help detect heat in swine as the females in estrus will attempt to locate a boar. Signs of estrus in the mare are elevation of the tail, contractions of the vulva (winking), spreading of the legs, and frequent urination.

Proper Timing of Insemination

The duration of estrus and ovulation time are quite variable in farm animals. This variability poses difficulty in determining the best time for insemination. An additional challenge is that sperm are short-lived when put into the female reproductive tract. Also, estrus is sometimes expressed without ovulation occurring; sows, for example, typically show estrus 3–5 days after farrowing, but ovulation does not occur. Sows should not be bred at this time either artificially or naturally.

Insemination time should be as close to ovulation time as possible; otherwise, sperm stay in the female reproductive tract too long and lose their fertilizing capacity. Cows found in estrus in the morning are usually inseminated the evening of the same day, and cows in heat in the evening are inseminated the following morning. Insemination, therefore, should occur toward the end or after estrus has been expressed in the cow. Ewes are usually inseminated in the second half of estrus, and goats are inseminated 10–12 hours after the beginning of estrus. Sows ovulate from 30–38 hours after the beginning of estrus, so insemination is recommended at the end of the first day and at the beginning of the second day of estrus. When sows are inseminated both days, conception rate is improved compared to insemination on only one day.

Insemination of dairy cows occurs while the cow is standing in a stall or stanchion. Beef cows are penned and inseminated in a chute that restrains the animal. The most common insemination technique in cattle involves the inseminator having one arm in the rectum to manipulate the insemination tube through the cervix (Fig. 11.8). The insemination tube is passed just through the cervix, and the semen is deposited into the body of the uterus. The insemination procedure for sheep and goats is similar to that for cattle; however, a speculum (a tube approximately 1.5 in. in diameter and 6 in. long) allows the inseminator to observe the cervix in sheep and goats. The inseminating tube is passed through the speculum into the cervix, where the semen is deposited into the uterus or cervix.

The sow is usually inseminated without being restrained. The inseminating tube is easily directed into the cervix because the vagina tapers into the cervix. The semen is expelled into the body of the uterus. The mare is hobbled or adequately restrained prior to insemination. The vulva area is washed, and the tail

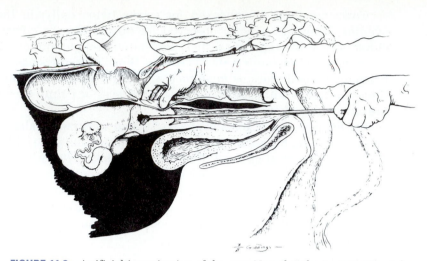

FIGURE 11.8   Artificial insemination of the cow. Note that the insemination tube has been manipulated through the cervix. The inseminator's forefinger is used to determine when the insemination rod has entered into the uterus. Courtesy of Dennis Giddings.

is wrapped or put into a plastic bag. The plastic-covered arm of the inseminator is inserted into the vagina, and the index finger is inserted into the cervix. The insemination tube is passed through the cervix, and the semen is deposited into the uterus.

The turkey hen is inseminated by first applying pressure to the abdominal area to cause eversion of the oviduct (Fig. 11.9). The insemination tube is

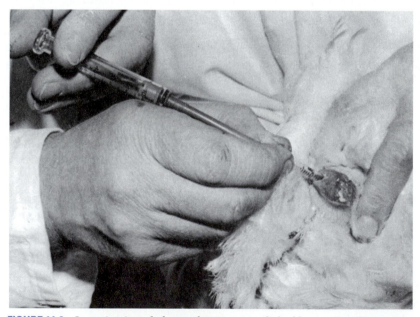

FIGURE 11.9   Insemination of a hen turkey is accomplished by everting the opening of the oviduct through the opening of the cloaca. The tube is inserted 1 or 2 inches, pressure on the oviduct is relaxed, and the proper amount of semen is deposited as the syringe is slowly withdrawn. Courtesy of P. E. Lake and J. M. Stewart, *Artificial Insemination in Poultry,* 1978, Ministry of Agriculture, Fisheries, and Food, No. 213, London: Her Majesty's Stationery Office, British Crown Copyright.

inserted into the oviduct approximately 2 in., and the semen is released. The first insemination is made when 5–10% of the flock have started laying eggs. A second insemination a week later ensures a high level of fertility. Thereafter, insemination is done at 2-week intervals. Fertility in the turkey usually persists at a high level for 2–3 weeks after insemination. This is possible because sperm are stored in special glands of the hen, where they are nourished and retain their fertilizing capacity. Most of these glands are located near the junction of the uterus and vagina.

EXTENT OF ARTIFICIAL INSEMINATION

The number of farm animals artificially inseminated each year is not well documented. AI is used primarily in the dairy industry, where 70% of the cows and 25% of the heifers are bred artificially. Most of the inseminations are in the Holstein breed (Table 11.1). Less than 5% of the beef cows are inseminated, where semen from Angus and Simmental bulls is used most extensively (Table 11.1). The greater use of AI in dairy compared to beef is reflected in Table 11.1, where in 1995 13.5 million units of dairy semen was produced compared to 2.9 million units of beef semen.

It is estimated that approximately 15,000 sows are artificially bred in the United States each year, which is less than 1% of the total sows bred in this country. Satisfactory techniques to freeze boar semen were accomplished in 1971, but frozen semen yields smaller litter sizes than fresh semen. More than 400,000 sows are artificially inseminated each year in certain European countries, representing 20–30% of the sows bred in these countries.

AI in horses is still limited because of difficulty in providing extended semen storage. AI in sheep and goats in the United States is limited because herds and flocks are dispersed over wide areas and the cost per unit of semen is high. Little AI is done in chickens; however, AI is rather extensive in turkeys. It is especially important in the broad-breasted turkey, which, because of the size of its breast, has difficulty mating naturally.

TABLE 11.1   Units of Semen Sold for Domestic Use, Custom Frozen, and Exported, 1995

Breed	Domestic Sales	Custom Frozen	Total	Export Sales No. Units	Export Sales $Mil
			1,000 Units		
Holstein	11,930	611	12,541	7,416	—
Jersey	595	26	621	422	—
Brown Swiss	118	69	187	277	—
Other dairy	143	8	151	34	—
Total dairy	12,787	714	13,501	8,149	$61.3
Angus	598	469	1,067	126	—
Simmental	103	112	215	77	—
Polled Hereford	82	62	144	62	—
Red Angus	52	104	156	105	
Limousin	31	139	170	28	
Maine-Anjou	8	146	154	—	
Other beef	123	850	973	92	—
Total beef	997	1,882	2,879	490	$1.8

Source: National Association of Animal Breeders.

ESTROUS SYNCHRONIZATION

Estrous synchronization is controlling or manipulating the estrous cycle so that females express estrus at approximately the same time. Estrous synchronization is a useful part of an AI program because checking heat and breeding animals under extensive management conditions is time-consuming and expensive. Also, estrous synchronization is a successful tool in making embryo transfer programs successful.

Prostaglandin

In 1979, a prostaglandin was cleared for use in cattle. *Prostaglandins* are naturally occurring fatty acids that have important functions in several of the body systems. The prostaglandin that has a marked effect on the reproductive system is prostaglandin F_2 alpha ($PGF_{2\alpha}$).

It is pointed out in Chapter 10 that the corpus luteum (CL) controls the estrous cycle in the cow by secreting the hormone progesterone. Progesterone prevents the cow from expressing heat and ovulation. The prostaglandin destroys the CL, thus destroying the source of progesterone. About 3 days after the injection of prostaglandin, the cow will be in heat. For prostaglandin to be effective, the cow must have a functional CL. It is ineffective in heifers that have not reached puberty and in noncycling mature cows. Also, prostaglandin is ineffective if the CL is immature or has already started to regress. Prostaglandin is, then, only effective in heifers and cows that are in days 5–18 of their estrous cycle. Because of this relationship, prostaglandin is given on either a one-injection or a two-injection system.

One-Injection System

The one-injection system requires one prostaglandin injection and an 11-day AI breeding season. The first 5 days are a conventional AI program of heat detection and insemination. After 5 days, a calculation can be made to determine what percentage of the females in the herd have been in heat. If a smaller than expected percentage are cycling, the cost of the prostaglandin may not be justified. If the decision is to proceed with the injection, the remaining animals not previously bred are injected on day 6. Then the AI breeding season continues 5 more days, for a total of 11 days.

Field trial results in herds where a high percentage of females are cycling show a 50–60% pregnancy rate at the end of 11 days of AI breeding for those receiving prostaglandin versus a 30–40% pregnancy rate for those not receiving it. The amount of prostaglandin required per calf that is produced by AI will be between 1.2 and 2.0 units.

Two-Injection System

In the two-injection system, all cows are injected with prostaglandin at two different times. Counting the first injection as day 1, the second injection is administered on day 11, 12, or 13. All cows capable of responding to the drug should be in heat during the first 5 days after the second injection of prostaglandin. Heat detection and insemination can occur each of these 5 days, or a fixed-time insemination can be performed 76–80 hours after the second injection.

Field trial results in herds in which a high percentage of the females are cycling show a 35–55% pregnancy rate at the end of the fifth day or 76- to 80-hour one-time insemination versus a 10–12% rate in the cows not receiving prostaglandin. The two-injection system will require 4–7 units of prostaglandin per AI calf.

Prostaglandin is not a wonder drug. It will only work in well-managed herds where a high percentage of the females are cycling. Biologically, it has been well demonstrated that prostaglandin can synchronize estrus. However, producers must weigh the cost against the economic benefit. Caution should be exercised in administering prostaglandin to pregnant cows, as it may cause abortion. Estrous synchronization in cattle may not be advisable in areas where inadequate protection is given to young calves during severe blizzards. Also, excellent herd health programs must be utilized to prevent high losses from calf scours and other diseases that become more serious where large numbers of newborn calves are grouped together.

MGA and Prostaglandin

MGA (melengestrol acetate) is a feed additive that suppresses estrus in heifers and is widely used in the feedlot industry. MGA and prostaglandin have been used in combination in highly successful estrous synchronization programs. In one such program, MGA is fed at 0.5 mg per head per day for 14 days. Seventeen days after the last MGA feeding, a single injection of prostaglandin is given (Fig. 11.10). Most heifers show estrus approximately 48–72 hours after the prostaglandin injection. In a 30-day breeding program, more than 80% of the heifers usually conceive.

Syncro-Mate-B

Another estrous synchronization product, Syncro-Mate-B, has been available for beef and dairy heifers since 1983. Heifers are injected with 5 mg of estradiol valerate and 3 mg of norgestomet and implanted with 6 mg norgestomet implant subcutaneously on top of the ear. Nine days later, the implant is removed. Most heifers will show estrus approximately 24–48 hours after implant removal. Heifers can be inseminated 12 hours after being detected in estrus or time-inseminated 48–54 hours after implant removal. Syncro-Mate-B will cause many noncycling heifers to show heat; however, conception rates in these heifers at this estrus are low (~20%). In cycling heifers treated with Syncro-Mate-B, conception rates are usually 40–60%.

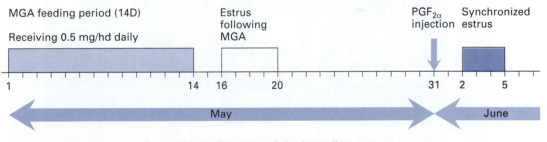

June 2 is the first day of the breeding season.

FIGURE 11.10 The MGA-prostaglandin ($PGF_{2\alpha}$) system for synchronizing estrus. Courtesy of Colorado State University.

In a sense, there are some natural occurrences of estrous synchronization. The weaning process in swine is an example; the sow will typically show heat 3–8 days after the pigs are weaned. Estrus is suppressed through the suckling influence, so when the pigs are removed, the sow will show heat.

EMBRYO TRANSFER

Embryo transfer is sometimes referred to as *ova transplant* or *embryo transplant.* In this procedure, an embryo in its early stage of development is removed from its own mother's (the donor's) reproductive tract and transferred to another female's (the recipient's) reproductive tract. The first successful embryo transfer was accomplished with rabbits in 1890. In the past several decades, successful embryo transfers have been reported in sheep, goats, swine, cattle, and horses.

In recent years, commercial embryo transfer companies have been established in the United States and several other foreign countries. Most commercial work is done with beef and dairy cattle. It is estimated that 40,000–50,000 beef calves were born in the United States and Canada in 1995. This represents approximately 1 embryo transfer calf per 700–800 beef calves born. This compares to 20 done in 1972. During the 1990s, bovine embryos were exported from the United States, primarily to Western Europe and South America. Approximately 90% of the embryos were from dairy cattle.

Superovulation is the production of a greater than normal number of eggs. Females that are donors of eggs for embryo transfer are injected with hormones to stimulate increased egg production. In cattle, an average response is 5–12 usable embryos per superovulation treatment; however, a range of 0–25 embryos can be expected (Fig. 11.11). On average, two to four calves will result per superovulated donor if fertile donors are utilized in a well-managed embryo transfer program. This procedure gives embryo transfer its greatest advantage because it increases the number of offspring that a superior female can produce. The key to justification of embryo transfer is identification of

FIGURE 11.11   Ten Holstein embryo transfer calves resulting from one superovulation and transfer from the Holstein cow in the background. The ten recipient cows are shown on the left-hand side of the fence. Courtesy of Colorado State University.

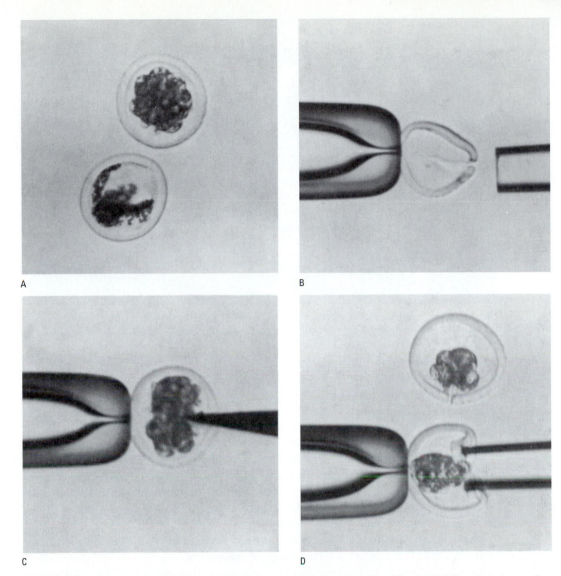

FIGURE 11.12   The process of embryo splitting. (A) A morula (40–70 cells of the fertilized egg) and an unfertilized egg. (B) The unfertilized egg with the cellular contents sucked out of it. (C) The morula is divided into two groups of cells with the microsurgical blade. (D) One-half of the cells are left in the morula while the other one-half of the cells are placed inside the other unfertilized egg. The result is two genetically identical embryos ready for transfer. Courtesy of Williams et al., *1983 NAAB Conference, Beef AI* and *Embryo Transfer.*

genetically superior females. Embryo transfer is usually confined to seedstock herds, where genetically superior females can be more easily identified and where high costs can be justified.

Earlier, embryo transfer was done surgically, but nonsurgical techniques are now used. The procedure after superovulation is to breed the donor 12 and 24 hours after she comes into heat with two doses of semen each time. The fertilized eggs are recovered about a week later by flushing the uterus with a buffered solution. This is essentially the reverse of the AI process and does not require surgery. The embryos are located and evaluated using a microscope and

FIGURE 11.13 "Question" and "Answer" were the world's first split-embryo foals resulting from nonsurgical embryo transfer. Courtesy of Colorado State University.

loaded into an AI rod. Nonsurgical transfer of an embryo into a recipient is done with the AI rod. Conception rates are usually higher if the transfer is done surgically. Embryos are usually transferred shortly after being collected, so recipient females need to be in the same stage of the estrous cycle (within 24 hours) for successful transfer to occur. Estrous synchronization of females is necessary, so large numbers of females must be kept for this purpose.

Embryos can now be frozen in liquid nitrogen and remain dormant for months or years. Conception rates are lower when frozen embryos are used than when fresh embryos are used, but research is bringing about improvement. In cattle, approximately 85% of the frozen embryos will be normal after thawing. Forty to fifty-five percent of these normal embryos will result in confirmed pregnancies in 60–90 days. This compares to a pregnancy rate of 55–65% from fresh embryos transferred the same day of collection.

Recent advances in embryo transfer research have permitted the embryo to be mechanically divided so that identical twins can be produced from a single embryo (Figs. 11.12 and 11.13). Perhaps in the future an embryo of a desired mating and sex could be selected from an inventory of frozen embryos. This embryo would then be thawed and transferred nonsurgically in the cow when she comes into heat. Embryo splitting and embryo transfer are areas of biotechnology that continue to advance rapidly.

CHAPTER SUMMARY

- Artificial insemination (AI) is the deposition of semen in the female reproductive tract by methods other than natural mating.
- The advantages of AI are the extended use of sires, controlling reproductive diseases, and using sires that have been injured or are dangerous when used naturally.

■ AI is widely used in dairy cattle and turkeys and to a lesser extent in other farm animal species.

■ Estrous synchronization is controlling or manipulating the estrous cycle so that females express estrus at approximately the same time.

■ Prostaglandin F_2 alpha, Melengestrol Acetate (MGA), and Syncro-Mate-B are commonly used products to synchronize estrus in cattle.

■ Embryo transfer (ET) involves removing embryos from females (donors) and transferring them into the reproductive tracts of other females (recipients).

REVIEW QUESTIONS

1. What is artificial insemination?
2. What is the primary advantage of artificial insemination?
3. In relation to the time of ovulation, when should animals be inseminated?
4. When does ovulation usually occur?
5. In relation to estrus, when should insemination occur?
6. Why does fertility in turkey hens persist for 2 to 3 weeks after insemination?
7. Why is the use of artificial insemination especially important for turkeys?
8. What is estrous synchronization?
9. What are three methods of estrous synchronization in cattle?
10. How can estrus be synchronized in sows?
11. What is embryo transfer?
12. What is superovulation and why is it important?

SELECTED REFERENCES

Publications

Ernst, R. A., Ogasawara, F. X., Rooney, W. F., Schroeder, J. P., and Ferebee, D. C. 1970. *Artificial Insemination of Turkeys.* University of California Agric. Ext. Publ. AXT-338.

Foote, R. H. 1994. Embryo Transfer in Domestic Animals. *Encyclopedia of Agricultural Science.* San Diego: Academic Press, Inc.

Herman, H. A., et al. 1994. *The Artificial Insemination and Embryo Transfer of Dairy and Beef Cattle* (including information pertaining to goats, sheep, horses, swine, and other animals).

Hafez, E. S. E. 1987. *Reproduction in Farm Animals.* Philadelphia: Lea and Febiger.

Kiessling, A. A., Hughes, W. H., and Blankevoort, M. R. 1986. Superovulation and embryo transfer in the dairy goat. *J. Am. Vet. Med. Assoc.* 188:829.

Lake, P. E., and Steward, J. M. 1978. *Artificial Insemination in Poultry.* Scotland Ministry of Agriculture, Fisheries, and Food, no. 213.

Pickett, B. W., and Back, D. G. 1973. *Procedures, Collection, Evaluation and Insemination of Stallion Semen.* Fort Collins: Colorado State University Exp. Sta., Animal Reprod. Lab., Gen. Series 935.

Seidel, G. E., Jr. 1981. Superovulation and embryo transfer in cattle. *Science* 211:351.

Seidel, G. E., Jr., Seidel, S. M., and Bowen, R. A. 1978. *Bovine Embryo Transfer Procedures.* Colorado State University Exp. Sta., Animal Reprod. Lab., Gen. Series 975.

Squires, E. L., Cook, V. M., and Voss, J. L. 1984. Collection and transfer of equine embryos. Colorado State University, Animal Reprod. Lab. Bull. no. 01.

Synchronization of Beef Cattle with Prostaglandin. 1979. DeForest, WI: American Breeders Service.

Visuals

Artificial Insemination in Poultry (slide-tape—57 slides; 10 min). Poultry Science Department, Ohio State University, 674 W. Lane Ave., Columbus, OH 43210.

Artificial Insemination: Striving for Perfection. 1993 (Video; 17 min.; swine focus). CEV, P.O. Box 65265, Lubbock, TX 79464

Embryo Transfer. 1995 (video; 40 min.). CEV, P.O. Box 65265, Lubbock, TX 79464.

Equine A.I. (sound filmstrips), covering "Introduction to Equine AI," "Teasing and Rectal Palpation," "Semen Collection and Evaluation," "Broodmare Health Care," and "Inseminating and Diagnosing Pregnancy." Vocational Education Productions, California Polytechnic State University, San Luis Obispo, CA 93407.

Fundamental Livestock Parturition. 1997 (video; 30 min). CEV, P.O. Box 65265, Lubbock, TX 79464.

Modern Cattle Breeding Techniques (sound filmstrips), covering "Artificial Insemination—An Overview" and "Embryo Transfer—The New Horizon." Prentice-Hall Media, 150 White Plains Road, Tarrytown, NY 10591.

CHAPTER 12

Genetics

Body tissues of animals and plants are composed of cells, whose structure can be observed microscopically. These cells, with certain exceptions, have an outer membrane, an internal cytoplasm, and a nucleus (see Chapter 18, Fig. 18.3). This nucleus contains rodlike bodies called ***chromosomes*** (Fig. 12.1). Body cells contain these chromosomes in pairs, and each chromosome contains genes, which are the functional units of inheritance. When cells divide to produce more body cells, the chromosomes replicate by a process called *mitosis* (Fig. 12.2). During mitosis, each member of each chromosome pair divides so that the two new cells (*daughter cells*) are identical to the original cell that divided.

Each species has a characteristic number of chromosomes (Table 12.1), which is maintained through meiosis and fertilization. Poultry have more chromosomes than livestock, whereas swine have the smallest number. For comparative purposes, humans have 23 pairs of chromosomes.

PRODUCTION OF GAMETES

The testicles of the male and the ovaries of the female produce sex cells (gametes) by a process called *gametogenesis.* The gametes produced by the testicles are called *sperm;* the gametes produced by the ovaries are called *eggs,* or *ova.* Specifically, the production of sex cells that become sperm is called ***spermatogenesis;*** the production of ova, ***oogenesis.*** The unique type of cell division in which gametes (sperm or ova) are formed is called ***meiosis.*** Each newly formed gamete contains only one member of each of the chromosome pairs present in the body cells.

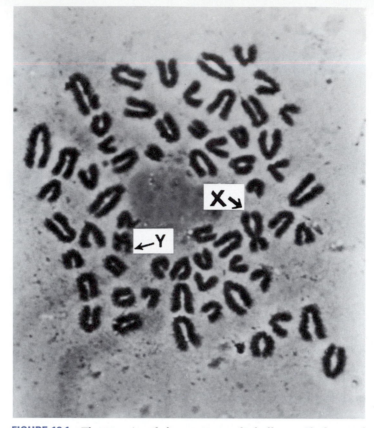

FIGURE 12.1   The 30 pairs of chromosomes of a bull magnified several hundred times. Note the *X* and *Y* chromosomes. Courtesy of Nat M. Kieffer, Texas A&M University.

Let us examine gametogenesis in a theoretical species in which there are only two pairs of chromosomes in each body cell. Meiosis occurs in the primordial germ cell (cells capable of undergoing meiosis) located near the outer wall of the **seminiferous tubules** of each testicle and near the surface of each ovary. The initial steps in meiosis are similar for the male and female. The chromosomes replicate themselves so that each chromosome is doubled. Then each pair of chromosomes comes together in extremely accurate pairing called *synapsis.* After chromosome replication and synapsis, the cell is called a *primary spermatocyte* in the male and a *primary oocyte* in the female. Because subsequent differences exist between spermatogenesis and oogenesis, the two are described here separately.

SPERMATOGENESIS

The process of spermatogenesis in our theoretical species is shown in Fig. 12.3. The primary spermatocyte contains two pairs of chromosomes in synapsis. Each chromosome is doubled following replication. Thus, the primary spermatocyte contains two bodies or structures formed by four parts. By two rapid cell divisions, in which no further replication of the chromosomes occurs, four

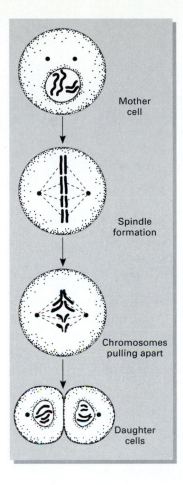

Mother
cell

Spindle
formation

Chromosomes
pulling apart

Daughter
cells

FIGURE 12.2   Mitosis. Drawing by Dennis Giddings.

cells, each of which contains two chromosomes, are produced. The spermatid cells each lose much of their cytoplasm and develop a tail. This process, spermatogenesis, results in formation of sperm. Four sperm are produced from each primary spermatocyte. Whereas four chromosomes in two pairs are present in the primordial germ cell, only two chromosomes (one-half of each chromosome pair) are present in each sperm. Thus, the number of chromosomes in the sperm has been reduced to half the number of the primordial germ cell.

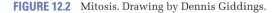

TABLE 12.1   Pairs of Chromosomes in Livestock and Poultry

Species	Number of Pairs
Turkeys	41
Chickens	39
Horses	32
Cattle	30
Goats	30
Sheep	27
Humans[a]	23
Swine	19

[a] Shown for comparison.

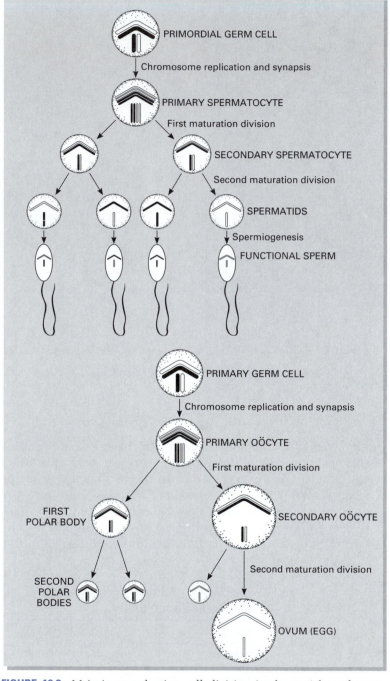

FIGURE 12.3 Meiosis or reduction cell division in the testicle and ovary (example with two pairs of chromosomes). Drawing by Dennis Giddings.

OOGENESIS

The process of oogenesis in our theoretical species is shown in Fig. 12.3. Like the primary spermatocyte of the male, the primary oocyte of the female contains a tetrad. The first maturation division produces one relatively large nutrient-containing cell, the secondary oocyte, and a smaller cell, the first polar body. Each of these two cells contains a dyad. The unequal distribution of nutrients that results from the first maturation division in the female serves to maximize the quantity of nutrients in one of the cells (the secondary oocyte) at the expense of the other (the first polar body). The second maturation division produces the egg (ovum) and the second polar body, each of which contains two chromosomes. The first polar body may also divide, but all polar bodies soon die and are reabsorbed. Note the egg, like the sperm, contains only one chromosome of each pair that was present in the primordial germ cell. The gametes each contain one member of each pair of chromosomes that existed in each primordial germ cell.

FERTILIZATION

When a sperm and an egg of our theoretical species unite to start a new life, each contributes one chromosome to each pair of chromosomes in the fertilized egg, now called a *zygote* (Fig. 12.4). *Fertilization* is defined as the union of the sperm and the egg along with the establishment of the paired condition of the chromosomes. The zygote is termed *diploid* (*diplo* means "double") because it has chromosomes in pairs; one member of each pair comes from the sire and

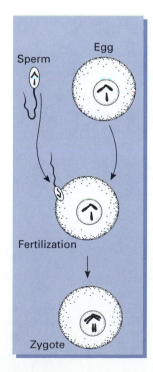

FIGURE 12.4 Combining of chromosomes through fertilization (two pairs of genes used for simplification of example). Drawing by Dennis Giddings.

one member comes from the dam. Gametes have only one member of each pair of chromosomes; therefore, gametes are termed ***haploid*** (*haplo* means "half"). Gametogenesis thus reduces the number of chromosomes in a cell to half the diploid number. Fertilization reestablishes the normal diploid number.

DNA AND RNA

The two members of each typical pair of chromosomes in a cell are alike in size and shape and carry **genes** that affect the same hereditary characteristics (Fig. 12.5). Such chromosomes are said to be ***homologous.*** The genes are points of activity found in each of the chromosomes that govern how traits develop. The genes form the coding system that directs enzyme and protein production. Thus, they control the development of traits.

Chromosomes in advanced organisms like poultry and livestock are composed basically of a protein sheath surrounding ***deoxyribonucleic acid*** (**DNA**). Segments of DNA are the genes. DNA is composed of deoxyribose sugar, phosphate, and four nitrogenous bases. The combination of deoxyribose, phosphate, and one of the four bases is called a ***nucleotide;*** when many nucleotides are chemically bonded to one another, they form a strand that composes one-half of the DNA molecule. (A molecule formed by many repeating sections is called a ***polymer.***) Two of these strands wind around each other in a double helix to form the DNA molecule.

The bases of DNA are the parts that hold the key to inheritance. The four bases are adenine (A), thymine (T), guanine (G), and cytosine (C). In the two strands of DNA, A is always complementary to (pairs with) T, and G is always complementary to C (Fig. 12.6). During meiosis and mitosis, the chromosomes

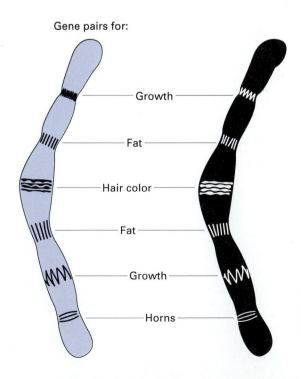

Gene pairs for:

Growth

Fat

Hair color

Fat

Growth

Horns

FIGURE 12.5 A simplified example showing a pair of chromosomes containing several pairs of genes. Drawing by Dennis Giddings.

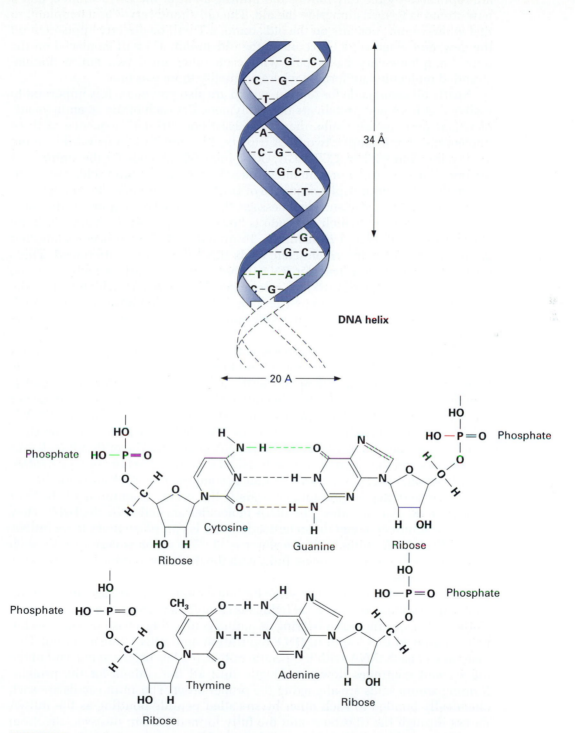

FIGURE 12.6 DNA helix and structure of nucleotides.

are replicated by the unwinding and pulling apart of the DNA strands, and a new strand is formed alongside the old. The old strand serves as a template, so that wherever an A occurs on the old strand, a T will be directly opposite it on the new, and wherever a C occurs on the old strand, a G will be placed on the new. Complementary bases pair with each other until two entire double-stranded molecules are formed where originally there was one.

Nearly all genes code for protein, which are also polymers. It is important to realize that DNA and protein are both polymers. For each of the 26 amino acids of which proteins are made, there is at least one "triplet" sequence of three nucleotides. For example, two DNA triplets, TTC and TTT, code for the amino acid lysine; four triplets, CGT, CGA, CGG, and CGC, all code for the amino acid alanine. If we think of a protein molecule as a word, and amino acids as the letters of the word, each triplet sequence of DNA can be said to code for a letter of the word, and the entire encoded message, the series of base triplets, is the gene.

The processes by which the code is "read" and protein is synthesized are called *transcription* and *translation.* To understand these processes, another group of molecules, the **ribonucleic acids** (**RNAs**), must be introduced. There are three types of RNA: **transfer RNA** (**tRNA**), which identifies both an amino acid and a base triplet in mRNA; **messenger RNA** (**mRNA**), which carries the information codes for a particular protein; and **ribosomal RNA** (**rRNA**), which is essential for ribosome structure and function. All three RNAs are coded by the DNA template.

The first step in protein synthesis is that of transcription. Just as the DNA molecule serves as a template for self-replication using the pairing of specific bases, it can also serve as a template for the mRNA molecule. Messenger RNA is similar to DNA but is single-stranded rather than shorter, coding for only one or a few proteins. The triplet sequence that codes for one amino acid in mRNA is called a *codon.* Through transcription, the encoded message held by the DNA molecule becomes transcribed onto the mRNA molecule. The mRNA then leaves the nucleus and travels to an organelle called a *ribosome,* where protein synthesis will actually take place. The ribosome is composed of rRNA and protein.

The second step in protein synthesis is the union of amino acids to their respective tRNA molecules. The tRNA molecules are coded by the DNA. They have a structure and contain an anticodon that is complementary to an mRNA codon. Each RNA unites with one amino acid. This union is very specific such that, as an example, lysine never links with the tRNA for alanine but only to the tRNA for lysine.

The mRNA attaches to a ribosome for translation of its message into protein. Each triplet codon on the mRNA (which is complementary to one of DNA) associates with a specific tRNA bearing its amino acid, using a base-pairing mechanism similar to that found in DNA replication and mRNA transcription. This matching of each tRNA with its specific mRNA triplets begins at one end of the mRNA and continues down its length until all the codons for the protein-forming amino acids are aligned in the proper order. The amino acids are each chemically bonded to each other by so-called peptide bonding as the mRNA moves through the ribosome, and the fully formed protein disassociates from the tRNA-mRNA complex and is ready to fulfill its role as a part of a cell or as an enzyme to direct metabolic processes. Figure 12.7 summarizes the steps of protein synthesis. It depicts the amino acids arginine (arg), leucine (leu), threonine (thr), and valine (val) being moved into position to join a chain that already includes the amino acids alanine (ala), proline (pro), and leucine (leu).

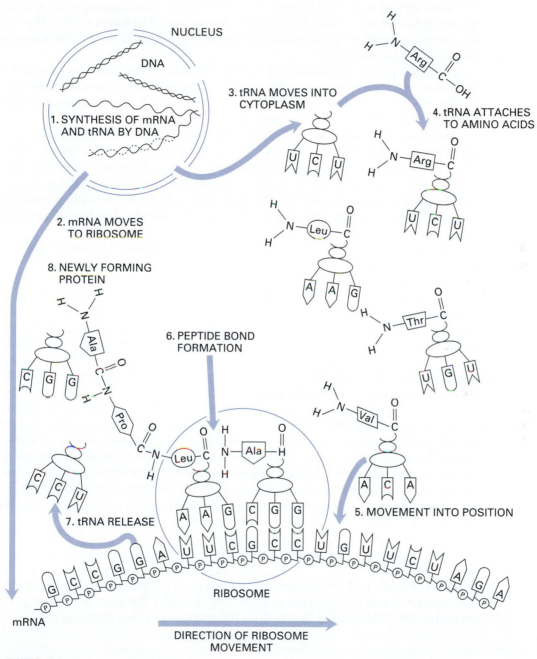

FIGURE 12.7 Protein synthesis in the cell.

GENES AND CHROMOSOMES

Because chromosomes are in pairs, genes are also in pairs. The location of a gene in a chromosome is called a ***locus*** (plural, *loci*). For each locus in one of the members of a pair of homologous chromosomes, a corresponding locus occurs in the other member of that chromosome pair. The transmission of genes from parents to offspring depends entirely on the transmission of chromosomes from parents to offspring.

A special pair of chromosomes, the so-called sex chromosomes (*X* and *Y*), exist as a pair in which one of the chromosomes does not correspond entirely to the other in terms of what loci are present. The *Y* chromosome is much shorter in length than the *X* chromosome.

The *X* and *Y* chromosomes determine the sex of mammals. A female has two *X* chromosomes, and the male has an *X* and a *Y* chromosome. The female, being *XX,* can contribute only an *X* chromosome to the offspring. The male contributes either an *X* or a *Y* (Fig. 12.1), thus determining the sex of the newborn.

In all bird species, including chickens and turkeys, the female determines the sex of the offspring. The sex chromosomes are identical in the sperm (*XX*) but they are different in the egg (*XY*).

The genes located at corresponding loci in **homologous chromosomes** may correspond to each other in the way that they control a trait, or they may contrast. If they correspond, the individual is said to be **homozygous** at the locus (*homo* means "alike"; *zygous* refers to the individual); if they differ, the individual is said to be **heterozygous** (*hetero* means "different"). Those genes that occupy corresponding loci in homologous chromosomes but that affect the same character in different ways (e.g., black or red color) are called ***alleles.*** Genes that are alike and that affect the character developing in the same way are called ***identical alleles.*** Some authorities speak of alike genes in homologous chromosomes as **identical alleles.**

The geneticist usually illustrates the chromosomes as lines and indicates the genes by alphabetical letters. When the genes at corresponding loci on homologous chromosomes differ, one of the genes often overpowers, or dominates, the expression of the other. This allele is called ***dominant.*** The allele whose expression is prevented is called ***recessive.*** The dominant allele is symbolized by a capital letter. The recessive allele is symbolized by a lowercase letter. For example, in cattle, black hair color is dominant to red hair color, so we let *B* = black and *b* = red. Three combinations of genes are possible:

$$\dashv B \dashv B \qquad \dashv B \dashv b \qquad \dashv b \dashv b$$

Both *BB* animals and *bb* animals are homozygous for the genes that determine hair color, but one is homozygous-dominant (*BB*) and the other is homozygous-recessive (*bb*). The animal that is *Bb* is heterozygous; it has allelic genes.

SIX FUNDAMENTAL TYPES OF MATING

With three kinds of individuals (homozygous-dominant, heterozygous, and homozygous-recessive) and one pair of genes considered, six types of mating are possible. Using the genes designated as *B* = black and *b* = red in cattle, the six mating possibilities are *BB* × *BB*, *BB* × *Bb*, *BB* × *bb*, *Bb* × *Bb*, *Bb* × *bb*, and *bb* × *bb*. Keep in mind that our discussion of these genes is applicable to any other pair of genes in any species.

Homozygous-Dominant × Homozygous-Dominant (*BB* × *BB*). When two homozygous-dominant individuals are mated, each of them can produce only one kind of gamete—the gamete carrying the dominant gene. In our particular example, this gamete carries gene *B*. The union of gametes from two homozygous-dominant parents results in a zygote that is homozygous-dominant (*B* × *B* = *BB*). Thus, homozygous-dominant parents produce only homozygous-dominant offspring (Fig. 12.8).

Homozygous-Dominant × Heterozygous (*BB* × *Bb*). The mating of a homozygous-dominant with a heterozygous individual results in an expected ratio of 1 homozygous-dominant:1 heterozygous. The homozygous-dominant parent produces only one kind of gamete, the one carrying the dominant gene (*B*). The heterozygous parent produces two kinds of gametes, one carrying the dominant gene (*B*) and one carrying the recessive gene (*b*). The latter two kinds of gametes are produced in approximately equal numbers. The chances are equal that a gamete from the parent producing only the one kind of gamete (the one having the dominant gene) will unite with each of the two kinds of gametes produced by the heterozygous parent; therefore, the number of homozygous-dominant offspring (*B* × *B* = *BB*) and heterozygous offspring (*B* × *b* = *Bb*) produced should be approximately equal (Fig. 12.9).

Homozygous-Dominant × Homozygous-Recessive (*BB* × *bb*). The homozygous-dominant individual can produce only gametes carrying the dominant gene (*B*). The recessive individual must be homozygous-recessive to show the recessive

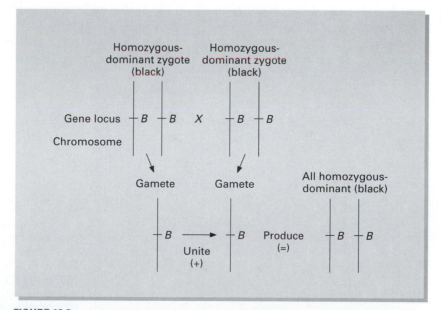

FIGURE 12.8   Mating of homozygous-dominant (*BB*) × homozygous-dominant (*BB*).

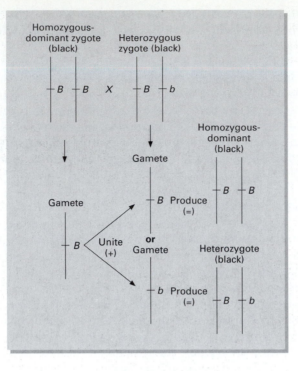

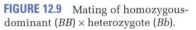

FIGURE 12.9 Mating of homozygous-dominant (*BB*) × heterozygote (*Bb*).

trait; therefore, it produces only gametes carrying the recessive gene (*b*). When the two kinds of gametes unite, all offspring produced receive both the dominant and the recessive gene (*B* × *b* = *Bb*) and are thus all heterozygous (Fig. 12.10).

Heterozygous × Heterozygous (*Bb* × *Bb*).   Each of the two heterozygous parents produces two kinds of gametes in approximately equal ratios: one kind of gamete carries the dominant gene (*B*); the other kind carries the recessive gene

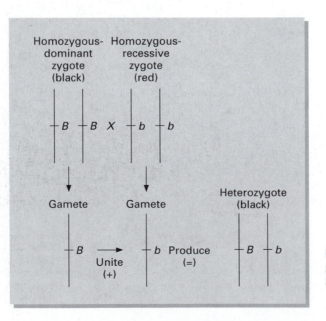

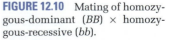

FIGURE 12.10 Mating of homozygous-dominant (*BB*) × homozygous-recessive (*bb*).

(*b*). The two kinds of gametes produced by one parent each have an equal chance of uniting with each of the two kinds of gametes produced by the other parent. Thus, four equal-chance unions of gametes are possible. If the gamete carrying the *B* gene from one parent unites with the gamete carrying the *B* gene from the other parent, the offspring produced are homozygous-dominant (*B* × *B* = *BB*). If the gamete carrying the *B* gene from one parent unites with the gamete carrying the *b* gene from the other parent, the offspring are heterozygous (*B* × *b* = *Bb*). The latter combination of genes can occur in two ways: the *B* gene can come from the male parent and the *b* gene from the female parent, or the *B* gene can come from the female parent and the *b* gene from the male parent. When the gametes carrying the *b* gene from both parents unite, the offspring produced are recessive (*b* × *b* = *bb*). The total expected ratio among the offspring of two heterozygous parents is 1*BB*:2*Bb*:1*bb* (Fig. 12.11). This 1:2:1 ratio is the genetic, or ***genotypic,*** ratio. The appearance of the offspring is in the ratio of 3 dominant:1 recessive because the **BB** and **Bb** animals are all black (dominant) and cannot be genetically distinguished from one another by looking at them. This 3:1 ratio, based on external appearance, is called the ***phenotypic*** ratio.

Heterozygous × Homozygous-Recessive (*Bb* × *bb*).

The heterozygous individual produces two kinds of gametes, one carrying the dominant gene (*B*) and the other carrying the recessive gene (*b*), in approximately equal numbers. The recessive individual produces only the gametes carrying the recessive gene (*b*). There is an equal chance that the two kinds of gametes produced by the heterozygous parent will unite with the one kind of gamete produced by the recessive parent (*B* × *b* = *Bb*; *b* × *b* = *bb*). The offspring produced when these gametes unite thus occur in the expected ratio of 1 heterozygous:1 homozygous-recessive (Fig. 12.12).

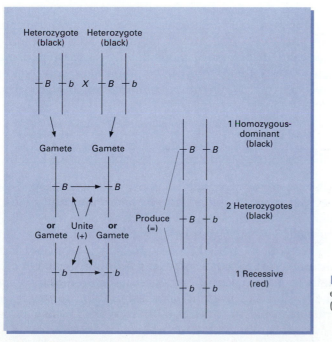

FIGURE 12.11 Mating of heterozygote (*Bb*) × heterozygote (*Bb*).

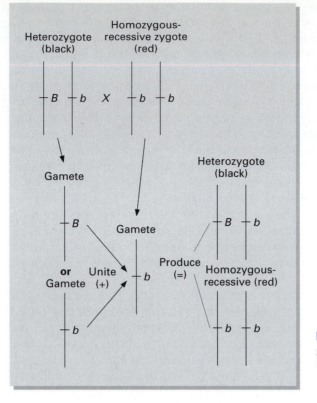

FIGURE 12.12 Mating of heterozygote (*Bb*) × homozygous-recessive (*bb*).

Homozygous-Recessive × **Homozygous-Recessive (*bb* × *bb*).**   The recessive individuals are homozygous; therefore, they can produce gametes carrying only the recessive gene (*b*). When these gametes unite (*b* × *b* = *bb*), all offspring produced will be recessive (Fig. 12.13). This example illustrates the principle that recessives, when mated together, breed true.

Knowledge of the resulting combinations from each of the six fundamental types of matings provides a background for understanding complex crosses. When more than one pair of genes is considered in a mating, one can understand the expected results; that is, one can combine each of the combinations of one pair of genes with each of the other combinations of one pair of genes to obtain the expected ratios.

MULTIPLE GENE PAIRS

Suppose that there are two pairs of genes to be considered, each pair independently affecting a particular trait. Let us consider two pairs of genes: one determines coat color in cattle; the other determines whether the animal is polled (hornless) or horned. The genes are designated as follows: *B* = black (dominant), *b* = red (recessive), *P* = polled (dominant), and *p* = horned (recessive).

If a bull that is heterozygous for both traits (*BbPp*) is mated to cows that are also heterozygous for both traits (*BbPp*), one can determine the expected phenotypic and genotypic ratios. Results of crossing *Bb* × *Bb* (giving a ratio of 3 black:1 red in the offspring, or ¾ black and ¼ red) has previously been shown

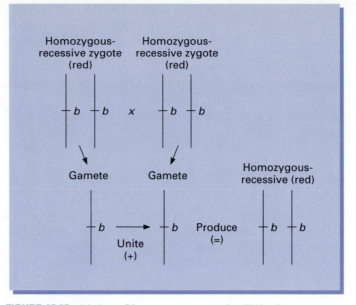

FIGURE 12.13 Mating of homozygous-recessive (*bb*) × homozygous-recessive (*bb*).

(Fig. 12.11). Similarly, a cross of *Pp* × *Pp* gives a ratio of 3 polled:1 horned. Combined, the *BbPp* × *BbPp* produces the following phenotypes:

$$BbPp \times BbPp$$

$$3 \text{ black} \times \begin{cases} 3 \text{ polled} = 9 \text{ black polled} \\ 1 \text{ horned} = 3 \text{ black horned} \end{cases}$$

$$1 \text{ red} \times \begin{cases} 3 \text{ polled} = 3 \text{ red polled} \\ 1 \text{ horned} = 1 \text{ red horned} \end{cases}$$

Of the ¾ that are black, ¾ will be polled and ¼ will be horned. Therefore, ¾ of ¾ (or ⁹⁄₁₆) will be black and polled; and ¼ of ¾ (or ³⁄₁₆) will be black and horned. Similarly, of the ¼ that are red, ¾ will be polled and ¼ will be horned, and ¾ of ¼ (or ³⁄₁₆) will be red and polled and ¼ of ¼ (or ¹⁄₁₆) will be red and horned. This distribution of phenotypes is more often expressed as a 9:3:3:1 ratio instead of a set of fractions. The following demonstrates the different types of genotypes and phenotypes from the *BbPp* × *BbPp* mating (continued on next page):

Genotypes (9)   *Phenotypes (4)*

$$1 \, BB \times \begin{bmatrix} 1 \, PP = 1 \, BBPP & \text{black, polled} \\ 2 \, Pp = 2 \, BBPp & \text{black, polled} \\ 1 \, pp = 1 \, BBpp & \text{black, horned} \end{bmatrix}$$

$$2 \, Bb \times \begin{bmatrix} 1 \, PP = 2 \, BbPP & \text{black, polled} \\ 2 \, Pp = 4 \, BbPp & \text{black, polled} \\ 1 \, pp = 2 \, Bbpp & \text{black, horned} \end{bmatrix}$$

$$1\ bb \times \begin{bmatrix} 1\ PP = 1\ bbPP & \text{red, polled} \\ 2\ Pp = 2\ bbPp & \text{red, polled} \\ 1\ pp = 1\ bbpp & \text{red, horned} \end{bmatrix}$$

Table 12.2 shows another method of demonstrating the different phenotypes and genotypes that can be produced from a *BbBp* (bull) × *BbPp* (cow) mating. The four different sperm have the opportunity to combine with each of the four different eggs. Nine different genotypes and four different phenotypes are possible.

GENE INTERACTIONS

A gene may interact with other genes in the same chromosome (*linear interaction*), with its corresponding gene in a homologous chromosome (*allelic interaction*), or with genes in nonhomologous chromosomes (*epistatic interaction*). In addition, genes interact with the cytoplasm and with the environment. The environmental factors with which genes interact are internal, such as hormones, and external, such as nutrition, temperature, and amount of light.

Linear interactions are known to exist in lower animals (*Drosophila*), but have not been demonstrated in farm animals.

Allelic Interactions

When contrasting genes occupy corresponding loci in the same (homologous) pair of chromosomes, each gene exerts its influence on the trait, but the effects of each of the genes depend on the relationship of dominance and recessiveness. **Allelic interactions** may also be called *dominance interactions*. When unlike genes occupy corresponding loci, complete dominance might exist. In this situation, only the effect of the dominant gene is expressed. A good example, as previously discussed, is provided by cattle, in which the hornless (polled) condition is dominant to the horned condition. The heterozygous animal is polled and is phenotypically indistinguishable from the homozygous polled animal.

TABLE 12.2   Genotypes and Phenotypes with Two Heterozygous Gene Pairs

Sperm	Eggs			
	BP	Bp	bP	bp
BP	BBPP[a] black, polled[b]	BBPp black, polled	BbPP black, polled	BbPp black, polled
Bp	BBPp black, polled	BBpp black, horned	BbPp black, polled	Bbpp black, horned
bP	BbPP black, polled	BbPp black, polled	bbPP red, polled	bbPp red, polled
bp	BbPp black, polled	Bbpp black, horned	bbPp red, polled	bbpp red, horned

[a] Genotype. Number of different genotypes: 1 (*BBPP*), 2 (*BBPp*), 2 (*BbPP*), 4 (*BbPp*), 1 (*BBpp*), 1 (*bbPP*), 2 (*Bbpp*), 2 (*bbPp*), 1 (*bbpp*).
[b] Phenotype. Number of different phenotypes: 9 black, polled; 3 black, horned; 3 red, polled; 1 red, horned.

There may be lack of dominance, in which heterozygous animals show a phenotype that is different from either homozygous phenotype and is usually intermediate between them. A good example is observed in sheep, in which two alleles for ear length lack dominance to each other. Sheep that are *LL* have long ears, *Ll* have short ears, and *ll* are earless. Thus the heterozygous *Ll* sheep are different from both homozygotes and are intermediate between them in phenotype.

Lack of dominance can also be considered additive gene action. Additive gene action occurs when each gene has an expressed phenotypic effect. For example, if *D* gene and *d* gene influenced rate of gain (where *D* = 0.10 lb/day and *d* = 0.05 lb/day), then *DD* = increase in gain of 0.20 lb, *Dd* = increase in gain of 0.15 lb, and *dd* = increase of gain of 0.10 lb.

There is evidence that many pairs of genes affect production traits (like rate of gain) in an additive manner. For example, consider two pair of genes that influence daily gain, where *D* = 0.10 lb, *d* = 0.05 lb, *N* = 0.10 lb, and *n* = 0.05 lb:

$$DDNN = 0.40 \text{ lb}$$

$$\left.\begin{array}{l} DDNn \\ DdNN \end{array}\right\} = 0.35 \text{ lb}$$

$$\left.\begin{array}{l} DdNn \\ BBnn \\ bbNN \end{array}\right\} = 0.30 \text{ lb}$$

$$\left.\begin{array}{l} Bbnn \\ bbNn \end{array}\right\} = 0.25 \text{ lb}$$

$$bbnn = 0.20 \text{ lb}$$

Sometimes heterozygous individuals are superior to either of the homozygotes. This condition, in which the heterozygotes show overdominance, is an example of a selective advantage for the heterozygous condition. Overdominance means that heterozygotes possess greater vigor or are more desirable in other ways (such as producing more milk or being more fertile) than either of the two homozygotes that produced the heterozygote. Because the heterozygotes of breed crosses are more vigorous than the straightbred parents, they are said to possess **heterosis.** This greater vigor or productivity of crossbreds is also said to be an expression of **hybrid vigor** (same as heterosis).

The effects of dominance on the expression of traits that are important in livestock production (production traits like fertility, milk production, growth rate, feed conversion efficiency, carcass merit, and freedom from inherited defects) can be illustrated by bar graphs (Fig. 12.14). Figure 12.14A, which illustrates complete dominance, shows equal performance of the homozygous-dominant and heterozygous individuals, both of which are quite superior to the performance of the recessive individual.

With lack of dominance, one homozygote is superior, the heterozygote is intermediate, and the other homozygote is inferior (Fig. 12.14B). When traits are controlled by genes that show the effects of overdominance, the heterozygote is superior to either of the homozygous types (Fig. 12.14C).

A gene or a pair of genes in one pair of chromosomes may alter or mask the expression of genes in other chromosomes. A gene that interacts with other

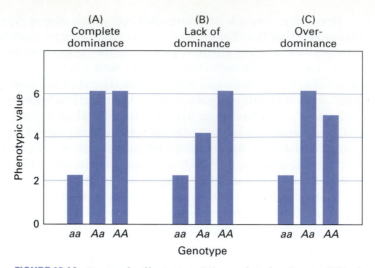

FIGURE 12.14   Bar graphs illustrating: (A) complete dominance; (B) lack of dominance; (C) overdominance.

genes that are not allelic to it is said to be *epistatic* to them. In other cases, a gene or one pair of genes in one pair of chromosomes may influence many other genes in many pairs of chromosomes.

Several coat colors in horses are due to epistatic interactions. For example, the two basic coat colors in horses are black and chestnut. The recessive gene *b* produces the chestnut color; thus, *BB* and *Bb* are black, while *bb* is chestnut. Some horses possess a gene for white color that masks all other genes for color that exist in the genotype. This gene is called the dominant white gene *W*. The recessive gene *w* allows other color genes in the genotype to express themselves. Thus, the genotypes and phenotypes are:

BBWW or *BBWw:* white   *BBww:* black

BbWW or *BbWw:* white   *Bbww:* black

bbWW or *bbWw:* white   *bbww:* chestnut

Epistasis can result from dominant or recessive genes or a combination of both. For example, in White Rock chickens, *C* = color, *c* = albino, *W* = color, and *w* = white. Homozygous *cc* prevents *W* from showing, and *ww* prevents *CC* from showing. In this example, if the mating *CcWw* × *CcWw* was made, there would be only two genotypes (colored or white). This contrasts with four phenotypes if *C* and *W* showed complete dominance, or nine phenotypes if there was no dominance expressed in the two pairs of genes. Thus, epistasis changes the phenotypic ratio, but the number of different genotypes remains the same.

INTERACTIONS BETWEEN GENES AND ENVIRONMENT

Genes interact with both external and internal environments. The external environment includes temperature, light, altitude, humidity, disease, and feed supply. Some breeds of cattle (Brahman) can withstand high temperatures better than others. Some breeds (Scottish Highland) can withstand the rigors of extreme cold better than others.

Perhaps the most important external environmental factor is feed supply. Some breeds or types of cattle can survive when feed is in short supply for considerable periods of time, and they may consume almost anything that can be eaten. Other breeds of cattle select only highly palatable feeds, and these animals have poor production when good feed is not available.

Allelic, epistatic, and environmental interactions all influence the genetic improvement that can be made through selection. When the external environment has a large effect on production traits, genetic improvement is quite low. For example, if animals in a population are maintained at different nutritional levels, those that are best fed obviously grow faster. In such a case, much of the difference in growth may be due to the nutritional status of the animals rather than to differences in their genetic composition. The producer has two alternatives: (1) standardize the environment so that it causes less variation among the animals, or (2) maximize the expression of the production trait by improving the environment. The first approach is geared to increasing genetic improvement so that selection is more effective. Improvements made by this approach are permanent. The latter approach does not improve the genetic qualities of animals, but it allows animals to express their genetic potentials. The most sensible approach is to expose the breeding animals to an environment similar to one in which commercial animals are expected to perform economically.

BIOTECHNOLOGY

Biotechnology is defined as the use of living organisms to improve, modify, or produce industrially important products or processes. Microorganisms, for instance, have been used for centuries in the production of food and fermented substances, such as some dairy products. **Genetic engineering** of animals by selection and hybridization was first implemented at the turn of the century. Superovulation, sexing of semen, embryo splitting (cloning), and embryo transfer (discussed in Chapter 11) are some of the more recent advances in the field of biotechnology.

During the past several years, research has provided the tools to identify and manipulate DNA (genes) at the molecular and cellular levels. These genetic-engineering methods, which are used to alter hereditary traits, have created a renewed excitement in biotechnology. The American Association for the Advancement of Science ranks genetic engineering as among the four major scientific advancements of the twentieth century, similar in significance to unlocking the atom, escaping the earth's gravity, and the computer revolution.

Applications

Enzymes are used as "genetic scissors" to cut the DNA at designated places. These genetic components can then be reconstructed into unique combinations not easily achieved with natural selection. In addition to altering existing genes, synthetic genes can be constructed.

Genes can be injected from other animals of the same or different species (Fig. 12.15). The technology mimics nature in that genes can be moved around

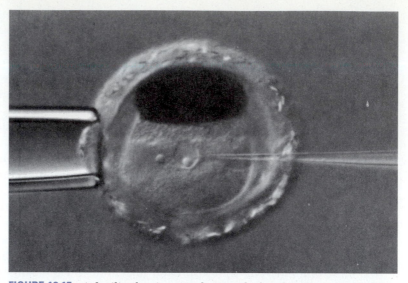

FIGURE 12.15   A fertilized swine egg photographed at the moment it is micro-injected with new genetic material. The vacuum in the large pipette at the bottom anchors the cell while a mixture containing the genes is forced through the smaller pipette into one of the egg's pronucleii. Courtesy of R. E. Hammer and R. L. Brinster, University of Pennsylvania School of Veterinary Medicine.

within and occasionally between species by some types of virus infections. The manipulation of the genetic material does not occur at the whole animal level, but it is performed **in vitro.** The genes are then inserted into either embryos, which are transferred to recipient animals or inserted into cells that are grafted into living animals.

Bovine somatotropin (BST), a bovine growth hormone, is a natural protein produced in the pituitary gland of cattle. It helps direct the energy in feed to meet the animal's need for growth and milk production. BST can be produced outside the animal's body by the genetic-engineering process shown in Fig. 12.16. Providing supplemental BST to the dairy cow increases feed consumption and milk production. BST directs more of the additional feed to increase milk production than to body maintenance.

Porcine somatotropin (PST) has also been produced by a process similar to that used for BST (Fig. 12.16). Research has shown that PST, given as a supplement, increases litter size, reduces the time needed for pigs to reach market weight, improves feed efficiency, and reduces carcass fat production.

Several calves born in Canada, Germany, and the United States (in 1986, 1988, and 1989, respectively) have had genes introduced into them from other species, including genes from humans. A human gene that produces the protein interferon was introduced to improve disease resistance. Other genes have been introduced to stimulate growth and increase milk production. All physiological responses to these introduced genes have not been clearly elucidated.

Figure 12.17 shows a **transgenic** pig. While normal in appearance, the boar contains a new gene composed of the mouse metallothionien promotor/regulator fused to the bovine growth hormone structural gene. This gene, transmitted by the boar in Fig. 12.17, causes elevated levels of growth hormone in the blood and results in a 10 to 15% increase in growth rate. The pigs are also more efficient in feed conversion to body weight and have significantly less body fat.

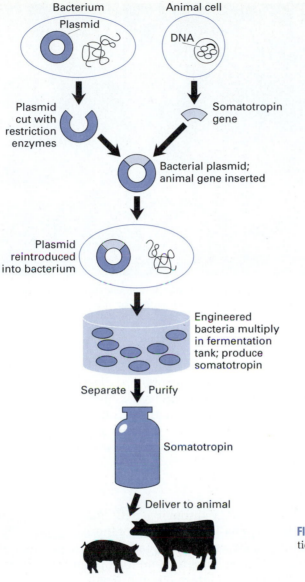

FIGURE 12.16 Somatotropin production for use in cows and pigs.

Nuclear fusion is the union of nuclei from two sex cells (sperm and eggs). In terms of biotechnology, this means not only can nuclei from a sperm and egg be united but also the union of nuclei from two eggs or two sperm can occur. The union of nuclei from two females results in all female offspring, while the union of two sperm nuclei produces ½ females (*XX*), ¼ surviving males (*XY*), and ¼ lethal males (*YY*). Thus, nuclei from two outstanding males (or females) can be united and then transferred to a recipient female. This has been accomplished in laboratory animals but not in larger, domestic farm animals. Even self-fertilization, once considered possible only in plants, may become a choice in genetically engineered farm animals.

Gene therapy involves inserting genes into a patient's cells to treat or cure certain diseases. In September, 1990, the first federally approved gene therapy

FIGURE 12.17 This normal-appearing boar is a transgenic pig. He received a growth gene (from both mouse and cattle origin) by the process shown in Fig. 12.15. Courtesy of R. L. Brinster and R. E. Hammer, School of Veterinary Medicine, University of Pennsylvania.

for humans took place. A girl having a hereditary disorder of the immune system had some of her white blood cells removed, genetically altered, then transfused back into her blood stream. The genetically altered cells were to produce a needed enzyme. Early results of the therapy appear promising. Similar approaches have been used to treat advanced melanoma (skin cancer).

Future Expectations and Concerns

Future advances in biotechnology through genetic manipulation appear to be mind-boggling and ethically challenging. Many of these changes will take years to accomplish because genetic engineering tasks are complicated and expensive. The present state of the art has a high number of failures. Research in the following areas gives insight into future possibilities.

1. Identification of the *genome* (all the genes) of domestic animals may become possible (Table 12.3).
2. Animal productivity (e.g., growth, changing the fat-to-lean ratio, milk production) could be significantly increased and the cost of animal production could be reduced.
3. Disease-resistant animals could be genetically engineered.
4. Parasites could be controlled by genetic interference with their immune systems.
5. Domestic and international markets could be expanded through new genetically engineered products.

TABLE 12.3   Estimated Number of Total Genes and Number Identified

Species	Total Number of Genes (estimated)	Number of Loci Identified (mapped)
Cattle	100,000+	169
Horse	100,000+	21
Human	100,000+	3500
Poultry	—	400+
Sheep	100,000+	26
Swine	100,000+	376

Source: Hetzel, 1989. *Proc. New Zealand Soc. of Anim. Prod.* 49:53.

6. USDA and Colorado State University scientists have developed technology to sex semen by separating X (female-producing) sperm from Y (male-producing) sperm, then artificially inseminating cows with sperm for the desired sex of calf. A 90% accuracy appears to be achievable. This process is currently not commercially feasible because pregnancy rates are lower with sexed semen when compared with typical AI. Costs also have to be carefully assessed. Sexed semen would appeal to the dairy industry because heifer calves as replacements are higher valued than bull calves. Sexed semen used in combination with superovulation, in vitro fertilization, embryo transfer, or just with artificial insemination could significantly affect management practices in the livestock industry.

Future advances in biotechnology, such as the cloning of a sheep in 1996, will likely bring many significant genetic changes and challenges that will benefit both humans and domestic animals. However, genetic engineering of animals raises questions of ethics, food-product safety, and environmental safety. These issues are discussed in more detail in Chapter 35.

CHAPTER SUMMARY

- Cells contain several pair of chromosomes which are composed of deoxyribonucleic acid (DNA). Genes are segments of DNA that are the basic units of inheritance.

- Two chromosomes (X and Y) determine the sex of animals. In livestock, XX is female while XY is male. In poultry, XX is male and XY is female. The parent carrying the XY chromosome determines the sex of the offspring.

- Genetic variation and genetic change occur through chromosome (gene) segregation and chromosome (gene) recombination resulting from sex cell formation and fertilization.

- Genes produce their effects through dominance, overdominance (heterosis), lack of dominance (additive), or epistasis (interactions between pairs of genes).

- Biotechnology and genetic engineering are bringing changes and challenges to both humans and domestic animals.

REVIEW QUESTIONS

1. The functional units of inheritance are called _____ which are located on rodlike bodies called _____ located within the nucleus of each cell.

2. Cell division that yields two daughter cells that have identical chromosomes as compared to the parent cell is termed?

3. Production of sex cells or gametes is termed? Specifically, what is this process called for the production of ova? For the production of sperm?

4. During gametogenesis, what is the unique type of cell division called which reduces the chromosome number of gametes to one-half of that possessed by all other body cells?

5. Where does meiosis occur in the testicle? In the ovary?

6. What is the union of the sperm and the ovum called?

7. What is a newly fertilized ovum called?

8. *True or False:* A zygote has the haploid number of chromosomes.

9. *True or False:* All cells in the body except gametes have the diploid number of chromosome.

10. Within all cells except gametes, chromosomes are paired, being alike in size and shape and carrying genes that affect the same hereditary traits. Paired chromosomes are said to be _____ .

11. What is the chemical component of genes?

12. DNA carries the code or instructions for control of enzymes and _____ .

13. The DNA code of a gene is "read" by what process?

14. Genes are transcribed to produce what type of chemical compound?

15. Through what process does the information in RNA transcripts result in protein synthesis?

16. What is the location of a gene on a chromosome called?

17. What are a pair of genes which occupy corresponding loci on homologous chromosomes called?

18. An individual possessing alleles for a given trait which are identical is said to be _____ . When the alleles for the trait are different, the individual is said to be _____ .

19. When a pair of genes or alleles for a trait differ, and the individual is heterozygous for that trait, one of the genes in the pair is expressed and

often overpowers the expression of the other gene. What is the term used to describe the gene which is expressed? What is the gene whose expression is prevented called?

20. What is the actual gene makeup for a trait called? What is the expression of the gene makeup or appearance of the individual called?

21. Crossbreeding animals of different breeds results in offspring with _____ or hybrid vigor, which means that the offspring are more vigorous than the parents.

SELECTED REFERENCES

Publications

American Health Institute. 1988. *Bovine Somatotropin (BST).* Alexandria, VA: AHI.

Berkowitz, D. B. 1994. Transgenic Animals. *Encyclopedia of Agricultural Science.* San Diego: Academic Press, Inc.

Bourdon, R. M. 1997. *Understanding Animal Breeding.* Upper Saddle River, NJ: Prentice-Hall, Inc.

Cundiff, L. V., Van Vleck, L. D. Young, L. D., and Dickerson, G. D. 1994. Animal Breeding and Genetics. *Encyclopedia of Agricultural Science.* San Diego: Academic Press, Inc.

Genetic Engineering: A Natural Science. St. Louis, MO: Monsanto Company.

Genetically modified livestock: Progress, prospects, and issues. 1993. *Journal of Animal Science,* vol. 71 (supplement 3).

Legates, J. E. 1990. *Breeding and Improvement of Farm Animals.* 4th ed. New York: McGraw-Hill.

Petters, R. M. 1986. Recombinant DNA, gene transfer and the future of animal agriculture. *J. Anim. Sci.* 62:1759.

Pursel, V. G., et al. 1989. Genetic engineering of livestock. *Science* 244:1281.

USDA. 1986. *Biotechnology: An Overview. Biotechnology: Its Application to Animals. 1986 Yearbook of Agriculture (Research for Tomorrow).* Washington, DC: GPO.

Van Vleck, L. D., Oltenacu, E., and Pollack, J. 1986. *Genetics for the Animal Sciences.* New York: W. H. Freeman.

Visuals

Bowling, A. 1996. *Horse Genetics.* Oxon, United Kingdom: CAB International.

Basic Animal Microgenetics. 1996. (Video; 43 min.). CEV, P.O. Box 65265, Lubbock, TX 79464.

Cattle Abnormalities (8861, 16-mm color film; 26 min). Bureau of Audio-Visual Instruction, University of Wisconsin Extension Service, Box 2093, Madison, WI 53701.

Genetics of Coat Color of Horses (SL 501 slide set). Educational Aids, National 4-H Council, 7100 Connecticut Ave., Chevy Chase, MD 20815.

The Nature of Change (videotape 17376 or 16-mm film 17373). Modern Talking Picture Service Film Scheduling Center, 5000 Park St., N., St. Petersburg, FL 33709. A companion booklet, "Genetic Engineering: A Natural Science," can be obtained from Direct Mail Corp., 1533 Washington Ave., St. Louis, MO 63103.

DNA and Genes, VHS video; *Biotechnology: Science for the Future,* VHS video. VEP, California Polytechnic State University, San Luis Obispo, CA 93407.

CHAPTER 13

Genetic Change Through Selection

CONTINUOUS VARIATION AND MANY PAIRS OF GENES

Most economically important traits in farm animals, such as milk production, egg production, growth rate, and carcass composition, are controlled by hundreds of pairs of genes; therefore, it is necessary to expand one's thinking beyond inheritance involving one and two pairs of genes. The exact number of total genes in a species is not known, although estimates of 100,000+ genes in mammalian species have been made. A multibillion dollar project is underway to map the human **genome,** all the genes on the 23 pairs of chromosomes. If this is accomplished by the projected year 2005, it is anticipated that many genes of livestock and poultry will also be identified by that time.

Consider even a simplified example of 20 pairs of heterozygous genes (one gene pair on each pair of 20 chromosomes) affecting yearling weight in sheep, cattle, or horses. The estimated numbers of genetically different gametes (sperm or eggs) and genetic combinations are shown in Table 13.1. Remember that for one pair of heterozygous genes, there are three different genetic combinations (e.g., *AA, Aa,* and *aa*), and for two pairs of heterozygous genes, there are nine different genetic combinations (Table 12.2 in Chapter 12).

Most farm animals are likely to have some gene pairs heterozygous and some homozygous, depending on the mating system being utilized. Table 13.2 shows the number of gametes and genotypes where eight pairs of genes are either heterozygous or homozygous and each gene pair is located on a different pair of chromosomes.

Many economically important traits in farm animals show continuous variation primarily because they are controlled by many pairs of genes. As these

TABLE 13.1   Number of Gametes and Genetic Combinations with Varying Numbers of Heterozygous Gene Pairs

No. of Pairs of Heterozygous Genes	No. of Genetically Different Sperm or Eggs	No. of Different Genetic Combinations (genotypes)
1	2	3
2	4	9
n	2^n	3^n
20	$2^n = 2^{20} = $ ~1 million	$3^n = 3^{20} = $ ~3.5 billion

many genes express themselves, and the environment also influences these traits, producers usually observe and measure large differences in the performance of animals for any one trait. For example, if a large number of calves were weighed at weaning (~205 days of age) in a single herd, there would be considerable variation in the calves' weights. Distribution of weaning weights of the calves would be similar to the examples shown in Fig. 13.1 or 13.2. The bell-shaped curve distribution demonstrates that most of the calves are near the average for all calves, with relatively few calves having extremely high or low weaning weights when compared at the same age.

Figure 13.2 shows how the statistical measurement of standard deviation (SD) is used to describe the variation or differences in a herd where the average weaning weight is 440 lb and the calculated standard deviation is 40 lb. Using herd average and standard deviation, the variation in weaning weight is shown in Fig. 13.2 and can be described as follows:

1 SD: 440 lb ± 1 SD (40 lb) = 400–480 lb
         (68% of the calves are in this range)
2 SD: 440 lb ± 2 SD (80 lb) = 360–520 lb
         (95% of the calves are in this range)
3 SD: 440 lb ± 3 SD (120 lb) = 320–560 lb
         (99% of the calves are in this range)

One percent of the calves (4 calves in a herd of 400) would be on either side of the 320–560-lb range. Most likely, two calves would be below 320 lb and two calves would weigh more than 560 lb.

TABLE 13.2   Number of Gametes and Genetic Combinations with Eight Pairs of Genes with Varying Amounts of Heterozygosity and Homozygosity

Genes in Sire: Genes in Dam:	Aa aa		Bb Bb		Cc CC		Dd Dd		Ee Ee		FF FF		GG gg		Hh Hh		Total
No. different sperm for sire	2	×	2	×	2	×	2	×	2	×	1	×	1	×	2	=	64
No. different eggs for dam	1	×	2	×	1	×	2	×	2	×	1	×	1	×	2	=	16
No. different genetic combinations possible in offspring	2	×	3	×	2	×	3	×	3	×	1	×	1	×	3	=	324

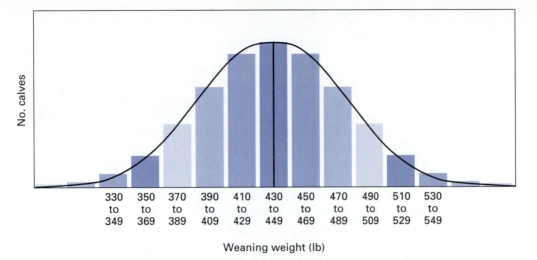

FIGURE 13.1 Variation or differences in weaning weight in beef cattle. The variation shown by the bell-shaped curve could be representative of a breed or a large herd. The dark vertical line in the center is the average or the mean and, in this example, is 440 lb.

Weaning weight is a phenotype, since the expression of this characteristic is determined by the genotype (genes received from the sire and the dam) and the environment to which the calf is exposed. The genetic part of the expression of weaning weight is obviously not simply inherited. There are many pairs of genes involved, and, at the present time, the individual pairs of genes cannot be identified similarly to traits controlled by one to two pair of genes.

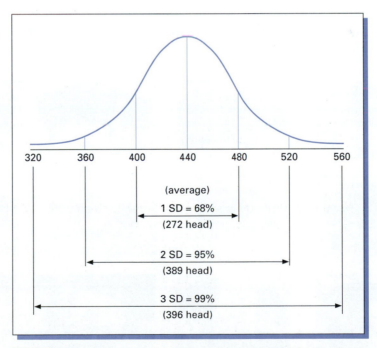

FIGURE 13.2 A normal bell-shaped curve for weaning weight showing the number of calves in the area under the curve (400 calves in the herd).

Consider the following formula:

$$phenotype = genotype \pm environment$$

The phenotype will more closely predict the genotype if producers expose their animals to a similar environment; however, the latter must be within economic reason. This resemblance between phenotype and genotype, where many pairs of genes are involved, is predicted with the estimate of heritability. Use of **heritability,** along with selecting animals with superior phenotypes, is the primary method of making genetic improvement in traits controlled by many pairs of genes. The application of this method is discussed in more detail later in the chapter.

Traits influenced little by the environment can also show considerable variation. An example is the white belt, a breed-identifying characteristic in Hampshire swine (Fig. 13.3). Note that the white belt can be nonexistent in some Hampshires whereas others can be almost white. To be eligible for pedigree registration, Hampshire pigs must be black with the white belt entirely circling the body, including both front legs and feet. Too much white can also limit registration. It is difficult to select a herd of purebred Hampshires that will breed true for desired belt pattern. Apparently, some of the genes for this trait exist in heterozygous or epistatic combinations.

In a beef herd where the average weaning weight is 440 lb, or a dairy herd where annual milk production averages 12,500 lb per cow, the managers of these herds may want to increase the herd average. This assumes the increased production is economically feasible—that the economic benefits of improvement will cover the additional costs.

Selection

Selection is differential reproduction—preventing some animals from reproducing while allowing other animals to become parents of numerous offspring.

FIGURE 13.3   Variation in belt pattern in Hampshire swine. Courtesy of *The Hampshire Herdsman*.

In the latter situation, the selected parents should be genetically superior for the economically important traits. Factors affecting the rate of genetic improvement from selection include selection differential, heritability, and generation interval.

Selection Differential

Selection differential, sometimes called *reach,* shows the superiority (or inferiority) of the selected animals compared to the herd average. To improve weaning weights in beef cattle, producers cull as many below-average-producing cows as economically feasible. Then replacement heifers and bulls are selected that are above the herd average for weaning weight. For example, if the average weaning weight of the selected replacement heifers is 480 lb in a herd averaging 440 lb, then the selection differential for the heifers would be 40 lb. Part of this 40-lb difference is due to genetic differences, and the remaining part is due to differences caused by the environment.

Heritability

A *heritability* estimate is a percentage figure that shows what part of the total phenotypic variation (phenotypic differences) is due to genetics. Heritability can also be defined as that portion of the selection differential that is passed on from parent to offspring.

Realized heritability is the portion actually obtained compared to what was attempted in selection. To illustrate realized heritability, let us suppose a farmer has a herd of pigs whose average postweaning gain is 1.80 lb/day. If the farmer selects from this original herd a breeding herd whose members have an average gain of 2.3 lb/day, the farmer is selecting for an increased daily gain of 0.5 lb/day.

If the offspring of the selected animals gain 1.95 lb, then an increase of 0.15 lb (1.95 − 1.80) instead of 0.5 lb (2.3 − 1.8) has been obtained. To find the portion obtained of what was reached for in the selection, one divides 0.15 by 0.5, which gives 0.3. This figure is the realized heritability. If one multiplies 0.3 by 100%, the result is 30%, which is the percentage obtained of what was selected for in this generation.

Table 13.3 shows heritability estimates for several species of livestock. Traits having heritability estimates 40% and higher are considered highly heritable, 20–39% are classified as having medium heritability, and lowly heritable traits are below 20%.

PREDICTING GENETIC CHANGE

The rate of genetic change that can be made through selection can be estimated using the following equation:

$$Genetic\ change\ per\ year = \frac{heritability \times selection\ differential}{generation\ interval}$$

Consider the following example of selecting for weaning weight in beef cattle and predicting the genetic change. The herd average is 440 lb and bulls and

TABLE 13.3   Heritability Estimates for Selected Traits in Several Species of Farm Animals

Species Trait	Percent Heritability	Species Trait	Percent Heritability
Beef Cattle		**Horses** (*Continued*)	
Age at puberty	40%	Riding performance	
Weight at puberty	50	Dressage (earnings)	20%
Scrotal circumference	50	Cutting ability	5
Birth weight	40	Thoroughbred racing	
Gestation length	40	Log of earnings	50
Body condition score	40	Time	15
Calving interval	10	Pacer: best time	15
Percent calf crop	10	Trotter	
Weaning weight	30	Log earnings	40
Postweaning gain	45	Time	30
Yearling weight	40		
Yearling hip ht. (frame size)	40	**Poultry**	
Mature weight	50	Age at sexual maturity	35
Carcass quality grade	40	Total egg production	25
Yield grade	30	Egg weight	50
Tenderness of meat	50	Broiler weight	40
Longevity	20	Mature weight	50
		Egg hatchability	10
Dairy Cattle		Livability	10
Services per conception	5		
Birth weight	40	**Sheep**	
Milk production	25	Multiple births	20
Fat production	25	Birth weight	30
Protein	25	Weaning weight	30
Solids-not-fat	25	Postweaning gain	40
Type score	30	Mature weight	50
Feet and leg score	10	Fleece weight	40
Teat placement	30	Face covering	50
Udder score	20	Loin eye area	40
Mastitis (susceptibility)	10	Carcass fat thickness	50
Milking speed	20	Weight of retail cuts	40
Mature weight	50		
Excitability	25		
		Swine	
Goats		Litter size	10
Milk production	30	Birth weight	10
Mohair production	20	Litter weaning weight	15
		Postweaning gain	30
Horses		Feed efficiency	30
Withers height	45	Backfat (live animal)	40
Pulling power	25	Carcass fat thickness	50
Riding performance		Loin eye area	50
Jumping (earnings)	20	Percent lean cuts	45

heifers are selected from within the herd. The selection differential measures the phenotypic superiority (or inferiority) of the selected animals compared to the herd average—e.g., bulls (535 lb – 440 lb = 95 lb); heifers (480 lb – 440 lb = 40 lb). Heritability times the selection differential measures the genetic superiority of the selected animals compared to the herd average.

Bulls	*Weight*
Average of selected bulls	535 lb
Average of all bulls in herd	440 lb

Selection differential	95 lb
Heritability	0.30
Total genetic superiority	28 lb
Only half passed on (28 ÷ 2)	14 lb

Heifers	*Weight*
Average of selected heifers	480 lb
Average of all heifers in herd	440 lb
Selection differential	40 lb
Heritability	0.30
Total genetic superiority	12 lb
Only half passed on (12 ÷ 2)	6 lb

Fertilization combines the genetic superiority of both parents, which in this example is 14 lb + 6 lb = 20 lb. The calves obtain half of their genes from each parent; therefore ½ (28) + ½ (12) = 20 lb.

The selected heifers represent only approximately 20% of the total cow herd, so the 20 lb is for one generation. Since the heifer replacement rate in the cow herd is 20%, it would take 5 years to replace the cow herd with selected heifers. Because of this generation interval, the genetic change per year from selection would be 20 lb ÷ 5 = 4 lb/year.

Generation Interval

This trait is defined as the average age of the parents when the offspring are born. This is calculated by adding the average age of all breeding females to the average age of all breeding males and dividing by 2. The generation interval is approximately 2 years in swine, 3–4 years in dairy cattle, and 5–6 years in beef cattle.

The previous example of 4 lb/year shows genetic change if selection was for only one trait. If selection is practiced for more than one trait, genetic change is $1/\sqrt{n}$, where n is the number of traits in the selection program. If four traits were in the selection program, the genetic change per trait would be $1/\sqrt{4}$ = ½. This means that only ½ the progress would be made for any one trait compared to giving all the selection to one trait. This reduction in genetic change per trait should not discourage producers from multiple-trait selection, as herd income is dependent on several traits. Maximizing genetic progress in a single trait may not be economically feasible, and it may lower productivity in other economically important traits.

Sometimes traits are genetically correlated. A genetic correlation between two traits means that some of the same genes affect both traits. In swine, the genetic correlation between rate of gain and feed per pound of gain (from similar beginning weights to similar end weights) is negative. This is desirable from a genetic improvement standpoint because animals that gain faster require less feed (primarily less feed for maintenance). This relationship is also desirable because rate of gain is easily measured, whereas feed efficiency is expensive to measure.

Yearling weight or mature weight in cattle is positively correlated with birth weight, which means that as yearling or mature weight increases, birth weight also increases. This weight increase may pose a potential problem, since birth weight to a large extent reflects calving difficulty and calf death loss.

EVIDENCE OF GENETIC CHANGE

The previously shown examples of selection are theoretical. Does selection really work, or is the observed improvement in farm animals a result of improving only the environment? Let's evaluate several examples.

There have been marked visual, meat-to-bone ratio changes in the thick-breasted modern turkey selected from the narrow-breasted wild turkey, which is the only animal native to the United States. Estimates for genetic change for meat-to-bone ratio have been 0.5% per year for the modern turkey. Genetic selection and improving the environment, particularly through improved nutrition and health, have produced the modern turkey. Tom turkeys can produce 27 lb of liveweight (22 lb dressed weight) in 5 months; the wild turkey weighs 10 lb in 6 months.

The modern turkey is not likely to survive in the same environment as the wild turkey. The modern turkey needs an environment with intensive management. For example, turkeys are mated artificially because the heavy muscling in the breast prevents them from mating naturally. Selection has been effective in changing rate of growth and meatiness in turkeys, but it has necessitated a change in the environment for these birds to be productive. Therefore, the improvement in turkeys has resulted from improving both the genetics and the environment.

There are tremendous size differences in horses. Draft horses have been selected for large body size and heavy muscling to perform work. The light horse is more moderate in size, and ponies are small. Miniature horses have genetic combinations that result in a very small size. The miniature horse is considered a novelty and a pet. Table 13.4 shows the height and weight variations in different types of horses. Apparently these different horse types have been produced from horses that originally were quite similar in size and weight. The extreme size differences in horses result primarily from genetic differences, as the size differences are apparent when the horses are given similar environmental opportunities.

One of the best documented examples of genetic change in farm animals comes from a backfat selection experiment in swine. Figure 13.4 shows the result of 13 generations of selection for high and low backfat thickness. Also, a line of pigs was maintained in which no selection for backfat thickness was

TABLE 13.4   Height and Weight Differences in Mature Draft Horses, Light Horses, Ponies, and Miniature Horses

| Horse Type | Approximate Height at Withers | | Approximate Mature Weight (lb) |
	Hands	Inches	
Draft	17	68	1,600
Light	15	60	1,100
Pony	13	52	700
Miniature	8	32	250

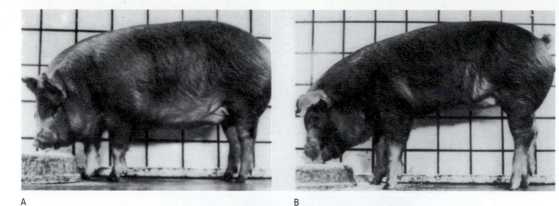

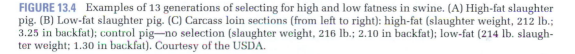

FIGURE 13.4 Examples of 13 generations of selecting for high and low fatness in swine. (A) High-fat slaughter pig. (B) Low-fat slaughter pig. (C) Carcass loin sections (from left to right): high-fat (slaughter weight, 212 lb.; 3.25 in backfat); control pig—no selection (slaughter weight, 216 lb.; 2.10 in backfat); low-fat (214 lb. slaughter weight; 1.30 in backfat). Courtesy of the USDA.

practiced. This nonselected line showed the environmental changes in backfat thickness from year to year. In this experiment, backfat thickness was the only trait for which selection was practiced. Selection for and against backfat thickness was effective, since backfat thickness is a highly heritable trait (50%).

In past years, the backfat probe was used to measure differences in backfat thickness in live pigs (Fig. 13.5). Use of the backfat probe improved the effectiveness of selection by measuring backfat thickness in breeding animals without having to slaughter them. Today, ultrasound has replaced the backfat probe in measuring fat thickness in live animals.

The effectiveness of selection against excessive backfat thickness in swine is shown in Table 13.5. The reduction in lard production per hog has decreased dramatically, even though the average slaughter weight has increased approximately 19 lb since 1960.

Selection for milk production in dairy cattle has an additional challenge in a genetic improvement program because the bull does not express the trait. This is an example of a **sex-limited** trait.

Genetic evaluation of a bull is based primarily on how his daughters' milk production compares to that of their **contemporaries.** Milk production is a

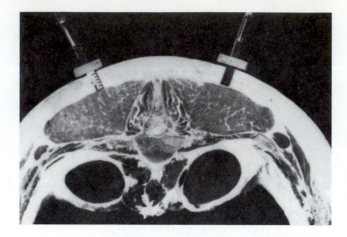

FIGURE 13.5 A pork carcass cross section with steel rulers measuring the backfat over the loin eye muscle. Courtesy of Iowa State University.

moderately heritable trait (25%), and an increase of 10,000 pounds per cow over the past 50 years is most noteworthy, even though the trait is not expressed in the bull (Table 13.6). The genetic change in milk production in Holsteins has been estimated at 168 lb per year from 1960 to 1988. The total (phenotypic change) was 208 lb and the management (environmental) change is also noted in Fig. 13.6.

TABLE 13.5   Trends in U.S. Lard Production

Year	Total Hog Slaughter (mil head)	Total Lard Production (mil lb)	Average Slaughter Weight (lb)	Lard Production per Hog (lb)
1950	79.3	2,631	—	33.2
1960	84.1	2,562	236	30.4
1970	87.0	1,913	240	22.0
1980	97.2	1,217	242	12.5
1985	79.6	852	247	10.7
1990	85.1	910	252	10.6
1995	95.7	1,034	255	10.8

Source: USDA.

TABLE 13.6   Changes in Milk Production in the United States, 1940–1995

Year	No. Cows (mil)	Average Milk Per Cow (lb)	Total Milk (bil lb)
1940	23.7	4,622	109.4
1950	21.9	5,314	116.6
1960	17.5	7,029	123.1
1970	12.0	9,751	117.0
1980	10.8	11,875	128.4
1990	10.1	14,645	148.3
1995	9.5	16,443	155.6

Source: USDA.

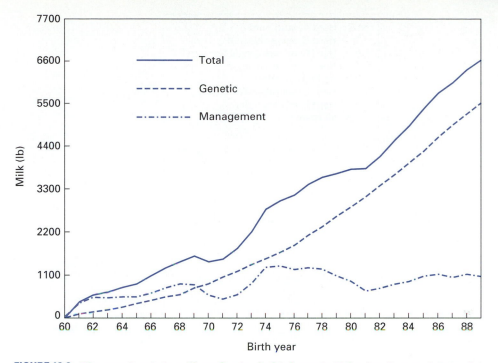

FIGURE 13.6   The genetic gain in milk production by birth year in Holsteins. Source: *J. Dairy Sci.* 76:3143.

GENETIC IMPROVEMENT THROUGH ARTIFICIAL INSEMINATION

The primary reason for using artificial insemination (AI) is that genetically superior sires can be used more extensively. AI is used more in dairy cattle than any other species of farm animals. Using AI dairy bulls has resulted in higher milk production than using natural service bulls.

Figure 13.7 shows the rate of genetic improvement for yearling weight in beef cattle under different selection schemes. The most significant points are these:

1. Marked difference between herd A (no AI, males and females selected for yearling weight) and herd D (bulls obtained from a herd where no selection is practiced for yearling weight).

2. The small difference between herd B (selects bulls from herd A on yearling weight and selects for yearling weight in females) and herd C (selects bulls from herd A but practices no selection for yearling weight on females). Effective sire selection accounts for 80–90% of the genetic change in a herd.

3. Greatest change in herd F where the top 1% of the sons of the top 1% of the bulls were used to build generation upon generation of AI selection. The rate of improvement using AI in herd F is 1.5 times that of herd A with natural service. The primary reasons for the greater improvement in herd F is a larger selection differential.

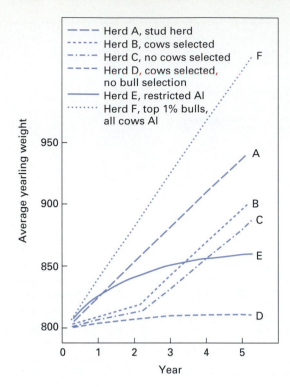

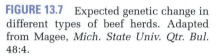

FIGURE 13.7 Expected genetic change in different types of beef herds. Adapted from Magee, *Mich. State Univ. Qtr. Bul.* 48:4.

SELECTION METHODS

The three methods of selection are (1) tandem, (2) independent culling level, and (3) selection index, which is based on an index of net merit.

Tandem is selection for one trait at a time. When the desired level is achieved in one trait, then selection is practiced for the second trait. The tandem method is rather ineffective if selection is for more than two traits or if the desirable aspect of one trait is associated with the undesirable aspect of another trait. The tandem method is not generally recommended because it is the least effective of the three methods of selection.

Independent culling level establishes minimum culling levels for each trait in the selection program. Even though it is the second most effective type, it is the most prevalent one used. Independent culling level is most useful when the number of traits being considered is small and when only a small percentage of offspring is needed to replace the parents. For example, with poultry, 100 or more chicks might be produced from each hen and 1,500 or more chicks from each rooster. Here only 1 female out of 50 or 1 male out of 750 is needed to replace the parents. It is simple using this method to select from the upper 10% of the females and the upper 1% of the males.

Table 13.7 shows an example of using independent culling levels in selecting yearling bulls. Birth weight is correlated to calving ease. Weaning weight and postweaning gain measure growth. Puberty and semen production are measured with scrotal circumference. Any bull that does not meet the minimum or maximum level is culled. Birth weights are evaluated on a maximum level (upper limit) since high birth weights result in increased calving difficulty.

TABLE 13.7 Independent Culling Level Selection in Yearling Bulls

Trait	Culling Level	Bull				
		A	B	C	D	E
Birth weight (lb)	85 (max)	(105)	82	85	83	80
Weaning weight (lb)	500 (min)	550	(495)	505	570	600
Postweaning gain (lb/day)	3.0 (min)	3.7	3.6	3.1	3.3	3.5
Scrotal circumference (cm)	30 (min)	34	37	31	(29)	35

The circled records in Table 13.7 indicate bulls that did not meet the independent culling levels. Bulls A, B, and D would be culled.

In the *selection index* method, each animal is rated numerically in comparison to every other animal for each trait being considered. A simplified example of the selection index method is to rank each of the five bulls in Table 13.7 on a scale of 1 to 5, with 1 representing the best performance and 5 the poorest. Table 13.8 gives the ranks of the bulls based on the 1-to-5 rating scale.

In Table 13.8, the total score (sum of rankings for each of the four traits) is shown for each bull. The bull with the lowest total score would represent the best bull. The bull with the highest total score would be the poorest bull.

A comparison of Tables 13.7 and 13.8 identifies some strengths and weaknesses of independent culling level versus index method of selection. Bull E is evaluated as the best bull under both selection methods.

Bulls B and C identify major weaknesses of independent culling levels. Under the selection index method, bull B was the second-best bull; using independent culling levels, the bull was culled (and likely castrated) when weaned at about 7 months of age. Bull B is superior in four traits but slightly below the culling level for weaning weight.

Bull C barely survives the four culling levels; however, he is ranked second to last in the index method. Therefore, mediocre animals may be selected using independent culling levels but culled using the index method.

An advantage of independent culling levels is that selection can occur during different productive stages during the animal's lifetime (e.g., at weaning). This is more cost-effective than the index method, where no culling should occur until records are recorded for all traits. A combination of the two methods may be most useful and cost-effective.

The example in Table 13.8 is oversimplified. Equal emphasis is given to each of the four traits. A selection index could be computed for the traits in

TABLE 13.8 Simplified Selection Index Method of Selecting Yearling Bulls Based on Total Score

Trait	Bulls				
	A	B	C	D	E
Birth weight	5	2	4	3	1
Weaning weight	3	5	4	2	1
Postweaning gain	1	2	5	4	3
Scrotal circumference	3	1	4	5	2
Total score	12	10	17	14	7

Table 13.8 by putting the heritabilities, economic values, genetic correlations, and variabilities of these four traits into a single formula. This process is rather complicated, as it requires the use of complex statistical methods. An example of such an index, used in boar selection, is given in Chapter 26.

BASIS FOR SELECTION

The three primary bases of selection—(1) **pedigree** (names and records of relatives), (2) **individual appearance** or **individual performance,** and (3) **progeny testing**—are shown in Fig. 13.8.

Pedigree information is most useful before animals have expressed their own individual performance or before their progeny's performance is known. Pedigree is also useful for assessing genetic abnormalities and traits expressed much later in life (longevity), as well as for selecting for traits expressed in only one sex. In evaluating pedigree information, only the information from the animal's closest relatives should be used because of the genetic relationship involved. For example, the genetic relationship of an animal to each parent is ½ (50%), to each grandparent ¼ (25%), and to each great-grandparent ⅛ (12.5%).

Selection on individual appearance and performance should be practiced on traits that are economically important and have high heritabilities. One primary advantage of selection on individual performance is that it permits rapid generation turnover and thus shortens the general interval.

Progeny testing is more accurate than the other two methods, provided the progeny test is accurate and extensive. Progeny tests are particularly useful in selecting for carcass traits (where good carcass indicators are not available in the live animal), sex-limited traits such as milk production, and traits with low heritabilities.

Disadvantages of progeny testing include: (1) too few animals can be progeny-tested, (2) longer generation interval required to obtain progeny information, and (3) decreased accuracy in poorly conducted tests. Progeny testing is usually limited to sires because a sufficient number of offspring cannot be obtained on a dam to give an adequate progeny test.

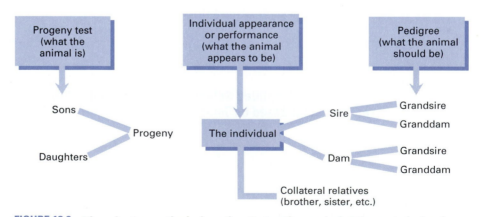

FIGURE 13.8   The selection methods, how they tie together, and a brief appraisal of each.

In a proper progeny test, sires are assigned females at random after the dams have been classified by age, line of breeding, and known performance. A sire, bred to a selected group of females, can give biased information that does not accurately evaluate his breeding ability. An adequate progeny test involves a minimum number of progeny. The greater the number, the greater the validity of the test. Progeny test data are biased and their accuracy is greatly reduced if dams and offspring are not fed and managed uniformly until all data are collected.

All three bases of selection can be used effectively in animal-breeding programs. Outstanding young sires can best be identified by using individual performance and pedigree information on the sire, dam, and half-sibs. These outstanding young sires then need to prove themselves genetically, based on a valid progeny-testing program. Results of progeny tests determine which sires to use extensively in later years.

CHAPTER SUMMARY

- Phenotype (what is seen or measured) is determined by the genotype (genetic makeup) and the environment to which the animal is exposed.

- Heritability measures the proportion (0–100%) of the total phenotypic variation that is due to genetics. Traits high in heritability are ≥40%, while lowly heritable traits are ≤20%.

- Selection differential is the superiority (or inferiority) of the selected animals compared to the average of the group from which they came. Generation interval is the average age of the parents when the offspring are born.

- $\text{Genetic change per year} = \text{heritability} \times \dfrac{\text{selection differential}}{\text{generation interval}}$

- Independent culling level is the most common method of selection. Selection can be based on pedigree, individual performance, progeny testing, or a combination of all three.

REVIEW QUESTIONS

1. Phenotype = genotype ± _____.
2. Phenotypic variation that is due to differences in an animal's genotype is called _____.
3. What is the superiority (or inferiority) of animals selected to be parents compared to the average of the herd called?
4. The amount of genetic change made in a herd of animals during a year can be predicted by knowing what three variables?
5. What is the average age of parents when their offspring are born called?
6. What is a sex-limited trait?
7. What are three methods of genetic selection?
8. Which selection procedure involves selection for only one trait at a time and is least effective?

9. Which selection procedure involves ranking each individual animal against all others in the herd for the traits being selected and then obtaining a total score of the ranks for each animal, and is most effective?

10. What are the three primary bases of selection?

SELECTED REFERENCES

Publications

Bourdon, R. M. 1997. *Understanding Animal Breeding.* Upper Saddle River, NJ: Prentice-Hall, Inc.

Cundiff, L. V., Van Vleck, L. D., Young, L. D., and Dickerson, G. D. 1994. Animal Breeding and Genetics. *Encyclopedia of Agricultural Science.* San Diego: Academic Press, Inc.

Freeman, A. E., and Lindberg, G. L. 1993. Challenges to Dairy Management: Genetic Considerations. *Journal of Dairy Science* 76:3143.

Genetics and Goat Breeding. Proceedings of the Third International Conference on Goat Production and Disease. 1982. Scottsdale, AZ: *Dairy Goat Journal.*

Hetzer, H. O., and Harvey, W. R. 1967. Selection for high and low fatness in swine. *J. Anim. Sci.* 26:1244.

Hintz, R. L. 1980. Genetics of performance in the horse. *J. Anim. Sci.* 51:582.

Legates, J. E. 1990. *Breeding and Improvement of Farm Animals.* 4th ed. New York: McGraw-Hill.

Van Vleck, L. D., Oltenacu, E., and Pollack, J. 1986. *Genetics for the Animal Sciences.* New York: W. H. Freeman.

Visuals

Development of the Modern Chicken (79 slides and audiotape; 15 min); and *Development of the Modern Turkey* (76 slides and audiotape; 14 min). Poultry Science Department, Ohio State University, 674 W. Lane Ave., Columbus, OH 43210.

CHAPTER 14

Mating Systems

Purebred breeders (sometimes called *seedstock producers*) and commercial breeders (producers) are the two general classifications of animal breeders. **Purebred** livestock typically come from the pure breeds for which their ancestry is recorded on a pedigree by a breed association (Fig. 14.1). Most commercial slaughter livestock are crossbreds resulting from crossing two or more breeds or lines of breeding.

A knowledge of breeding or mating systems is important because producers can use mating systems to preserve genetic superiority and utilize hybrid vigor. Genetic improvement can be optimized in most herds and flocks by utilizing a combination of selection and mating systems.

Mating systems are identified primarily by genetic relationship of the animals being mated. The two major systems of mating are **inbreeding** and **outbreeding.** Inbreeding is the mating of animals more closely related than the average of the breed or population. Outbreeding is the mating of animals not as closely related as the average of the population.

Since mating systems are based on the *relationship* of animals being mated, it is important to understand more detail about genetic relationship. Proper pedigree evaluation also involves understanding the genetic relationship between animals. Relationship is best described in knowing which genes two animals have in common and whether the genes in an animal or animals exist primarily in a heterozygous or homozygous condition. Figure 14.2 shows the mating systems and their relationship to homozygosity and heterozygosity.

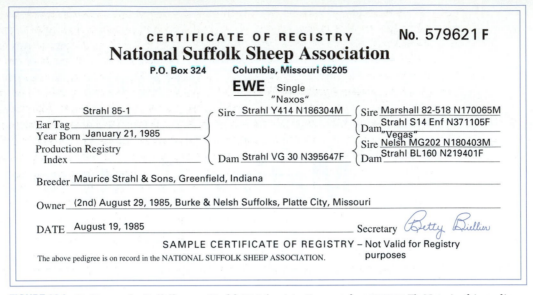

FIGURE 14.1 Pedigree of a Suffolk ewe, Strahl 85-1 (registration number 579621 F). Note in this pedigree that the sire, dam, and four grandparents are listed, with a registration number following each of their names. Courtesy of the National Suffolk Sheep Association.

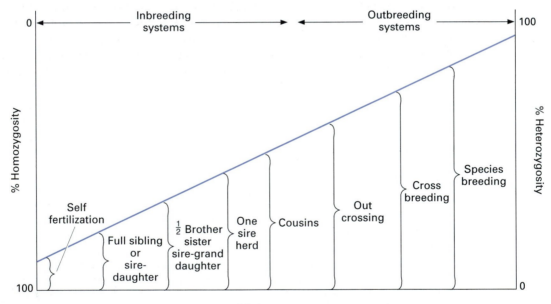

FIGURE 14.2 Relationship of the mating system to the amount of heterozygosity or homozygosity. Self-fertilization is currently not an available mating system in animals.

INBREEDING

Because inbreeding is the mating of related animals, the resulting inbred off-spring have an increased homozygosity of gene pairs compared to noninbred animals in the same population (breed or herd). An example of inbreeding is shown in Fig. 14.3, where animal A has resulted from mating B (the sire) to C

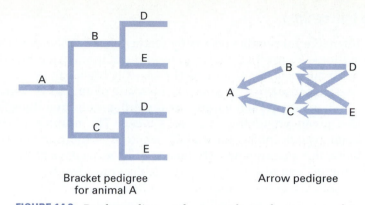

Bracket pedigree
for animal A

Arrow pedigree

FIGURE 14.3 Bracket pedigree and arrow pedigree showing animal A resulting from a full brother–sister mating.

(the dam). Note that B and C have the same parents (D and E). B and C are genetically related because they have a brother–sister relationship. B and C do not have the same identical genetic makeup because each has received a sample half of genes from each parent. The arrow in the arrow pedigree represents a sample half of genes from the parent to the offspring.

Animal A is inbred because it has resulted from the mating of related animals. Although the calculations are not shown in the figure, animal A has 25% more homozygous genes as compared with a noninbred animal in the same population.

Figure 14.4 shows another example of inbreeding, where animal X has resulted from a sire–daughter mating. Sire A was mated to dam D, with D being sired by A. The increased homozygosity of animal X is 25%, which is the same as animal A in Fig. 14.3.

The two different forms of inbreeding are these:

1. *Intensive inbreeding*—mating of closely related animals whose ancestors have been inbred for several generations.

2. *Linebreeding*—a mild form of inbreeding where inbreeding is kept relatively low while maintaining a high genetic relationship to an ancestor or line of ancestors.

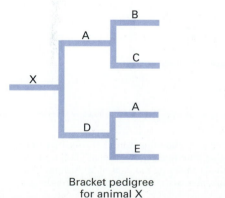

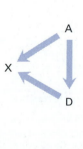

FIGURE 14.4 Bracket pedigree and arrow pedigree showing animal X resulting from a sire–daughter mating.

Bracket pedigree
for animal X

Arrow pedigree

Intensive Inbreeding

Intensive inbreeding occurs by mating closely related animals for several generations. Figure 14.5 shows an example of intensive inbreeding. The increase in the homozygosity of animal H's genes is higher than 25% (compare with Fig. 14.4) because both the sire and grandsire of animal H were also inbred.

There are numerous, genetically different inbred lines that can be produced in a given population such as a breed. The number of different, completely homozygous inbred lines is 2^n, where n is the number of heterozygous gene pairs. For example, with two pairs of heterozygous gene pairs, there are 2^2 or 4 different inbred lines that are possible. For example, with *BbPp* genes in a herd completely heterozygous (see Chapter 12), the different, completely homozygous lines that can result from inbreeding are *BBPP, BBpp, bbPP,* and *bbpp.* Each of these four inbred lines is genetically pure (homozygous) for those two pair of genes and each will breed true when animals are mated within the same homozygous line.

There have been research projects with swine and beef cattle in which inbred lines have been produced, anticipating that the crossing of these inbred lines would produce results similar to hybrid corn (Fig. 14.6). One continuing research project is at the San Juan Basin Research Center (SJBRC) in Hesperus, Colorado, where cattle have been inbred for more than 30 years. Figure 14.7 shows the arrow pedigree of an inbred bull (Royal 4160) produced in the Royal line of Hereford cattle. Note that College Royal Domino 3 is the sire of 10 animals in Royal 4160's pedigree, while Royal 3016 has sired 6 animals in the same pedigree. The increased homozygosity of this bull is 58%, which represents some of the most highly inbred cattle in the world.

Information obtained from inbreeding studies with farm animals demonstrates the following results and observations:

1. Increased inbreeding is usually detrimental to reproductive performance and preweaning and postweaning growth. Also, inbred animals are more susceptible to environmental stresses. Whereas 60–70% of the inbred

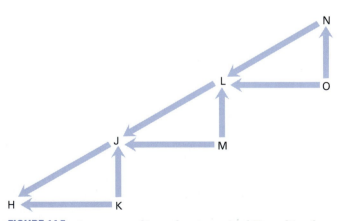

FIGURE 14.5 An arrow pedigree showing animal H resulting from three generations of sire–daughter matings. Note that J is the sire of H and also the sire of K, the latter being the dam of H. Animal J has resulted from two previous, successive generations of sire–daughter matings.

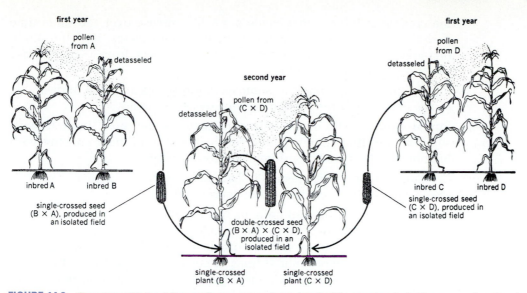

FIGURE 14.6 Crossing of inbred lines of corn to produce the double-crossing hybrid corn seed utilized widely today. Note the relative size of the corn ears from the inbred lines compared with the resulting hybrid cross. Courtesy of the USDA.

lines show the detrimental effects to increased inbreeding, there are 30–40% of the lines that show no detrimental effect, with some lines demonstrating improved productivity.

2. In a Colorado research beef herd, the inbred lines showed a yearly genetic increase of 2.6 lb in weaning weight over a 26-year period, while

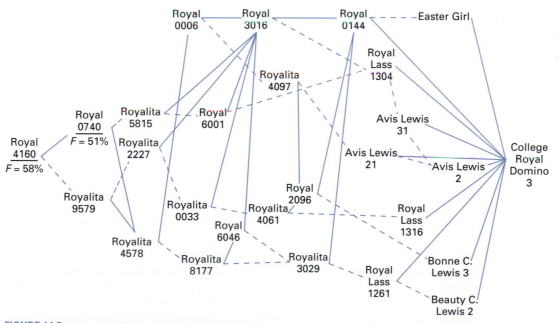

FIGURE 14.7 Arrow pedigree of Royal 4160, which has an inbreeding coefficient of 58%. Solid lines represent the genetic contribution of the bulls, whereas the cow's contribution is represented by a broken line. Courtesy of the CSU Expt. Sta. (San Juan Basin Research Center) General Series 982.

the crosses of inbred lines made a 4.6 lb increase over the same time period. Heterosis is demonstrated in the line crosses, and the 4.6-lb increase is typical of what breeders might expect from using intense selection in an outbred herd.

3. Inbreeding quickly identifies some desirable genes and also undesirable genes, particularly the hidden, serious recessive genes.

4. Inbred animals with superior performance are the most likely to have superior breeding values, which result in more uniform progeny with high levels of genetically influenced productivity.

5. Crossing of inbred lines results in heterosis; however, in most cases heterosis compensates for inbreeding depression.

6. Crossing of inbred lines of animals has not yielded the same results as crossing inbred lines of corn. The reasons appear to be: (1) inbreeding animals is slower (cannot self-fertilize as in corn); (2) it is easier and less costly to produce more inbred lines of corn; (3) inbred lines of animals are eliminated because of extremely poor reproductive performance and being less adapted to environmental stress.

7. There is merit in using some inbreeding in developing new lines of poultry and swine. (This is discussed in more detail later in the chapter.)

It is not logical for breeders to develop their own lines of highly inbred beef cattle. Inbreeding depression usually affects the economics of the operation. Both purebred and commercial producers can take advantage of highly productive inbred lines by crossing these inbred bulls with unrelated cows.

Inbreeding such as sire–daughter matings are logical ways to test for undesirable recessive genes. Also, seedstock producers use inbreeding in well-planned linebreeding programs. Consideration should be given to implementing linebreeding when breeders have difficulty introducing sires from other herds that are genetically superior to those they are producing.

Linebreeding

Linebreeding is a mild form of inbreeding used to maintain a high genetic relationship to an outstanding ancestor, usually a sire. This mating system is best used by seedstock producers who have a high level of genetic superiority in their herd. They find it difficult to locate sires that are superior to the ones they are raising in their herds.

Occasionally a breeder may produce a sire with a superior combination of genes, where the sire consistently produces high-producing offspring. Some of these sires may not be outproduced by younger sires. This is observed in some dairy bulls, when they remain competitively superior as long as they produce semen. These sires warrant use in a linebreeding program.

Figure 14.8 gives an example of linebreeding. Impressive, an outstanding quarter horse stallion, is linebred to his ancestor, Three Bars, by three separate pathways. The inbreeding of Impressive is approximately 9%, whereas the genetic relationship of Impressive to Three Bars is approximately 44%. Inbreeding below 20% is considered low, whereas a genetic relationship is high when it approaches 50%.

Impressive has nearly the same genetic relationship as if Three Bars had been his sire (44 versus 50%). Progeny of Impressive have produced outstand-

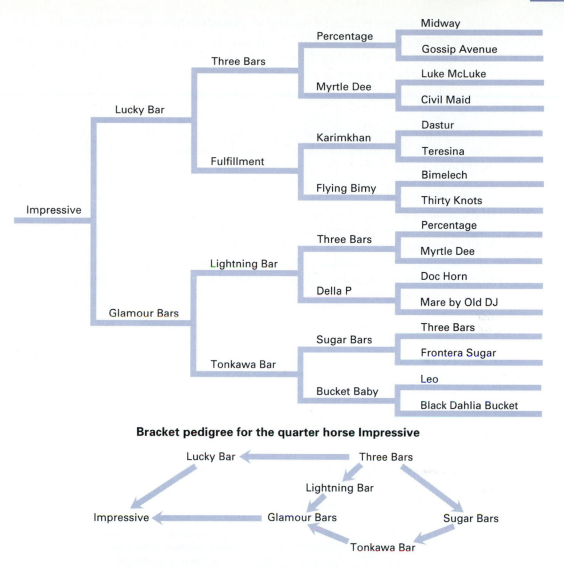

Bracket pedigree for the quarter horse Impressive

**Arrow pedigree showing the genetic pathways by which
Three Bars contributes to the inbreeding and linebreeding of Impressive**

FIGURE 14.8 Horse pedigree showing linebreeding. Courtesy of the American Quarter Horse Association.

ing records, particularly as halter-point and working-point winners in the show ring.

OUTBREEDING

The four types of outbreeding are as follows:

1. *Species cross*—crossing of animals of different species (e.g., horse to donkey or cattle to bison).

2. *Crossbreeding*—mating of animals of different established breeds.

3. *Outcrossing*—mating of unrelated animals within the same breed.

4. *Grading up*—mating of purebred sires to commercial grade females and their female offspring for several generations. Grading up can involve some crossbreeding or it can be a type of outcrossing system.

Species Cross

A species designation is part of the zoological classification (Fig. 14.9) used in taxonomy (a branch of zoology) to classify animals on similarities in body structure. These differences and similarities in body structure translate to genetic differences and similarities. For example, some animals of different species but the same genus can be crossed to produce viable offspring. Animals of different genus cannot be successfully crossed because chromosome number and other genes are different. Therefore, a **species cross** is the widest possible kind of outbreeding that can be achieved.

One of the most common species cross is the **mule,** resulting from crossing the jack of the ass species and the mare of the horse species (*Equus asinus* × *Equus caballus*). Mules existed in large numbers as work animals before the advent of the tractor. The **hinny** is the reciprocal cross of the mule (*Equus caballus* stallion × *Equus asinus* jennet). The hinny never became as popular as the mule.

Mare mules are usually sterile, which gives verification to genetic differences between the ass and horse. There have been a few reports of fertile mare mules.

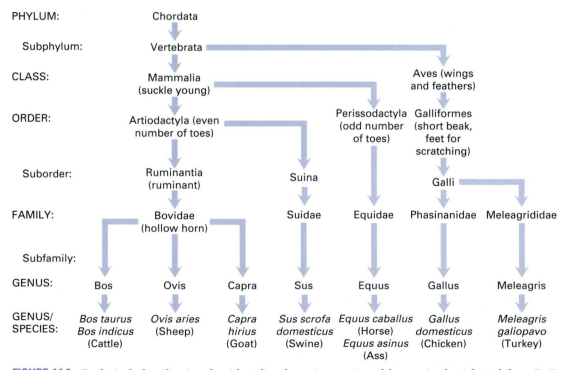

FIGURE 14.9 Zoological classification that identifies the major species of farm animals. Adapted from R. F. Plimpton, and J. F. Stephens, *Animal and Science for Man: A Study Guide* (Minneapolis: Burgess, 1979).

Crossing of the zebu or humped cattle on European-type cattle (*Bos indicus* × *Bos taurus*) is common in the southeastern part of the United States. These crosses are more adaptable and productive in hot, humid environments than either of the straight species. Some authorities raise questions about *Bos indicus* and *Bos taurus* being separate species, and their crosses are usually referred to as *crossbreds* rather than species crosses.

Numerous crosses of American bison and cattle have been made. Some of these crosses have been designated as separate breeds called **Cattalo** or **Beefalo.** These crosses are intended to be more adaptable to harsh environments (cold temperatures and limited forage). Fertility problems have existed in some of these crosses, and these numbers are limited.

Sheep and goats have been crossed even though they have different genus classification. Fertilization occurs but embryos die in early gestation.

There are other species crosses that have occurred. Most species crosses have little commercial value, although insight into the evolutionary process is interesting. Recent advances in genetic engineering might make some genetic combinations between species more feasible. *Gene-splicing* (inserting a gene or genes from one animal to another) has occurred between species. In the future, opportunity to combine desirable genes, both within a species and between species, could become a reality.

Crossbreeding

There are two primary reasons for using **crossbreeding:** (1) breed complementation and (2) heterosis (hybrid vigor). *Breed complementation* implies crossing breeds so their strengths and weaknesses complement one another. There is no one breed superior in all desired production characteristics; therefore, planned crossbreeding programs that use breed complementation can significantly increase herd productivity.

Crossbreeding, if properly managed, allows for the effective use of heterosis, which has a marked effect on productivity in swine, poultry, and beef cattle. *Heterosis* is defined as the increase in productivity in the crossbred progeny above the average of breeds or lines that are crossed. An example of calculating heterosis is shown in Table 14.1. The calculated heterosis for calf-crop percent in this example is 5%, whereas heterosis for weaning weight is 4%.

TABLE 14.1   Computation of Heterosis for Percent Calf Crop and Weaning Weights

	Calf Crop (%)	Weaning Weight (lb)	Lb Calf Weaned per Cow Exposed
Breed A	82	460	377
Breed B	78	540	421
Average of the two breeds (without heterosis)	80	500	399
Average of crossbreds (with heterosis)	84	520	447
Superiority of crossbreds over average of two breeds	4	20	48
Percent heterosis	5% (4 ÷ 80)	4% (20 ÷ 500)	12% (48 ÷ 399)

TABLE 14.2   Relationship of Heritability and Heterosis for Most Traits

Traits	Heritability	Heterosis
Reproduction	Low	High
Growth	Medium	Medium
Carcass	High	Low

Crossbreeding is sometimes questioned when the crossbred performance is less than the parent breed. In Table 14.1, for example, the weaning weight of the crossbreds is 520 lb, while one parent (Breed B) is 540 lb. However, for calf-crop percent, the crossbreds are 84%, which is higher than either Breed A or Breed B. The value of crossbreeding, in this example, is best demonstrated by combining calf-crop percent and weaning weight (calf-crop percent × weaning weight = lb calf weaned per cow exposed). Note in Table 14.1 that lb of calf weaned per cow exposed is 447 lb for the crossbreds, whereas Breed A is 377 lb and Breed B is 421 lb.

The amount of heterosis expressed is related to the heritability of the trait. Table 14.2 shows that heterosis is highest for lowly heritable traits and lowest for highly heritable traits. These relationships are helpful to commercial producers in selecting and crossbreeding to enhance genetic improvement. Figure 14.10 shows the relative importance of selection and crossbreeding in an improvement program. This figure demonstrates that selecting genetically superior animals is more important than crossbreeding. However, using the two in combination gives the highest level of performance.

Crossbreeding is most commonly used in swine, beef cattle, and sheep. Little crossbreeding is done in dairy cattle because of the primary emphasis on one trait (milk production) and the superiority of the Holstein breed for that trait. Poultry breeders utilize heterosis primarily through crossing lines that have been developed from crossing breeds and inbreeding separate and distinct lines.

Figures 14.11, 14.12, and 14.13 show the crossbreeding systems most frequently used in swine, beef cattle, and sheep. More specific detail in using these crossbreeding systems for these species is given in Chapters 22, 26, and 28.

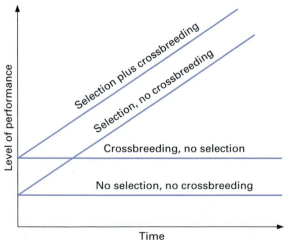

FIGURE 14.10 Improvement in performance with various combinations of selection and crossbreeding.

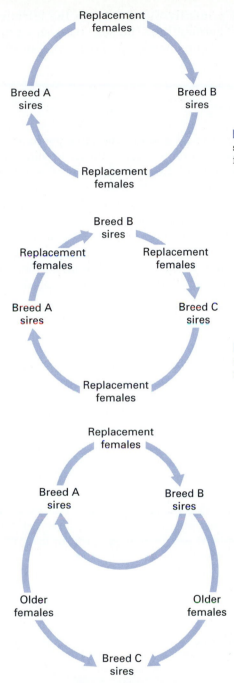

FIGURE 14.11 Two-breed rotation cross. Females sired by breed A are mated to breed B sires, and females sired by breed B are mated to breed A sires.

FIGURE 14.12 Three-breed rotation cross. Females sired by a specific breed are bred to the breed of sire next in rotation.

FIGURE 14.13 Terminal (static) or modified-terminal crossbreeding system. It is terminal or static if all females in herd (A × B) are then crossed to breed C sires. All male and female offspring are sold. It is a modified-terminal system if part of females are bred to A and B sires to produce replacement females. The remainder of the females are terminally crossed to breed C sires.

Outcrossing

The most widely used breeding system for most species is **outcrossing**. As unrelated animals within the same breed are mated, the gene pairs are primarily heterozygous, although there is a slight increase in homozygosity over time. Homozygosity for several breeds has been estimated between 10 and 20%. This slight increase in homozygosity occurs because the animals mated are somewhat related in that they are members of the same breed.

Usefulness of outcrossing is primarily dependent on the effectiveness of selection (selection differential × heritability). The gene pairs stay primarily in a heterozygous condition, as there is no attempt to maximize homozygosity or heterozygosity.

Grading Up

The continuous use of purebred sires of the same breed in a grade herd or flock is called *grading up.* In this situation, grading up is similar to outcrossing. The accumulated percentage of inheritance of the desired purebred is 50% (½), 75% (¾), 84.5% (⅞), and 94% (¹⁵⁄₁₆) for four generations when grading up is practiced. The fourth generation resembles the purebred sires so closely in genetic composition that it approximates the purebred level.

The grading-up system is useful in the breeding of cattle and horses, but it has little value in breeding sheep, swine, or poultry. High-producing purebred sheep, swine, and poultry breeding stock are available at reasonable prices; therefore, the breeder can buy them for less than he or she can produce them by grading up. The use of production-tested males that are above average in performance in a commercial herd can grade up the herd not only to a general purebred level but also to a high level of production.

A recent use of grading up on a large scale has occurred with the introduction of many beef cattle breeds. Most of the introduction has been with males (bulls or semen), as females have been less available and more expensive because of the numbers needed.

Imported bulls (or their semen) have been used on commercial cows or purebred cows of another breed. Grading up, as used here, is a type of crossbreeding, although the intent is not to maintain heterosis but to increase the frequency of genes from the introduced breed.

After several successive generations of mating the new breed to cows carrying a certain percentage of the new breed, the resulting offspring have been designated purebreds. In most breeds, this designation has been given when the calves had ⅞ or ¹⁵⁄₁₆ of the genetic composition of the new breed. Figure 14.14 shows how these matings are made. It would require a minimum of 7 years to produce the first ¹⁵⁄₁₆ calves of the new breed.

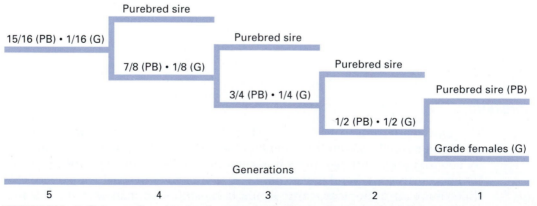

FIGURE 14.14   Utilizing grading up to produce purebred offspring from a grade herd.

FORMING NEW LINES OR BREEDS

New breeds have been formed and are currently being formed by crossing several breeds. These are sometimes given a general classification of **synthetic breeds** and **composite breeds.** In beef cattle, the Brangus, Barzona, Beefmaster, and Santa Gertrudis breeds are composite (synthetic) breeds formed several years ago, whereas MARC I (crosses of Charolais, Brown Swiss, Limousin, Hereford, and Angus breeds) and RX3 (crosses of Red Angus, Hereford, and Red Holstein breeds) are examples of composites currently being developed. Columbia, Targhee, and Polypay are examples of synthetic breeds of sheep.

Crossing several swine breeds, a practice associated with some inbreeding, has been used to develop the hybrid boars now being merchandised by several companies. Hybrid boars are used extensively in the swine industry.

In poultry, breeding for egg production differs from breeding for broiler production. Different traits are emphasized in the production of these two products. Both inbreeding and heterosis are utilized in the production of specific lines and strains of birds that are highly productive in the production of either eggs or broilers.

Figure 14.15 shows the change from poultry breeds to crossmated lines and strains of birds. The incrossmated represents many of the synthetic egg-type strains or lines. These are primarily two- and three-way crosses of primarily Mediterranean breeds (e.g., White Leghorns). The crossmated chickens are the commercial broiler chickens that are primarily cornish-type males or White Plymouth females. The area of the chart depicting crossmated shows the tremendous changes of breeding methods utilized by the broiler industry.

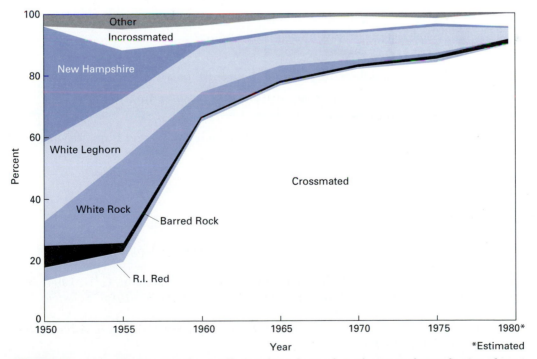

FIGURE 14.15 In poultry breeding the specific breeds are losing their identity in the production of crossmated and incrossmated lines. Data is from over 30 million birds recorded in the National Poultry Improvement Plan. Courtesy of the USDA.

CHAPTER SUMMARY

■ Genetic relationship estimates the genes two animals have in common because the same ancestors appear in the first six generations of their pedigrees.

■ Mating systems are identified by the genetic relationship of the animals being mated.

■ Inbreeding is the mating of animals more closely related than the average of the population, while outbreeding is the mating of animals not as closely related as the average of the population.

■ Inbreeding increases the genetic homozygosity while outbreeding increases the genetic heterozygosity.

■ Linebreeding is a mild form of inbreeding while maintaining a high genetic relationship to an outstanding ancestor.

■ Outcrossing is the mating of unrelated animals within the same breed.

■ Crossbreeding is the mating of animals from different breeds resulting in heterosis (hybrid vigor).

REVIEW QUESTIONS

1. Animals which are derived from matings within a single breed are called _____.
2. What is inbreeding?
3. How does inbreeding affect homozygosity of traits?
4. What is a major disadvantage of inbreeding?
5. What is a major advantage of inbreeding?
6. What is line breeding?
7. What is crossbreeding?
8. What are the two primary reasons for crossbreeding?
9. What is outcrossing?
10. What are newly developed breeds which are created from crossing several established breeds called?
11. What is grading up?

SELECTED REFERENCES

Bourdon, R. M. 1997. *Understanding Animal Breeding.* Upper Saddle River, NJ: Prentice-Hall, Inc.

Cundiff, L. V., Van Vleck, L. D., Young, L. D., and Dickerson, G. D. 1994. Animal Breeding and Genetics. *Encyclopedia of Agricultural Science.* San Diego: Academic Press, Inc.

Legates, J. E. 1990. *Breeding and Improvement of Farm Animals.* 4th ed. New York: McGraw-Hill.

CHAPTER 15

Nutrients and Their Function

A *nutrient* is any feed constituent that functions in the support of life. There are many different feeds available to animals to provide these nutrients.

Most animal feeds are classified as **concentrates** and **roughages.** Concentrates include cereal grains (e.g., corn, wheat, barley, oats, and milo), oil meals (e.g., soybean meal, linseed meal, and cottonseed meal), molasses, and dried milk products. Concentrates are high in energy, low in fiber, and highly (80–90%) digestible. Roughages include legume hays, grass hays, and straws, the latter being by-products from the production of grass, seed, and grain. Additional roughages are silage, stovers (dried corn, cane, or milo stalks and leaves with the grain portion removed), soilage (cut green feeds), and grazed forages. Roughages are less digestible than concentrates. Roughages are typically 50–65% digestible, but the digestibility of some straws is significantly lower.

NUTRIENTS

The six basic classes of nutrients—water, carbohydrates, fats, proteins, vitamins, and minerals—are found in varying amounts in animal feeds. Nutrients are composed of at least 20 of the more than 100 known chemical elements. These 20 elements and their chemical symbols are: calcium (Ca), carbon (C), chlorine (Cl), cobalt (Co), copper (Cu), fluorine (F), hydrogen (H), iodine (I), iron (Fe), magnesium (Mg), manganese (Mn), molybdenum (Mo), nitrogen (N), oxygen (O), phosphorus (P), potassium (K), selenium (Se), sodium (Na), sulfur (S), and zinc (Zn).

Water

Water contains hydrogen and oxygen. The terms **water** and **moisture** are used interchangeably. Typically, water refers to drinking water, whereas moisture is used in reference to the amount of water in a given feed or ration. The remainder of the feed, after accounting for moisture, is referred to as *dry matter.* Moisture is found in all feeds, ranging from 10% in air-dry feeds to more than 80% in fresh green forage. Livestock and poultry consume several times more water than dry matter each day and will die from lack of water more quickly than from lack of any other nutrient. Thus, water is considered to be the most important nutrient. Water in feed is no more valuable than water from any other source. This knowledge is important to assessing feeds that vary in their moisture content.

Water has important body functions. It enters into most of the metabolic reactions, assists in transporting other nutrients, helps maintain normal body temperature, and gives the body its physical shape (water is the major component within cells).

Carbohydrates

Carbohydrates contain carbon, hydrogen, and oxygen in either simple or complex forms. The more simple carbohydrates, such as starch found in grains, supply the major energy source for cattle diets, particularly feedlot diets. The more complex carbohydrates, such as cellulose found in roughages such as hay, are the major components of the cell walls of plants. These complex carbohydrates are not as easily digested as simple carbohydrates and they require host or microbial interaction for effective utilization.

Fats

Fats and oils, also referred to as **lipids,** contain carbon, hydrogen, and oxygen, although there is more carbon and hydrogen in proportion to oxygen than with carbohydrates. Fats are solid and oils are liquid at room temperature due generally to relative saturation. Fats contain 2.25 times more energy per pound than carbohydrates.

Most fats are comprised of three fatty acids attached to a glycerol backbone. **Example:**

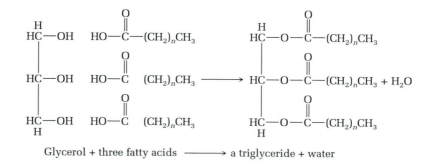

Glycerol + three fatty acids ⟶ a triglyceride + water

There are saturated and unsaturated fats, depending on their particular chemical composition. Saturated fatty acids have single bonds tying the carbon atoms together (e.g., —C—C—C—C—), whereas unsaturated fatty acids have one or more double bonds (e.g., =C=C—C=C—). The term *polyunsaturated fatty acids* is applied to those having more than one double bond. More than 100 fatty acids have been identified, but only four have been determined to be dietary essential (linoleic acid, linolenic acid, arachidonic acid, and alpha linolenic acid). The two apparent functions of the essential fatty acids are (1) as precursors of prostaglandins and (2) as structural components of cells.

Proteins

Proteins always contain carbon, hydrogen, oxygen, and nitrogen and sometimes iron, phosphorus or sulphur, or both. Protein is the only nutrient class that contains nitrogen. Proteins in feeds, on average, contain 16% nitrogen. This is why feeds are analyzed for the percent nitrogen in the feed, with the percent multiplied by 6.25 (100% ÷ 16% = 6.25) to convert it to percent protein. If, for example, a feed is 3% nitrogen, 100 g of the feed contains 3 g nitrogen. Multiplying 6.25 × 3 gives 18.75%, meaning that 100 g of this feed contains 18.75 g protein.

Proteins are composed of various combinations of some 25 amino acids. Amino acids are called the *building blocks* of the animal's body. The building blocks for growth (including growth of muscle, bone, and connective tissue), milk production, and cellular and tissue repair are amino acids that come from proteins in feed. The interstitial (between cells) fluid, blood, and lymph require amino acids to regulate body water and to transport oxygen and carbon dioxide. All enzymes are proteins, so amino acids are also required for enzyme production. Amino acids have an amino group (NH_2) in each of their chemical structures. There are many different combinations of amino acids that can be structured together. The chemical, or peptide, bonding of amino acids is illustrated using alanine and serine, which results in the formation of a dipeptide:

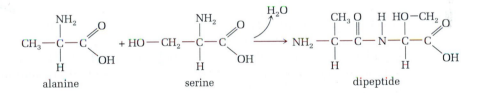

It can be seen that amino acids have both a basic portion, NH_2, and an acid portion,

and it is because of these that they can combine into long chains to make proteins. When digestion occurs, the action is at the peptide linkage to free amino acids from one another.

Some amino acids are provided in abundance in many proteins but others are quite limited. Thus, monogastric animals might have difficulty obtaining scarce amino acids that they cannot synthesize. Of the 15 nonessential amino acids, some (alanine, aspartic acid, glutamic acid, hydroxyproline, proline, and serine) can be synthesized by monogastric animals if carbon, hydrogen, oxygen, and nitrogen are available. Other amino acids, however, either cannot be synthesized at all by monogastric animals (isoleucine, leucine, lysine, methionine, phenylalanine, threonine, tryptophan, and valine) or are produced so slowly that they are called *semiessential* and must be provided in the feed of growing animals (arginine, histidine—also glycine for the chick).

All amino acids are needed by all animals; the terms *essential* and *nonessential* merely refer to whether or not they must be supplied through the diet. The 10 dietary essential amino acids are phenylalanine, valine, threonine, tryptophan, isoleucine, methionine, histidine, arginine, leucine, and lysine. The shortage of any particular amino acid can prevent an animal from using other amino acids for needed functions, and a protein deficiency results. Ruminants do not need a dietary supply of amino acids because the amino acids are synthesized in the ruminant stomach. However, the rumen microbes do have a requirement for nitrogen—e.g., urea.

Minerals

Chemical elements other than carbon, hydrogen, oxygen, and nitrogen are called *minerals.* They are inorganic because they contain no carbon; organic nutrients do contain carbon. Some minerals are referred to as *macro* (required in larger amounts) and others are *micro* or trace minerals (required in smaller amounts).

The macrominerals include calcium, chlorine, magnesium, phosphorus, potassium, sodium, and sulfur. Calcium and phosphorus are required in certain amounts and in a certain ratio (2 to 1) to each other for bone growth and repair and for other body functions. The blood plasma contains sodium chloride; the red blood cells contain potassium chloride. The osmotic relations between the plasma and the red blood cells are maintained by proper concentrations of the three electrolytes—sodium, potassium, and chloride. Potassium is found primarily within the cell, while sodium and chloride are outside the cell. Sodium chloride which is required by all animals is provided by salt. Sodium chloride may be depleted by excessive sweating that results from heavy physical work in hot weather. It is essential that salt and plenty of water be available under such conditions. The acid-base balance of the body is maintained at the proper level by electrolytes.

The essential microminerals for farm animals include cobalt, copper, fluorine, iodine, iron, manganese, molybdenum, selenium, and zinc. Microminerals may become a part of the molecule of a vitamin (e.g., cobalt is a part of vitamin B_{12}) and they may become a part of a hormone (e.g., thyroxine, a hormone made by the thyroid gland, requires iodine for its synthesis).

Hemoglobin of the red blood cells carries oxygen to tissues and carbon dioxide from tissues. Iron is required for production of hemoglobin because it is a part of the hemoglobin molecule. A small quantity of copper is also necessary for protection of hemoglobin (even though it does not normally become a part of the hemoglobin molecule in farm animals) because it apparently is necessary for normal iron absorption from the digestive tract and for release of iron to the blood plasma.

Certain important metabolic reactions in the body require the presence of minerals. Selenium and vitamin E both appear to work together to help prevent white muscle disease, which is a disease of the striated muscles, the smooth muscles, and cardiac muscles. Both vitamin E and selenium are more effective if the other is present. Excesses of certain minerals may be quite harmful. For example, excess amounts of fluorine, molybdenum, and selenium are highly toxic.

Vitamins

Vitamins are organic nutrients needed in very small amounts to provide for specific body functions in the animal. There are 16 known vitamins that function in animal nutrition. Vitamins may be classed as either *fat-soluble* or *water-soluble.*

The fat-soluble vitamins are vitamins A, D, E, and K. Vitamin A helps maintain proper repair of internal and external body linings and helps regulate growth. Because the eyes have linings, lack of vitamin A adversely affects the eyes. Vitamin A is also a part of the visual pigments of the eyes. Vitamin D is required for proper use of calcium and phosphorus in bone growth and repair. A major function of vitamin D is to regulate the absorption of calcium and phosphorus from the intestine. Vitamin D is produced by the action of sunlight on steroids of the skin; therefore, animals that are exposed to sufficient sunlight make all the vitamin D they need. Vitamin K is important in blood clotting; hemorrhage might occur if the body is deficient in vitamin K.

The water-soluble vitamins are ascorbic acid (vitamin C), choline, and the B-complex vitamins which include cyanacobalamin (B_{12}), folic acid, biotin, niacin, pantothenic acid, pyridoxine (B_6), riboflavin (B_2), and thiamin (B_1). More diseases caused by inadequate nutrition have been described in the human than in any other animal, and among the best known are those caused by a lack of certain vitamins: beri-beri (lack of thiamin); pellagra (lack of niacin); pernicious anemia (lack of vitamin B_{12}); rickets (lack of vitamin D); and scurvy (lack of vitamin C).

In ruminant animals, all of the water-soluble vitamins except vitamin C and choline are made by microorganisms in the rumen. Vitamin C is produced from glucose in the liver of farm animals. The B vitamins are also readily available to horses; produced by microbial fermentation in the cecum. Most fat-soluble vitamins are not synthesized by either ruminants or monogastrics and must be supplied in the diets of both groups (exceptions are vitamin K, which is synthesized by rumen bacteria in ruminants, and vitamin D through the skin). Many vitamins are supplied through feeds normally given to animals.

PROXIMATE ANALYSIS OF FEEDS

The nutrient composition of a feed cannot be determined accurately by visual inspection. A system has been devised by which the value of a feed can be approximated. **Proximate analysis** separates feed components into groups according to their feeding value. This analysis is based on a feed sample and analysis, and therefore is accurate only if a representative the sample is of the entire feed source.

The inorganic and organic components resulting from a proximate analysis are water, crude protein, crude fat (sometimes referred to as *ether extract*),

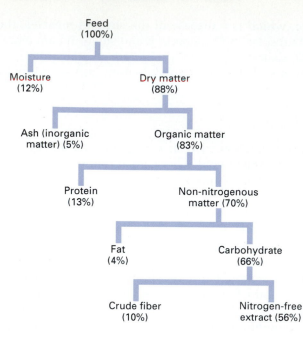

FIGURE 15.1 Proximate analysis showing the inorganic and organic components of a feed (similar to wheat) on a natural or air-dry basis. Courtesy of CSU Graphics.

crude fiber, nitrogen-free extract, and ash (minerals). Figure 15.1 shows these components resulting from a feed that would have a laboratory analysis of 88% dry matter, 13% protein, 4% fat, 10% crude fiber, and 56% nitrogen-free extract (NFE) on a natural or air-dry basis. The analysis might be reported on a dry-matter basis (no moisture) as 14.8% protein, 4.5% fat, 11.4% crude fiber, and 63.6% nitrogen-free extract. Therefore, caution needs to be exercised in interpreting the proximate analysis results because different laboratories may report their analytical values on either an air-dry (or as-fed) basis or a dry-matter basis.

The proximate analysis for the six basic nutrients does not distinguish the various components of a nutrient. For example, ash content of a feed does not tell the amount of calcium, phosphorus, or other specific minerals. Figure 15.2 gives the chemical analysis for organic and inorganic nutrients. There are specific chemical analyses for each of these nutrients in a feed if such an analysis is needed.

DIGESTIBILITY OF FEEDS

Digestibility refers to the amount of various nutrients in a feed that are digested in the digestive tract. Different feeds and nutrients vary greatly in their digestibility. Many feeds have been subjected to digestion trials, in which feeds of known nutrient composition have been fed to livestock and poultry. Feces have been collected and the nutrients in the feces analyzed. The difference between nutrients fed and nutrients excreted in the feces is the apparent digestibility of the feed.

For example, the digestibility of protein is obtained by determining the digestibility of nitrogen in a feed. Digestibility is expressed as a percentage of nitrogen, for example, as follows:

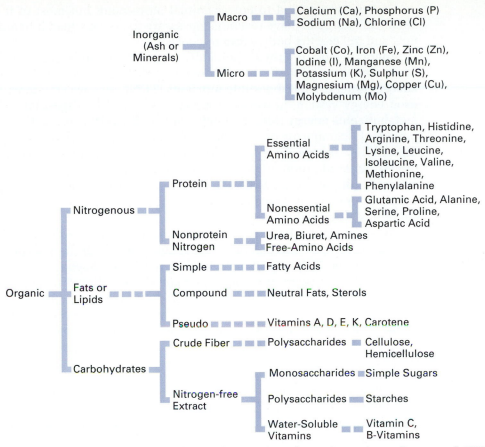

FIGURE 15.2 Chemical analysis scheme of inorganic and organic nutrients. Courtesy of CSU Graphics.

$$\frac{\text{Nitrogen in feed} - \text{Nitrogen in feces}}{\text{Nitrogen in feed}} \times 100 = \text{Percentage digestibility}$$

As an example, if 100 grams feed contains 3.2 grams nitrogen and 100 grams feces contains 0.8 grams nitrogen, the percent digestibility of nitrogen is:

$$\frac{3.2 - 0.8}{3.2} \times 100 = 75\%$$

Note that the determination of 3.2 grams nitrogen in 100 grams feed enables one to estimate the percentage of protein in the feed as 20% ($3.20 \times 6.25 = 20$).

ENERGY EVALUATION OF FEEDS

Nutrients that contain carbon provide the energy for animals. Carbohydrates, fats, and proteins can all be used to provide energy; however, carbohydrates supply most of the energy, as they are a more economical energy source than proteins. Complete oxidation (burning or taking on oxygen) of carbon releases the energy. Energy is used to drive a variety of body systems.

Energy can be used to power animal movement, but most of it is used as chemical energy to drive reactions necessary to convert feed into animal products and to keep the body warm or cool.

Energy needs of animals generally account for the largest portion of feed consumed. Several systems have been devised to evaluate feedstuffs for their energy content. **Total digestible nutrients (TDN)** once was the most commonly used energy system; however, it is being replaced by **digestible energy (DE)**, **metabolizable energy (ME)**, and **net energy (NE)**. TDN is typically expressed in pounds, kilograms, or percentages after obtaining the proximate analysis and digestibility figures for a feed. The formula for calculating TDN is TDN = (digestible crude protein) + (digestible crude fiber) + (digestible nitrogen-free extract) + (digestible crude fat × 2.25). The factor of 2.25 is used to equate fat to a carbohydrate basis, since fat has 2.25 times as much energy as an equivalent amount of carbohydrate.

An example of calculating TDN in 100 grams of feed is shown in Table 15.1.

TDN is roughly comparable to digestible energy (DE) but it is expressed in different units. TDN and DE both tend to overvalue roughages.

Even though there are some apparent shortcomings in using TDN as an energy measurement of feeds, it works well in balancing rations for cows and growing cattle. There is more precision in the energy measurement of feeds by using the net energy (NE) system. This system usually measures energy values in megacalories per pound or kilogram of feed. The calorie basis, which measures the heat content of feed, is as follows:

Calorie (cal)—amount of energy or heat required to raise the temperature of 1 gram of water 1°C.

Kilocalorie (kcal)—amount of energy or heat required to raise the temperature of 1 kilogram of water 1°C.

Megacalorie (Mcal)—equal to 1,000 kilocalories or 1,000,000 calories.

Figure 15.3 shows various ways that the energy of feeds is utilized by animals and the various energy measurements of feeds. Note that some energy in the feed is lost in the feces (not digested), the urine (digested but not used by the body cells), gases from microbial fermentation of the feed, and heat, resulting from digestion and metabolism of the feed.

TABLE 15.1   An Example of Calculating Total Digestible Nutrients (TDN)

Nutrient	Amount of Nutrient (g)	Digestibility (%)	Amount of Digestible Nutrient
Protein	20	75	15.00
Carbohydrates			
Soluble (NFE)	55	85	46.75
Insoluble (fiber)	10	20	2.00
Fat (5 × 2.25)	11.15	85	9.50
			TDN = 73.25

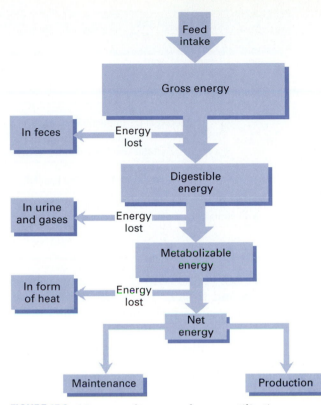

FIGURE 15.3 Measures of energy and energy utilization.

Feeds provide energy that the animal uses to supply two basic functions: (1) maintenance energy and (2) production energy (e.g., growth and lactation). Maintenance energy is used to keep the basal metabolism functioning, to provide for the voluntary activity of the animal, to generate heat to keep the body warm, and to provide energy to cool the body.

Production energy becomes stored energy and energy needed for work. The stored energy exists in fetus development, semen production, growth, fat deposition, and milk, eggs, and wool production.

Gross energy (*GE*) is the quantity of heat (calories) released from the complete burning of the feed sample in an apparatus called a *bomb calorimeter.* GE has little practical value in evaluating feeds for animals because the animal does not metabolize feeds in the same manner as a bomb calorimeter. For example, oat straw has the same GE value as corn grain. Digestible energy (DE) is GE of feed minus fecal energy. Metabolizable energy (ME) is GE of feed minus energy in the feces, urine, and gaseous products of digestion. Net energy (NE) is the ME of feed minus the energy used in the consumption, digestion, and metabolism of the feed. This energy lost between ME and NE is called **heat increment.**

Another way to illustrate the several measures of feed energy and how they are utilized is shown in Fig. 15.4. In this example, 2,000 kcal of GE (in approximately 1 lb of feed) is fed to a laying hen. The DE shows that 400 kcal was lost

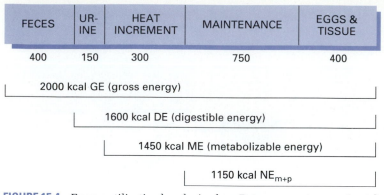

FECES	UR-INE	HEAT INCREMENT	MAINTENANCE	EGGS & TISSUE
400	150	300	750	400

2000 kcal GE (gross energy)

1600 kcal DE (digestible energy)

1450 kcal ME (metabolizable energy)

1150 kcal NE_{m+p}

FIGURE 15.4   Energy utilization by a laying hen. Data represent approximately 1 lb of feed containing 2,000 kcal of gross energy. Courtesy of CSU Graphics.

in the feces. The ME (1,450 kcal) is used for heat increment, maintenance, and production (eggs and tissue). There are 300 kcal lost in the heat increment, which leaves 1,150 kcal for maintenance and production. For maintenance, 750 kcal are needed, which leaves 400 kcal for egg production and tissue growth. Therefore only 20% (440/2,000) of the GE is used for production.

Net energy for maintenance (NEm) and *net energy for gain* (NEg) are more commonly used for formulating diets for feedlot cattle than any other energy system. *Net energy for lactation* (NEl) is used in dairy cow ration formulation. NEm in animals is the amount of energy needed to maintain a constant body weight. Animals of known weight, fed for zero energy gain, have a constant level of heat production.

The NEg measures the increased energy content of the carcass after feeding a known quantity of feed energy. All feed fed above maintenance is not utilized at a constant level of efficiency. Higher rates of gain require more feed per unit of gain as composition of gain varies with rate of gain.

The TDN and NE systems are compared in Table 15.2. Information in the table is based on feeding a simple diet of ground ear corn (90%) and supplement (10%) to a yearling steer (772 lb) for different rates of gain.

TABLE 15.2   Comparison of TDN and NE Systems for a Yearling Steer with Varying Rates of Gain

	Rate of Gain (lb/day)		
Energy System	0	2	2.9
TDN (lb)	6.40	12.80	15.00
NEm (Mcal)	6.24	6.24	6.24
NEg (Mcal)		4.29	6.48
Total lb feed			
TDN basis	8.2	16.5	19.3
NE basis	7.2	15.0	19.0
Feed per lb gain (NE)		7.5	6.6
Feed per lb gain (TDN)		8.2	6.6

FEEDS AND FEED COMPOSITION

Classification of Feeds

Feeds are naturally occurring ingredients in diets of farm animals used to sustain life. The terms *feeds* and *feedstuffs* are generally used interchangeably; however, *feedstuffs* is a more inclusive term. Feedstuffs can include certain nonnutritive products such as additives to promote growth and reduce stress, for flavor and palatability, to add bulk, or to preserve other feeds in the ration.

The National Research Council (NRC) classification of feedstuffs is as follows:

1. Dry roughages and forages

   Hay (legume and nonlegume)
   Straw
   Fodder
   Stover
   Other feeds with greater than 18% fiber (hulls and shells)

2. Range, pasture plants, and green forages

3. Silages (corn, legume, and grass)

4. Energy feeds (cereal grains, mill by-products, fruits, nuts, and roots)

5. Protein supplements (animal, marine, avian, and plant)

6. Mineral supplements

7. Vitamin supplements

8. Nonnutritive additives (antibiotics, coloring materials, flavors, hormones, preservatives, and medicants)

Roughages and forages are used interchangeably, although roughage usually implies a bulkier, coarser feed. In the dry state, roughages have more than 18% crude fiber. The crude fiber is primarily a component of cell walls that is not highly digestible. Roughages are also relatively low in TDN, although there are exceptions; for example, corn silage has over 18% crude fiber and approximately 70% TDN.

Feedstuffs that contain 20% or more protein, such as soybean meal and cottonseed meal, are classified as protein supplements. Feedstuffs having less than 18% crude fiber and less than 20% protein are classified as energy feeds or concentrates. Cereal grains are typical energy feeds, which is reflected by their high TDN values.

Nutrient Composition of Feeds

Feeds are analyzed for their nutrient composition, as discussed earlier. The ultimate goal of nutrient analysis of feeds is to predict the productive response of animals when they are fed rations of a given composition.

The nutrient composition of some of the more common feeds is shown in Tables 15.3 (ruminants) and 15.4 (monogastric animals). The information in Table 15.3 represents averages of numerous feed samples. Feeds are not constant in composition, and an actual analysis should be obtained whenever economically feasible. The actual analysis is not always feasible or possible because of lack of available laboratories and insufficient time to obtain the

TABLE 15.3 Nutrient Composition of Selected Feeds Commonly Used in Diets of Ruminants

Feed	Dry Matter (%)	TDN (%)	ME (Mcal/lb)	NEm (Mcal/lb)	NEg (Mcal/lb)	NEl (Mcal/lb)	Crude Protein (%)	Vitamin A (carotene) (mg/lb)	Calcium (%)	Phosphorus (%)
				On a Dry-Matter Basis (moisture-free)						
Alfalfa hay (early bloom)	90.0	57.0	0.94	0.55	0.25	0.59	18.4	57.8	1.25	0.23
Barley	89.0	83.0	1.36	0.97	0.64	0.85	13.0	—	0.09	0.47
Bermuda grass (grazed)	31.0	66.0	—	0.67	0.38	0.61	14.6	150.2	0.49	0.27
Bluegrass (grazed)	30.6	58.0	0.95	0.56	0.27	0.73	17.0	86.4	0.39	0.39
Bone meal (steamed)	95.0	16.0	—	—	—	—	12.7	—	30.51	14.31
Brome (grazed early)	32.5	63.0	1.04	0.62	0.34	0.62	20.3	208.8	0.59	0.37
Buffalo grass	47.7	56.0	0.92	0.54	0.23	0.60	9.2	42.6	0.52	0.16
Corn (no. 2 dent)	89.0	91.0	1.50	1.04	0.67	0.96	10.0	0.91	0.02	0.35
Corn (ground ear)	87.0	90.0	1.48	1.01	0.63	0.85	9.3	—	0.05	0.31
Corn silage	27.9	70.0	1.15	0.71	0.45	0.80	8.4	—	0.28	0.21
Corn stover (no ears, husks)	87.2	59.0	0.97	0.55	0.25	0.52	5.9	—	0.49	0.09
Cottonseed meal	93.5	74.0	1.22	0.75	0.49	0.77	42.4	—	0.20	1.09
Dicalcium phosphate	96.0	—	—	—	—	—	—	—	23.10	18.65
Grama (early vegetation)	41.0	64.0	1.05	0.63	0.36	0.50	13.1	—	0.54	0.19
Limestone (ground)	100.0	—	—	—	—	—	—	—	33.84	0.02
Meadow hay (native)	92.9	51.0	0.84	0.50	0.14	0.56	9.1	—	0.57	0.17
Milk	12.0	130.0	2.14	2.09	0.91	1.52	25.8	—	0.93	0.75
Milo (sorghum)	89.0	80.0	1.31	0.84	0.56	0.81	12.4	—	0.04	0.33
Molasses (cane)	75.0	72.0	1.25	2.27	1.48	0.77	4.3	—	1.19	0.11
Oats (grain)	89.0	76.0	1.25	0.79	0.52	0.80	13.2	—	0.11	0.39
Prairie hay (midbloom)	91.0	51.0	0.84	0.50	0.14	—	8.1	9.1	0.34	0.21
Sorghum, Sudan grass (grazed)	22.7	63.0	1.04	0.62	0.35	0.72	8.7	—	0.43	0.35
Soybean meal (solvent)	89.0	81.0	1.33	0.88	0.59	0.84	51.5	—	0.36	0.75
Wheat (hard, red winter)	89.1	88.0	1.45	0.98	0.64	0.92	14.6	—	0.06	0.57
Wheat (grazed early)	21.5	73.0	1.20	0.75	0.49	0.79	28.6	236.4	0.42	0.40
Wheat (straw)	90.1	48.0	0.79	0.47	0.09	0.37	3.6	1.0	0.17	0.08
Wheatgrass, crested (early)	30.8	67.0	1.10	0.66	0.40	0.57	23.6	197.1	0.46	0.35

Sources: Adapted from the National Research Council, 1989, and Preston, 1990.

TABLE 15.4 Nutrient Composition of Selected Feeds Commonly Used in Rations of Monogastric Animals (air-dry basis)

Feed	ME (Mcal/lb)	Protein (%)	Calcium (%)	Phosphorus (%)	Iron (mg/lb)	Manganese (mg/lb)	Zinc (mg/lb)	A (IU/lb)	Niacin (mg/lb)	Pantothenic Acid (mg/lb)	Riboflavin (mg/lb)	Choline (mg/lb)	B$_{12}$ (mg/lb)	Lysine (%)	Methionine (%)	Tryptophan (%)
			Minerals					Vitamins						Amino Acids		
Alfalfa meal (dehydrated)	1032	17.5	1.44	0.22	141	13	8	12,272	17	13	7	497	0.002	0.73	0.2	0.28
Barley	1304	11.6	0.05	0.36	23	4	8	—	29	4	1	450	—	0.40	0.2	0.14
Blood meal	876	85.0	0.30	0.25	1364	3	139	—	10	1	1	340	0.20	8.10	1.5	1.10
Bone meal	—	—	28.00	13.00	—	2	—	—	—	—	—	—	—	—	—	—
Corn	1511	8.8	0.02	0.28	16	2	4	454	15	3	0.5	241	—	0.24	0.2	0.05
Dicalcium phosphate	—	—	26.0	20.0	—	—	—	—	—	—	—	—	—	—	—	—
Feather meal	1032	86.4	0.20	0.80	—	10	—	—	12	4	1	405	0.27	1.10	0.4	0.50
Fish meal (menhaden)	1014	60.5	5.11	2.88	200	15	67	—	25	4	2	1389	0.07	4.83	1.8	0.68
Limestone	—	—	39.0	—	—	—	—	—	—	—	—	—	—	—	—	—
Meat and bone meal	1106	50.4	10.1	4.96	223	6	42	—	21	2	2	907	0.03	2.60	0.7	0.28
Milo (sorghum)	1468	8.9	0.28	0.32	18	6	6	—	19	5	0.5	308	—	0.22	0.1	0.10
Oats	1213	11.4	0.06	0.27	32	20	—	—	7	13	0.5	500	—	0.40	0.2	0.16
Skim milk (dried)	1527	33.5	1.28	1.02	23	1	18	—	5	15	10	568	—	2.40	0.9	0.44
Soybean meal (solvent)	1404	48.5	0.27	0.62	54	12	20	—	10	7	1	1295	—	3.18	0.7	0.67
Wheat, hard (red winter)	1464	14.1	0.05	0.37	23	28	6	—	25	6	2	495	—	0.40	0.2	0.18
Whey (dried)	1450	13.6	0.97	0.76	59	3	—	—	5	20	12	900	—	0.97	0.2	0.19

Source: Adapted from the National Research Council, 1989.

analysis. Therefore, feed-analysis tables become the next-best source of reliable information on nutrient composition of feeds. It is not uncommon to expect the following deviations of actual feed analysis from the table values for several feed constituents: crude protein (±15%), energy values (±10%), and minerals (±30%).

Digestible protein is included in some feed composition tables, but because of the large contribution of body protein to the apparent protein in the feces, digestible protein is more misleading than crude protein. For this reason, crude protein is more commonly found in feed-composition tables and used in formulating diets for ruminants.

Digestible protein (DP) can be calculated from crude protein (CP) content by using the following equation (%DP and %CP are on a dry-matter basis):

$$\% \ DP = 0.9 \ (\% \ DP) - 3$$

Five measures of energy values—TDN, ME, NEm, NEg, and NEl—are shown in Table 15.3. TDN is shown because there are more TDN values for feeds and because TDN has been a standard system of expressing the energy value of feeds. Some individuals seek ME (metabolizable energy) values for feed because these values are in calories rather than pounds. NEm and NEg values are used primarily to formulate feedlot diets and diets for growing replacement heifers, as these values offset the major problem associated with the TDN energy system. NEl is used in formulating diets for dairy cows.

CHAPTER SUMMARY

- Nutrients are feed constituents that support or sustain life.
- The six classes of nutrients are water, carbohydrates, fats, proteins, vitamins, and minerals.
- Carbohydrates and fats are the primary energy sources in feed.
- Proteins are composed of some 25 amino acids; the latter are known as the building blocks of the animal's body.
- Proteins contain 16% nitrogen and they comprise most of the muscle mass.
- Vitamins and minerals are required in smaller amounts than the other nutrient classes; however, they are necessary for certain metabolic reactions to occur.
- The proximate analysis of a feed sample identifies components that reflect their feeding value—that is, water, crude protein, fat (ether extract), crude fiber, nitrogen-free extract, and ash (minerals).
- Energy evaluation of feeds is measured by total digestible nutrients (TDN), digestible energy (DE), metabolizable energy (ME), and net energy (NE).

REVIEW QUESTIONS

1. Any feed constituent that functions in the support of life is a _____.
2. What are the two classes into which most animals feeds fall?

3. What are the six basic classes of nutrients?

4. What is dry matter of feed?

5. Fats and oils are referred to as _____.

6. What are the three types of carbohydrates?

7. What are the building blocks of protein?

8. What are the two classes of amino acids?

9. What is an essential amino acid?

10. What are the two general classes of mineral nutrients?

11. What are the two groups of vitamins?

12. What are the fat-soluble vitamins?

13. What are the water-soluble vitamins?

14. What is proximate analysis of feeds?

15. The amount of various nutrients in a feed which can be absorbed from the digestive tract is referred to as _____.

16. How is energy obtained from feed nutrients?

SELECTED REFERENCES

Publications

Cheeke, P. J. 1991. *Applied Animal Nutrition.* Englewood Cliffs, NJ: Prentice-Hall.

Church, D. C. 1991. *Livestock Feeds and Feeding.* 3d ed. Englewood Cliffs, NJ: Prentice-Hall.

Ensminger, M. E., Oldfield, J. E., and Heinemann, W. W. 1990. *Feeds and Nutrition.* 2d ed. Clovis, CA: Ensminger Publishing Co.

Jurgens, M. H. 1993. *Animal Feeding and Nutrition.* 5th ed. Dubuque, IA: Kendall-Hunt.

Kunkle, W. E. 1994. Feeds and Feeding. *Encyclopedia of Agricultural Science.* San Diego: Academic Press, Inc.

National Research Council. *Nutrient Requirements of Beef Cattle,* 1996; *of Dairy Cattle,* 1978; *of Goats,* 1981; *of Horses,* 1989; *of Poultry,* 1986; *of Sheep,* 1985; and *of Swine,* 1988. Washington, D.C.: National Academy Press.

Perry, T. W. 1994. Animal Nutrition, Principles. *Encyclopedia of Agricultural Science.* San Diego: Academic Press, Inc.

Pond, W. G., et al. 1995. *Basic Animal Nutrition and Feeding.* New York, NY: John Wiley and Sons.

Visuals

Introduction to Feeds and Feeding (VHS video), VEP, California Polytechnic State University, San Luis Obipso, CA 93407.

CHAPTER 16

Digestion and Absorption of Feed

Animals obtain substances needed for all body functions from the feeds they eat and the liquids they drink. Before the body can absorb and use them, feeds must undergo a process called *digestion.* Digestion includes mechanical action, such as chewing and contractions of the intestinal tract; chemical action, such as the secretion of hydrochloric acid (HCl) in the stomach and bile in the small intestine; and action of enzymes, such as maltase, lactase, and sucrase (which act on disaccharides), lipase (which acts on lipids), and peptidases (which act on proteins). Enzymes are produced either by the various parts of the digestive tract or by microorganisms. The role of digestion is to reduce feed particles to molecules so they can be absorbed into the blood and eventually support body functions.

CARNIVOROUS, OMNIVOROUS, AND HERBIVOROUS ANIMALS

Animals are classed as carnivores, omnivores, or herbivores according to the types of feed they normally eat. Carnivores, such as dogs and cats, normally consume animal tissues as their source of nutrients; herbivores, such as cattle, horses, sheep, and goats, primarily consume plant tissues. Humans and pigs are examples of omnivores, who eat both plant and animal products.

Carnivores and omnivores are monogastric animals, meaning that the stomach is simple in structure, having only one compartment. Some herbivores, such as horses and rabbits, are also monogastric. Other herbivores, such as cattle, sheep, and goats, are ruminant animals, meaning that the stomach is complex in structure, containing four compartments. Animals classified as carnivores, omnivores, and herbivores can utilize certain feeds they do not normally consume. For example, animal products can be fed to herbivorous ruminants, and certain cereal products can be fed to carnivores.

The digestive tracts of pigs and humans are similar in anatomy and physiology; therefore, much of the information gained from studies on pig nutrition and digestive physiology can be applied to the human. Both the pig and the human are omnivores and both are monogastric animals. Neither can synthesize the B-complex vitamins or amino acids to a significant extent. Both pigs and humans tend to eat large quantities, which can result in obesity. Humans can control obesity by controlling food intake and exercising as a means of using, rather than storing, excess energy. Obesity in swine can be controlled by limiting the amount of feed available to them and through genetic selection of leaner animals. The latter has received the greater emphasis as pigs are typically fed ad libitum.

DIGESTIVE TRACT OF MONOGASTRIC ANIMALS

The anatomy of the digestive tract varies greatly from one species of animal to another. The basic parts of the digestive tract are mouth, stomach, small intestine, and large intestine (or colon). The primary function of the parts preceding the intestines is to reduce the size of feed particles. The small intestine functions in splitting food molecules and in nutrient absorption, whereas the large intestine absorbs water and forms indigestible wastes into solid form called *feces.* In a mammal having a simple stomach (such as the pig), the mouth has teeth and lips for grasping and holding feed that is masticated (chewed), and salivary glands that secrete saliva for moistening feed so it can be swallowed.

Feed passes from the mouth to the stomach through the esophagus. A sphincter (valve) is at the junction of the stomach and esophagus. It can prevent feed from coming up the esophagus when stomach contractions occur. The stomach empties its contents into that portion of the small intestine known as the *duodenum.* The pyloric sphincter, located at the junction of the stomach and the duodenum, can be closed to prevent feed from moving into or out of the stomach. Feed goes from the duodenum to the jejunum portion and then to the ileum portion of the small intestine. It then passes from the small intestine to the large intestine, or colon. The ileocecal valve, located at the junction of the small intestine and the colon, prevents material in the large intestine from moving back into the small intestine.

The small intestine actually empties into the side of the colon near, but not at, the anterior end of the colon. The blind anterior end of the colon is the cecum, or, in some animals, the vermiform appendix. The large intestine empties into the rectum. The anus has a sphincter, which is under voluntary control so that defecation can be prevented by the animal until it actively engages in the process. The structures of the digestive system of the pig are shown in Fig. 16.1.

Animals such as pigs, horses, and poultry are classed as monogastric animals, but they differ markedly in certain ways. For example, the horse (Fig. 16.2) has a large structure called the **cecum,** where feed is fermented. Because the cecum is posterior to the area where most feed is absorbed, horses do not

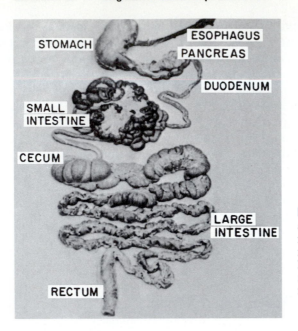

FIGURE 16.1 Digestive tract of the pig as an example of the digestive tract of a monogastric animal. Reprinted with permission from D. C. Church and W. G. Pond, *Basic Animal Nutrition and Feeding* (Corvallis, OR: D. C. Church, copyright © 1974).

obtain all of the nutrients made by microorganisms in the cecum. Digestive tract sizes and capacities of monogastric animals are contrasted in Table 16.1.

The digestive tracts of most poultry species differ from the pig in several respects. Because they have no teeth, poultry break their feed into a size that can be swallowed by pecking with their beaks or by scratching with their feet. Feed goes from the mouth through the esophagus to the crop, which is an enlargement of the esophagus where feed can be stored. Some fermentation may occur in the crop, but it does not act as a fermentation vat. Feed passes

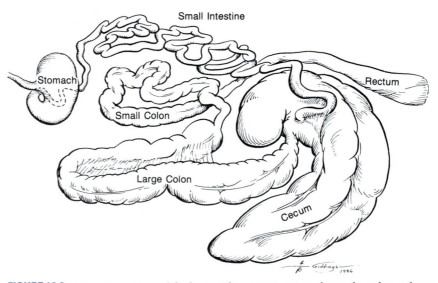

FIGURE 16.2 Digestive system of the horse. The posterior view shows the colon or large intestine proportionally larger than the rest of the digestive tract. Note particularly the location of the cecum at the anterior end of the colon. Drawing by Dennis Giddings.

TABLE 16.1 **Digestive Tract Sizes and Capacities of Selected Monogastric Animals**

Part of Digestive Tract	Species			
	Human	Pig	Horse	Chicken
Esophagus	—	—	4 ft	Total length of digestive
Stomach	1 qt	2 gal	4 gal	tract in mature chickens
Small intestine	1 gal	2 gal	12 gal	is approximately 7 ft
		60 ft	70 ft	(beak to crop, 7 in.; beak to
Large intestine	1 qt	3 gal	11 gal	proventriculus, 14 in.;
		12 ft	20 ft	duodenum, 8 in.;
Cecum			8 gal	ileum and jejunum,
			4 ft	48 in.; and cecum, 7 in.)

from the crop to the proventriculus, which is a glandular stomach in birds that secretes gastric juices and HCl but does not grind feed. Feed then goes to the gizzard, where it is ground into finer particles by strong muscular contractions. The gizzard apparently has no function other than to reduce the size of feed particles; birds from which it has been removed digest a finely ground ration. Feed moves from the gizzard into the small intestine. Material from the small intestine empties into the large intestine. At the junction of the small and large intestines are two ceca, which contribute little to digestion. Material passes from the large intestine into the cloaca, into which urine also empties. Material from the cloaca is voided through the vent (Fig. 16.3).

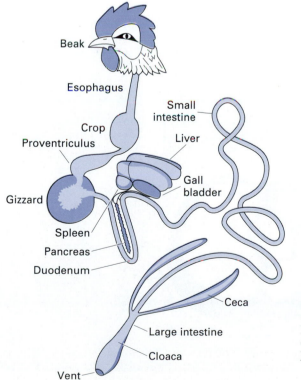

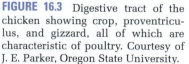

FIGURE 16.3 Digestive tract of the chicken showing crop, proventriculus, and gizzard, all of which are characteristic of poultry. Courtesy of J. E. Parker, Oregon State University.

STOMACH COMPARTMENTS OF RUMINANT ANIMALS

In contrast to the single stomach of monogastric animals, stomachs of cattle, sheep, and goats have four compartments—**rumen, reticulum, omasum,** and **abomasum** (Figs. 16.4 and 16.5). The rumen is a large fermentation vat where bacteria and protozoa thrive and break down roughages to obtain nutrients for their use. It is lined with numerous papillae, which give it the appearance of being covered with a thick coat of short projections. The papillae increase the surface area of the rumen lining. The microorganisms in the rumen can digest cellulose and can synthesize amino acids from nonprotein nitrogen as well as the B-complex vitamins. Later, these microorganisms are digested in the small intestine to provide these nutrients for the ruminant animal's use.

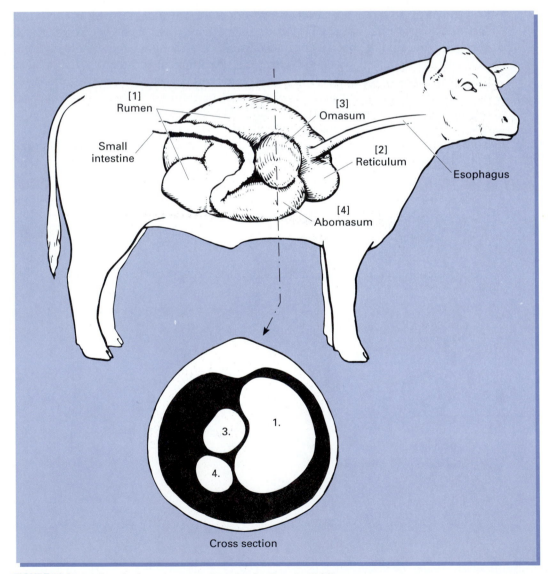

FIGURE 16.4   Beef cattle digestive tract. Drawing by Dennis Giddings.

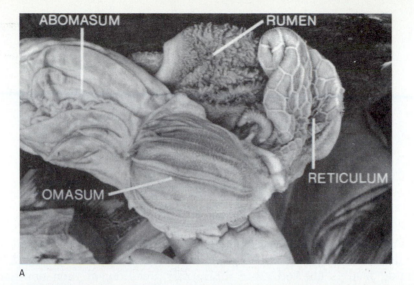

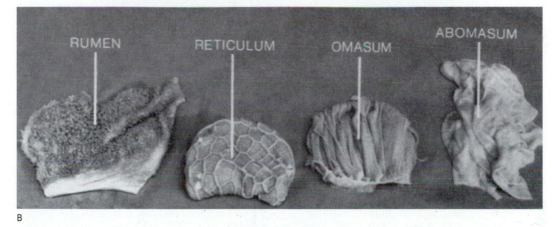

FIGURE 16.5 Lining of the four compartments of the ruminant stomach (goat). (A) Compartments connected. (B) Compartments separated. Courtesy of George F. W. Haenlein, University of Delaware.

The reticulum has a lining with small compartments similar to a honeycomb; thus it is occasionally referred to as the *honeycomb.* Its function is to interact with the rumen in initiating the mixing activity of the rumen and providing an additional area for fermentation. The omasum has many folds, so it is often called the *manyplies.* The omasum may not have a major digestive function, although some believe that the folds produce a grinding action on the feed. The abomasum, or true stomach, corresponds to the stomach of monogastric animals and performs a similar digestive function.

The size and capacity of the ruminant stomach and intestinal tract are given in Table 16.2. The data in Table 16.2 are for mature ruminants, as the relative proportions of the stomach compartments are considerably different in the young lamb and calf. At birth, the abomasum comprises 60% of the total stomach capacity, whereas the rumen is only 25% of the total.

TABLE 16.2   Digestive Tract Sizes and Capacities of Mature Ruminant Animals

Part of Digestive Tract	Species	
	Cow	Ewe
Stomach		
Rumen	40 gal	5 gal
Reticulum	2 gal	2 qt
Omasum	4 gal	1 qt
Abomasum	4 gal	3 qt
Small intestine	15 gal (130 ft)	2 gal (80 ft)
Large intestine	10 gal	6 qt

Animals having the four-compartment stomach eat forage rapidly, and later, while resting, they regurgitate each bolus of feed known as the **cud.** The regurgitated feed is chewed more thoroughly, swallowed, then another bolus is regurgitated and chewed. This process continues until the feed is thoroughly masticated. The regurgitation and chewing of undigested feed is known as **rumination.** Animals that ruminate are known as *ruminants.* As feeds are fermented by microorganisms in the rumen, large amounts of gases (chiefly methane and carbon dioxide) are produced. The animal normally can eliminate the gases by controlled belching, also called **eructation.**

DIGESTION IN MONOGASTRIC ANIMALS

Feed that is ingested (taken into the mouth) stimulates the secretion of saliva. Chewing reduces the size of ingested particles and saliva moistens the feed. The enzyme amylase, present in saliva of some species including pigs and humans, acts on starch. Ruminants do not secrete salivary amylase. However, little actual breakdown of starch into simpler compounds occurs in the mouth, primarily because feed is there for a short time.

An **enzyme** is an organic catalyst that speeds a chemical reaction without being altered by the reaction. Enzymes are rather specific; that is, each type of enzyme acts on only one or a few types of substances. Therefore, it is customary to name enzymes by giving the name of the substance on which it acts and adding the suffix *-ase,* which, by convention, indicates that the molecules so named are enzymes (Table 16.3). For example, lipase is an enzyme that acts on lipids (fats); maltase is an enzyme that acts on maltose to convert it into two molecules of glucose; lactase is an enzyme that acts on lactose to convert it into one molecule of glucose and one molecule of galactose; and sucrase is an enzyme that acts on sucrose to convert it into one molecule of glucose and one molecule of fructose. Some lipase is present in saliva but little hydrolysis of lipids into fatty acids and glycerides occurs in the mouth.

As soon as it is moistened by saliva and chewed, feed is swallowed and passes through the esophagus to the stomach. The stomach secretes HCl, mucus, and the digestive enzymes pepsin and gastrin. The strongly acidic environment in the stomach favors the action of pepsin. Pepsin breaks proteins down into polypeptides. The HCl also assists in coagulation, or curdling, of

TABLE 16.3   Important Enzymes in the Digestion of Feed

Enzyme	Substrate	Substances Resulting from Enzyme Action
Amylase	Starch	Disaccharides, dextrin
Chymotrypsin	Peptides	Amino acids and peptides
Lactase	Lactose	Glucose and galactose
Lipase	Lipids	Fatty acids and glycerides
Maltase	Maltose	Glucose and glucose
Pepsin	Protein	Polypeptides
Peptidases	Peptides	Amino acids
Sucrase	Sucrose	Glucose and fructose
Trypsin	Protein	Polypeptides

milk. Little breakdown of proteins into amino acids occurs in the stomach. Mucous secretions help to protect the stomach lining from the action of strong acids.

In the stomach, feed is mixed well and some digestion occurs; the mixture that results is called *chyme.* The chyme passes next into the duodenum, where it is mixed with secretions from the pancreas, bile, and enzymes from the intestine.

Secretion from the pancreas and discharge of bile from the gall bladder are stimulated by secretin, pancreozymin, and cholecystokinin, three hormones that are released from the duodenal cells. The enzymes from the pancreas are lipase, which hydrolyzes fats into fatty acids and glycerides; trypsin, which acts on proteins and polypeptides to reduce them to small peptides; chymotrypsin, which acts on peptides to produce amino acids; and amylase, which breaks starch down to disaccharides, after which the disaccharides are broken down to monosaccharides. The liver produces bile that helps emulsify fats; the bile is strongly alkaline and so helps to neutralize the acidic chyme coming from the stomach. Some minerals that are important in digestion also occur in bile.

By the time they reach the small intestine, amino acids, fatty acids, and monosaccharides (simple sugars or carbohydrates) are all available for absorption. Thus, the small intestine is the most important area for both digestion and absorption of feed. Absorption of feed molecules may be either passive or active. Passive passage results from diffusion, which is the movement of molecules from a region of high concentration of those molecules to a region of low concentration. Active transport of molecules across the intestinal wall may be accomplished through a process in which cells of the intestinal lining (**villi,** shown in Fig. 16.6) engulf the molecules and then actively transport these molecules to either the bloodstream or the lymph. Energy is expended in accomplishing the active transport of molecules across the gut wall.

When molecules of digested feed enter the capillaries of the blood system, they are carried directly to the liver. Molecules may enter the lymphatic system, after which they go to various parts of the body, including the liver. The liver is an extremely important organ both for metabolizing useful substances and for detoxifying harmful ones.

In some monogastric animals, such as the horse, postgastric (cecal) fermentation of roughages occurs. In these animals, the feed that can be digested by a monogastric animal is digested and absorbed before the remainder reaches the

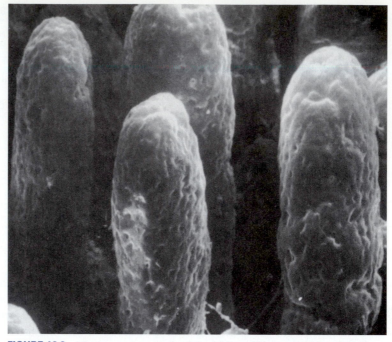

FIGURE 16.6   Electron micrograph of the lining of the small intestine. These projections (villi) increase the surface area and are covered with cells that digest and absorb nutrients from the feed. Magnification is approximately ×200. Courtesy of Dr. G. L. Waxler, 1972. *Am. J. Vet. Res.* 33:1323.

cecum. These animals are perhaps more efficient than ruminants in their use of feeds such as concentrates. In the ruminant animals, the concentrates given along with roughages are used by the bacteria and protozoa. Because the microorganisms in ruminants use starches and sugars, little glucose is available to ruminants for absorption. The microorganisms do provide volatile fatty acids, which are absorbed by the ruminant and converted to glucose as an energy source. The postgastric fermentation that occurs in horses breaks down roughages, but this takes place posterior to the areas where nutrients are most actively absorbed; consequently, all nutrients in the feed are not obtained by the animal in postgastric fermentation.

DIGESTION IN RUMINANT ANIMALS

In mature ruminant animals (cattle, sheep, and goats), predigestive fermentation of feed occurs in the rumen and reticulum. The bacteria and protozoa in the rumen and reticulum use roughages consumed by the animal as feed for their growth and multiplication; consequently, billions of these microorganisms develop. The rumen environment is ideal for microorganisms because moisture, a warm temperature, and a constant supply of nutrients are present. Excess microorganisms are continuously removed from the rumen and reticulum along with small feed particles that escape microbial fermentation and pass through the omasum into the abomasum. When feed passes into the abomasum, strong acids destroy the bacteria and protozoa. The ruminant animal then digests the microorganisms in the small intestine and uses them as a source of nutrients. The digested microbial cells provide the animal with most of its amino acid needs and some energy. Thus, ruminant animals and microorganisms mutually benefit each other. All digestive processes in ruminants are the same as those in monogastric animals after the feed reaches the abomasum, which corresponds to the stomach of monogastric animals.

The rumen fermentation process also produces **volatile fatty acids** (acetic, propionic, and butyric acids), which are waste products of microbial fermentation of carbohydrates. The animal then uses these volatile fatty acids (VFA) as its major source of energy. In the process of fermenting feeds, methane gas is also produced by the microorganisms. The animal releases the gas primarily through belching. Occasionally, the gas-releasing mechanism does not function properly and gas accumulates in the rumen, causing **bloat** to occur. Death will occur owing to suffocation if gas pressure builds to a high level and interferes with adequate respiration.

A young, nursing ruminant consumes little or no roughage. Consequently, at this early stage of life, its digestive tract functions similar to a monogastric animal. Milk is directed immediately into the abomasum in young ruminants by the **esophageal groove** (Fig. 16.7). The sides of the esophageal groove extend upward by a reflex action and form a tube through which milk passes directly from the esophagus to the abomasum. This allows milk to bypass fermentation in the rumen. Rumen fermentation is an inefficient use of energy and protein in a high-quality feed such as milk.

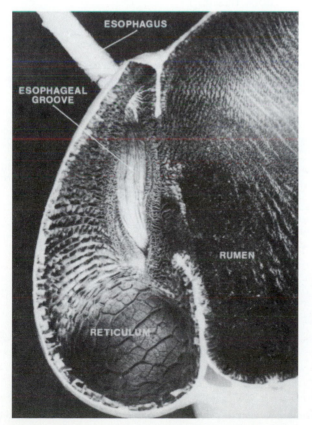

FIGURE 16.7 The esophageal groove with its location relative to the esophagus, reticulum, and rumen. Courtesy of N. J. Benevenga et al., 1969. Preparation of the ruminant stomach for classroom demonstration. *J. Dairy Sci.* 52:1294.

When roughage is consumed it is directed into the rumen, where bacteria and protozoa break it down into simple forms for their use. The rumen starts to develop functionally as soon as roughage enters it, but some time is required before it is completely functional. Complete development of the rumen, reticulum, and abomasum requires about 2 months in sheep and about 3–4 months in cattle. One can influence development of the rumen by the type of feed one gives to the animal. If only milk and concentrated feeds are given, the rumen shows little development. If very young ruminants are forced to live on forage, the rumen develops much more rapidly.

Energy Pathways

Figure 16.8 shows the digestion and utilization of carbohydrates and fats contained in the ingested forages and grains. The primary energy end products of glucose and fatty acids supply energy in the body tissues and become milk fat and lactose in the lactating ruminant. Excess energy is stored as body fat in the body tissues.

The primary organs and tissues in energy metabolism are shown in Fig. 16.8. These are the rumen, abomasum, small intestines, liver, blood vessels, mammary gland, and body tissues. Undigested carbohydrates (primarily complex carbohydrates such as lignin) are excreted through the large intestine.

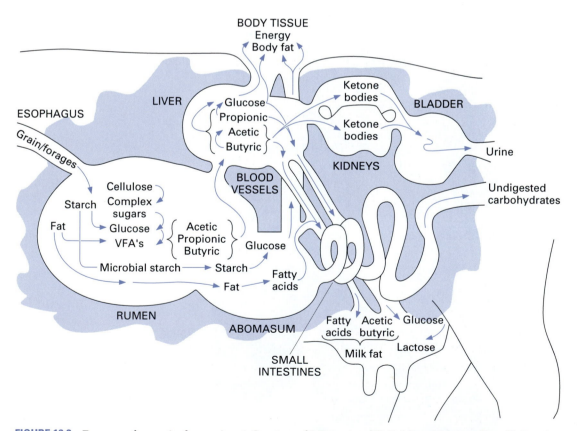

FIGURE 16.8   Energy pathways in the ruminant. Courtesy of J. Bryant and B. R. Moss, Montana State University.

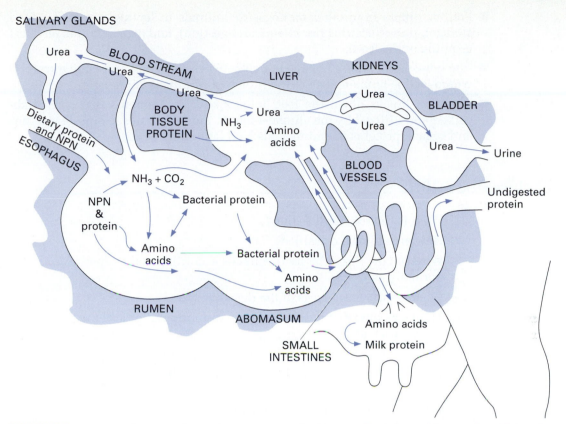

FIGURE 16.9 Protein pathways in the ruminant. Courtesy of J. Bryant and B. R. Moss, Montana State University.

Other energy waste products such as ketone bodies are excreted through the kidneys in the urine.

Protein Pathways

The digestion, utilization, and excretion of dietary protein and **nonprotein nitrogen** (**NPN**) are shown in Fig. 16.9. The end products of protein and NPN are amino acids, ammonia (NH_3), and synthesized amino acids. Excess NH_3 can be formed into urea in the liver, then excreted through the urine, with some urea returning to the rumen as a component of saliva.

CHAPTER SUMMARY

- The digestive tract parts of a monogastric animal are the mouth, esophagus, stomach, small intestine, and large intestine (colon).
- Stomachs of ruminant animals have four compartments—rumen, reticulum, omasum, and abomasum. The latter functions like the stomach of a monogastric animal.
- Microorganisms in the rumen digest cellulose, synthesize amino acids, and synthesize B vitamins.

- Poultry differs from other monogastric animals in having a crop (for food storage), proventriculus (for chemical digestion), and a gizzard (for physical grinding of the feed).
- The horse has a cecum (a large pouch between the small and large intestine) where feed is fermented as in the rumen of ruminant animals.
- Digestion breaks down the feed into smaller components that can be absorbed and utilized by the animal—that is, carbohydrates to simple starches and sugars, proteins to amino acids, and fat to fatty acids.

REVIEW QUESTIONS

1. What is the role of digestion?
2. How are animals classified?
3. What are the three classes of animals according to the types of feed they normally eat?
4. What are the basic parts of the digestive tract?
5. The large intestine absorbs water and forms indigestible wastes into a solid form called _____.
6. What are the three regions of the small intestine?
7. *True or False:* The stomach of ruminants such as cattle, sheep, and goats has four compartments.
8. What is the function of the rumen?
9. What structure of the colon, in some species such as horses, performs the function of microbial fermentation of feed?
10. What feed component is most effectively digested by fermentation in the rumen and cecum?
11. What is the significance of fermentative digestion of cellulose by microorganisms in the rumen and cecum?
12. Why are ruminants more efficient at utilizing roughages than are animals, such as horses, which possess a large active cecum?
13. What are the three regions of the large intestine?
14. What is the glandular stomach in birds called?
15. What is the crop in birds?
16. *True or False:* Mechanical digestion in birds occurs in the mouth, just as it does in mammals.
17. What are the four stomach compartments in ruminants?
18. Ruminants regurgitate undigested roughage which is referred to as _____ , or chewing the _____.
19. _____ are organic catalysts that speed up a chemical reaction without being altered by the reaction and are important in digestion of specific nutrients.
20. What is the function of the esophageal groove?
21. What are the products of fermentative digestion in the rumen which ruminant animals utilize as energy?

SELECTED REFERENCES

Publications

Cheeke, P. J. 1991. *Applied Animal Nutrition.* Englewood Cliffs, NJ: Prentice-Hall.

Church, D. C. 1991. *Livestock Feeds and Feeding.* Englewood Cliffs, NJ: Prentice-Hall.

Ensminger, M. E., Oldfield, J. E., and Heinemann, W. W. 1990. *Feeds and Nutrition.* 2d ed. Clovis, CA: Ensminger Publishing Co.

Jurgens, M. H. 1993. *Animal Feeding and Nutrition.* 5th ed. Dubuque, IA: Kendall/Hunt.

Kunkle, W. E. 1994. Feeds and Feeding. *Encyclopedia of Agricultural Science.* San Diego: Academic Press, Inc.

Pond, W. G., et al. 1995. *Basic Animal Nutrition and Feeding.* New York, NY: John Wiley and Sons.

Visuals

Introduction to Livestock Nutrition (VHS video). VEP, California Polytechnic State University, San Luis Obispo, CA 93407.

CHAPTER 17

Providing Nutrients for Body Functions

Feeding animals is of fundamental importance to any livestock or poultry production program because animals must be healthy to function efficiently and yield maximum profits to the producer. The basic task of the producer, then, is to supply animals with feed that satisfies their body functions for maintenance, growth, fattening, reproduction, lactation, egg laying, wool production, and work. Each of these functions has a unique set of nutrient requirements, and they are additive when more than one function is occurring.

Profits derived from any feeding program must be assessed against production costs. Knowledgeable producers can increase their profits by feeding their animals both adequately and economically.

NUTRIENT REQUIREMENTS FOR BODY MAINTENANCE

Body maintenance requires that nutrients are supplied to keep the body functioning in a state of well-being. There is no gain or loss of weight or production. Maintenance functions that have a high priority for nutrients are (1) body tissue repair, (2) control of body temperature, (3) energy to keep all vital organs (respiratory, digestive, and so on) functioning, and (4) water balance maintenance.

Nutrients first meet maintenance needs before supplying any of the other body functions. Approximately half of all feed fed to livestock and poultry is used to fill the maintenance requirement. Feedlot animals on full feed may only use 30–40% of their nutrients for maintenance, while some mature breeding animals may need 90% of their feed for maintenance. Highly efficient dairy cows, producing over 100 lb of milk per day, have a daily feed consumption four to five times their maintenance requirement.

Body Size and Maintenance

Maintenance needs are related to body size. A large animal obviously needs more feed than a small one, but maintenance requirements are not linearly related to body weight. Small animals require more feed per unit of body weight for maintenance than large ones. The approximate maintenance requirement in relation to weight is expressable as $Wt^{0.75}$ rather than $Wt^{1.00}$. Thus, if a 500-lb animal requires 15 lb of feed per day for maintenance, a 1,000-lb animal of the same type does not require twice as much feed even though the latter animal weighs twice as much as the first. The quantity of $1,000^{0.75}$ can be determined and applied to show that the 1,000-lb animal requires approximately 1.7 times as much feed for maintenance as the 500-lb animal. The 1,000-lb animal requires, therefore, approximately 25.5 lb daily (15 lb × 1.7 = 25.5 lb). Table 17.1 shows how the TDN requirement changes with increasing weight. Where corn is 91% TDN, on an air-dry basis, it would take approximately 12.5 lb (12.5 lb × 0.91 = 11.4 lb TDN) of corn to fill the maintenance requirement of a 1,000-lb animal.

NUTRIENT REQUIREMENTS FOR GROWTH

Growth occurs when protein synthesis is in excess of its breakdown. Growth occurs when cells increase in number or size, or when a combination of both takes place. Growth at the tissue level is accomplished primarily through the building of muscle, bone, and connective tissue.

There are several important nutrient requirements for growth, including energy, protein, minerals, vitamins, and energy. The dry matter of muscle and connective tissue is composed largely of protein; therefore, young, growing animals that need feed to sustain growth in addition to maintenance have greater protein requirements. A young, growing animal is like a muscle-building factory, and protein in the feed is the raw material for the manufacturing process. If provided with only a maintenance amount of feed for an extended period, a young animal may be permanently stunted.

Monogastric animals not only need a certain quantity of protein but they also must have certain amino acids for proper growth. The protein needs of hogs, for example, are usually supplied by feeding them soybean meal as a supplemental source of amino acids. Young ruminant animals cannot consume enough roughage to make maximum growth. If young ruminant animals are

TABLE 17.1 TDN Needed for Maintenance of Cattle in the Growing-Finishing Period

Weight of Cattle (lb)	TDN Needed Daily for Maintenance (lb)
400	5.7
600	7.7
800	9.7
1,000	11.4
1,100	12.3

being nursed by dams that produce adequate amounts of milk, the young will do well on good pastures, good quality hay, or both together.

The mineral needs of a young, growing animal include calcium and phosphorus for proper bone growth, salt or a normal sodium level in the body, and any mineral that may be deficient in the area in which the animal lives. Calcium is usually plentiful in legume forages, and phosphorus is usually plentiful in grains, so a combination of hay and grain should provide all the calcium and phosphorus that young ruminant animals need. Animals fed on hay alone may need additional phosphorus, and those fed diets high in concentrates may need additional calcium. Some producers feed a mixture of steamed bone meal or dicalcium phosphate and salt at all times to ensure that their animals have the necessary calcium and phosphorus.

Two minerals, iodine and selenium, require special consideration. Some areas may be deficient in one or both of these elements. An insufficient amount of iodine in the ration of pregnant females might cause an iodine deficiency in the fetus, which prevents thyroxine from being produced and thus causes a goiter in the newborn. Young animals with goiters die shortly after they are born. Iodized salt can be easily provided to the pregnant females to avoid the iodine deficiency.

A lack of selenium might cause the young to be born with white muscle disease; it can be prevented by giving the pregnant female an injection of selenium. The injectable selenium is distributed commercially and directions for proper dosages that are supplied by the distributor should be followed closely because an overdose in natural feeds or supplements can kill the animal.

Vitamins are needed by young, growing animals. Young ruminant animals are usually on pasture with their dams and thus are exposed to sunshine. The action of ultraviolet rays from the sun converts steroid in the skin into vitamin D, providing the animal with this vitamin. Vitamin D is needed for the proper use of calcium and phosphorus in bone growth. Because pigs, poultry, and rabbits are often raised inside, where sunshine is limited or lacking, they need a dietary source of vitamin D.

Most vitamins must be supplied to pigs and poultry through feeds. The only vitamin commonly fed to ruminant animals is vitamin A, and then only when they are on dry pasture or they are fed hay that is quite mature or hay that has been moistened in processing and has, therefore, been dried in the sun for several days. Vitamin A is easily lost in sunlight and during extended dry storage. The activity of vitamin A in silage is often quite high because this vitamin is usually preserved by the acid fermentation that takes place when silage is made. Silage, however, is usually quite low in vitamin D because the plants used in making silage do not make this vitamin through the action of sunlight and silages are not exposed to sunlight for long periods after they are cut.

Young animals need sufficient energy to sustain their growth, high metabolic rate, and activities. They obtain some energy from their mother's milk, and additional energy is supplied as carbohydrates (starch and sugar) and fats from grazed forage or supplemental feeds. Feed grains are high in carbohydrates and also contain some fats. Young ruminants on good pasture typically obtain sufficient energy from pasture and from milk.

The energy needs of young pigs and poultry are generally supplied by feeding them grains such as corn, barley, or wheat. Young horses can usually obtain their energy needs when they are on pasture with their dams since most mares produce much milk; however, for adequate growth after weaning, they need some grain in addition to good pasture or good quality hay.

NUTRIENT REQUIREMENTS FOR FATTENING

Fattening is the storing of surplus feed energy as fat both within and around body tissues. Fattening is desirable to give meat some of its palatability characteristics and to provide energy reserves for postpartum reproductive performance.

Fattening is the result of excess energy from carbohydrates, fats, or protein above that required for maintenance and growth. Usually fattening animals are full-fed high-energy rations during the last phase of the growing-finishing feeding program.

Gain from growth is usually less costly than gain from fattening. It takes 2.25 times the energy to produce a pound of fat compared with a pound of protein tissue.

NUTRIENT REQUIREMENTS FOR REPRODUCTION

The requirements for **reproduction** fall into two categories—those for (1) gamete production and (2) fetal growth in the uterus. In general, healthy males and females are capable of producing gametes. The energy needs for germ cell production are no greater than those needed to keep animals in a normal, healthy condition. For example, ruminant animals grazing on pastures of mixed grass and legumes are generally neither deficient in phosphorus nor lacking in fertility. A lack of phosphorus may cause irregular estrous cycles and impaired breeding in females.

Animals that are losing weight rapidly because of poor feed conditions and animals that are overly fat may be low in fertility. To attain optimum fertility from female animals, they should be in moderate body fat condition (BCS 5) as breeding season approaches, but should, ideally, be increasing in condition (e.g., gaining weight) for 2–4 weeks before and during the breeding season.

The nutrients required by the growing **fetus** are much greater in the last trimester of pregnancy than earlier, as little fetal growth occurs during the first two trimesters of pregnancy. Because the fetus is growing, its requirements are the same as those for growth of a young animal after it is born. Healthy females can withdraw nutrients from their bodies to support the growing fetus temporarily while the amount or quality of their feed is low, but reproductive performance will be lower if nutrition is inadequate for 2–3 months in cattle and a few weeks in swine and sheep.

NUTRIENT REQUIREMENTS FOR LACTATION

Among common farm animals, dairy cows and dairy goats produce the most milk; however, most all females are expected to produce milk for their young. Milk production requires considerable protein, minerals, vitamins, and energy. The need for protein is greater because milk contains more than 3% protein. As an example, a cow that weighs 1,500 lb and produces 50 lb of milk per day needs at least 30 lb of feed per day, which contains 15% protein; this gives her 4.5 lb protein for her body and the milk she produces. If the protein she eats is 60% digestible, there is 2.7 lb of digestible protein, of which 1.5 lb is present in her milk. If a cow of this weight is to produce 100 lb of milk per day, she must now consume 50 lb of feed containing 15% protein to compensate for the 3 lb of protein in the 100 lb of milk that she gives. Generally, during peak milk production, feed consumption cannot compensate for nutrient output and the cow does mobilize some body protein. Actually, more body energy is mobilized than body protein to meet the nutrient deficit.

Calcium and phosphorus are the two most important minerals needed for lactation. Milk is rich in these minerals; their absence or imbalance may result in decreased lactation or even cause death. The dairy cow may develop milk fever shortly after calving if there is an exceptionally heavy drain of calcium from her system. Cows afflicted by milk fever might become comatose and die if not treated. An intravenous injection of calcium gluconate usually helps the cow recover in less than a day. Milk fever rarely occurs in species or animals that produce relatively small quantities of milk.

Dairy cows produce milk that contains considerable quantities of vitamin A and most B-complex vitamins. Because cows are ruminants, it is unnecessary to feed them B-complex vitamins, and they require vitamin D supplementation only if confined indoors.

Exceptions occur with beef cows that give large amounts of milk. If they do not have adequate feed while they are nursing their calves, they may not conceive for the next calf crop. In those parts of the world where sheep and dairy goats are the principal dairy animals, the energy needs of these animals are quite similar to those of dairy cows that produce much milk.

The requirements for milk production in sows are usually provided by increasing the percentage of protein in the ration, by increasing the amount of feed allowed, and by providing a mineral mix (a combination of minerals that usually contains calcium, phosphorus, salt, and some trace minerals).

Energy is perhaps the most vital requirement for the production of much milk. The energy need is based on the amount of milk being produced. A lactating cow needs energy for body maintenance while she also produces milk and provides the energy stored in it. She cannot eat enough hay to obtain the quantity of energy needed so she must receive high-energy feeds such as concentrates; even then her production may be limited by the amount of feed she can eat. A high-producing dairy cow may need three to four times the energy of a nonlactating cow of the same size. Even when fed large amounts of concentrates, a cow that is producing much milk often loses weight and body condition because she cannot consume enough feed to produce at her maximum level; therefore, she draws on her body reserves to supply part of her energy needs.

In the dairy cow, the roughage-to-concentrate ratio should be approximately 40:60, as a certain amount of roughage is needed to maintain the desired fat content in milk. Therefore, simply feeding more concentrates is not the only answer to increased milk production.

NUTRIENT REQUIREMENTS FOR EGG LAYING

The nutrient requirements of poultry are dependent on the specific purposes of production. For example, broilers need nutrients primarily for growth with less emphasis on egg production; Leghorn-type hens (layers) need nutrients with primary emphasis on eggs and less on growth. Nutrient requirements for growth are discussed earlier in this chapter.

Leghorn-type chickens are smaller in body size than broilers so their maintenance requirements are less. They are prolific in egg production so they are usually fed **ad libitum** during the growing and laying period. Because layers eat ad libitum to satisfy their energy needs, rations need to have adequate concentrations of energy, protein (amino acids), vitamins, and minerals.

NUTRIENT REQUIREMENTS FOR WOOL PRODUCTION

Nutrient requirements for wool production are in addition to nutrients needed for maintenance, growth, and reproduction. Insufficient energy owing to amount or quality of feed is usually the most limiting nutritional factor affecting wool production. As wool fibers are primarily protein in composition, the ration should be adequate in protein content.

Shearing removes the natural insulation and may cause an increase in energy requirements owing to heat loss. This is especially true when periods of cold weather occur shortly after shearing.

NUTRIENT REQUIREMENTS FOR WORK

Animals used for work, either for pulling heavy loads or for being ridden, require large amounts of energy in addition to the needs for maintenance. Horses are the primary work animals in the United States, but elsewhere, donkeys, cattle, and water buffaloes are used.

Horses, mules, and donkeys rely partly on perspiration to remove nitrogenous wastes. If a horse is used for hard work for five days of the week and is not allowed to exercise the next two days, a strain is placed on the kidneys and illness may result.

The primary requirement is energy above that needed for maintenance and growth. If energy in the ration is not sufficient to meet the work needs, then body fat stores will provide the additional energy needs.

RATION FORMULATION

This chapter is not intended to cover the details of ration formulation. Attempting to expose the reader to this area without providing the details can be misleading. Books are written and courses taught on feeds and feeding or animal nutrition that provide an in-depth coverage of this topic. The reader who desires to pursue ration formulation should refer to the references at the end of this chapter.

Chapters 15, 16, and 17 give a brief background on nutrition, which leads into ration formulation. The primary objective of ration formulation is economically matching the animal's nutrient requirements (Tables 17.2–17.8) with the available feeds, taking into consideration the nutrient content of the feeds. Additional considerations are the palatability of the ration, physical form of the feed, and other factors that affect feed consumption.

TABLE 17.2   Daily Nutrient Requirements of Sheep (dry-matter basis)

Weight (lb)	Gain (lb)	Dry Matter (lb)	TDN (lb)	ME (Mcal)	Crude Protein (lb)	Ca (g)	P (g)	Vitamin A (IU)	Vitamin E (IU)
Ewes (maintenance)									
110	0.02	2.2	1.2	2.0	0.21	3.0	2.8	2,350	15
176	0.02	2.9	1.6	2.6	0.27	3.3	3.1	3,760	20
Ewes (nonlactating and first 15 weeks of gestation)									
110	0.07	2.6	1.5	2.4	0.25	3.0	2.8	2,350	18
176	0.07	3.3	1.8	3.0	0.31	3.3	3.1	3,760	22
Ewes (last 4 weeks of gestation with 200% lambing rate)									
110	0.5	3.7	2.4	4.0	0.43	4.1	3.9	4,250	26
176	0.5	4.4	2.9	4.7	0.49	4.8	4.5	6,800	30
Ewes (first 8 weeks of lactation, suckling singles)									
110	−0.06	4.6	3.0	4.9	0.67	10.9	7.8	4,250	32
176	−0.06	5.7	3.7	6.1	0.76	12.6	9.0	6,800	39
Ewes (first 8 weeks of lactation, suckling twins)									
110	−0.13	5.3	3.4	5.6	0.86	12.5	8.9	5,000	36
176	−0.13	6.6	4.3	7.0	0.96	14.4	10.2	8,000	45
Lambs (finishing)									
66	0.65	2.9	2.1	3.4	0.42	4.8	3.0	1,410	20
88	0.60	3.5	2.7	4.4	0.41	5.0	3.1	1,880	24
110	0.45	3.5	2.7	4.4	0.35	5.0	3.1	2,350	24

Source: National Research Council, *Nutrient Requirements of Sheep,* 1985.

TABLE 17.3   Daily Nutrient Requirements (NRC) for Breeding Heifers and Cows

Weight (lb)	Daily Gain (lb)	Minimum Dry-Matter Consumption (lb)	Crude Protein (lb)	TDN (lb)	ME (Mcal)	Ca (g)	P (g)
Pregnant Heifers—Last Month of Pregnancy							
750	0.7	20.0	1.6	10.8	10.4	29	21
900	0.9	23.0	1.8	12.4	11.7	32	24
Cows Nursing Calves—Average Milking Ability[a]—First 3 Months Postpartum							
1,000	0	23.0	1.9	12.5	20.9	24	17
1,200	0	26.0	2.1	13.9	23.4	27	19
1,400	0	28.9	2.3	15.4	25.7	30	21
Cows Nursing Calves—Superior Milking Ability[b]—First 3 Months Postpartum							
1,000	0	25.4	2.6	14.9	24.9	35	24
1,200	0	28.4	2.8	16.4	27.3	37	25
1,400	0	31.3	3.0	17.8	29.7	40	27
Dry Pregnant Mature Cows—Month 3 of Pregnancy							
1,000	0	16.7	1.2	8.2	13.4	14	14
1,200	0	19.5	1.4	9.5	15.6	17	17
1,400	0	23.3	1.6	11.4	18.7	21	21
Dry Pregnant Cows—Month 8 of Pregnancy							
1,000	0.9	21.0	1.6	10.9	18.3	23	14
1,200	0.9	24.1	1.8	12.6	20.9	27	17
1,400	0.9	27.0	2.1	14.2	23.9	32	21

[a] Ten pounds of milk per day (equivalent of approximately 450 lb of calf at weaning if there is adequate forage).
[b] Twenty pounds of milk per day (equivalent of approximately 650 lb of calf at weaning if there is adequate forage).
Source: National Research Council, *Nutrient Requirements of Beef Cattle,* 1996.

TABLE 17.4 Daily Nutrient Requirements for Growing-Finishing Heifers and Steers

Weight[a] (lb)	Daily Gain (lb)	Minimum Dry-Matter Consumption (lb)	Protein (%)	Crude Protein (lb)	NEm (Mcal)	NEg (Mcal)	TDN (lb)	Ca (g)	P (g)
Growing-Finishing Heifer Calves (medium-frame)									
500	1.0	11.8	9.4%	1.10	4.84	3.81	10.1	21	17
500	2.0	11.8	11.4	1.35	4.84	5.37	11.9	27	21
800	1.0	16.7	8.1	1.36	6.24	2.52	11.2	15	15
800	2.0	16.8	9.0	1.51	6.24	6.91	15.2	21	20
1,000	2.0	19.8	8.1	1.61	7.52	6.71	16.3	19	19
Growing-Finishing Steer Calves (medium-frame)									
500	1.0	12.3	9.5%	1.16	4.84	2.53	8.8	18	16
500	2.0	13.1	11.4	1.49	4.84	3.33	9.9	22	19
500	3.0	11.8	14.4	1.69	4.84	4.17	10.4	26	21
800	2.0	18.6	9.2	1.72	6.89	5.33	15.0	21	20
800	3.0[b]	16.8	10.8	1.81	6.89	7.80	17.0	26	23
1,000	2.0	22.0	8.4	1.85	8.14	7.73	18.1	21	21
1,000	3.0[b]	19.8	9.5	1.88	8.14	8.47	19.2	22	22

[a] Average weight for a feeding period.
[b] Most steers of the weight indicated, and not exhibiting compensatory growth, fail to sustain the energy intake necessary to maintain this rate of gain for an extended period.
Source: National Research Council, *Nutrient Requirements of Beef Cattle*, 1984.

TABLE 17.5 Daily Nutrient Requirements of Dairy Cows

Body Weight (lb)	NE₁ (Mcal)	TDN (lb)	Crude Protein (lb)	Ca (lb)	P (lb)	Vitamin A (IU)
Mature Lactating Cows (maintenance)						
800	7.16	6.9	0.70	0.029	0.024	30
1,100	8.46	8.1	0.80	0.044	0.031	38
1,300	9.70	9.3	0.89	0.053	0.037	46
1,550	10.89	10.5	0.99	0.062	0.044	53
Mature Dry Cows (last 2 months of gestation)						
800	9.30	9.1	1.96	0.057	0.035	30
1,100	11.00	10.8	2.32	0.073	0.044	38
1,300	12.61	12.4	2.66	0.086	0.053	46
1,550	14.15	13.9	2.98	0.101	0.062	53
Milk Production (meal or lb of nutrient per lb of milk for various fat percentages)						
Percentage fat						
3.0	0.291	0.282	0.077	0.0025	0.00170	
3.5	0.313	0.304	0.082	0.0026	0.00175	
4.0	0.336	0.326	0.087	0.0027	0.00180	
4.5	0.354	0.344	0.092	0.0028	0.00185	
5.0	0.377	0.365	0.098	0.0029	0.00190	

Source: National Research Council, *Nutrient Requirements of Dairy Cattle*, 1989.

TABLE 17.6 Daily Nutrient Requirements (or percent of ration) for Swine

| | Growing-Finishing (fed ad libitum) | | | Breeding Swine | |
				Bred Gilts and Sows (4 lb daily air-dry feed intake)	Lactating Gilts and Sows (10.5 lb daily air-dry feed intake)
Live Weight (lb)	11–22	48–110	110–242		
Expected Daily Gain (lb)	0.55	1.54	1.81		
Expected FE (feed/gain)	1.84	2.71	3.79		
ME (kcal, daily)	1,490	6,200	10,185	5,836	15,320
Crude protein, lb(%)[a]	0.20(20)	0.63(15)	0.89(13)	(12)	(13)
Indispensable amino acids					
Lysine g (%)	5.3(1.15)	14.3(0.75)	18.7(0.60)	(0.43)	(0.60)
Arginine g (%)	2.3(0.50)	4.3(0.05)	3.1(0.10)	(0)	(0.40)
Histidine g (%)	1.4(0.31)	4.2(0.22)	5.6(0.18)	(0.15)	(0.25)
Isoleucine g (%)	3.0(0.15)	8.7(0.46)	11.8(0.38)	(0.30)	(0.39)
Leucine g (%)	3.9(0.85)	11.4(0.60)	15.6(0.50)	(0.30)	(0.48)
Methionine plus cystine g (%)	2.7(0.58)	7.8(0.41)	10.6(0.34)	(0.23)	(0.36)
Phenylalanine plus tyrosine g (%)	4.3(0.94)	12.5(0.66)	17.1(0.55)	(0.45)	(0.70)
Threonine g (%)	3.1(0.68)	9.1(0.48)	12.4(0.40)	(0.30)	(0.43)
Tryptophan g (%)	0.8(0.17)	2.3(0.12)	3.1(0.10)	(0.09)	(0.12)
Valine g (%)	3.1(0.68)	9.1(0.48)	12.4(0.40)	(0.32)	(0.60)
Minerals (selected)					
Calcium g (%)	3.7(0.80)	11.4(0.60)	15.6(0.50)	(0.75)	(0.75)
Phosphorus g (%)	3.0(0.65)	9.5(0.50)	12.4(0.40)	(0.60)	(0.60)
Sodium g (%)	0.5(0.10)	1.9(0.10)	3.1(0.10)	(0.15)	(0.20)
Chlorine g (%)	0.4(0.08)	1.5(0.08)	2.5(0.08)	(0.12)	(0.16)
Magnesium g (%)	0.2(0.04)	0.8(0.04)	1.2(0.04)	(0.04)	(0.04)
Iron mg	46	114	124	144	380
Zinc mg	46	114	155	90	238
Manganese mg	2	4	6	18	48
Vitamins (selected)					
A IU	1,012	2,470	4,043	7,200	9,500
D IU	101	285	466	360	950
E IU	7	21	34	18	47.5
Riboflavin mg	1.61	4.57	6.22	5.4	14.2
Niacin mg	6.90	19	21.77	18	47.5
Pantothenic acid mg	4.60	15.20	21.77	21.6	57
B_{12} mg	8.05	19	15.55	27	71.2

[a] Pounds per day or percentage of the ration.
Source: National Research Council, *Nutrient Requirements of Swine,* 1988.

Additional material on feeding farm animals can be found in the following chapters on individual species: beef cattle (Chapter 23), dairy cattle (Chapter 25), swine (Chapter 27), sheep (Chapter 29), horses (Chapter 31), poultry (Chapter 32), and goats (Chapter 33).

NUTRIENT REQUIREMENTS OF RUMINANTS

The following are some major comparisons that demonstrate changes in nutrient requirements for maintenance growth, lactation, and reproduction:

1. For ewes (maintenance), note the increased requirement in dry matter, energy (TDN or ME), protein, calcium, phosphorus, vitamin A, and vitamin D as body weight changes from 110 to 176 lb (Table 17.2). More nutrients are needed to maintain a heavier body weight.

TABLE 17.7   Daily Nutrient Requirements for Horses

	Weight (lb)	Daily Gain (lb)	Daily Feed[a] (lb)	Digestible Energy (Mcal)	Crude Protein (g)	Ca (g)	P (g)	Vitamin A (1000 IU)
Ponies (approximate mature weight, 440 lb)								
Maintenance	440	0	8.2	7.4	296	8	6	6
Mares, last 90 days of gestation	440	0.60	8.1	8.2	361	16	12	12
Lactating mares, first 3 months (18 lb of milk per day)	440	0	11.0	13.7	688	27	18	12
Yearling (12 months of age)	308	0.44	6.4	8.7	392	12	7	6
Horses (approximate mature weight, 880 lb)								
Maintenance	880	0	13.9	13.4	536	16	11	12
Mares, last 90 days of gestation	880	1.17	13.7	14.9	654	28	21	24
Lactating mares, first 3 months (26 lb of milk per day)	880	0	17.1	22.9	1,141	45	29	24
Yearling (12 months of age)	583	0.88	10.9	15.6	700	23	13	12
Horses (approximate mature weight, 1,320 lb)								
Maintenance	1,320	0	18.8	19.4	776	24	17	18
Mares, last 90 days of gestation	1,320	1.47	18.5	21.5	947	41	31	36
Lactating mares, first 3 months (40 lb of milk per day)	1,320	0	26.0	33.7	1,711	67	43	36
Yearling (12 months of age)	847	1.32	14.8	22.7	1,023	36	20	17

[a] Air-dry basis.

Source: National Research Council, *Nutrient Requirements of Horses,* 1989.

2. In Table 17.2, compare the requirements under ewes (last 6 weeks of gestation) with the requirements of ewes (maintenance) and ewes (nonlactating and first 15 weeks of gestation). During the latter part of gestation, there are greater nutrient demands owing to rapid fetal growth. During the latter part of gestation compared with maintenance only, energy and protein requirements almost double, with mineral and vitamin requirements showing significant increases.

3. During lactation, nutrient requirements are even higher than those during gestation. Even with larger amounts of dry matter being supplied, the ewes lose weight. Compare the requirements of ewes nursing single lambs versus twins and note the even higher requirements.

4. The nutrient requirements for lambs being finished for slaughter show the gains approximately the same with increased nutrient requirements due to increased body weight. The increased nutrient requirements are primarily due to an increased maintenance requirement.

CHAPTER SUMMARY

- Nutrients are used by animals for maintenance (maintain body functions), growth, fattening, reproduction, egg laying, wool production, and work.

- Maintenance involves no gain or loss of body weight. Maintenance needs increase as body weight increases.

TABLE 17.8   Nutrient Requirements for Leghorn-Type Chickens and Broilers (as percentages or as milligrams or units per kilogram [2.2 lb] of diet)

| | Leghorn-Type Chickens | | | | Broilers | |
| | Growing | | Laying | | | |
Energy Base kcal ME/kg Diet[a]	0–6 Weeks 2,900	14–20 Weeks 2,900	2,900	Daily Intake per Hen (mg)[b]	0–3 Weeks 3,200	6–8 Weeks 3,200
Protein (%)	18	12	14.5	16,000	23.0	18.0
Arginine (%)	1.00	0.67	0.68	750	1.44	1.00
Glycine and serine (%)	0.70	0.47	0.50	550	1.50	0.70
Histidine (%)	0.26	0.17	0.16	180	0.35	0.26
Isoleucine (%)	0.60	0.40	0.50	550	0.80	0.60
Leucine (%)	1.00	0.67	0.73	800	1.35	1.00
Lysine (%)	0.85	0.45	0.64	700	1.20	0.85
Methionine plus cystine (%)	0.60	0.40	0.55	600	0.93	0.60
Methionine (%)	0.30	0.20	0.32	350	0.50	0.32
Phenylalanine plus tyrosine (%)	1.00	0.67	0.80	880	1.34	1.00
Phenylalanine (%)	0.54	0.36	0.40	440	0.72	0.54
Threonine (%)	0.68	0.37	0.45	500	0.80	0.68
Tryptophan (%)	0.17	0.11	0.14	150	0.23	0.17
Valine (%)	0.62	0.41	0.55	600	0.82	0.62
Linoleic acid (%)	1.00	1.00	1.00	1,100	1.00	1.00
Calcium (%)	0.80	0.60	3.40	3,750	1.00	0.80
Phosphorus, available (%)	0.40	0.30	0.32	350	0.45	0.35
Potassium (%)	0.40	0.25	0.15	165	0.40	0.30
Sodium (%)	0.15	0.15	0.15	165	0.15	0.15
Chlorine (%)	0.15	0.12	0.15	165	0.15	0.15
Magnesium (mg)	600	400	500	55	600	600
Manganese (mg)	60	30	30	3.30	60.0	60.0
Zinc (mg)	40	35	50	5.50	40.0	40.0
Iron (mg)	80	60	50	5.50	80.0	80.0
Copper (mg)	8	6	6	0.88	8.0	8.0
Iodine (mg)	0.35	0.35	0.30	0.03	0.35	0.35
Selenium (mg)	0.15	0.10	0.10	0.01	0.15	0.15
Vitamin A (IU)	1,500	1,500	4,000	440	1,500	1,500
Vitamin D (ICU)	200	200	500	55	200	200
Vitamin E (IU)	10	5	5	0.55	10	10
Vitamin K (mg)	0.50	0.50	0.50	0.055	0.50	0.50
Riboflavin (mg)	3.60	1.80	2.20	0.242	3.60	3.60
Pantothenic acid (mg)	10.0	10.0	2.20	0.242	10.0	10.0
Niacin (mg)	27.0	11.0	10.0	1.10	27.0	11.0
Vitamin B_{12} (mg)	0.009	0.003	0.004	0.00044	0.009	0.003
Choline (mg)	1,300	500	?	?	1,300	500
Biotin (mg)	0.15	0.10	0.10	0.011	0.15	0.10
Folacin (mg)	0.55	0.25	0.25	0.0275	0.55	0.25
Thiamin (mg)	1.8	1.3	0.80	0.088	1.80	1.80
Pyridoxine (mg)	3.0	3.0	3.0	0.33	3.0	2.5

[a] These are typical dietary energy concentrations.
[b] Assumes an average daily intake of 110 g of feed/hen.
Source: National Research Council, *Nutrient Requirements of Poultry,* 1986.

■ Growth occurs when protein synthesis is in excess of protein breakdown and loss.

■ Fattening occurs when the energy in feeds (primarily from carbohydrates and fats) exceeds the body's need for maintenance and growth.

■ Rations are formulated to match the animal's nutrient requirements with the nutrient content of available feeds.

REVIEW QUESTIONS

1. What are nutrients needed for?

2. What body maintenance functions have a high priority for nutrients?

3. *True or False:* Small animals require less nutrients for body maintenance than large animals.

4. *True or False:* Diets of primarily roughage are adequate to meet nutrient requirements for growth in young ruminants.

5. What are the two categories of nutrient requirements for reproduction?

6. When is the nutrient requirement for fetal growth greatest?

7. What are the two most important minerals needed for lactation?

8. What is the most vital requirement for production of large quantities of milk during lactation?

SELECTED REFERENCES

Publications

Cheeke, P. J. 1991. *Animal Nutrition.* Englewood Cliffs, NJ: Prentice-Hall.

Church, D. C. 1991. *Livestock Feeds and Feeding.* Englewood Cliffs, NJ: Prentice-Hall.

Ensminger, M. E., Oldfield, J. E., and Heinemann, W. W. 1990. *Feeds and Nutrition.* 2d ed. Clovis, CA: Ensminger Publishing Co.

Jurgens, M. H. 1993. *Animal Feeding and Nutrition.* 5th ed. Dubuque, IA: Kendall/Hunt.

National Research Council. *Nutrient Requirements of Beef Cattle,* 1996; *of Dairy Cattle,* 1989; *of Goats,* 1981; *of Horses,* 1989; *of Poultry,* 1986; *of Sheep,* 1985; and *of Swine,* 1989. Washington, DC: National Academy Press.

National Research Council. 1982. *United States—Canadian Tables of Feed Composition.* Washington DC: National Academy Press.

Pond, W. G. 1995. *Basic Animal Nutrition and Feeding.* New York, NY: John Wiley and Sons.

Visuals

Decision Making: Energy Utilization for Feedlot Cattle (videotape). CEV, P.O. Box 65265, Lubbock, TX 79464.

Feeds for Horses. 1996. CEV, P.O. Box 65265, Lubbock, TX 79464.

Introduction to Livestock Feeding (video). CEV, Inc., 5147-A 69th St., Lubbock, TX 79464.

CHAPTER 18

Growth and Development

Profitable and efficient production of livestock and poultry involves understanding their growth and development. Manipulation of genetic and environmental factors can change growth patterns in farm animals.

Many aspects of growth and development are contained in other chapters of this book. The material on reproduction, genetics, nutrition, and products should be integrated with this chapter to understand more fully animal growth and development.

Generally, **growth** is an increase in body weight until mature size is reached. This growth is an increase in cell size and cell numbers with protein deposition resulting. More specifically, growth is an increase in the mass of structural tissue (bone, muscle, and connective tissue) and organs accompanied by a change in body form and composition.

Development is defined as the directive coordination of all diverse processes until maturity is reached. It involves growth, cellular differentiation, and changes in body shape and form. In this chapter, growth and development are combined and discussed as one entity.

PRENATAL (LIVESTOCK)

The three phases of prenatal life—sex cells, the embryo, and the fetus—are briefly discussed in Chapter 10. Embryological development is a fascinating process; a spherical mass of cells differentiates into specific cell types and eventually into recognizable organs (Figs. 18.1 and 18.2). The endoderm (Fig. 18.2D) differentiates into the digestive tract, lungs, and bladder; the mesoderm (Fig. 18.2D) into the skeleton, skeletal muscle, and connective tissue; the ecto-

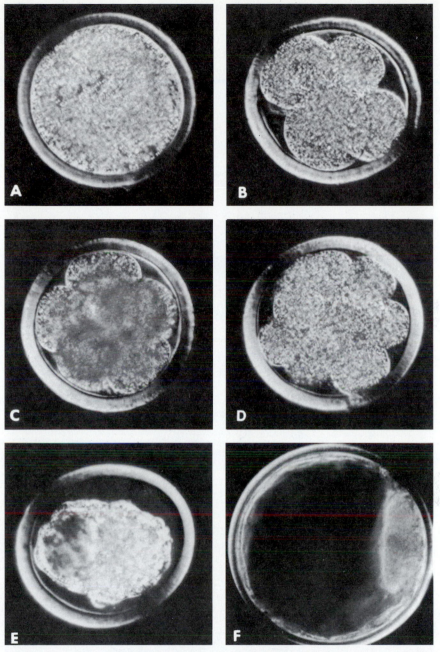

FIGURE 18.1 Embryonic development during the first 8 days of pregnancy in cattle. (A) Unfertilized egg. (B) Four-cell embryo on day 2 of pregnancy (estrus = day 0). (C) Eight-cell embryo on day 3 of pregnancy. (D) An 8- to 16-cell embryo on day 4 of pregnancy. (E) Very early blastocyst stage (approximately 60 cells). (F) Expanded blastocyst (>100 cells) recovered from the uterus on day 8 of pregnancy. Magnification approximately ×300. Courtesy of G. E. Seidel, Jr., 1981, Superovulation and embryo transfer in cattle, *Science* 211:351. Copyright © 1981 by the American Association for the Advancement of Science.

derm (Fig. 18.2D) into the skin, hair, brain, and spinal cord. The growth, development, and differentiation processes, involving primarily protein synthesis, are directed by DNA chains of chromosomes and the organizers in the developing embryo (see Chapter 12). Thus the nucleus is a center of activity for different types of cells, directing the growth and development process (Fig. 18.3).

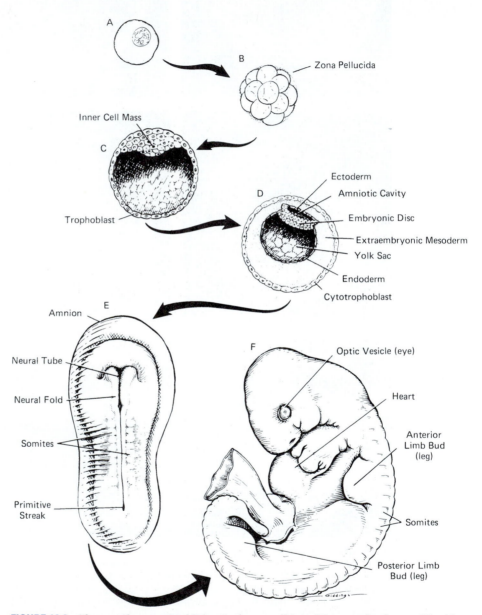

FIGURE 18.2   The morphogenesis of (A) a single egg cell into (B) a morula, then to (C) a blastocyst. (D) The stage at which the two cavities have formed in the inner cell mass; an upper (amniotic) cavity and a lower cavity yolk sac. The embryonic disc containing the ectoderm and endoderm germ layers is located between cavities. (E) A cattle embryo showing the neural tube and somites. (F) The development of the 14-day cattle embryo. Drawing by Dennis Giddings.

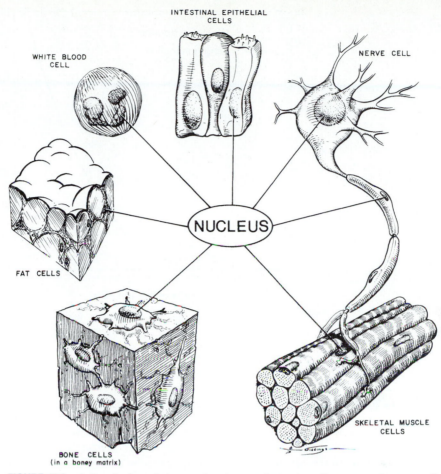

FIGURE 18.3 Cells of selected tissues showing similarity in cell structure but not cell shape. The nucleus gives direction to differentiation of cells and their function. Drawing by Dennis Giddings.

The fetus undergoes marked changes in shape and form during prenatal growth and development. Early in the prenatal period, the head is much larger than the body. Later, the body and limbs grow more rapidly than other parts. The order of tissue growth follows a sequential trend determined by physiological importance, starting with the central nervous system and progressing to bones, tendons, muscles, intermuscular fat, and subcutaneous fat.

During the first two-thirds of the prenatal period, most of the increase in muscle weight is due to hypertrophy (increase in size of fibers). During the last 3 months of pregnancy, hyperplasia (increase in number of fibers) represents most of the muscle growth. Individual muscles vary in their rate of growth, with larger muscles (those of the legs and back) having the greatest rate of postnatal growth. Water content of fetal muscle declines with fetal age, and this decline in water content continues through postnatal growth as well.

The relative size of the fetus changes during gestation, with the largest increase in weight occurring during the last trimester of pregnancy (Fig. 18.4).

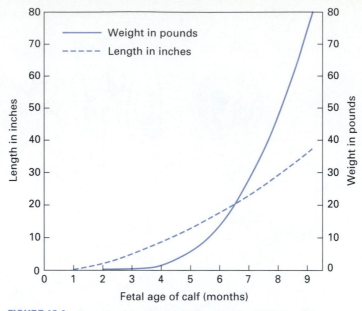

FIGURE 18.4   Growth of the fetal calf. Drawing by CSU Graphics.

BIRTH (LIVESTOCK)

After birth, the number of muscle fibers does not appear to increase significantly; therefore, postnatal muscle growth is primarily by hypertrophy. In red meat animals, all muscle fibers appear to be red at birth, but shortly thereafter some of them differentiate into white and intermediate muscle types.

At birth, the various body parts have considerably different proportions when compared with mature body size and shape. At birth, the head is relatively large, the legs are long, and the body is small; in the mature animal, the head is relatively small, the legs are relatively short, and the body is relatively large. Birth weight represents approximately 5–7% of the mature weight, while leg length at birth is approximately 60%; height at withers is approximately 50% of those same measurements at maturity. Hip width and chest width at birth are approximately one-third of the same measurements at maturity. This shows that the distal parts (leg and shoulder height) are developed earlier than proximal parts (hips and chest).

POULTRY

Embryonic Development

The development of a chick differs from that of mammals because there is no connection with its mother. The chick develops in the egg, entirely outside the hen's body. Embryonic development is much more rapid in chicks than in farm mammals.

Every egg whether fertile or nonfertile has a germ spot called the *blastoderm* (see Chapter 6, Fig. 6.1). This is where the chick embryo develops if the fertile egg is properly incubated.

A controlled environment must be maintained during incubation to produce live chicks from the fertile eggs. The major components of the controlled environment are (1) temperature (99.5–100°F); (2) 60–75% relative humidity; (3) turning the egg every 1–8 hours; and (4) providing adequate oxygen.

Three hours after fertilization, the blastoderm divides to form two cells. Cell division occurs until maturity except during the holding period before incubating the eggs.

There are four membranes that are essential to the growth of the chick embryo (Fig. 18.5). The **allantois** is the membrane that allows the embryo to breathe. It takes oxygen through the porous shell and oxygenates the blood of the embryo. The allantois removes the carbon dioxide, receives excretions from the kidneys, absorbs albumen used as food for the embryo, and absorbs calcium from the shell for use by the embryo. The **amnion** is a membrane filled with a colorless fluid that serves as a protection from the mechanical shock. The **yolk sac** is a layer of tissue growth over the surface of the yolk. This tissue has special cells that digest and absorb the yolk material for the developing embryo. The **chorion** surrounds both the amnion and yolk sac.

Table 18.1 identifies some of the primary changes in the growth and development of the chick embryo. Many of these phenomenal changes occur rapidly, sometimes in only hours.

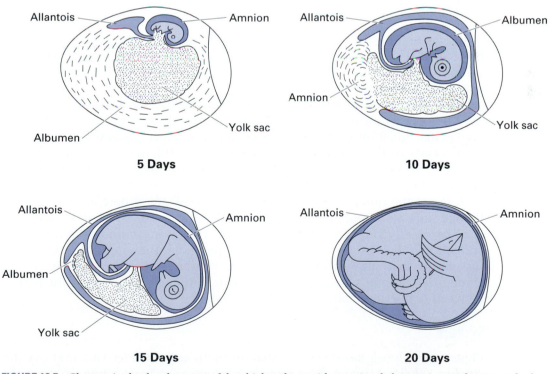

FIGURE 18.5 Changes in the development of the chick embryo with associated changes in membranes and other contents of the egg.

TABLE 18.1   Major Changes in Weight, Form, and Function of the Chick Embryo (White Leghorn) During Incubation

Day	Weight (g)	Developmental Changes
1	00.0002	Head and backbone are formed; central nervous system begins
2	00.0030	Heart forms and starts beating; eyes begin formation
3	00.0200	Limb buds form
4	00.0500	Allantois starts functioning
5	00.1300	Formation of reproductive organs
6	00.2900	Main division of legs and wings; first movements noted
7	00.5700	
8	01.1500	Feather germs appear
9	01.5300	Beak begins to form; embryo begins to look birdlike
10	02.2600	Beak starts to harden; digits completely separated
11	03.6800	
12	05.0700	Toes fully formed
13	07.3700	Down appears on body; scales and nails appear
14	09.7400	Embryo turns its head toward blunt end of egg
15	12.0000	Small intestines taken into body
16	15.9800	Scales and nails on legs and feet are hard; albumen is near gone; yolk is main food
17	18.5900	Amniotic fluid decreases
18	21.8300	
19	25.6200	Yolk sac enters body through umbilicus
20	30.2100	Embryo becomes a chick; it breaks amnion, then breathes air in air cell
21	36.3000	Chick breaks shell and hatches

Source: Rollins, 1984.

BASIC ANATOMY AND PHYSIOLOGY

In the developmental process, cells become grouped in appearance and function. Specialized groups of cells that function together are called *tissues.* The primary types of tissues are (1) muscle, (2) nerves, and (3) connective and epithelial, with examples shown in Fig. 18.3.

Organs are groups of tissues that perform specific functions. For example, the uterus is an organ that functions in the reproductive process. A group of organs that function in concert to accomplish a larger, general function comprise a *system.* The reproductive system (Chapter 10), digestive system (Chapter 16), and mammary system (Chapter 19) are discussed in their respective chapters.

It is not the intent of this chapter to discuss all the different systems even though they are important in growth and development. A few additional systems are briefly surveyed—those considered important in understanding farm animals and their productivity. Figures are provided, showing selected systems for one or two species. The systems are similar and generally comparable among the various species of farm animals, although some large differences between poultry and farm mammals do exist.

Skeletal System

Figure 18.6 shows the skeletal system of the horse and Fig. 18.7 portrays the chicken's skeletal system. Even though only bones and some joints are shown in these figures, teeth and cartilage are also considered part of the skeletal system.

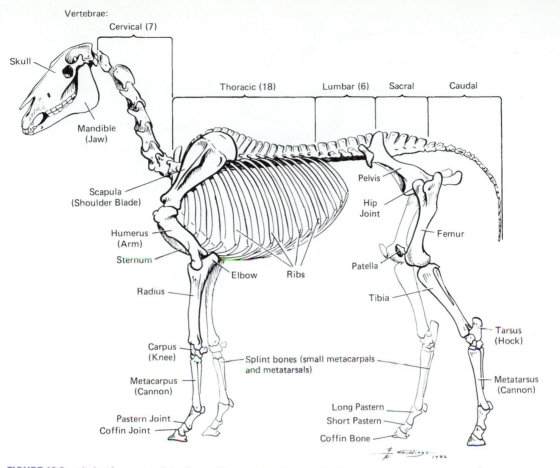

FIGURE 18.6 Skeletal system of the horse. Drawing by Dennis Giddings.

The skeleton protects other vital organs and gives a basic form and shape to the animal's body. Bones function as levers, store minerals, and the bone marrow is the site of blood cell formation.

Chicken bones are more pneumatic (bone cavities are filled with air spaces), harder, thinner, more brittle, and have a different ossification process than those for mammals.

Muscle System

There are three types of muscle tissue—skeletal, smooth, and cardiac. Skeletal muscle is the largest component of red meat animal products. Smooth muscle is located in the digestive, reproductive, and urinary organs. The heart is composed of cardiac muscle.

Figure 18.8 identifies some of the major muscles similar in name and location in the meat animal species and horses. Of special note is the longissimus dorsi, which is discussed in Chapter 3. The size of this muscle and the marbling it contains are important factors in determining yield grades and quality grades of meat animals.

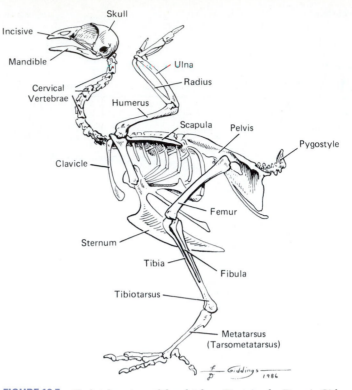
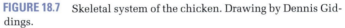

FIGURE 18.7   Skeletal system of the chicken. Drawing by Dennis Giddings.

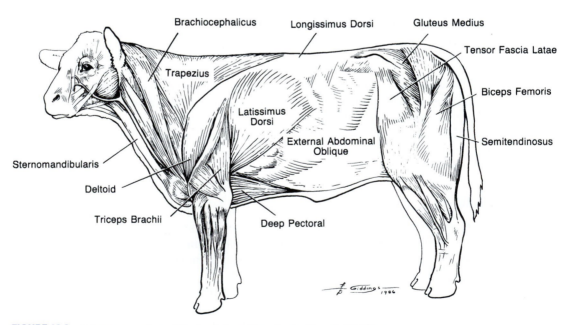

FIGURE 18.8   Primary muscles of the beef steer. Drawing by Dennis Giddings.

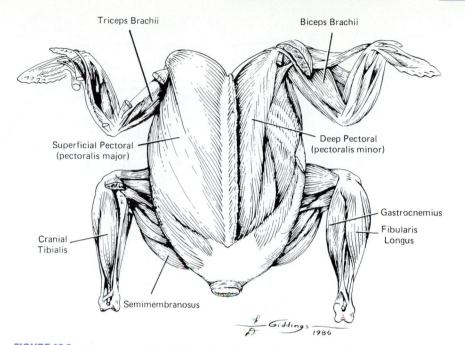

FIGURE 18.9 Primary muscles of the turkey. Drawing by Dennis Giddings.

The primary muscles of the turkey are shown in Fig. 18.9. The muscles of poultry are referred to as *dark meat* (legs and thighs) and *white meat* (breast and wings).

Circulatory System

Figure 18.10 shows the major components of the dairy cow's circulatory system. The circulatory systems of other farm animal species are similar to that of the dairy cow.

The heart, acting as a pump, and the accompanying vessels comprise the circulatory system. *Arteries* are vessels transporting blood away from the heart, while the vessels carrying blood to the heart are called *veins.* The lymph vessels transport lymph (intercellular fluid) from tissues to the heart.

The circulatory system is important in growth and development; the blood transports oxygen, nutrients, cellular waste products, and hormones.

Milk production in all species is dependent on the nutrients in milk arriving through the circulatory system. Dairy cows producing 20,000–40,000 lb milk per year (over 100 lb per day) must consume large amounts of feed, with these feed nutrients circulating in high concentration through the mammary blood supply. Approximately 400–500 lb of blood circulates through the dairy cow's mammary gland for each pound of milk produced.

Endocrine System

Growth and development is highly dependent on the endocrine system. This system consists of several endocrine (ductless) glands that secrete **hormones** into the circulatory system. Hormones are chemical substances that affect a gland (or organ) or, in some cases, all body tissues.

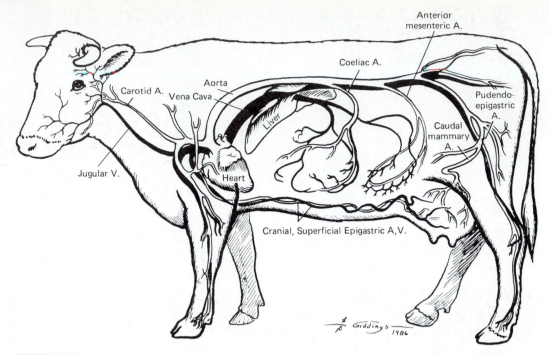

FIGURE 18.10   Circulatory system of the dairy cow (A = artery, V = vein). Drawing by Dennis Giddings.

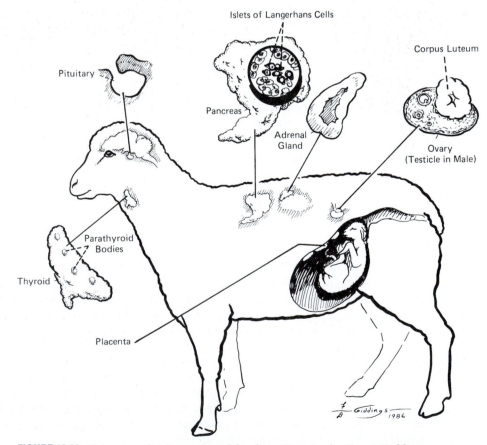

FIGURE 18.11   Primary endocrine organs of the sheep. Drawing by Dennis Giddings.

TABLE 18.2   Major Hormones Affecting Growth and Development in Farm Animals

Hormone	Source	Major Effect
Growth (Somatotrophic, STH)	Pituitary (anterior)	Body cell growth—especially muscle and bone cells
Adrenocorticotropic (ACTH)	Pituitary (anterior)	Stimulates adrenal cortex to produce adrenal cortical steroid hormones
Glucocorticoids	Adrenal (cortex)	Conversion of proteins to carbohydrates
Mineralocorticoids	Adrenal (cortex)	Regulates sodium and potassium balance; water balance
Thyroid-stimulating (TSH)	Pituitary (anterior)	Stimulates thyroid gland to produce thyroid hormones
Thyroid	Thyroid	Regulates metabolic rate
Testosterone	Testicles (interstitial cells)	Libido; accessory sex gland development; male secondary sex characteristics; spermatogenesis
Follicle-stimulating (FSH)	Pituitary (anterior)	Development of ovarian follicles; spermatogenesis
Luteinizing (LH)	Pituitary (anterior)	Ovum maturation; ovulation; formation of corpus luteum (CL); stimulates interstitial cells (testicle) to produce testosterone
Prolactin (luteotropic, LTH)	Pituitary (anterior)	Milk secretion (initiation and maintenance); CL maintenance during pregnancy
Estrogen	Ovary (follicle); placenta	Female reproductive organ growth; female secondary sex characteristics; mammary gland duct growth
Progesterone	Corpus luteum; placenta	Uterine growth; maintenance of pregnancy; alveoli growth in mammary gland growth
Antidiuretic (ADH) (Vasopressin)	Pituitary (posterior)	Controls water loss in kidney
Oxytocin	Pituitary (posterior) (released with nursing reflex)	Causes uterine contractions during parturition; causes milk letdown in mammary gland
Relaxin	Ovary; placenta	Relaxes pelvic ligaments during parturition
Epinephrine	Adrenal (medulla)	Increases blood glucose concentration
Norepinephrine	Adrenal (medulla)	Maintains blood pressure
Insulin	Pancreas	Lowers blood sugar
Parathormone (PTH)	Parathyroid	Calcium and phosphorus metabolism
Glucagon	Pancreas	Raises blood sugar

Source: Compiled from several sources.

The major endocrine glands are shown in Fig. 18.11; the hormones produced and their major effects are identified in Table 18.2. Figure 19.3 (Chapter 19) shows the origin of most of the hormones noted in Table 18.2. The important roles of the pituitary and hypothalmus should also be observed in Fig. 19.3.

GROWTH CURVES

Postweaning growth is a curved-line function regardless of how it is expressed mathematically. If growth is considered an increase in body weight, then most of the early growth follows a straight line or is linear in the age—weight relationship (Fig. 18.12). As the animal increases in age and approaches puberty,

Age

Weight

FIGURE 18.12   Early growth in pigs is linear (straight line), whereas later in life the rate of growth decreases. Courtesy of Elanco Products Co.

rate of growth usually declines, and true growth ceases when the animal reaches maturity.

After an animal reaches maturity, it may have large fluctuations in body weight simply by increasing or decreasing the amount of fat or water that is stored. This increase in weight owing to fattening is not true growth because no net increase in body protein occurs. In fact, animals tend to lose body protein as they grow older. The loss of body protein is one of the phenomena in the aging process.

Figures 18.13 and 18.14 show typical growth curves for most farm animals. The different curves for large and small breeds is primarily a function of differences in skeletal frame size. Maturity is reached at heavier weights in larger breeds, which have larger skeletal frame sizes than smaller-framed animals.

The relatively straight-lined growth shown for turkeys and broilers (Fig. 18.14) does curve and level off as shown for the other species. This occurs when the birds become older; for broilers, this occurs at approximately 11 weeks of age.

CARCASS COMPOSITION

Products from meat animals and poultry are composed primarily of fat, lean, and bone. A superior carcass is characterized by a low proportion of bone, a high proportion of muscle, and an optimum amount of fat. Understanding the growth and development of these animals is important in knowing when animals should be slaughtered to produce the desirable combinations of fat, lean, and bone.

Figure 18.15 shows the expected changes in fat, muscle, and bone as animals increase in liveweight during the linear-growth phase. As the animal moves from the linear-growth phase into maturity, the growth curves shown in Fig. 18.15 are extended similar to those shown in Fig. 18.13.

At light weights, fat deposition begins rather slowly, then increases geometrically when muscle growth slows as the animal approaches physiological maturity. Muscle comprises the greatest proportion and its growth is linear during the production of young slaughter animals. Bone has a smaller relative growth rate than either fat or muscle. Because of the relative growth rates, the ratio of muscle to bone increases as the animal increases in weight.

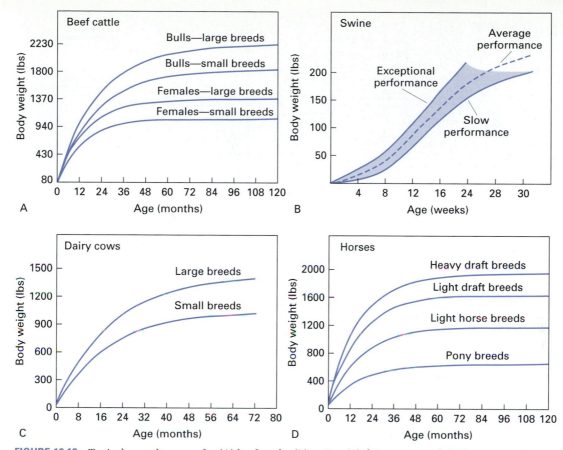

FIGURE 18.13   Typical growth curves for (A) beef cattle, (B) swine, (C) dairy cows, and (D) horses. Courtesy of R. A. Battaglia and V. B. Mayrose, *Handbook of Livestock Management Techniques* (New York: Macmillan, 1981). Copyright © Macmillan Publishing Co.

Effects of Frame Size

Different maturity types of animals (sometimes referred to as *breed types* or *biological types*) have a marked influence on carcass composition at similar liveweights (Fig. 18.16). The earlier-maturing types increase fat deposition at lighter weights than either the average or later-maturing types.

Skeletal frame size, measured as hip height, is a more specific way of defining maturity type. Figure 18.17 shows the difference in fat, lean; and bone composition for small-, medium-, and large-framed beef steers at different liveweights. Note that the three different frame sizes have a similar composition at 900 lb (small frame), 1,100 lb (medium frame), and 1,300 lb (large frame).

Effect of Sex

The effect of sex is primarily on the fat component, although there are differences among species. Heifers deposit fat earlier than steers or bulls, with bulls being leaner at the same slaughter weight (Fig. 18.18A). Heifers are typically

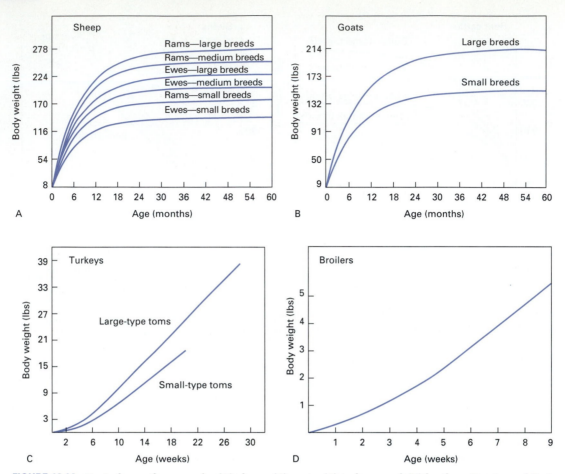

FIGURE 18.14 Typical growth curves for (A) sheep, (B) goats, (C) turkeys, and (D) broilers. Courtesy of R. A. Battaglia and V. B. Mayrose, *Handbook of Livestock Management Techniques* (New York: Macmillan, 1981). Copyright © Macmillan Publishing Co.

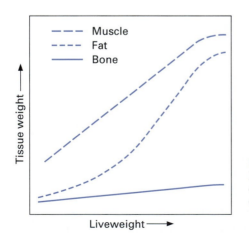

FIGURE 18.15 Tissue growth relative to increased liveweight. Adapted from Berg and Walters (1983), courtesy of the American Society of Animal Science.

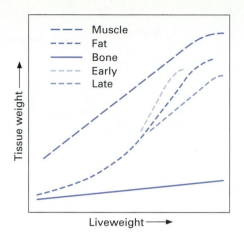

FIGURE 18.16 Effects of different maturity types on carcass composition. Adapted from Berg and Walters (1983), courtesy of the American Society of Animal Science.

slaughtered at lighter weights (100–200 lb less) than steers to have a similar fat-to-lean composition.

Swine are different from cattle, as barrows are fatter than gilts or boars at similar slaughter weights (Fig. 18.18B). The reason for this species difference is not known.

At typical slaughter weights, the sex differences in sheep do not have a marked effect on fat and lean composition. Apparently, sheep are slaughtered at an earlier stage of development (before puberty) than cattle and swine. Compositional differences in sheep are similar to cattle if sheep are fed beyond their usual slaughter weights.

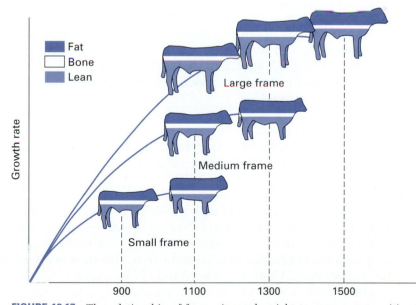

FIGURE 18.17 The relationship of frame size and weight to carcass composition in beef steers. Drawing by CSU Graphics.

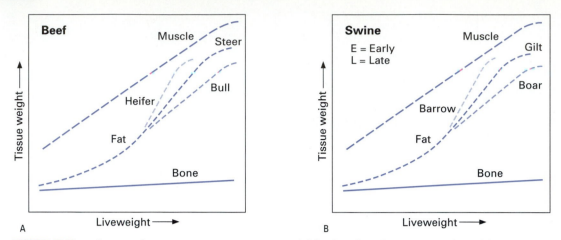

FIGURE 18.18   Influence of sex on carcass composition in (A) beef cattle and (B) swine. Adapted from Berg and Walters (1983), courtesy of the American Society of Animal Science.

Effect of Muscling

Relatively large differences exist in muscling among the various species of farm animals. The proportion of muscle in the carcass varies indirectly with fat, and a higher proportion of fat is associated with a lower proportion of muscle. Muscle has a much faster relative growth rate than bone. Muscle weight relative to liveweight or muscle-to-bone ratio can be used as a valuable measurement of muscle yield.

Most of the current cattle population in the United States does not have large differences in muscling. There are, however, muscling differences in 10–15% of the cattle, which are best expressed by a muscle-to-bone ratio (pounds of muscle ÷ pounds of bone). Muscle-to-bone ratios can range from 3 lb of muscle to 1 lb of bone in thinly muscled slaughter steers and heifers to 5 lb of muscle to 1 lb of bone in more thickly muscled cattle. Muscle-to-bone ratios higher than 5:1 are found in double-muscled cattle. These differences in muscle-to-bone ratio are economically important, particularly if more boneless cuts are merchandised.

Some of the muscling differences for cattle are shown in Fig. 18.19: lighter-muscled cattle (Fig. 18.19A), heavy-muscled cattle (Fig. 18.19B), and double-muscled cattle (Fig. 18.19C). Double-muscled cattle do not have a duplicate set of muscles, but they do have an enlargement (hypertrophy) of the existing muscles.

Extreme muscling, as observed in double-muscled or heavy-muscled types of animals, can be associated with problems in total productivity. Some heavy-muscled swine might die when subjected to stress conditions or produce undesirable meat that is pale, soft, and exudative (see porcine stress syndrome (PSS) and PSE pork in Chapter 26). Reproductive efficiency is lower in some heavy-muscled animals. There is some latitude in increasing muscle-to-bone ratio in slaughter animals without encountering negative side effects.

Individual muscle dissection has shown a similarity of proportion of individual muscles to the total body muscle weight. Small differences in muscle distribution have been shown, but the differences do not seem to be economi-

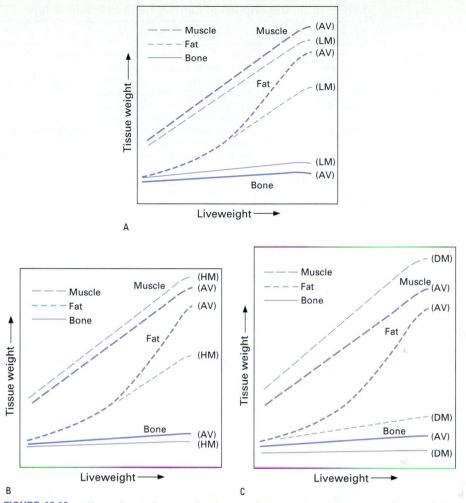

FIGURE 18.19 Effect of muscling on fat, lean, and bone composition in cattle: (A) light-muscled (LM), (B) heavily muscled (HM), (C) double-muscled (DM), compared to average muscling (AV). Adapted from Berg and Walters (1983), courtesy of the American Society of Animal Science.

cally significant for the industry. Thus, it appears that increasing muscle weight in areas of high-priced cuts at the expense of muscle weight in lower-priced cuts is not commercially feasible. Even though muscle distribution is not important, muscling as previously defined—as muscle-to-bone ratio—has high economic importance.

AGE AND TEETH RELATIONSHIP

For many years, producers have observed teeth of farm animals to estimate their age or their ability to graze effectively with advancing age. Some animals are **mouthed** to classify them into appropriate age groups for showring classification. The latter has not always proven exact because of animal variation in teeth condition and the ability of some exhibitors to manipulate the condition of the teeth.

TABLE 18.3   **Description of Cattle's Teeth to Estimate Age**

Approximate Age	Description of Teeth
Birth	Usually only one pair of middle incisors
1 month	All eight, temporary incisors
1.5–2 yr	First pair of permanent (middle incisors)
2.5–3 yr	Second pair of permanent incisors
3.25–4 yr	Third pair of permanent incisors
4–4.5 yr	Fourth (corner) pair of permanent incisors
5–6 yr	Middle pair of incisors begins to level off from wear; corner teeth might also show some wear
7–8 yr	Both middle and second pair of incisors show wear
8–9 yr	Middle, second, and third pair of incisors show wear
10 yr and over	All eight incisors show wear

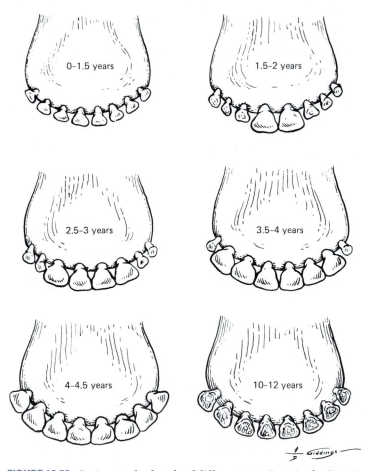

FIGURE 18.20   Incisor teeth of cattle of different ages. Drawing by Dennis Giddings.

Table 18.3 gives some of the guidelines used in estimating the age of cattle by observing their teeth. The **incisors** (front cutting teeth) are the teeth used. There are no upper incisors, and the molars (back teeth) are not commonly used to determine age.

Figure 18.20 shows the incisor teeth of cattle of different ages. If the grazing area is sandy and the grass is short, wearing of the teeth progresses faster than previously described. Also, some cows may be **broken-mouthed,** which means some of their permanent incisor teeth have been lost.

The age relationship to teeth condition for sheep and horses is given in Chapters 29 and 30, respectively.

CHAPTER SUMMARY

■ Growth is an increase in body weight until mature size is reached. Development is the directive coordination of all physiological processes until maturity is obtained.

■ Prenatal growth is a unique series of events starting from the union of two sex cells, which then grow and differentiate into a fetus with different organs and tissues.

■ Embryonic development of the chick differs from mammals as the chick develops more rapidly in the egg outside the hen's body.

■ Muscle growth occurs by an increase in fiber size (hypertrophy) and an increase in fiber number (hyperplasia).

■ Anatomy and physiology of farm animals involve understanding the structure and function of cells, tissues, organs, and systems. The most common systems are skeletal, nervous, muscular, circulatory, and endocrine.

■ Growth curves are influenced by nutrition, sex, frame size, and muscling. They are important in determining when animals should be harvested to produce desirable combinations of fat, lean, and bone.

REVIEW QUESTIONS

1. What is the difference between growth and development?
2. What are the three phases of prenatal life?
3. When does the greatest increase in prenatal weight occur?
4. *True or False:* At birth the distal parts of the body (legs) are more developed that the proximal parts (hips and chest).
5. What is the germ spot on a fertilized or nonfertilized poultry egg called?
6. What are the four membranes essential for growth of the chick embryo?
7. Specialized groups of cells are called _____.
8. Groups of tissues that perform specific functions are _____.
9. What is a group of organs that function together to accomplish a larger general function called?
10. What are the two functions of the skeletal system?

11. What are the three types of muscle tissue?

12. _____ are blood vessels transporting blood away from the heart, whereas _____ are those vessels transporting blood back to the heart.

13. Chemical substances secreted by a ductless gland into the circulatory system to affect a distant gland or organ are called?

14. When does the greatest postnatal growth occur?

15. What are the three primary tissues that a carcass is composed of?

16. When is fat deposition in the body greatest?

17. *True or False:* Females typically are leaner than intact or castrated males.

18. *True or False:* An animal's age can often be estimated by examining its teeth.

SELECTED REFERENCES

Publications

Berg, T. T., and Butterfield, R. M. 1976. *New Concepts of Cattle Growth.* Sydney, Australia: Sydney University Press.

Berg, R. T., and Walters, L. E. 1983. The meat animal. Changes and challenges. *J. Anim. Sci.* 57:133 (Suppl. 2).

Bone, J. F. 1988. *Animal Anatomy and Physiology.* 4th ed. Reston, VA: Reston Publishing Co.

Currie, W. B. 1988. *Structure and Function of Domestic Animals.* Boston, MA: Butterworths Publishers.

Frandson, R. D. 1986. *Anatomy and Physiology of Farm Animals.* 4th ed. Philadelphia: Lea & Febiger.

Hammond, J., Jr., Robinson, T., and Bowman, J. 1983. *Hammond's Farm Animals.* Baltimore: Edward Arnold University Park Press.

Prior, R. L., and Lasater, D. B. 1979. Development of the bovine fetus. *J. Anim. Sci.* 48:1546.

Rollins, F. D. 1984. Development of the embryo. *Arizona Coop. Ext. Serv. Publication* 8427.

Swatland, H. J. 1984. *Structure and Development of Meat Animals.* Englewood Cliffs, NJ: Prentice-Hall.

Trenkle, A. H., and Marple, D. N. 1983. Growth and development of meat animals. *J. Anim. Sci.* 57:273 (Suppl. 2).

Visuals

Anatomy of the Fowl (sound filmstrip). Vocational Education Productions, California Polytechnic State University, San Luis Obispo, CA 93407.

CHAPTER 19

Lactation

Lactation, the production of milk by the mammary gland, is a distinguishing characteristic of mammals whose young, at first, feed solely on milk from their mothers. Even after they start to eat other feeds, the young continue to nurse until they are weaned (separated from their mothers so they cannot nurse).

The mammary gland serves two functions: (1) it provides nutrition to animal offspring and (2) it is a source of passive immunity to the offspring. The importance of milk as a nutritional source to perpetuate each mammalian species has been known since the beginning of history. Only during the past few decades have the basic mechanisms of immunity through milk been determined.

Humans consume milk and recognize it as a palatable source of nutrients. Milk consumed in the United States comes primarily from dairy cows and, to a much lesser extent, from goats and sheep. In other countries, the human milk supply also comes from water buffalo, yak, reindeer, camel, donkey, and sow.

MAMMARY GLAND STRUCTURE

The mammary gland is an **exocrine gland** that produces the external secretion of milk transported through a series of ducts. The cow has four separate mammary glands that terminate into four teats; sheep and goats have two glands and two teats; and the mare has four mammary glands that terminate into two teats. The mammary glands of these species are located in the groin area (Fig. 19.1).

Sows have 6–20 mammary glands located in two rows along the abdomen, with each gland having a teat. Typically there are 10–14 of the sow's mammary glands that are functional (Fig. 19.1). There is evidence that teat number in swine is not related to litter size or litter-weaning weight.

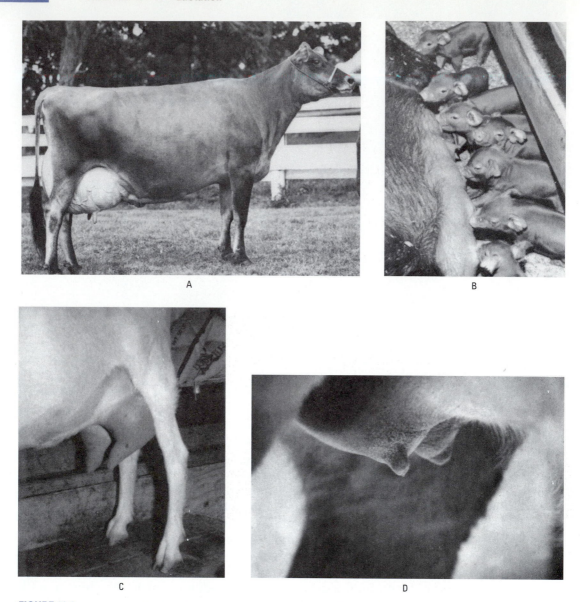

FIGURE 19.1   Mammary glands of (A) cow, (B) sow, (C) goat, and (D) mare. Courtesy of (A) *Hoards Dairyman,* (B) University of Nebraska, (C) George F. W. Haenlein, and (D) Colorado State University.

Figure 19.2 shows the basic structure of the cow's udder. The udder is supported horizontally and laterally by suspensory ligaments (Fig. 19.2A). The internal structure of the udder is similar for all farm animal species, except for the number of glands and teats.

Secretory tissue of the mammary gland (Fig. 19.2C) is composed of millions of grapelike structures called ***alveoli.*** Each **alveolus** has its separate blood supply from which milk constituents are obtained by epithelial cells lining the alveolus. Milk collects in the alveolus lumen and, during milk letdown, travels through ducts to a larger collection area called the *gland cistern.* During milk letdown and the milking process, milk is forced into the teat cistern and through the streak canal to the outside of the teat.

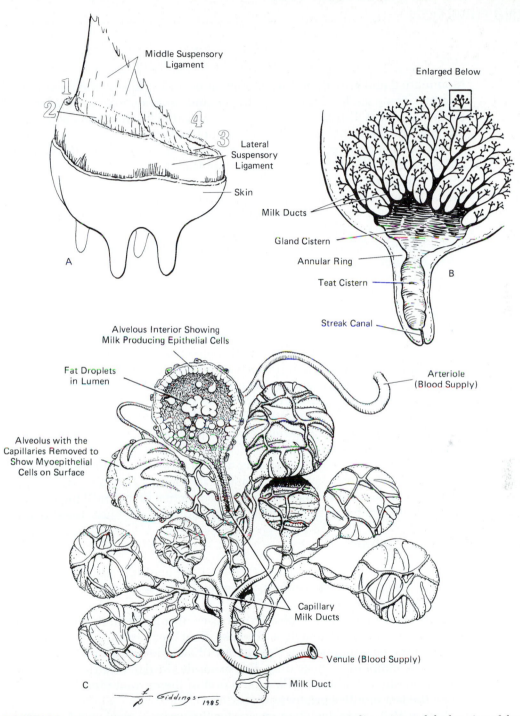

Middle Suspensory Ligament

1
2
4
3

Lateral Suspensory Ligament

Skin

A

Enlarged Below

Milk Ducts

Gland Cistern

Annular Ring

Teat Cistern

Streak Canal

B

Alvelous Interior Showing Milk Producing Epithelial Cells

Fat Droplets in Lumen

Alveolus with the Capillaries Removed to Show Myoepithelial Cells on Surface

Arteriole (Blood Supply)

Capillary Milk Ducts

Venule (Blood Supply)

Milk Duct

C

Giddings 1985

FIGURE 19.2 (A) Basic structure of the cow's udder, showing suspensory ligaments and the location of the four separate quarters. (B) A section through one of the quarters, showing secretory tissue, ducts, and milk-collecting cisterns. (C) An enlarged lobe with several alveoli and their accompanying blood supply. Drawing by Dennis Giddings.

MAMMARY GLAND DEVELOPMENT AND FUNCTION

Development

Mammary gland growth and development occur rapidly as the female reaches puberty. The ovarian hormones (estrogen and progesterone) have a large effect on development. Estrogen is primarily responsible for duct and cistern growth, whereas progesterone stimulates growth of the alveoli.

Estrogen and progesterone are produced by the ovary under the stimulation of FSH and LH from the anterior pituitary (Fig. 19.3). The pituitary also has a direct influence on mammary growth through the production of growth hormone. Growth hormone stimulates general cell growth of the mammary gland.

Figure 19.3 also identifies the additional indirect effect (other than FSH and LH) on mammary growth and development. Thyroid hormones are produced under the influence of TSH, and the adrenal gland produces the corticosteroids when stimulated by ACTH from the pituitary. All of these hormones work in concert to produce mammary growth and function.

Milk Secretion

Growth hormone, adrenal corticoids, and prolactin are primarily responsible for the initiation of lactation. These hormones become effective as parturition nears and when estrogen and progesterone hormone levels decrease.

Through milking or nursing, the milk in the gland cistern is soon removed. There remains a large amount of milk in the alveoli that is forced into the ducts by the contractions of the myoepithelial cells (Fig. 19.2). These cells contract under the influence of the hormone oxytocin, which is secreted by the posterior pituitary (Fig. 19.3).

Oxytocin, released by the suckling reflex, can also be released by other stimuli. The milk letdown can be associated with feeding the cows or washing the udder. The latter is a typical management practice before milking cows. Oxytocin release can be inhibited by pain, loud noises, and other stressful stimuli.

Maintenance of Lactation

Lactation is maintained primarily through hormonal influence. Prolactin, thyroid hormones, adrenal hormones, and growth hormone are all important in the maintenance of lactation.

Daily milk production typically increases during the first few weeks of lactation, peaks at approximately 4–6 weeks, then decreases over the next several weeks of the lactation period. Persistency of lactation measures how milk production is maintained over time. For example, in dairy cattle, persistency is determined by calculating milk production of the current month as a percentage of the last month's production.

Figure 19.4 shows lactation curves for the various species of farm animals. The curve for dairy cows (Fig. 19.4A) represents approximately 14,000 lb produced in 305 days.

Decreased milk production during the lactation period is due primarily to a decreased number of active alveoli and less secretory tissue (epithelial cells) in the alveoli. These and other changes are associated with hormonal changes.

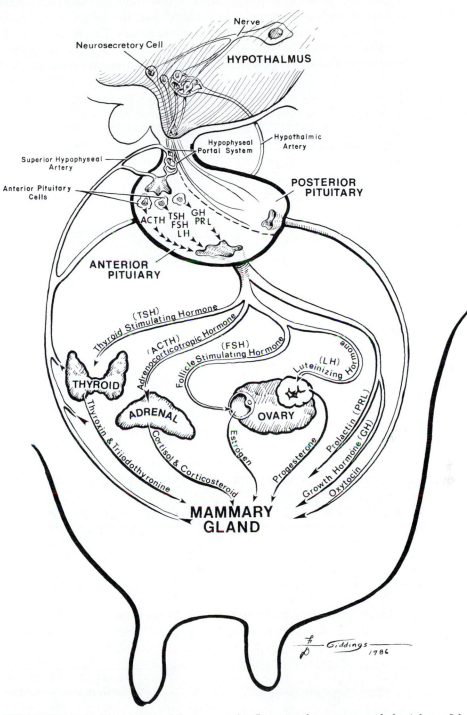

FIGURE 19.3 Major hormones (and their sources) influencing the anatomy and physiology of the mammary gland. Refer to Figure 18.11 for the location of the endocrine glands throughout the body. Drawing by Dennis Giddings.

When milking or nursing is stopped, the alveoli are distended and the capillaries are filled with blood. After a few days the secretory tissue becomes involuted (reduced in size and activity) and the lobes of the mammary gland consist primarily of ducts and connective tissue. The female then becomes **dry** as milk secretion is not occurring. After the dry period (approximately 2 months in the dairy cow) and when parturition approaches, hormones and

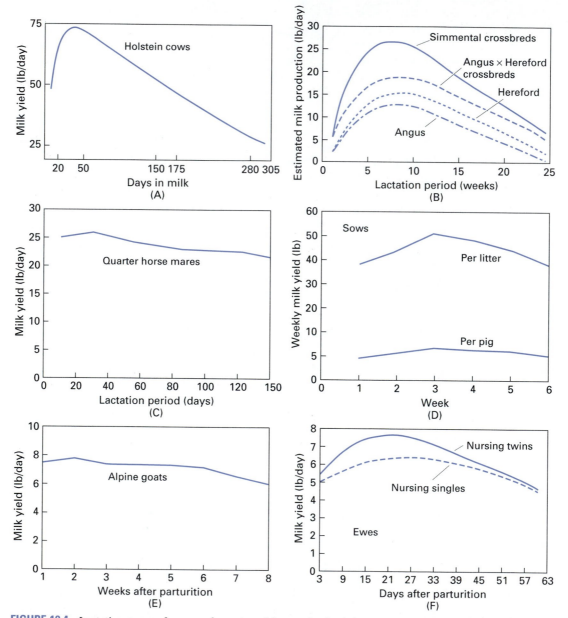

FIGURE 19.4 Lactation curves for several species of farm animals. (A) Dairy cows. Courtesy of Colorado State University (CO, ID, and UT DHIA records). (B) Beef cows. Courtesy of Jenkins and Ferrell, 1984 Beef Cow Efficiency Forum. (C) Quarter horse mares. Courtesy of *J. Anim. Sci.* 54:496. (D) Sows. Courtesy of *Missouri Agric. Research Bull.* 712. (E) Alpine goats. Courtesy of *J. Dairy Sci.* 63:1677. (F) Ewes. Courtesy of *Michigan State Agric. Expt. Sta. Research Report* 491.

other influences prepare the mammary gland to resume its secretion and production of milk.

FACTORS AFFECTING MILK PRODUCTION

Inheritance determines the potential for milk production, but feed and management determine whether or not this potential is attained. The best feed and care will not make a high-producing female out of one that is genetically a low producer. Likewise, a female with high genetic potential will not produce at a high level unless she receives proper feed and care. Production is also influenced by the health of the animal. For example, **mastitis** (an inflammation of the udder) in dairy cows can reduce production by 30% or more. Proper management of dairy cows, particularly adherence to routine milking and feeding schedules, contribute to a high level of production.

In lactating farm animals, the level of milk production is important because milk from these females provides much of the required nutrients for optimal growth in the young. If beef cows, for example, are inherently heavy milkers, their milk-producing levels will be established by the ability of the young to consume milk. Normally, if the cow is not milked, she will adjust her production to the consumption level of the calf. However, if the cow is milked early in lactation, her milk flow may be far greater in amount than the calf can consume. This situation makes it necessary either to continue milking the cow until the calf becomes large enough to consume the milk or to put another calf onto the cow.

Anything that causes a female to reduce her production of milk is likely to cause some regression of the mammary gland, thereby preventing resumption of full production. If the young animal lacks vigor, it does not consume all the milk that its mother can supply, and her level of production is lowered accordingly. This is one reason inbred animals grow more slowly and crossbred animals more rapidly during the nursing period. Crossbreds are usually larger and more vigorous at birth and can stimulate a high level of milk production by their dams, whereas inbred animals are smaller and less vigorous and cause their dams to produce milk at a comparatively low rate.

Females with male offspring produce more milk than females with female offspring. Male offspring are usually heavier at birth and have a faster growth rate than females. This puts a greater demand on the milk-producing ability of the dam.

Females with multiple births usually produce more total milk than females with single births. Age of the female also affects milk yield; younger and older females produce less milk compared to females who have had several lactations. For example, sows are usually at peak milk yields during third or fourth lactation, beef cows produce the most milk at 5–9 years of age, and dairy cows have the highest milk yields between 5–8 years of age.

Large amounts of nutrients are needed to supply the requirements of lactating females producing high levels of milk because much energy is required for milk secretion and because milk contains large quantities of nutrients. Adequate nutrition is much greater for lactation than for gestation; for example, a cow producing 100 lb of milk daily could yield 4.0 lb of butterfat, 3.3 lb of protein, and almost 5.0 lb of lactose in that milk. If nutrition is inadequate in quality or quantity, a cow that has the inherent capacity to produce 100 lb of milk daily will draw nutrients from her own body as the body attempts to sustain a high level of milk production. Withdrawal of nutrients reduces the body stores. This withdrawal usually occurs to a limited extent in high producers even when they are well fed. Cows reduce their milk production in response to inadequate amounts of quality of feed, but they usually do so only after the loss of nutrients is sufficiently severe to cause a loss in body weight. In gestation, a cow becomes relatively efficient in digesting her feed, so feed restrictions in gestation are less harmful than they might be at other times.

MILK COMPOSITION

Species Differences

Milk composition is markedly different for the mammalian species (Table 19.1). Milk from reindeer and aquatic mammals is exceptionally high in total solids, whereas the mare's milk is low in this regard. Large differences in milk fat percentages exist in mammals; donkeys are low (1.4%) and fur seals are high (53.3%).

Except in protein, milk from cows and goats is similar in composition to milk from humans. Milk from cows or goats is much higher in protein than milk from humans. This is not a problem in bottle-feeding babies as the additional protein is not harmful.

In dairy cows, the amount or percentage of fat is easily changed through changing the diet fed. Carbohydrates (lactose) remain relatively constant even with dietary fluctuations. Varying the protein content of the ration has little change on the protein content of the milk.

COLOSTRUM

The fetus develops in a sterile environment. Its immune system has not yet been challenged by the microorganisms existing in the external environment. Before or shortly after birth, the fetal immune system must be made functional or death is imminent.

TABLE 19.1   Average Milk Composition of Several Species

Species	Total Solids (%)	Fat (%)	Protein (%)	Lactose (%)	Minerals (%)
Human	13.3	4.5	1.6	7.0	0.2
Cow (*Bos taurus*)	12.7	3.9	3.3	4.8	0.7
Cow (*Bos indicus*)	13.5	4.7	3.4	4.7	0.7
Goat	12.4	3.7	3.3	4.7	0.8
Water buffalo	19.0	7.4	6.0	4.8	0.8
Ewe	18.4	6.5	6.3	4.8	0.9
Sow	19.0	6.8	6.3	5.0	0.9
Mare	10.5	1.2	2.3	5.9	0.4
Reindeer	33.7	18.7	11.1	2.7	1.2

Immunoglobulins (Ig) are involved in the passive immunity transfer maternally to the offspring. In some animals, the Ig are transferred in utero through the bloodstream, whereas in other animals immunoglobulins are transferred through colostrum. Certain animals utilize both methods of Ig transfer; however, the colostrum method is most common for larger, domestic animals.

These Ig antibodies give the newborn protection from harmful microorganisms that invade the body and cause illness. The intestinal wall of the newborn is quite porous, permitting absorption of colostrum antibodies to enter the bloodstream. Within a few hours (no more than 24), the gut wall becomes less porous, allowing little absorption of the antibodies to occur. Thus, passive immunity of the newborn is dependent on an adequate supply of antibodies in the colostrum and on consuming the colostrum within a few hours after parturition.

CHAPTER SUMMARY

- Lactation is the production of milk from the mammary gland.
- Milk from mammals provides nutrients and passive immunity to their young and it is a nutrient source for humans.
- Colostrum, the first milk produced after the young are born, is the source of immunoglobulins which provide early immunity to the offspring.
- The mammary gland is an exocrine gland that produces an external secretion (milk) from structures called alveoli.
- Mammary gland development, milk secretion, and the maintenance of lactation are affected by several hormones—namely, estrogen, progesterone, thyroid hormone, growth hormone, adrenal hormone, and prolactin.
- Lactation curves for most farm animals peak a few weeks after lactation is initiated then decrease throughout the lactation period.
- Milk is composed primarily of water with lesser amounts of protein, fat, lactose, and minerals.

REVIEW QUESTIONS

1. What are the two functions of the mammary gland?
2. What are the secretory units of the milk-secreting tissue within the mammary gland?
3. What two reproductive hormones directly influence growth and development of the mammary gland?
4. How is milk released from the alveoli during milking or suckling?
5. What hormone is responsible for milk letdown?
6. What determines an animal's potential for milk production?
7. What determines whether an animal's genetic potential for milk production is attained?
8. An inflammation of the udder which can reduce milk production by 30% or more is called _____ .
9. *True or False:* Milk from cows and goats is similar to human milk in all components except protein?

SELECTED REFERENCES

Publications

Ackers, R. M. 1994. Lactation. *Encyclopedia of Agricultural Science.* San Diego: Academic Press, Inc.

Allen, A. D., Lasley, J. F., and Tribble, L. F. Milk production and related factors in sows. *Missouri Agric. Research Bull.* 712.

Gibbs, P. G., Potter, G. D., Blake, R. W., and McMullan, W. C. 1982. Milk production of quarter horse mares during 150 days of lactation. *J. Anim. Sci.* 54:496.

Henry, M. S., and Benson, M. E. 1990. Milk production and efficiency of lactating ewe lambs. *Michigan State Agric. Expt. Sta. Report* 491.

Keown, J. F., Everett, R. W., Empet, N. B., and Wadell, L. H. 1986. Lactation curves. *J. Dairy Sci.* 69:769.

Larson, B. L. (ed.). 1985. *Lactation.* Ames, IA: Iowa State University Press.

Offedal, O. T., Hintz, H. F., and Schryver, H. F. 1983. Lactation in the horse: Milk composition and intake by foals. *J. Nutr.* 113:2196.

Sakul, H. and Boylau, W. J. 1992. Lactation curves for several U.S. sheep breeds. *Animal Production* 54:229.

Visuals

External Features of the Udder; Internal Features of the Udder; Removal of Milk and Proper Milking Practices; Abnormalities in Mammary Glands; and *Milk Composition and Factors Affecting Milk Yield* (videotapes). Agricultural Products and Services, 2001 Killebrew Dr., Suite 333, Bloomington, MN 55420.

CHAPTER 20

Adaptation to the Environment

Livestock throughout the world are expected to produce under an extremely wide range of environments. Variations in temperature, humidity, wind, light, altitude, feed, water, and exposure to parasites and disease organisms are some of the environmental conditions to which livestock are exposed (Fig. 20.1).

Under **intensive management,** such as integrated poultry enterprises, confinement swine operations, and large dairies, many environmental conditions are highly controlled. Feed and water are plentiful, rations are carefully balanced, excellent health programs are carefully monitored, and, in many cases, temperature, humidity, and other weather influences are controlled.

Extensive management involves less producer control over the environmental conditions in which animals are expected to produce. Many ruminant animals are extensively managed while also exposed to numerous climatic conditions as they graze the forage, often sparse. Livestock that produce and survive under these conditions must have physiological mechanisms for adaptability.

Economics dictate to a large extent whether animals will be managed intensively or extensively. Under intensive management, many environmental conditions are changed to fit the animals. Under extensive management, animals are changed (selected for adaptability) to fit the environment.

It is important to understand the major environmental effects on animal productivity. Management decisions can then be made to determine whether changing the environment or changing the adaptability of animals is the most economically feasible alternative. In some situations, environmental changes and changes in animal adaptability can produce optimal results.

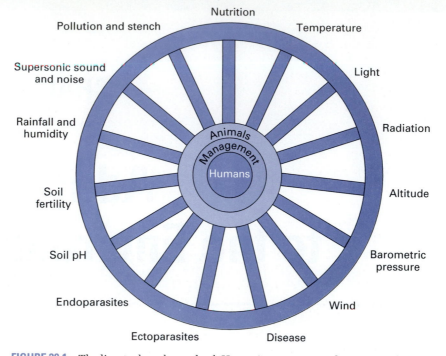

FIGURE 20.1   The livestock ecology wheel. Human's management decisions with farm animals play a role in determining how the animals will interact with their environment. Courtesy of J. Bonsma, *Livestock Production* (Cody, WY: Agi Books, 1983). Copyright © Agi Books.

MAJOR ENVIRONMENTAL EFFECTS

Adjusting to Environmental Changes

As the seasons change, two major kinds of changes occur in the environment: changes in temperature and changes in length of daylight. As summer approaches, temperature and length of daylight increase, thus increasing the amount of heat available to the animal. As autumn approaches, length of daylight and temperature decrease. Hormone changes help the animal respond physiologically to these seasonal changes.

The thyroid gland, the lobes of which are located on both sides of the trachea, regulates the metabolic activity of the animal. When the environmental temperature is cool, the air inhaled tends to cool the thyroid gland. The thyroid responds by increasing its production of the hormone thyroxine; the thyroxine, in turn, stimulates metabolic activity and heat is generated. The growing and shedding of insulating coats of fiber are slow changes that are also influenced by hormones. Sudden warm periods before shedding occurs in spring, and sudden cool periods before a warm coat has been produced in autumn can have severe effects on an animal. These sudden temperature changes are the greatest predisposers for the onset of respiratory diseases such as pneumonia, influenza, and shipping fever.

Temperature Zones of Comfort and Stress

The **comfort zone,** or **thermoneutral zone (TNZ),** identifies a range of temperatures where heat production and heat loss from the body are about the same. When the temperature is below the TNZ, animals increase their feed intake and reduce blood flow to the surface and the extremities. Mammals generate heat by shivering. Fowl fluff their feathers to create a large space of dead air about themselves to prevent rapid heat transfer from their bodies. Some animals hunch to expose less body surface. Cattle congregate in an area that provides protection from blizzards, and they huddle closely so that each animal receives heat from the others. During a severe blizzard, some animals can be killed from crowding, trampling, and suffocation.

To avoid excessive heat loss, animals maintain the temperature of their extremities at a level below that of the rest of the body through a unique mechanism called *countercurrent blood flow action.* Blood in arteries coming from the core of the body is relatively warm and blood in veins in the extremities is relatively cool. The blood in veins in the extremities cools the warmer arterial blood so that the extremities are kept at a cool temperature. With the extremities thus kept at a relatively cool temperature, loss of heat to the environment is reduced.

Between 65 and 80°F, dilation of blood vessels near the skin and in the extremities occurs so that the surface of the animal becomes warmer, water consumption increases, respiration increases, and, in animals that can sweat, perspiration increases. Above 80°F, animals that have the capacity of sweating keep their body surfaces wet with sweat so that evaporation can help cool them.

When the environmental temperature exceeds 90°F, farm animals may die, have lower daily gains and poor feed conversion, and have lower reproductive performance through higher embryonic deaths. During these high temperatures, animals become less active to reduce the amount of heat they generate, and lie down in the shade to reduce their exposure to the sun. Animals typically increase their consumption of water and excretion of urine. If the water consumed is cooler than the temperature of the animal, it can help considerably to cool the body.

CHANGING THE ENVIRONMENT

Adjusting Rations for Weather Changes

Nutrient requirements and predicted gains of farm animals are published from research studies in which animals are protected from environmental extremes. The most common environmental factor that alters both performance and nutrient requirements is temperature. Thus, livestock producers should be aware of critical temperatures that affect performance of their animals and should consider making changes in their feeding and management programs if economics so dictate.

The TNZ is the range in effective temperature where rate and efficiency of performance is optimum. Critical temperature is the lower limit of the TNZ and is typified by the ambient temperature below which the performance of animals begins to decline as temperatures become colder. Figure 20.2 shows that the maintenance energy requirement increases more rapidly during cold weather than does rate of feed intake. This results in a reduction of gain and more feed required per pound of gain, which typically causes cost per pound of gain to be higher.

Effective ambient temperatures below the lower critical temperature (below the TNZ) constitute cold stress, and those above the TNZ constitute heat stress. The term *effective ambient temperature* is an index of the heating or cooling power of the environment in terms of dry-bulb temperature. It includes an environmental factor—such as radiation, wind, humidity, or precipitation—that can alter environmental heat demand.

The critical temperatures and optimal temperature ranges are shown for some animals in Fig. 20.3. Effective ambient temperatures have not been calculated for all farm animals, although the effects of some combined environmental factors are known. For example, the windchill index for cattle is shown in Table 20.1. An example in Table 20.1 shows that with a temperature of 20°F and a wind speed of 30 mph, the effective ambient temperature is −16°F.

The *lower critical temperature* (LCT) for cattle depends on how much insulation is provided by the hair coat, whether the animal is wet or dry, and how much feed the cow consumes. Table 20.2 shows some lower critical temperatures for beef cattle. For example, a cow being fed a maintenance ration may have a low critical temperature of 32°F when dry but a low critical temperature of 60°F when wet.

Coldness of a specific environment is the value that must be considered when adjusting rations for cows. *Coldness* is simply the difference between effective temperature (windchill) and lower critical temperature. Using this definition for coldness, instead of using temperature on an ordinary thermometer, helps explain why wet, windy days in March might be colder for a cow than extremely cold but dry, calm days in January.

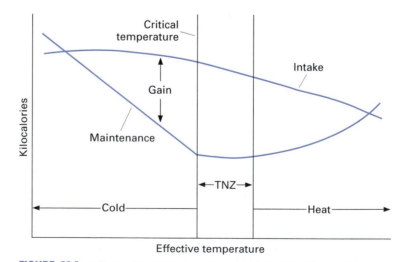

FIGURE 20.2 Effect of temperature on rate of feed intake, maintenance energy requirement, and gain. Adapted from Ames, 1980.

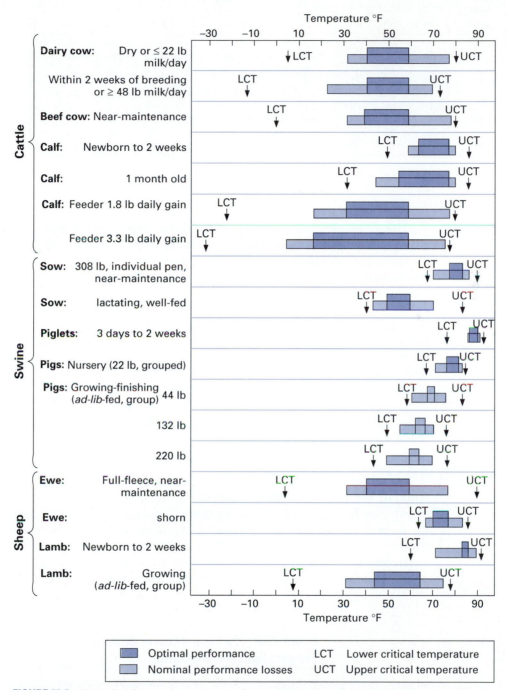

FIGURE 20.3 Critical ambient temperatures and temperature zones for optimal performance in cattle (*Bos taurus*), swine, and sheep. Adapted from Yousef, 1985.

TABLE 20.1   Windchill Factors for Cattle with Winter Coat

Wind Speed (mph)	Temperature (°F)												
	−10	−5	0	5	10	15	20	25	30	35	40	45	50
Calm	−10	−5	0	5	10	15	20	25	30	35	40	45	50
5	−16	−11	−6	−1	3	8	13	18	23	28	33	38	43
10	−21	−16	−11	−6	−1	3	8	13	18	23	28	33	38
15	−25	−20	−15	−10	−5	0	4	9	14	19	24	29	34
20	−30	−25	−20	−15	−10	−5	0	4	9	14	19	24	29
25	−37	−32	−27	−22	−17	−12	−7	−2	2	7	12	17	22
30	−40	−41	−36	−31	−26	−21	−16	−11	−6	−1	3	8	13
35	−60	−55	−50	−45	−40	−35	−30	−25	−20	−15	−10	−5	0
40	−78	−73	−68	−63	−58	−53	−48	−43	−38	−33	−28	−23	−18

The major effect of cold on nutrient requirements of cows is increased need for energy, which usually means increasing the total amount of daily feed.

Feeding tables recommend that a 1,200-lb cow receive 16.5 lb of good mixed hay to supply energy needs during the last one-third of pregnancy. How much feed should the cow receive if she is dry and has a winter hair coat but the temperature is 20°F with a 15-mph wind? The coldness is calculated by subtracting the windchill or effective temperature (4°F) from the cow's lower critical temperature (32°F). Thus, the magnitude of coldness is 28°F. A rule of thumb (more detailed tables are available) is to increase the amount of feed 1% for each degree of coldness. A 28% increase of the 16.5 lb (the original requirement) would mean that 21.1 lb of feed must be fed to compensate for the coldness.

This example is typical of many feeding situations; however, if the cow was wet, the same increase in feed would be required at 31°F windchill (28°F of coldness).

Inability of Animals to Cope with Heat

Animals are sometimes unable to control their body temperatures in conditions of extreme heat or cold. When the body temperature exceeds normal because the animal cannot dissipate its heat, a condition known as *fever* results. Fever is often caused by systemic infectious diseases and is often most severe when environmental temperatures are extremely high or low. Keeping the ill animal comfortable while medication is given will assist recovery.

TABLE 20.2   Estimated Lower Critical Temperatures for Beef Cattle

Coat Description	Critical Temperature (°F)
Summer coat or wet	59
Fall coat	45
Winter coat	32
Heavy winter coat	18

When the temperature becomes excessively high, some animals (notably pigs) might lose normal control of their senses and do things that aggravate the situation. If nothing is done to prevent them from doing so, pigs that get too hot will often run up and down a fence line, squealing, until they collapse and die. Pigs seen doing this should be cooled with water and encouraged to lie down on damp soil.

Most dairy cows will have a reduction in milk production when the environmental temperature exceeds 80°F. This occurs primarily because appetite is depressed and feed consumption declines. Milk yields will improve when thermal stress is alleviated by providing shade, fans, or refrigerated air.

CHANGING THE ANIMALS

One of the most impressive examples of adaptability under extensive management are the small Black Bedouin goats that graze the Negev desert of Israel and the eastern Sinai. Owing to the arid and desert conditions, they are watered only once in 2–4 days, which greatly increases their foraging range.

Ninety Bedouin goats (2,970 lb total body mass) can be herded successfully 2 days away from water points, where only one fat-tail Awassi sheep (110 lb) or 1.5 Black Mediterranean goats (165 lb) would be able to live because of more frequent watering needs.

During water deprivation for 4 days, the Bedouin goats lose 25–30% of their body weight from reduction of total body water and blood plasma volume. Nevertheless, these goats have relatively high milk yields, as 35-lb goats produce over 4.5 lb of milk per day. The Bedouin goats also have relatively low daily caloric demands for maintenance per unit of metabolic weight. Milk production efficiencies have exceeded 33% of energy consumed. The Bedouin goats have developed this adaptability over several centuries of natural selection as they have survived and produced in these arid, desert environments.

Inability of Animals to Cope with Other Stresses

Some animals are unable to adjust to environmental stresses. This inability may be genetically controlled.

When cattle are grazed at high altitudes, 1–5% develop high mountain (brisket) disease. This disease usually occurs at altitudes of 7,000 feet or more. It is characterized as right-heart failure (deterioration of the right side of the heart). It was considered that this disease was caused by chronic **hypoxia** (lack of oxygen), which leads to pulmonary **hypertension** (high blood pressure). As the animal becomes more affected, marked subcutaneous edema (filling of tissue with fluid) develops in the brisket area. Some recent studies at the U.S. Poisonous Plant Center at Logan, Utah, however, have shown that certain poisonous plants that grow only at high altitudes can cause brisket disease when fed to cattle maintained at low altitudes.

Some young cattle and sheep whose dams are fed plants in selenium-deficient areas develop **white muscle disease.** The disease is characterized by calcification of the heart and voluntary muscles, giving the muscles a white appearance. White muscle disease can be prevented by administering selenium to the mother during pregnancy and to the young at birth. In the absence of treatment, some animals are affected and some are not; therefore, resistance and susceptibility appear to be inherited.

CHAPTER SUMMARY

■ Extensively managed livestock are expected to be productive under a wide range of environmental conditions.

■ Variations in environmental conditions are primarily due to changes in temperature, humidity, wind, daylight, altitude, feed, water, and exposure to parasites and disease-causing organisms.

■ Producers may change animals genetically to adapt to different environmental conditions or change the environment when economically feasible.

■ The thermoneutral zone (TNZ) or comfort zone identifies a range of temperatures where heat production and heat loss from the animal's body is about the same. Temperatures above or below the TNZ significantly affect feed intake and the animal's performance.

REVIEW QUESTIONS

1. Livestock operations, such as integrated poultry enterprises and confinement swine operations, which have highly controlled environmental conditions are said to be under _____ management.

2. Livestock operations with less producer control over the environmental conditions in which animals are expected to produce are under _____ management.

3. What are the two major kinds of changes that occur in the environment during changes in season that influence productivity of livestock?

4. What is the range of temperatures called where heat production and heat loss from the body are about the same?

5. *True or False:* When temperatures are below the lower limit of an animal's thermoneutral zone, additional feed will be required to maintain body temperature.

6. *True or False:* Temperatures exceeding the upper limit of an animal's thermoneutral zone will have no effect on animals.

SELECTED REFERENCES

Ames, D. R. 1980. Livestock nutrition in cold weather. *Anim. Nutr. and Health,* Oct.

Bonsma, J. 1983. *Livestock Production.* Cody, WY: Ag Books.

Hahn, G. L. 1985. Weather and climate impacts on beef cattle. *Beef Research Progress Report* no. 2, MARC, USDA, ARS-42.

Johnson, H. D. (ed.). 1987. *Bioclimatology and the Adaptation of Livestock.* Elsevier Science Publishers.

McDowell, L. R. (ed.). 1985. *Nutrition of Grazing Ruminants in Warm Climates.* New York: Academic Press.

National Research Council. 1981. *Effect of Environment on Nutrient Requirements of Domestic Animals.* Washington, DC: National Academy Press.

Yousef, M. K. 1985. *Stress Physiology in Livestock. Vol. II: Ungulates.* Boca Raton, FL: CRC Press.

CHAPTER 21

Animal Health

Productive animals are typically healthy animals. Death loss (**mortality rate**) is the most dramatic sign of health problems; however, lower production levels and higher costs of production which are due to sickness (**morbidity**) are economically more serious.

Disease is defined as any deviation from normal health in which there are marked physiological, anatomical, or chemical changes in the animal's body. Noninfectious diseases and infectious diseases are the two major disease types. Noninfectious diseases result from injury, genetic abnormalities, ingestion of toxic materials, and poor nutrition. Microorganisms are not involved in noninfectious diseases. Examples of noninfectious diseases are plant poisoning, bloat, and mineral deficiencies.

Infectious diseases are caused by microorganisms such as bacteria, viruses, and protozoa. A *contagious disease* is an infectious disease that spreads rapidly from one animal to another. Brucellosis, ringworm, and transmissible gastroenteritis (TGE) are examples of infectious diseases.

PREVENTION

The old adage, "An ounce of prevention is worth a pound of cure," is an important part of herd health. Preventative herd health programs can eliminate or reduce most of the health-related livestock losses. Most major animal disease problems are associated with health management.

The components of a herd health-management program include (1) veterinarian-assisted planning, (2) sanitation, (3) proper nutrition, (4) record analysis, (5) physical facilities, (6) source of livestock, (7) proper use of biologics and pharmaceuticals, (8) minimizing stress, and (9) personnel training. All of these components require cooperative efforts between the producer and veterinarian.

Planning with the Veterinarian

Veterinarians are professionally educated and trained in animal health. Veterinarians located near animal producers are familiar with the disease and health problems common in the producer's area. Producers having the most successful herd health programs use veterinarians in planning such programs. Important parts of the plan include regularly scheduled visits by the veterinarian throughout the year to assess preventative programs as well as to provide animal treatments. The veterinarian can also train farm and ranch personnel in simple herd health-management practices and serve as a reference for new products.

In most livestock and poultry operations, it is more economically feasible to have veterinarians assist in planning and implementing preventative health programs than to use their services in crisis situations only.

Producers should keep cost-effective health records as recommended by the veterinarian. These records are needed to assess problem areas and to develop a more effective herd health plan. The records can include such information as what was done when in the preventative and treatment programs, descriptions of the conditions during health problems, and deaths. Possible causes of death also should be recorded, with **necropsy** records obtained when economically feasible.

Sanitation

The severity of some diseases is dependent on the number and virulence of microorganisms entering the animal's body. Many microorganisms live and multiply outside the animal, so the number of microorganisms can be reduced by implementing sanitation practices. These practices, in turn, reduce the incidence of disease outbreaks.

Manure and other organic waste materials are ideal environments for the proliferation of microorganisms. A good sanitation program includes cleaning of organic materials from buildings, pens, and lots. This allows the effective destruction of microorganisms from high temperatures and drying. Buildings, pens, and pastures should be well drained, preventing prolonged wet areas or mud holes. These sanitation practices help both in disease prevention and controlling parasites.

Antiseptics and **disinfectants** are carefully selected and effectively utilized in a good sanitation program. Antiseptics are substances, usually applied to animal tissue, that kill or prevent the growth of microorganisms. Disinfectants are products that destroy pathogenic microorganisms. They are usually agents used on inanimate objects. In the absence of disinfectants, sanitizing with clean water is helpful.

Other important sanitation measures include the prompt and proper disposal of dead animals, either to rendering plants or by burial.

Proper Nutrition

Well-nourished animals receive an adequate daily supply of essential nutrients. Undernourished animals usually have a weak immune system, thus making them more vulnerable to invading microorganisms.

Nutrition principles are presented in Chapters 15, 16, and 17. Feeding programs for the various species of farm animals are covered in Chapters 23, 25, 27, 29, 31, 32, and 33.

Record Analysis

Herd health must be approached from an economic standpoint. Does a health problem exist, or is it just a perception? Proper records permit health problems to be identified, determine what is causing them, and thereby allow alternative methods of prevention and treatment to be assessed.

Physical Facilities

Physical facilities contribute to animal health problems by causing physical injury or stress, or by allowing dissemination of pathogens through a group of animals. They can contribute to the spread of disease by not preventing its transmission (e.g., venereal disease transmission owing to poor or inadequate fences). Even proper facilities that are misused, such as at a feedlot, can provide an easier transmission of disease owing to crowding, stress, and poor sanitation.

Source of Livestock

Producers can reduce the spread of infectious diseases into their herds and flocks by (1) purchasing animals (entering their herds) from other producers who have effective herd health-management programs; (2) controlling exposure of their animals to other people and vehicles; (3) providing clothing, boots, and disinfectant to people who must be exposed to the animals and facilities; (4) isolating animals to be introduced into their herds so that disease symptoms can be observed for several weeks; (5) controlling insects, birds, rodents, and other animals that can carry disease organisms; and (6) keeping their animals out of drainage areas that run through their farm from other farms.

Use of Biologics and Pharmaceuticals

Drugs are classified as biologicals or pharmaceuticals. **Biologics** are used primarily to prevent diseases, whereas **pharmaceuticals** are used mainly to treat diseases. Both are needed in a successful herd health-management program.

Most biologicals are used to stimulate immunity against specific diseases. Vaccines are biological agents that stimulate active immunity in the animal.

Other biologicals stimulate the body's immune system to produce antibodies that fight diseases. This is similar to immunity through natural infection. An antibody is a protein molecule that circulates in the bloodstream and neutralizes disease-causing microorganisms. An animal is immunized when sufficient antibodies are produced to prevent the disease from developing. Specific antibodies must be produced for specific diseases; thus vaccination for several different diseases may be necessary. Also, periodic revaccination might be needed to maintain antibodies in the blood at an adequate level.

Immunity from vaccination occurs only if the immune system properly responds. In some situations, vaccination is not effective. In most animal species, for example, the immune systems of newborn young are not yet developed enough to respond. This is why their immunity is dependent on the maternal antibodies ingested from their mother's milk. Also, undernourished animals, stressed animals, or those animals already exposed to the disease rarely give a positive response to vaccination.

The amount of vaccine, frequency, route of administration, and duration of immunity vary with the specific vaccine. The manufacturer's directions should be followed carefully.

Vaccines (biologicals) and drugs (pharmaceuticals) can be administered as follows:

1. Topically—applied to the skin.
2. Orally—through the mouth by feeding, drenching, or using balling guns. The latter is used to give pills such as capsules or boluses (Fig. 21.1).
3. Injected directly—into the animal's body by using a needle and syringe.

Types of injections include (1) subcutaneous (under the skin but not into the muscle), (2) intramuscular (directly into the muscle), (3) intravenous (into a vein), (4) intramammary (through a cannula in the teat canal and injecting the drug into the milk cistern), (5) intraperitoneal (into the peritoneal cavity), and (6) intrauterine (using an infusion pipette or a tube through the cervix). Some injection sites are shown in Figs. 21.2–21.4.

Pharmaceuticals are used to kill or reduce the growth of microorganisms in the treatment of diseases and infections. Pharmaceuticals come in a variety of forms: liquid, powder, boluses, drenches, and feed additives. They should be administered only after a specific diagnosis has been made. Examples of pharmaceuticals are antibiotics, steroids, sulfa compounds, hormones, and nitrofurans.

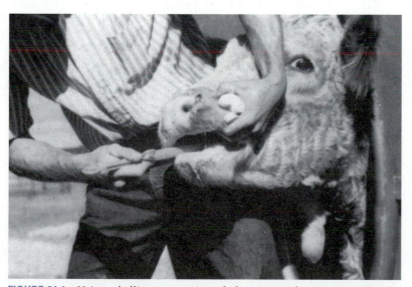

FIGURE 21.1   Using a balling gun to give a bolus or capsule to a cow. After the instrument is placed on the back of the tongue, the plunger is pushed to deposit the pill onto the back of the tongue. The animal then swallows the pill. Courtesy of Norden Laboratories.

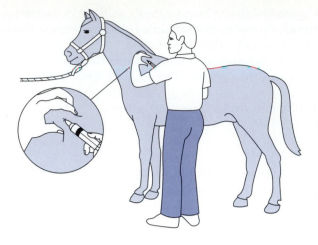

FIGURE 21.2 Subcutaneous injection in the horse. Courtesy of R. A. Battaglia, and V. B. Mayrose, *Handbook of Livestock Management Techniques* (New York: Macmillan, 1981). Copyright © Macmillan Publishing Co.

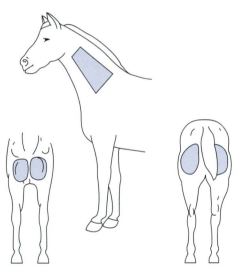

FIGURE 21.3 Intramuscular injection sites in the horse. Courtesy of R. A. Battaglia, and V. B. Mayrose, *Handbook of Livestock Management Techniques* (New York: Macmillan, 1981). Copyright © Macmillan Publishing Co.

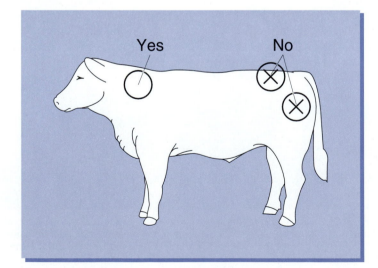

FIGURE 21.4 Selection of injection site is an important management decision in preventing blemishes to valuable cuts of meat. Courtesy of NCA.

Other pharmaceuticals are used to control external and internal parasites. Losses owing to external and internal parasites are usually reduced weight gains, poor feed conversion, lower milk and egg production, reduced hide value, and excessive carcass trim, rather than high death losses.

Common external parasites are flies, lice, mites, and ticks, which live off the flesh and blood of animals. These parasites can also transmit infectious diseases from one animal to another. Insecticides are used to control external parasites. They can be applied as systemics, spraying, fogging, dipping, back rubbers, ear tags, or dust bags.

Roundworms and tapeworms are the most common internal parasites, with flukes causing problems in certain environments. Internal parasites are controlled by interrupting their life cycle by (1) the presence of unfavorable climatic conditions (wet, warm weather favors the proliferation of many internal parasites); (2) destroying intermediate hosts; (3) managing animals so they will not ingest the parasites or their eggs; and (4) giving therapeutic chemical treatment. *Anthelmintics* are drugs that are given to kill the internal parasites.

Stress

Stress is any environmental factor that can cause a significant change in the animal's physiological processes. Physical sources of stress are temperature, wind velocity, nutritional deficiency or oversupply, mud, snow, dust, fatigue, weaning, ammonia buildup, transportation, castration, dehorning, and abusive handling. Social or behavior-related stress can result from aggression or overcrowding.

Prolonged stress can impair the body's immune system, causing a reduced resistance to disease. Stressful conditions in animals should be minimized within economic constraints.

Personnel Training

Maintaining good animal health requires continuous training of personnel. This is one of the most difficult areas to accomplish, particularly in a livestock operation that employs many people. While the owner understands what needs to be done, this may differ from what the employees actually accomplish. Cooperative training programs and informative sessions with the veterinarian can assist in the implementation of effective herd health-management programs.

DETECTING UNHEALTHY ANIMALS

Visual Observation

Detecting sick animals and separating them from the healthy animals is an important key to a successful treatment program. Animals treated in the early stages of sickness usually respond more favorably to treatment than do animals whose illness has progressed to advanced stages.

Early signs of sickness are not easily detected; however, the following are some of the observable signs, several of which are shown in Fig. 21.5.

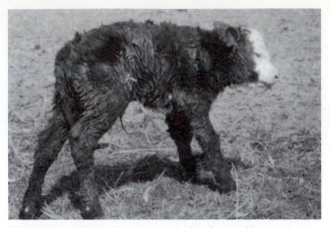

FIGURE 21.5 This calf is sick with diarrhea (calf scours). Note the visual signs of illness. Courtesy of Norden Laboratories.

1. Loss of appetite is observed.
2. Animal appears listless and depressed.
3. Ears may be droopy or not held in an alert position.
4. Animal has a hump in the back and holds its head in a lower position.
5. Animal stays separated from the rest of the herd or flock.
6. Coughing, wheezing, or labored breathing occurs.

Vital Signs

Body temperature, respiration rate, and heart rate are the vital signs of the animal. Health problems are usually evident when one or more of the vital signs deviate from the normal range. The average and normal range for the vital functions are given in Table 21.1.

TABLE 21.1   Vital Functions of Selected Livestock and Poultry Species

Animal	Rectal Temperature (°F)	Respiration Rate (per min)	Heart Rate (per min)
Cattle	101.5[a] (100.4–102.8)[b]	30[a]	50[a]
Swine	102.5 (101.6–103.6)	16	60
Sheep	102.3 (100.9–103.8)	19	75
Goat	102.3 (101.3–103.5)	15	90
Horse	100.0 (99.1–100.8)	12	45
Chicken	107.1 (105.0–109.4)	12–36[b]	275
Turkey	105	29–49[b]	165

[a] Average.
[b] Normal range.
Source: R. A. Battaglia, and V. B. Mayrose, *Handbook of Livestock Management Techniques* (New York: Macmillan, 1981). Copyright © Macmillan Publishing Co.

Body temperature is taken rectally. An elevated body temperature occurs in overheated animals or with most infectious diseases. A subnormal temperature indicates chilling or a critical condition of the animal.

By evaluating the animal's vital signs and visually observing its appearance, health problems can be identified in their early stages. This allows for early treatment, which usually prevents serious losses.

MAJOR DISEASES OF FARM ANIMALS

Table 21.2 identifies some of the major diseases and health problems of beef cattle, dairy cattle, swine, horses, poultry, and sheep.

TABLE 21.2   Selected Major Diseases and Health Problems of Beef Cattle, Dairy Cattle, Swine, Horses, Poultry, and Sheep

Species/Disease/Cause	Signs	Prevention	Treatment
Beef Cattle			
Bovine viral diarrhea (BVD) Virus. Laboratory diagnosis imperative for accuracy	(Feeder cattle) Ulcerations throughout digestive tract Diarrhea (often containing mucus or blood)	Vaccination prior to exposure Avoid contact with infected animals	Symptomatic treatment; antibiotics; sulfonamids; "Hyper" serums; force feed
	(Breeding cattle) Abortions	Vaccination prior to breeding	None Symptomatic treatment
BRSV (bovine respiratory syncytical virus)	Labored breathing; pneumonia	Vaccination	Antihistamines, corticosteroids
Brucellosis Bacteria	Abortions	Calfhood vaccination (in some states) between the ages of 2–12 months Test and slaughter reactors	None
Campylobacteriosis (vibriosis) Bacteria	Repeat services Abortions (1–2%)	Vaccination of females and bulls prior to breeding Use of artificial insemination Untried bulls on virgin heifers Avoid sexual contact with infected animals	None
Infectious bovine rhinotracheitis (IBR) Virus. Laboratory diagnosis imperative for accuracy	Respiratory involvement Fever, vaginitis, abortion, and infertility in females Abortions or stillborn calves Preputial infections in males	Vaccinate cows 30 days prior to breeding Vaccinate feeder cattle prior to exposure Semen from reputable bulls	Chlortetracycline, 350 g/day Sulfamethazine, 350 mg to keep down secondary infections
Leptospirosis Leptospira spp Several bacteria	(Breeding cattle) Fever, off feed, abortions, icterus, discolored urine	Vaccination at least annually (in high-risk areas, more frequently) Control rodents Avoid contact with wildlife and other infected animals	Streptomycin Penicillin Dihydrostreptomycin
	(Feeder cattle)	Vaccination on arrival	Dihydrostreptomycin

TABLE 21.2   (*Continued*)

Species/Disease/Cause	Signs	Prevention	Treatment
Beef Cattle (*continued*)			
Scours, calf colibacil-losis "septicemia" *E. coli* K99 (pathogenic bacteria) Several viruses plus stress factors	Diarrhea, weakness, and dehydration Rough hair and coat	Dry, clean calving areas Adequate colostrum	Fluid therapy Antibiotics to prevent secondary infections
Scours (viral) Virus (reovirus and corona virus)	Acute diarrhea, high mortality Affects calves shortly after birth (See Figs. 21.5 and 21.6).	Vaccination with specific viral vaccines	Fluid therapy not effective in severe cases
Trichomoniasis Protozoa	Repeat services Abortion (2–4 mo) Pyometra	Use of artificial insemi-nation Untried bulls on virgin heifers Avoid sexual contact with infected animals	Cull carrier animals
Dairy Cattle			
Mastitis Primarily infectious bacteria (*Streptococci, Staphylococci,* or *E. coli*)	Inflammation of the udder Decreased milk production	Use CMT (California mastitis test) or SCC (somatic cell counts)—checks for white blood cells in milk Avoid injury to udder Use proper milking techniques	Determine specific organism causing the problem, then select most effective antibiotic. For example, cloxacillin, penicillin, erythromycin, and neomycin are effective against several streptococcus and staphylococcus organisms
Milk fever (parturient paresis) Metabolic disorder (low blood calcium level associated with a deficiency of vitamin D and phosphorus)	Occurs at onset of lactation Muscular weakness; drowsiness; cow lies down in curled position	Proper nutrition and management of cows during dry period	Intravenous injection of calcium borogluconate
Respiratory diseases (See *Infectious Bovine Rhinotracheitis, "Shipping Fever" parainfluenza,* under **Beef Cattle**)			
Uterine infections Usually results from bacterial contamination at the time of calving	Infections classified according to tissues involved Animal may be system-ically or only locally affected Endometritis: inflamma-tion of lining of uterus Metritis: infection of all tissues of uterus Pyometra: pus in the uterus	Sanitation; cleanliness when giving calving assistance	Early detection; antibiotics, hormone therapy, prostaglandins
Venereal diseases (See *BVD, IBR, leptospirosis, vibriosis* and *trichomoniasis,* under **Beef Cattle**)			

TABLE 21.2 (*Continued*)

Species/Disease/Cause	Signs	Prevention	Treatment
Swine			
Atrophic rhinitis Bordetella bronchiseptica (bacteria); Pasteurella (secondary invader)	Sneezing (most common); sniffling; snorting; coughing; nose shape (twisted to one side) Nasal infection (inflammation of membranes in nose) Diagnosis confirmed by observing turbinate bone atrophy (in nose) during postmortem examination (Fig. 21.7)	Monitor performance; monitor contact with animals from outside the herd; correct environmental deficiencies (sanitation, temperature, humidity, ventilation, dust, drafts, excessive ammonia, and overcrowding)	Vaccinate against Bordetella and Pasteurella organisms; medicate sow feed with sulfamethazine or oxytetracycline; improve environment
Colibacillosis E. coli Bacteria	Diarrhea (pale, watery feces); weakness; depression Most serious in pigs under 7 days of age	Good sanitation; good management practices (nutrition, pigs suckle soon after birth, prevent chilling); vaccinate sows to increase protective value of colostrum	Effective treatment is limited; identify strain of *E. coli,* then use specific antibacterial drug
Mycoplasmal infections (Pneumonia and arthritis) Mycoplasma bacteria	Pneumonia: death loss low; dry cough; reduced growth rate; lesions in lungs Arthritis: inflammation in lining of chest and abdominal cavity; lameness; swollen joints; sudden death	Pneumonia: reduce animal contact; good nutrition, warm and dust-free environment, and parasite (ascarid and lung worm) control minimizes effects of the disease	Pneumonia: adequate treatment not available; sulfas and antibiotics prevent secondary pneumonia infections Arthritis: no satisfactory treatment; can depopulate and restock with disease-free animals
Pseudorabies Virus of herpesvirus group	Baby pigs (less than 3 weeks old): sudden death Pigs (3 weeks to 5 months): fever, loss of appetite, labored breathing, trembling and incoordination of hind legs Mature pigs: less severe fever, loss of appetite; abortion and reproductive failure	Prevent direct contact with infected swine; humans should wear clean clothes and disinfect boots when entering and leaving the premises Recovered swine are immune but should be considered carriers	Drugs and feed additives are not effective; quarantine; blood test and slaughter those infected; repopulate
Transmissible gastroenteritis (TGE) Virus of coronavirus group	Vomiting; diarrhea; weakness; dehydration; high death rate in pigs under 32 weeks of age (see Fig. 21.6)	Sanitation is most cost effective Prevent transfer of virus from infected animals through exposure to other pigs, birds, equipment, and people	No drugs are effective Fresh water and a draft-free environment will reduce losses Antibiotics may prevent some secondary infections
Horses			
Colic Noninfectious: a general term indicating abdominal pain	Looking at flank; kicking at abdomen; pawing; getting up and down; rolling; sweating	Parasite control, avoiding moldy or spoiled feeds; avoid overeating or drinking too much water when hot; have sharp points of teeth filed (floated) annually to assure proper chewing of feed	Quiet walking of horse; avoid undue stress; bran mash and aspirin; milk of magnesia; mild, soapy, warm water enema for impaction

TABLE 21.2   (*Continued*)

Species/Disease/Cause	Signs	Prevention	Treatment
Horses (*continued*)			
Lameness Most common are stone bruises, puncture wounds, sprains, and navicular disease	Departure from normal stance or gait; limping; head bobbing; dropping of the hip; alternate resting on front feet; pointing (extends one front foot in front of the other); stiffness Navicular disease: soft tissue bursitis in the foot	Bruise: avoid running horse on graveled roads or rocky terrain Puncture wounds: avoid areas where nails and other sharp objects may be present Sprains: avoid stress and strains on feet and legs Navicular disease: exact cause not known; avoid continual concussion on hard surfaces; provide proper shoeing, and trimming	Varies with cause of lameness Bruises: soaking foot in bucket of ice water or standing horse in mud; aspirin Puncture wound: tetanus protection; antibiotic; aspirin; soaking foot in hot epsom salts Navicular disease: no known cure; drug therapy and corrective shoes for temporary relief; surgery
Respiratory Three viral diseases: viral rhinopneumonitis; viral arteritis; influenza One bacterial disease: strangles	Nasal discharge; coughing; lung congestion; fever Rhinopneumonitis may cause abortion	Isolate infected animals; avoid undue stress; draft-free shelter; parasite control; avoid chilling Rhinopneumonitis, influenza, and strangles vaccines are available	Follow prevention guidelines Strangles: drain abscess
Parasites 150 different internal parasites with strongyles (bloodworms), ascarids (roundworms), bots, pinworms, and stomach worms most common External parasites (most common are lice, ticks, and flies)	Unthrifty appearance ("pot-bellied," rough hair coat); weakness; poor growth Lice, ticks, and flies can be visually observed on the horse	Good sanitation practices (clean feeding and watering facilities; regular removal of manure); periodic rest-periods for pastures; avoid over-crowding of horses; avoid spreading fresh manure on pastures	Internal parasites: worm horses twice a year (usually spring and fall); bot eggs can be shaved from the hairs of the legs External parasites: application of insecticides
Poultry			
Avian influenza Virus	Drop in egg production; sneezing; coughing; in the severe systemic form, deep drowsiness and high mortality are common	Vaccine (but has short immunity); select eggs and poults from clean flocks	No effective drug available
Coccidiosis Coccida (depends on type of coccidiosis—there are nine or more types. Some signs, prevention, and treatment of three more common types are identified)	Weight loss; unthriftiness; palor; blood in droppings; lesions in intestinal wall	Use coccidiostat (kills coccidia organism)	Sulfa drugs in drinking water
Lymphoid leukosis Virus	Combs and wattles may be shriveled, pale, and scaly; enlarged, infected liver Lesions common in liver and kidneys	Sanitation; development of resistant strains through breeding methods	None

TABLE 21.2 (*Continued*)

Species/Disease/Cause	Signs	Prevention	Treatment
Poultry (continued)			
Marek's disease Herpesvirus	Can cause high mortality in pullet flocks; paralysis; death might occur without any clinical signs	Vaccination	None
Mycoplasma infections Mycoplasma organisms (There are several diseases caused by mycoplasma organisms.) Most important are chronic respiratory disease (CRD) and infectious synovitis	CRD—difficult breathing; nasal discharge; rattling in windpipe; death loss may be high in turkey poults; swelling of face in turkeys Synovitis—swollen joints, (Fig. 21.8) tendons, tendon sheaths and footpads; ruffled feathers	For both CRD and synovitis: use chicks or poults from disease-free parent stock; sanitation (cleaning and disinfecting the premises)	Antibiotic in feed or drinking water
Newcastle disease	Gasping, coughing, hoarse chirping; paralysis; mortality may be high (Fig. 21.9)	Vaccination	None
Sheep			
C. pseudotuberculosis Bacteria	Enlarged lymph nodes	Sanitation; reduce opportunity for wound infection at docking, castration, and shearing	Antibiotics
Enterotoxemia (overeating disease) Bacteria	Often sudden death occurs without warning; sick lambs might show nervous symptoms; e.g., head drawn back, convulsions	Vaccination (most effective in young lambs under 6 weeks of age nursing heavy-milking ewes and in weaned lambs on lush pasture or in feedlots)	None; antitoxins can be used but they are expensive and immunity is temporary (2–3 weeks)
Epididymitis Bacteria	Most common in western United States; swelling of epididymis; poor semen quality; low conception rates	Rigid culling and vaccination	Rigid culling
Footrot Bacteria	Lameness; interdigital skin is usually red and swollen	Remove from wet pastures or stubble pastures; vaccines in some cases	Disinfectants such as 5% formalin or 10% copper sulfate; antibiotics
Pneumonia Many types; caused by viruses, stress, bacteria, and parasites	Sudden death; nasal discharge; depression; high temperature	Reduce stress factors	Antibiotics; sulfonamides
Pregnancy toxemia Undernutrition in late pregnancy; stress associated with poor body condition	Listlessness; loss of appetite; unusual postures; progressive loss of reflexes; hypoglycemia; coma; death	Prevent obesity in early pregnancy; good nutrition during last 6 weeks of pregnancy; reduce environmental stresses	In early stages the administration of some glucogenic materials may reduce mortality; in advanced stages no treatment improves survival

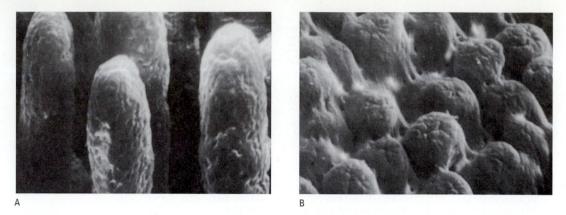

A   B

FIGURE 21.6   (A) Normal villi that line the intestinal wall. (B) De-nuded villi after virus infection in pigs and calves. Severe diarrhea results as the animal has lost the ability to digest and absorb nutrients. If the animal survives, the villi will regain their natural form in 1–2 weeks. Courtesy of Dr. G. L. Waxler, 1972. *Am. J. Vet. Res.* 33:1323.

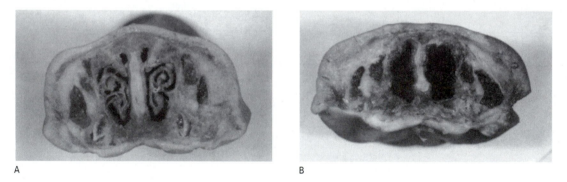

A   B

FIGURE 21.7   Cross sections through the hog's snout showing the effects of atrophic rhinitis. (A) Normal snout with turbinate bones. (B) Diseased snout showing severe atrophy or absence of turbinate bones. Courtesy of Elanco Products Company.

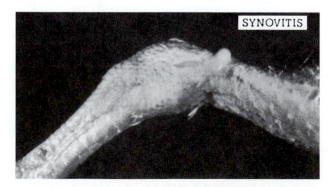

FIGURE 21.8   Infectious synovitis. A swollen hock and lameness are characteristic. When opened, the hock usually contains a creamy exudate. Courtesy of Salisbury Laboratories.

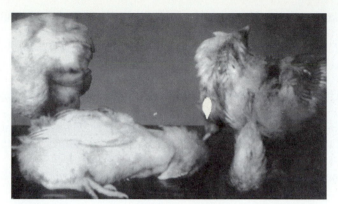

FIGURE 21.9   Newcastle disease. Birds may have complete or partial paralysis of legs and wings. The photo shows common signs when nerves are affected. Courtesy of Salisbury Laboratories.

CHAPTER SUMMARY

- Disease is any deviation from normal health in which there are marked physiological, anatomical, or chemical changes in the animal's body.

- Death loss (mortality rate) and sickness (morbidity) rate are minimized in an excellent herd health management program which includes (1) planning with veterinarian, (2) sanitation, (3) proper nutrition, (4) record keeping and analysis, (5) physical facilities, (6) source of livestock, (7) proper use of biologics and pharmaceuticals, (8) minimizing stress, and (9) training personnel who are working with the animals.

- Health problems are usually evident when one or more of the vital signs (temperature, respiration rate, and heart rate) deviate from the normal range.

- The detection, prevention, and treatment protocols are given for the major diseases and health problems for cattle, swine, horses, poultry, and sheep.

REVIEW QUESTIONS

1. Death loss or _____ rate is the most dramatic sign of health problems among animals.

2. How is disease defined?

3. What is a contagious disease?

4. The best herd health management programs involve _____ of disease.

5. What are the nine components of a herd health management program?

SELECTED REFERENCES

Publications

Battaglia, R. A., and Mayrose, V. B. 1981. *Handbook of Livestock Management Techniques.* New York: Macmillan (esp. Chap. 10, Animal health management).

Gaafar, S. M., Howard, W. E., and Marsh, R. E. (eds.). 1985. *Parasites, Pests and Predators.* New York: Elsevier Science.

Haynes, N. B. 1985. *Keeping Livestock Healthy.* Pownal, VT: Storey.

Keeler, R. F., VanKampen, K. R., and James, L. F. (eds.). 1978. *Effects of Poisonous Plants on Livestock.* New York: Academic Press.

Kirkbride, C. A. 1986. *Control of Livestock Diseases.* Springfield, IL: Charles C. Thomas.

Naviaux, J. L. 1985. *Horses in Health and Disease.* Philadelphia: Lea & Febiger.

Radostits, O. M. and Blood, D. C. 1985. *Herd Health.* Philadelphia: W. B. Saunders.

Stalheim, O. H. V. 1994. Animal Diseases. *Encyclopedia of Agricultural Science.* San Diego: Academic Press, Inc.

Visuals

Distinguishing Normal and Abnormal Cattle (videotape). Agricultural Products and Services, 2001 Killebrew Dr., Suite 333, Bloomington, MN 55420.

CHAPTER 22

Beef Cattle Breeds and Breeding

A ***breed*** of cattle is defined as a race or variety, the members of which are related by descent and are similar in certain distinguishable characteristics. More than 250 breeds of cattle are recognized throughout the world, and several hundred other varieties and types have not been identified with a breed name.

Some of the oldest recognized breeds in the United States were officially recognized as breeds during the middle to late 1800s. Most of these breeds originated from crossing and combining existing strains of cattle. When a breeder or group of breeders decided to establish a breed, distinguishing that breed from other breeds was of paramount importance; thus, major emphasis was placed on readily distinguishable visual characteristics, such as color, color pattern, polled or horned condition, and rather extreme differences in form and shape.

New cattle breeds, such as Brangus, Santa Gertrudis, and Beefmaster, have come into existence in the United States during the past 50–75 years. These breeds have been developed by attempting to combine the desirable characteristics of several existing breeds. Currently there are new breeds of cattle being developed. They are sometimes called ***composite breeds*** or ***synthetic breeds.*** In most cases, however, the same visual characteristics as previously mentioned are used to give the new breeds visual identity.

After some of these first breeds were developed, it was not long until the word ***purebred*** was attached to them. Herd books and registry associations were established to assure the "purity" of each breed and to promote and improve each breed. *Purebred* refers to purity of ancestry, established by the pedigree, which shows that only animals recorded in that particular breed have been mated to produce the animal in question. Purebreds, therefore, are cattle within the various breeds that have pedigrees recorded in their respective breed registry associations.

When viewing a herd of purebred Angus or Herefords, uniformity is noted, particularly uniformity of color or color pattern. Because of this uniformity of one or two characteristics, the word *purebred* has come to imply genetic uniformity (homozygosity) of all characteristics. Cattle within the same breed are not highly homozygous because high levels of homozygosity occur only after many generations of close inbreeding. This close inbreeding has not occurred in cattle breeds. If breeds were uniform genetically, they could not be improved or changed even if changes were desired.

WHAT IS A BREED?

The genetic basis of cattle breeds and their comparison is not well understood by most livestock producers. Often the statement is made, "There is more variation within a breed than there is between breeds." The validity of this claim needs to be examined carefully. Considerable variation does exist within a breed for most of the economically important traits. This variation is depicted in Fig. 22.1, which shows the number of calves of a particular breed that fall within certain weight-range categories at 205 days of age. A bell-shaped curve is formed by connecting the high points of each bar. The breed average is represented by the solid line that separates the bell curve into equal halves. Most of the calves are near the breed average; however, at the outer edge of the bell curve are high- and low-weaning weight calves. Note that they are fewer in number at these extremes.

Figure 22.2 allows us to compare, hypothetically, three breeds of cattle in terms of weaning weight. The breed averages are different; however, the variation within each breed is comparable among all three. The statement "There is more variation within a breed than between breed averages" is more correct than the statement "There is more variation within a breed than there is between breeds."

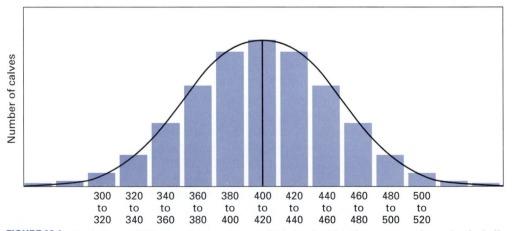

FIGURE 22.1   Variation or differences in weaning weight in beef cattle. The variation shown by the bell-shaped curve could be representative of a breed or a large herd. The vertical line in the center is the average or the mean, which is 410 lb. Note that the number of calves is greater around the average and is less at the extremely light and extremely heavy weights. Courtesy of Colorado State University.

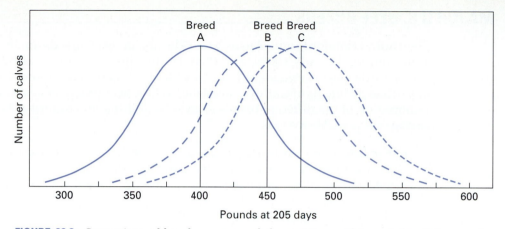

FIGURE 22.2 Comparison of breed averages and the variation within each breed for weaning weight in beef cattle. The vertical lines are the breed averages. Note that some individual animals in breeds B and C can be lower in weaning weight than the average of breed A. Courtesy of Colorado State University.

Figures 22.1 and 22.2 are hypothetical examples; however, they are based on realistic samples of data obtained from the various breeds of cattle. Figure 22.3 shows how breeds can differ greatly when only one trait is considered. Note how distinctly different the Charolais-sired calves are from the Jersey or Angus and Hereford calves. The statement "There is more variation within the Charolais breed than between the Simmental and Jersey breeds" is not valid in this case. Rather, "There is more variation within the Charolais breed than between breed averages" is a more correct statement.

The information in Fig. 22.3 supports Charolais as being the most superior breed based on breed averages. However, this is only for one trait, and other breeds show superiority to the Charolais breed when other economically important traits are considered.

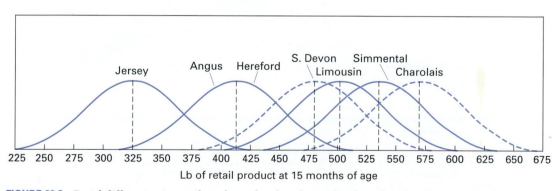

FIGURE 22.3 Breed differences in retail product when breeds are slaughtered at the same age. Courtesy of U.S. Meat Animal Research Center.

MAJOR U.S. BEEF BREEDS

Shorthorn, Hereford, and Angus were the major beef breeds in the United States during the early 1900s (Fig. 22.4). During the 1960s and 1970s, the number of cattle breeds remained relatively stable at between 15 and 20. Today, more than 60 breeds of cattle are available to U.S. beef producers. Why the large importation of the different breeds from several different countries? Following are several possible reasons:

1. Feeding more cattle larger amounts of grain, a practice started in the 1940s, resulted in many overfat cattle—cattle that had been previously selected to fatten on forage diets. Therefore, a need was established for grain-fed cattle that could produce a higher percentage of lean to fat at the desired slaughter weight.

2. Economic pressure to produce more weight in a shorter period of time demonstrated a need for cattle with more milk and more growth.

3. An opportunity was available for some promoters to capitalize on merchandising a certain breed as being the ultimate in all production traits. This opportunity was easily accomplished because there was little comparative information on breeds.

Color Plates M through P (following this page) identify the most numerous breeds of beef cattle in the United States. Table 22.1 gives some distinguishing characteristics and brief background information for more of these breeds.

The relative importance of the various breeds' contributions to the total beef industry is best estimated by the registration numbers of the breeds (Table 22.2).

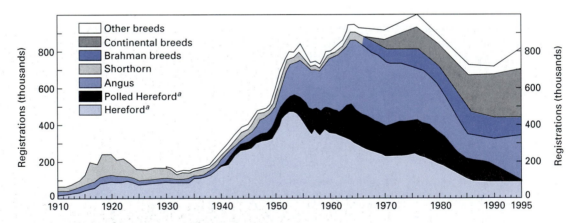

Prior to 1965: Brahman breeds were included in other breeds. Very few continental breeds were registered.

After 1965: Shorthorn ends, included in other breeds.
Brahman breeds (Brahman, Brangus, Beefmaster, Santa Gertrudis).
Continental breeds (Charolais, Gelbvieh, Limousin, Salers, Simmental).

[a]1995: Hereford and Polled Hereford breed associations were combined.

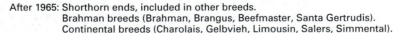

FIGURE 22.4 Registration numbers of beef cattle in the United States, 1910–1995. Courtesy of the USDA and breed associations.

Angus

Beefmaster

Brahman

Brangus

Charolais

Chianina

PLATE M.

PLATES M–P.
Some of the breeds of cattle used for beef production in the United States. Courtesy of the American Angus Association, BBU, American Brahman Breeders Association, International Brangus Breeders Association, Charolais Journal, American Breeders Service, North American Gelbvieh Association, North American Limousin Foundation, American Polled Hereford Association, Dickinson Ranch, Santa Gertrudis Breeders International, Leachman Cattle Co., American Scotch Highland Breeders Association, American Shorthorn Association, and American Simmental Association.

Devon

Galloway

Gelbvieh

Hereford

Limousin

Longhorn

PLATE N.

Maine Anjou

Murray Grey

Pinzgauer

Polled Hereford

Red Angus

Red Poll

PLATE O.

Salers

Santa Gertrudis

Scotch Highland

Shorthorn

Simmental

Tarentaise

PLATE P.

TABLE 22.1   Background and Distinguishing Characteristics of Major Beef Breeds in the United States

Breed	Distinguishing Characteristics	Brief Background
Angus	Black color; polled	Originated in Aberdeenshire and Angushire of Scotland. Imported into the United States in 1873.
Beefmaster	Has various colors; horned	Developed in the United States from Brahman, Hereford, and Shorthorn breeds. Selected for its ability to reproduce, produce milk, and grow under range conditions.
Brahman	Various colors, with gray predominant; they are one of the Zebu breeds, which have the hump over the top of the shoulder; most Zebu breeds also have large, drooping ears and excess skin in the throat and dewlap	Major importations to the United States from India and Brazil. Largest introductions in early 1900s. These cattle are heat tolerant and well adapted to the harsh conditions of the Gulf Coast region.
Brangus	Black and polled predominate, although Red Brangus is a separate breed	U.S. breed developed around 1912—⅜ Brahman, ⅝ Angus.
Charolais	White color with heavy muscling; horned or polled	One of the oldest breeds in France. Brought into the United States soon after World War I, but its most rapid expansion occurred in the 1960s.
Chianina	White color with black eyes and nose; extremely tall cattle	An old breed originating in Italy. Acknowledged to be the largest breed, with mature bulls weighing more than 3,000 lb.
Gelbvieh	Golden colored; horned or polled	Originated in Austria and Germany. Developed as a dual-purpose breed used for draft, milk, and meat.
Hereford	Red body with white face; horned	Introduced in the United States in 1817 by Henry Clay. Followed the Longhorn in becoming the traditionally known range cattle.
Limousin	Golden color with marked expression of muscling; polled or horned	Introduced into the United States in 1969, primarily from France.
Longhorn	Multicolored with characteristically long horns	Came to West Indies with Columbus. Brought to the United States through Mexico by the Spanish explorers. Longhorns were the noted trail-drive cattle from Texas into the Plains states.
Polled Hereford	Red body with white face; polled	Bred in 1901 in Iowa by Warren Gammon, who accumulated several naturally polled cattle from horned Hereford herds.
Red Angus	Red color; polled	Founded as a performance breed in 1954 by sorting the genetic recessives from Black Angus herds.
Salers	Uniform mahogany red with medium to long hair; horned	Raised in the mountainous area of France, where they were selected for milk, meat, and draft.
Santa Gertrudis	Red color; horned	First U.S. breed of cattle that was developed on the King Ranch in Texas. Combination of ⅝ Shorthorn and ⅜ Brahman.
Shorthorn	Red, white, or roan in color; horned and polled	Introduced into the United States in 1783 under the name *Durham*. Most prominent in the United States around 1920.
Simmental	Yellow to red-and-white color pattern; polled and horned	A prominent breed in parts of Switzerland and France. First bull arrived in Canada in 1967. Originally selected as a dual-purpose for milk and meat.
Tarentaise	Solid wheat-colored hair ranging from light cherry to dark blond; horned	Mountain cattle derived from an ancient Alpine strain in France. Originally a dairy breed in which maternal traits have been emphasized.

TABLE 22.2   Annual Registration Numbers for Major U.S. Beef Breeds

| Breed[a] | Annual Registrations (in thousands) | | | | | Year U.S. Association Formed |
	1995	1990	1985	1980	1975	
Angus	224.8	159.0	175.5	257.6	306.5	1883
Hereford[b]	120.0	170.5	180.1	353.2	419.2	1881
Limousin	79.3	78.4	44.5	28.0	32.4	1968
Simmental	71.0	79.3	73.0	66.1	75.5	1969
Charolais	55.6	44.8	23.2	23.0	45.2	1957
Beefmaster	47.1	38.4	32.1	30.0	12.0	1961
Gelbvieh	33.8	22.8	16.1	NA	NA	1971
Brangus	31.0	32.1	30.3	24.5	13.4	1949
Red Angus	30.0	15.4	12.0	12.5	10.1	1954
Salers	20.0	24.0	11.4	NA	0.2	1974
Shorthorn	20.0	18.0	16.7	19.4	29.2	1872
Brahman	15.4	13.0	29.9	36.4	27.5	1924

[a] Only breeds with more than 15,000 annual registrations are listed.
[b] Includes Polled Hereford.
Source: Beef breed associations.

Although registration numbers are for purebred animals, they reflect the commercial cow–calf producers' demand for the different breeds. Registration numbers show that Angus, Hereford/Polled Hereford, Limousin, and Simmental, are the most important breeds of the beef cattle industry in the United States.

The numbers of cattle belonging to various breeds in this country have changed over the past years, as shown in Fig. 22.4. No doubt some breeds will become more numerous in future years and others will decrease significantly in numbers. These changes will be influenced by economic conditions and how well the breeds meet the needs of the commercial beef industry.

IMPROVING BEEF CATTLE THROUGH BREEDING METHODS

Genetic improvement in beef cattle can occur by selection and by using a particular mating system. Significant genetic improvement by selection results when the selected animals are superior to the herd average and when the heritabilities of the traits are relatively high (40% and higher). It is important that the traits included in a breeding program be of economic importance and that they are measured objectively.

TRAITS AND THEIR MEASUREMENT

Most economically important traits of beef cattle are classified as follows: (1) reproductive performance, (2) weaning weight, (3) postweaning growth, (4) feed efficiency, (5) carcass merit, (6) longevity (functional traits), (7) conformation, and (8) freedom from genetic defects.

Reproductive Performance

Reproductive performance has the highest economic importance of all the traits. Most cow–calf producers have a goal for percent calf crop weaned (number of calves weaned compared to the number of cows in the breeding herd) of 85% or higher. Beef producers also desire each cow to calve every 365 days or less and to have a calving season for the entire herd of less than 90 days. All of these are good objective measures of reproductive performance.

The heritability of fertility in beef cattle is quite low (less than 20%), so little genetic progress can be made through selection for that specific trait. Heritabilities for birth weight and scrotal circumference are high; therefore, selection for these traits will improve the percent calf crop weaned. The most effective way to improve certain reproductive traits (such as rebreeding after calving) is to improve the environment, through adequate nutrition and good herd health practices, for example. Selecting bulls based on a reproductive soundness examination, which includes semen testing, will also improve reproductive performance in the herd.

Reproductive performance can be improved through breeding methods by crossbreeding to obtain heterosis for percent calf crop weaned, using bulls with relatively light birth weights (heritability of birth weight is 40%), which decreases calving difficulty, as well as by selecting bulls that have a relatively large scrotal circumference. Scrotal circumference has a high heritability (40%); bulls with a larger scrotal size (over 32 cm for yearling bulls) produce a larger volume of semen and have half-sister heifers that reach puberty at earlier ages than heifers related to bulls with a smaller scrotal size.

Weaning Weight

Weaning weight, as measured objectively by weighing calves on scales, reflects the milking and mothering ability of the cow and the preweaning growth rate of the calf. Weaning weight is commonly expressed as the adjusted 205-day weight, where the weaning weight is adjusted for the age of the calf and age of the dam. This adjustment puts all weaning-weight records on a comparable basis since older calves will weigh more than younger calves and mature cows (5–9 years of age) will milk heavier than younger cows (2–4 years of age) and older cows (10 years and older).

Weaning weights of calves are usually compared by dividing the calf's adjusted weight by the average weight of the other calves in the herd. This is expressed as a ratio. For example, a calf with a weaning weight of 440 lb, where the herd average is 400 lb, has a ratio of 110. This calf's weaning-weight ratio is 10% above the herd average. Ratios can be used primarily for selecting cattle within the same herd in which they have had similar environmental opportunity. Comparing ratios between herds is misleading from a genetic standpoint because most differences between herds are caused by differences in the environment. Weaning weight will respond to selection because it has a heritability of 30%.

Postweaning Growth

Postweaning growth measures the growth from weaning to a weight that approaches slaughter weight. Postweaning growth might take place in a pasture or in a feedlot. Usually, animals with relatively high postweaning gains make efficient gains at a relatively low cost to the producer.

Postweaning gain in cattle is usually measured in pounds gained per day after a calf has been on a feed test for 100–160 days. Weaning weight and postweaning gain are usually combined into a single trait, namely, adjusted 365-day weight (or yearling weight). It is computed as follows:

Adjusted 365-day weight = (160 × average daily gain) + adjusted 205-day weight

Average daily gain for 140 days and adjusted 365-day weight both have high heritabilities (40%); therefore, genetic improvement can be quite rapid when selection is practiced on postweaning growth or yearling weight.

Feed Efficiency

Feed efficiency is measured by the pounds of feed required per pound of liveweight gain. Specific records for feed efficiency can be obtained only by feeding each animal individually and keeping records on the amount of feed consumed. With the possible exception of some bull-testing programs, determining feed efficiency on an individual animal basis is seldom economically feasible.

Interpretation of feed-efficiency records can be somewhat difficult, depending on the endpoint to which the animals are fed. Feeding endpoint can be a certain number of days on feed (e.g., 140 days), a specified slaughter weight (e.g., 1,200 lb), or a carcass compositional endpoint (e.g., low Choice-quality grade or 0.4 inch of fat). Most differences in feed efficiency, shown by individual animals, are related to the pounds of body weight maintained through feeding periods and the daily rate of gain or feed intake of each animal. Cattle fed from a similar initial feedlot weight (e.g., 600 lb) to a similar slaughter weight (e.g., 1,100 lb) will demonstrate a high relationship between rate of gain and efficiency of gain. In this situation, cattle that gain faster will require fewer pounds of feed per pound of gain. Thus, a breeder can select for rate of gain and thereby make genetic improvement in feed efficiency. However, when cattle are fed to the same compositional endpoint (approximately the same carcass fat), there are small differences in the amount of feed required per pound of gain. This is true for different sizes and shapes of cattle, and even for various breeds that vary greatly in skeletal size and weight.

The heritability of feed efficiency at similar beginning and final weights is high (45%), so selection for more efficient cattle can be effective. It seems logical to use the genetic correlation between gain and efficiency where possible.

Carcass Merit

Carcass merit is presently measured by quality grades and yield grades, which are discussed in detail in Chapter 8. Many cattle-breeding programs have goals to produce cattle that will grade Choice and have yield grade 2 carcasses. Objective measurements using ultrasound backfat thickness on the live animal and hip height measurement can assist in predicting the yield grade at certain slaughter weights. Visual appraisal, which is more subjective, can be used to predict amount of fat or predisposition to fat and skeletal size. These visual estimates can be relatively accurate in identifying actual yield grades if cattle differ by as much as one yield grade.

Because quality grade cannot be evaluated accurately in the live animal, it is necessary to evaluate the carcass. Steer and heifer progeny of different bulls are slaughtered to identify the genetic superiority or inferiority of bulls for both quality grade and yield grade. The heritabilities of most beef carcass traits are high (over 40%), so selection can result in marked genetic improvement for these traits.

Longevity

Longevity measures the length of productive life. It is an important trait, particularly for cows, when replacement heifer costs are high or when a producer reaches an optimum level of herd performance and desires to stabilize it. Bulls are usually kept in a herd for 3–5 years, or inbreeding might occur. Some highly productive cows remain in the herd at age 15 years or older, whereas other highly productive cows have been culled before reaching 4 years of age. These cows may have been culled because of such problems as skeletal unsoundness, poor udders, eye problems (e.g., cancer eye), and lost or worn teeth. Little selection opportunity for longevity exists because few cows remain highly productive past the age of 10 years. Most cows that leave the herd early have poor reproduction performance.

Some producers need to improve their average herd performance as rapidly as possible rather than improve longevity. In this situation, a relatively rapid turnover of cows is needed. Some selection for longevity occurs because producers have the opportunity to keep more replacement heifers born to cows that are highly productive in the herd for a longer period of time than heifers born to cows that stay in the herd for a short time. Some beef producers attempt to identify bulls that have highly productive, relatively old dams. Certain conformation traits, such as skeletal soundness and udder soundness, may be evaluated to extend the longevity of production.

Conformation

Conformation is the form, shape, and visual appearance of an animal. How much emphasis to put on conformation in a beef cattle selection program has been, and continues to be, controversial. Some producers believe that putting a productive animal into an attractive package contributes to additional economic returns. It is more logical, however, to place more selection emphasis on traits that will produce additional numbers and pounds of lean growth for a given number of cows. Placing some emphasis on conformation traits such as skeletal, udder, eye, and teeth soundness is justified. Conformation differences such as fat accumulation or predisposition to fat can be used effectively to make meaningful genetic improvement in carcass composition.

Most conformation traits are medium to high (30–60%) in heritability, so selection for these traits will result in genetic improvement.

Genetic Defects

Genetic defects, other than those previously identified under longevity and conformation, need to be considered in breeding productive beef cattle. Cattle have numerous known hereditary defects; most of them, however, occur infrequently and are of minor concern. Some defects increase in their frequency, and selection needs to be directed against them. Most of these defects are determined by a single pair of genes that are usually recessive. When one of these hereditary defects occurs, it is a logical practice to cull both the cow and the bull.

Some commonly occurring genetic defects in cattle today are double muscling, syndactyly (mule foot), arthrogryposis (palate-pastern syndrome), osteopetrosis (marble bone disease), hydrocephalus, and dwarfism. **Double muscling** is evidenced by an enlargement of the muscles with large grooves between the muscle systems of the hind leg. Double-muscled cattle usually grow slowly and their fat deposition in and on the carcass is much less than that of the normal beef animal. **Syndactyly** is a condition in which one or more of the hooves are solid in structure rather than cloven. Mortality rate is high in calves with syndactyly. **Arthrogryposis** is a defect in which the pastern tendons are contracted, and the upper part of the mouth has not properly fused together. **Osteopetrosis** is characterized by the marrow cavity of the long bones being filled with bone tissue. All calves having osteopetrosis have short lower jaws, protruding tongue, and impacted molar teeth. A bulging forehead where fluid has accumulated in the brain area is typical of the defect of **hydrocephalus.** Calves with arthrogryposis, hydrocephalus, or osteopetrosis usually die shortly after birth. The most common type of **dwarfism** is snorter dwarfism, in which the skeleton is quite small and the forehead has a slight bulge. Some snorter dwarfs exhibit a heavy, labored breathing sound. This defect was most common in the 1950s, and it has decreased significantly since that time.

BULL SELECTION

Bull selection must receive the greatest emphasis if optimum genetic improvement of a herd is to be achieved. Bull selection accounts for 80–90% of the genetic improvement in a herd over a period of several years. This does not diminish the importance of good beef females, because genetically superior bulls have superior dams. However, most of the genetic superiority or inferiority of the cows will depend on the bulls previously used in the herd. Also the accuracy of records is much higher in bulls than in cows and heifers because bulls produce more offspring and are used in more herds (especially with AI).

All commercial and purebred producers should identify breeders who have honest, comparative records on their cattle. Most performance-minded seedstock producers record the birth weights, weaning weights, and yearling weights of their bulls. Some breeders also obtain feedlot and carcass data on their own bulls, or they use bulls from sires for whom these test data are known.

Purebred breeders should provide accurate performance data on their bulls, and commercial producers should request the information. Excellent performance records can be obtained and made available on the farm or ranch. The trait ratios are useful and comparative if the bulls have been fed and managed in similar environments.

Breeding Values

Phenotype is the appearance or performance of an animal determined by the *genotype* (genetic makeup) and the environment in which it was raised. Genotype is determined by two factors: (1) breeding value (what genes are present), and (2) nonadditive (how genes are combined). Environmental factors influence the phenotype through known and unknown effects. Age of dam, resulting in different levels of milk production, is a known environmental effect adjusted for in an adjusted 205-day weight (phenotype). Unknown effects are things like injury or health problems for which it is difficult to adjust the phenotypic record. Figure 22.5 shows how breeding value relates to the phenotype of an animal.

An example depicting breeding values is shown in Fig. 22.6. In the "brick wall concept," each brick represents a gene. Some genes have positive, additive effects, whereas other genes have negative effects. The different-sized bricks represent the magnitude of the gene's effect. In this example (Fig. 22.6), the 10 pairs of genes (20 bricks) are used to show a superior breeding value and a negative breeding value. The superior breeding value (higher brick wall) is where the sum of the additive gene effects (bricks) is above herd average.

Breed associations have developed computer programs that utilize performance records on the individual calf and its relatives. This information is used to calculate breeding values. Breeding values are most frequently reported on birth weight, maternal weaning weight, weaning weight, and yearling weight by using **expected progeny differences** (**EPDs**).

Sire Summaries

The development of sire summaries has made sire selection more effective. Most breed associations publish a sire summary that is updated annually or biannually.

Expected progeny difference (**EPD**) and **accuracy** (**ACC**) are the important terms used in understanding sire summaries. An EPD combines into one figure a measurement of genetic potential based on the individual's performance and the performance of related animals such as the sire, dam, and other relatives.

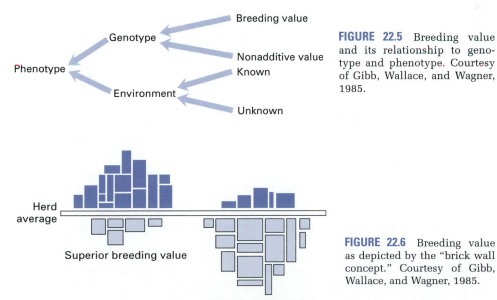

FIGURE 22.5 Breeding value and its relationship to genotype and phenotype. Courtesy of Gibb, Wallace, and Wagner, 1985.

FIGURE 22.6 Breeding value as depicted by the "brick wall concept." Courtesy of Gibb, Wallace, and Wagner, 1985.

EPD is expressed as a plus or minus value, reflecting the genetic-transmitting ability of a sire. The most common EPDs reported for bulls, heifers, and cows are birth-weight EPD, milk EPD, weaning-growth EPD, maternal EPD (includes milk and preweaning growth), and yearling-weight EPD. Some additional traits are included as well (Table 22.3).

Accuracy (ACC) is a measure of expected change in the EPD as additional progeny data become available. EPDs with ACC of 0.90 and higher would be expected to change very little, whereas EPDs with ACC below 0.70 might change dramatically with additional progeny data.

Table 22.4 shows sire summary data for several Angus bulls.

TABLE 22.3   Sire Summary Data (EPDs or ratios)

				Breed			
Trait	Angus	Charolais	Gelbvieh	Hereford[d]	Limousin	Red Angus	Simmental
Calving ease							
First calf[b]	0	0[a]	+[a]	0	0	0	+
Maternal[c]	0	0	+	0	0	0	+
Birth weight	+	+	+	+	+	+	+
Weaning weight	+	+	+	+	+	+	+
Yearling weight	+	+	+	+	+	+	+
Milk	+	+	+	+	+	+	+
Maternal	+	+	+	+	+	+	+
Milk + weaning							
Gestation length	0	0	+	0	+	0	0
Yearling height	+	0	0	0	0	0	0
Scrotal circumference	+	0	0	+	+	+	+
Mature cow weight	+	0	0	0	0	0	0
Stayability	0	0	0	0	0	+	0
Carcass traits	+	+	0	0	+	+	+

[a] 0 = absent from sire summary; + = appears in sire summary.
[b] Expressed as a ratio, not as an EPD. Measures calving ease of bull's own calves.
[c] Expressed as a ratio, not as an EPD. Measures calving ease of a bull's daughters.
[d] Includes both Hereford (horned) and Polled Hereford.

TABLE 22.4   Selected Data from an Angus Sire Evaluation Report

	Birth Weight		Weaning Weight Direct		Weaning Weight Maternal			Combination Index[c]	Yearling Weight	
					Milk[a]					
Sire	EPD	ACC	EPD	ACC	EPD	ACC	DTS[b]		EPD	ACC
A	−1	0.95	+20	0.95	+1	0.93	331	+11	+39	0.94
B	+9	0.95	+47	0.95	−17	0.89	88	6	+74	0.90
C	−7	0.83	−12	0.85	−10	0.79	29	−16	−4	0.85
D	+1	0.95	+22	0.95	+3	0.92	230	+14	+54	0.94
E	1	0.73	+15	0.70	+12	0.42	0	+20	+45	0.66

[a] Milk EPD is measured by comparing calf weaning weights from daughters of bulls.
[b] DTS is daughters.
[c] Calculated by taking ½ of weaning weight (direct) + milk EPD.

If bull B and bull C were used in the same herd (each on an equal group of cows), the expected performance on their calves would be:

Bull B's calves to be 16 lb heavier at birth

Bull B's calves to weigh 59 lb more at weaning

Bull B's calves to weigh 78 lb more as yearlings

Bull B's daughters to wean 7 lb less calf

Bull A and bull D have an optimum combination of calving ease (birth weight) and growth traits (weaning weight and yearling weight). Bull E is a promising young sire with an excellent combination of EPDs. However, the ACC is relatively low and could significantly change with more progeny and as daughters start producing.

Bull selection becomes more complex in that desired genetic improvement is for a combination of several traits. "Stacking pedigrees" for only one trait may result in some problems in other traits. A good example is selecting for yearling weight alone, which results in increased birth weight because the two traits are genetically correlated. Birth weight is associated with calving difficulty, so increased birth weight might be a problem. Large increases in birth weight EPD can occur and calving difficulty can be serious.

The most difficult challenge in bull selection is selecting bulls that will improve maternal traits. Frequently, too much emphasis is placed on yearling growth, frame size, and large mature size. These traits can be antagonistic to maternal traits such as birth weight, early puberty, and maintaining a cow size consistent with an economical feed supply. Figure 22.7 shows examples of stacking pedigrees for maternal traits for yearling bull selection where the bulls are used naturally on heifers. If an older bull can be used AI, then emphasis would be on accuracies higher than those available on young bulls.

Table 22.5 identifies bull-selection criteria that would apply to a large number of commercial producers. First, the breeding program goals and the maternal traits are identified. Second, the bull-selection criteria are identified, giving emphasis to the maternal traits. These traits would receive emphasis if a commercial producer was using only one breed or a rotational crossbreeding program or in producing the cows used in a terminal crossbreeding program. Finally, Table 22.5 identifies the traits emphasized in selecting terminal cross bulls. In selecting terminal cross sires, little emphasis is placed on maternal traits as no replacement heifers from this cross are kept in the herd.

SELECTING REPLACEMENT HEIFERS

Heifers, as replacement breeding females, can be selected for several traits at different stages of their reproductive life. The objective is to identify heifers that will conceive early in the breeding season, calve easily, give a flow of milk consistent with the feed supply, wean a heavy calf, and make a desirable genetic contribution to the calf's postweaning growth and carcass merit.

Beef producers have found it challenging to determine which of the young heifers will make the most productive cows. Table 22.6 shows the selection process that producers use to select the most productive replacement heifers. This selection process assumes that more heifers will be selected at each stage

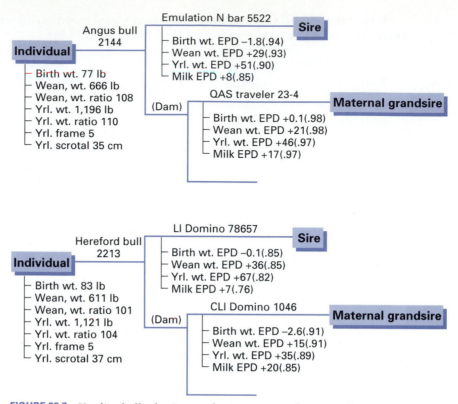

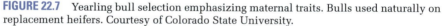

FIGURE 22.7   Yearling bull selection emphasizing maternal traits. Bulls used naturally on replacement heifers. Courtesy of Colorado State University.

of production than the actual numbers of cows to be replaced in the herd. The number of replacement heifers that producers keep is based primarily on their cost of production and their market value at various stages of production. More heifers than the number needed should be kept through pregnancy-check time. Heifers are selected on the basis that they become pregnant early in life primarily for economic reasons rather than expected genetic improvement from selection.

COW SELECTION

Cows should be culled from the herd based on the productivity of their calves and additional evidence that they can be productive the following year, such as early pregnancy and soundness of udders, eyes, skeleton, and teeth. Cow productivity is measured by pregnancy test, weaning and yearling weights (ratios) of their calves, and the EPDs of the cow. Table 22.7 shows the weaning and yearling values of the high- and low-producing cows in a herd. Cow 1 is a moderately high-producing cow, whereas cow 2 is the low-producing cow. Cow 2 should be culled and replaced with a heifer of higher breeding potential.

TABLE 22.5   Bull-Selection Criteria for Commercial Beef Producers

Goals and Maternal Traits	Bull-Selection Criteria for Maternal Traits[a]	Selection Criteria for Terminal Cross Bulls[b]
Breeding Program Goals 1. Select for an **optimum** combination of **maternal traits** to **maximize profitability** (Avoiding genetic antagonisms and environmental conflicts that come with maximum production or single-trait selection) 2. Provide genetic input so **cows** can be **matched with their environment**—cows that wean more lifetime pounds of calf without overtaxing the forage, labor, or financial resources 3. Stack pedigrees for maternal traits **Maternal Traits** **Mature weight:** Medium-sized cows (1,000–1,250 lb under average feed supply) **Milk production:** Moderate (wean 500–550-lb calves at 7 months of age under average feed) **Body condition** ("fleshing ability"): "5" condition score at calving without high-cost feeding **Early puberty/high conception:** Calve by 24 months of age **Calving ease:** Moderate birth weights (65–75-lb hfrs; 75–90-lb cows). Calf shape (head, shoulders, hips), which relates to unassisted births **Early growth and composition:** Rapid gains—relatively heavy weaning and yearling weights within medium (4–6) frame size. Yield grade 2 (steers slaughtered at 1,150 lb) **Functional traits (longevity):** Udders (shape, teats, pigment); eye pigment; disposition; structural soundness	**Yearling frame size:** 4.0–6.0 (smaller frame size will adapt better to harsher environments—e.g., less feed, more severe weather, less care) **Mature weight:** Under 2,000 lb in average condition (preference for future). Now we likely will have to consider bulls under 2,500 lb **Milk EPD:** 0 to +20 lb (prefer +10) **Maternal EPD:** +20 to +40 (prefer +25) **Body condition:** Backfat of 0.20–0.35 in. at yearling weights of 1,100–1,250 lb Monitor reproduction of daughters **Scrotal circumference:** 34–40 cm at 365 days of age. Passed breeding soundness examination **Birth weight EPD:** Preferably under +1.0 lb. Evaluate calving ease of daughters **Weaning weight EPD:** Approximately +20 lb **Yearling weight EPD:** Approximately +40 lb **Accuracies:** All EPDs of 0.90 and higher (older, progeny-tested sires). Young bulls with accuracies below 0.90—evaluate trait ratios and pedigree EPDs to predict mature bulls with above-average EPDs and accuracies. Select sons of bulls that meet EPDs listed above **Functional traits:** Visual evaluation of bull and his daughters **Visual:** Functional traits Sufficiently attractive to sire calves that would not be economically discriminated against in the marketplace Preference for "adequate middle" as medium-frame-size cattle need middle for feed capacity	**Yearling frame size:** 6.0–8.0 (should be evaluated with frame size of cows so slaughter progeny will average 5.0 and 6.0) **Mature weight:** No upper limit as long as birth weight and frame size are kept in desired range **Milk EPD:** Not considered **Maternal EPD:** Not considered **Body condition:** Evaluate with body condition of cows so that progeny will be between 0.25 and 0.45 in. at 1,100–1,300 lb slaughter weights **Scrotal circumference:** 32–38 cm at 365 days of age. Passed breeding soundness exam **Birth weight EPD:** Preferably no higher than +5.0; want calves' birth weights in 85–95-lb range with few, if any, over 100 lb **Weaning weight EPD:** +40 lb or higher **Yearling weight EPD:** +60 lb or higher **Accuracies:** All EPDs of 0.90 and higher (older, progeny-tested sires). Young bulls with accuracies below 0.90—evaluate trait ratios and pedigree EPDs to predict mature bulls with above-average EPDs and accuracies. Select sons of bulls that meet EPDs listed above **Visual:** Functional traits: Disposition and structural soundness Sufficiently attractive to sire calves that would not be economically discriminated against in the marketplace

[a] EPDs are an Angus base.
[b] EPDs are for Continental breeds.

TABLE 22.6 **Replacement Heifer Selection Guidelines at the Different Productive Stages**

Stage of Heifer's Productive Life	Emphasis on Productive Trait	
	Primary	Secondary
Weaning (7–10 months of age)	Cull only the heifers whose actual weight is too light to prevent them from showing estrus by 15 months of age. Also consider the economics of the weight gains needed to have puberty expressed. Cull heifers that are too large in frame and birth weight.	Weaning weight ratio Weaning EPD Milk EPD Predisposition to fatness Moderate skeletal frame Skeletal soundness
Yearling (12–15 months of age)	Cull heifers that have not reached the desired target breeding weight (e.g., minimum of 650–750 lb for small- to medium-sized breeds or cross, minimum of 750–850 lb for large-sized breeds and crosses). Cull heifers that are extreme in size and weight.	Milk and weaning weight EPDs Yearling weight ratio Yearling EPD Predisposition to fatness Moderate skeletal frame Skeletal soundness
After breeding (19–21 months of age)	Cull heifers that are not pregnant and those that will calve in the latter one-third of the calving season.	Milk and weaning weight EPDs Yearling weight ratio Yearling EPD Predisposition to fatness Moderate skeletal frame Skeletal soundness
After weaning first calf (31–34 months of age)	Cull to the number of first-calf heifers actually needed in the cow herd based on the weaning weight performance of the first calf. Preferably all the calves from these heifers have been sired by the same bull. Select for early pregnancy.	

TABLE 22.7 **Performance Data on High- and Low-Producing Cows in the Same Herd**

Cow Number	Number of Calves	Weaning Weight (lb)	Weaning Weight Ratio	Yearling Weight Ratio
1	9	572	105	102
2	9[a]	464	89	94

[a] One calf died before weaning (not computed in averages).

CROSSBREEDING PROGRAMS FOR COMMERCIAL PRODUCERS

Most commercial beef producers use crossbreeding programs to take advantage of heterosis in addition to the genetic improvement from selection. A crossbreeding system should be determined according to which breeds are available and then adapted to commercial producers' feed supply, market demands, and other environmental conditions. A good example of adaptability is the Brahman breed, which is more heat- and insect-resistant than most other breeds. Because of this higher resistance, the level of productivity (in the southern and

Gulf regions of the United States) is much higher for the Brahman and Brahman crosses.

Most commercial producers travel 150 or fewer miles to purchase bulls used in natural mating. Therefore, a producer should assess the breeders with excellent breeding programs in a 150-mile radius, as well as available breeds. This assessment, in most cases, should be determined before planning which breeds to use and in which combination to use them.

Breeds should be chosen for a crossbreeding system based on how well the breeds complement each other. Table 22.8 gives some comparative rankings of the major beef breeds for productive characteristics. Although the information in Table 22.8 is useful, it should not be considered the final answer for decisions on breeds to use. First, a producer needs to recognize that this information reflects breed averages; therefore, there are individual animals and herds of the same breed that are much higher or lower than the ranking given (Figs. 22.1–22.3). Producers need to use some of the previously described methods to identify superior animals within each breed. Second, it should be recognized

TABLE 22.8 Breed Crosses Grouped in Biological Types on Basis of Four Major Criteria

Breed Group[b]	Traits Used to Identify Biological Types[a]			
	Growth Rate and Mature Size	Lean-to-Fat Ratio	Age at Puberty	Milk Production
Jersey	X	X	X	XXXXX
Longhorn	X	XX	XXX	XX
Hereford-Angus	XXX	XX	XXX	XX
Red Poll	XX	XX	XX	XXX
Devon	XX	XX	XXX	XX
Shorthorn	XXX	XX	XXX	XXX
Galloway	XX	XXX	XXX	XX
South Devon	XXX	XXX	XX	XXX
Tarentaise	XXX	XXX	XX	XXX
Pinzgauer	XXX	XXX	XX	XXX
Brangus	XXX	XX	XXXX	XX
Santa Gertrudis	XXX	XX	XXXX	XX
Beefmaster	XXX	XX	XXXX	XX
Sahiwal	XX	XXX	XXXXX	XXX
Brahman	XXXX	XXX	XXXXX	XXX
Nellore	XXXX	XXX	XXXXX	XXX
Brown Swiss	XXXX	XXXX	XX	XXXX
Gelbvieh	XXXX	XXXX	XX	XXXX
Holstein	XXXX	XXXX	XX	XXXXXX
Simmental	XXXXX	XXXX	XXX	XXXX
Maine Anjou	XXXXX	XXXX	XXX	XXX
Salers	XXXXX	XXXX	XXX	XXX
Piedmontese	XXX	XXXXXX	XX	XX
Limousin	XXX	XXXXX	XXXX	X
Charolais	XXXXX	XXXXX	XXXX	X
Chianina	XXXXX	XXXXX	XXXX	X

[a] Breed crosses grouped into several biological types based on relative differences (X = lowest, XXXXXX = highest).
[b] Breed group based on these breeds of sires mated to Hereford and Angus cows.
Source: Adapted from *MARC Research Progress Reports.*

that these average breed rankings can change with time, depending on the improvement programs used by the leading breeders within the same breed. Obviously, those traits that have high heritabilities would be expected to change most rapidly, assuming the same selection pressure for each trait. A careful analysis of the information in Table 22.8 shows that no one breed is superior for all important productive characteristics. This gives an advantage to commercial producers using a crossbreeding program if they select breeds whose superior traits complement each other. An excellent example of breed complementarity is shown by the Angus and Charolais breeds, where they complement each other for both quality grade and yield grade.

Most of the heterosis achieved in cattle as a result of crossbreeding is expressed by weaning time. The cumulative effect of heterosis on pounds of calf weaned per cow exposed is shown in Fig. 22.8, in which maximum heterosis is obtained when crossbred calves are obtained from crossbred cows. The traits that express an approximate heterosis of a 20% increase in pounds of calf weaned per cow exposed to breeding are early puberty of crossbred heifers, high conception rates in the crossbred female, high survival rate of calves, increased milk production of crossbred cows, and a higher preweaning growth rate of crossbred calves.

Consistently high levels of heterosis (10–20%) can be maintained generation after generation if crossbreeding systems such as those shown in Figs. 22.9–22.11 are used. The crossbreeding system shown in Fig. 22.11 combines a two-breed rotation with a terminal cross. In this system, the two-breed rotation is used primarily to produce replacement females for the entire cow herd. In most cow herds, approximately 50% of the cows are bred to sires to produce

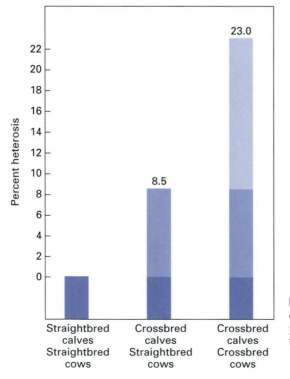

FIGURE 22.8 Heterosis, resulting from crossbreeding, for pounds of calf weaned per cow exposed to breeding. Courtesy of the USDA.

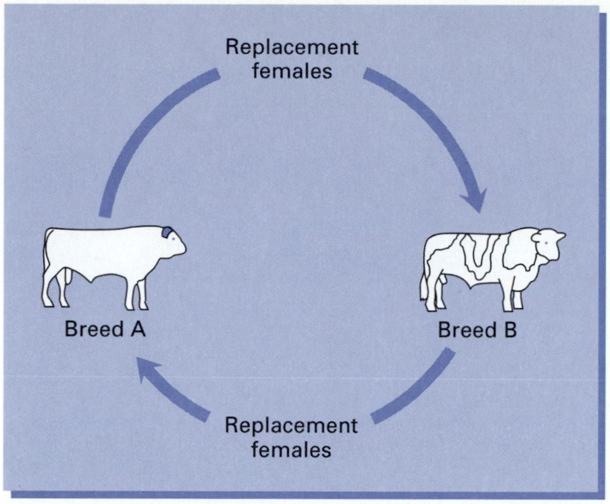

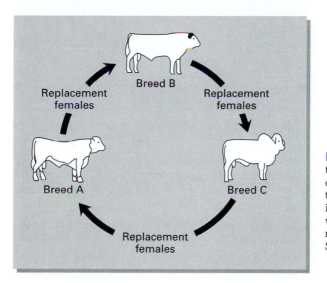

FIGURE 22.9 Two-breed rotation cross. Females sired by breed A are mated to breed B bulls, and heifers sired by breed B are mated to breed A bulls. This will increase the pounds of calf weaned per cow bred by approximately 15%. Courtesy of Colorado State University.

FIGURE 22.10 Three-breed rotation cross. Females sired by a specific breed are bred to the breed of the next bull in rotation. This will increase the pounds of calf weaned per cow by approximately 20%. Courtesy of Colorado State University.

replacement females, with the remaining 50% being bred to terminal cross sires. All terminal cross calves are sold. This crossbreeding system maintains heterosis as high as the three-breed rotation system.

In the rotational crossing, breeds with maternal-trait superiority (high conception, calving ease, and milking ability) would be selected. The terminal cross sire could come from a larger breed where growth rate and carcass cutability are emphasized. A primary advantage of the rotational-terminal cross system is that smaller- or medium-sized breeds can be used in rotational crossing and a larger breed could be used in terminal crossing. Terminal crossbreeding systems will maximize heterosis at approximately 25–30%. Some of the systems that will fit a one-breeding pasture program include (1) using a composite (synthetic) breed, (2) rotating two or three breeds of bulls (each breed every 3–5 years), and (3) putting multiple-sire breeds of bulls with crossbred cows. Each of these crossbreeding systems will maintain heterosis at approxi-

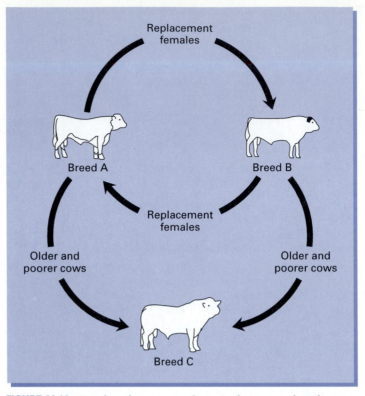

FIGURE 22.11 Two-breed rotation and terminal sire cross-breeding system. Sires are used in the two-breed rotation primarily to produce replacement heifers. Terminal cross sires are mated to the less productive females. This system will increase the pounds of calf weaned per cow bred by more than 20%. Courtesy of Colorado State University.

mately 10–15%. These systems may be more cost-effective and profitable than some of the more complex crossbreeding systems discussed earlier.

Table 22.9 shows the advantage a commercial producer has over a purebred breeder in being able to use more of the breeding methods for genetic improvement. Commercial producers can use crossbreeding, whereas purebred breeders cannot use crossbreeding if they maintain breed purity.

Traits with a low heritability respond little to genetic selection, but they show a marked improvement in a sound crossbreeding program. The commercial producer needs to select sires carefully to improve the traits with a high heritability.

TABLE 22.9 Heritability and Heterosis for the Major Beef Cattle Traits

Traits	Heritability	Heterosis
Reproduction	Low[a]	High
Growth	Medium	Medium
Carcass	High	Low

[a] Exceptions are age at puberty, scrotal circumference, and birth weight (dystocia). They are highly heritable.

CHAPTER SUMMARY

■ Phenotypic differences in breeds of beef cattle are due primarily to color (color pattern), polled or horned, mature size, and muscling.

■ There are more than 250 breeds of cattle in the world with more than 60 different breeds of beef cattle in the United States.

■ Based on annual registration numbers, the Angus breed is the most numerous breed in the United States, followed by Hereford/Polled Hereford, Limousin, Simmental, and Charolais.

■ Genetic improvement occurs by selecting for superiority (optimum combination) for the following traits: (1) reproductive performance, (2) weaning weight, (3) postweaning growth, (4) feed efficiency, (5) carcass merit (weight, yield grade, and quality grade), and (6) longevity.

■ Effective bull selection, primarily through EPDs, accounts for 80–90% of the genetic improvement in a herd over a period of several years.

■ It is economically important to select the biological type (mature weight, milk production, age at puberty, and carcass composition) that matches cows to a low-cost forage environment and their progeny to the marketplace.

■ Optimum levels of heterosis are obtained by using crossbreeding or composite breeds.

REVIEW QUESTIONS

1. What is a breed of cattle?
2. What does *purebred* refer to?
3. *True or False:* A herd of purebred cattle, such as Angus or Herefords, which are particularly uniform in one or two characteristics, such as color or color pattern, are highly homozygous.
4. What were the three major breeds of beef cattle in the United States during the early 1900s?
5. What are three reasons for the large increase in beef cattle breeds in the United States during the second half of the twentieth century?
6. What are the four most important breeds of beef cattle in the United States today according to breed registration numbers?
7. When does the most genetic improvement in beef cattle occur?
8. What are the eight economically important traits of beef cattle?
9. Which trait of beef cattle has the greatest economic importance?
10. To what segment of the beef industry is reproductive performance especially important?
11. *True or False:* Reproductive performance is lowly heritable so selection is relatively ineffective for making genetic progress in this trait.
12. What are the best ways to improve reproductive performance of beef cattle?
13. What does weaning weight reflect?

14. To what segment of the beef industry is weaning weight especially important?

15. What does postweaning growth measure?

16. To which segments of the beef industry is postweaning growth especially important?

17. Producers can make genetic improvement in feed efficiency by selecting for what related trait?

18. How is carcass merit presently measured?

19. *True or False:* Producers can make genetic improvement by selecting for carcass merit because the heritability of the traits which affect carcass meat is high.

20. What does longevity measure?

21. To which segment of the beef industry is longevity especially important?

22. What is conformation?

23. *True or False:* Bull selection must receive the greatest emphasis for optimum genetic improvement of the herd to be achieved.

24. Breeding values of beef cattle are most frequently reported as what?

25. *True or False:* Most heterosis in crossbred beef cattle is expressed in traits measured after the age of weaning.

26. What type of crossbreeding system will maximize heterosis in a herd of beef cattle?

27. What type of traits respond most to crossbreeding?

SELECTED REFERENCES

Publications

Beef Improvement Federation (BIF) Guidelines for Uniform Beef Improvement Programs. 1996. Ron Bolze, P.O. Box 786, Colby, KS 67701.

Boggs, D., 1992. *Understanding and Using Sire Summaries.* Beef Improvement Federation. BIF-FS3.

Bull and Heifer Replacement Workshops (proceedings). 1990. Fort Collins, CO: Colorado State University.

Gibb, J., Boggess, M. V., and Wagner, W. 1992. Understanding performance pedigrees. *Beef Improvement Federation,* BIF-FS2.

Gregory, K. E., and Cundiff, L. V. 1980. Crossbreeding in beef cattle: Evaluation of systems. *J. Anim. Sci.* 51:1224.

Hickman, C. G. (ed.). 1991. *Cattle Genetic Resources.* Amsterdam: Elsevier Science Publishers.

Jarrige, R., and Beranger, C. (ed.). 1992. *Beef Cattle Production.* Amsterdam: Elsevier Science Publishers.

Kress, D. D. 1989. Practical breeding programs for managing heterosis. *The Range Beef Cow Symposium XI.*

Legates, J. E. 1990. *Genetics of Livestock Improvement.* Englewood Cliffs, NJ: Prentice-Hall.

Middleton, B. K., and Gibb, J. B. 1991. An overview of beef cattle improvement programs in the United States. *J. Anim. Sci.* 69:3861.

Taylor, R. E. 1994. *Beef Production and Management Decisions.* New York: Macmillan Publishing Co.

Silcox, R., and McGraw, R. 1992. *Commercial Beef Sire Selection.* Beef Improvement Federation. BIF-FS9.

Visuals

Basic Genetics in Beef Cattle Selection; and *Selecting the Beef Heifer* (slide sets with audiotapes). Beef Improvement Federation, K. W. Ellis, Cooperative Extension Service, University of California, Davis, CA 95616.

Beef Breed Identification (slides). VEP, California Polytechnic State University, San Luis Obispo, CA 93407.

Breeding Cattle Selection, Yearling Bull Test (sound filmstrips). Vocational Education Productions, California Polytechnic State University, San Luis Obispo, CA 93407.

Cattle Breed Identification: British; Composites I and II; Continental; Zebu (videos). CEV, Inc., 5147-A 69th St., Lubbock, TX 79424.

Cows That Fit Montana (1985; videotape; 12 min). Department of Animal Science, Montana State University, Bozeman, MT 59715.

Like Begets Like (1989, videotape; 8.5 min). American Angus Association, 3201 Frederick Blvd., St. Joseph, MO 64501.

Using EPDs in Sire Selection (1988, videotape, 11 min). American Simmental Association, One Simmental Way, Bozeman, MT 59715.

CHAPTER 23

Feeding and Managing Beef Cattle

An overview of the beef industry and beef production is presented in Chapter 2. This chapter will identify the primary factors affecting beef cattle productivity and profitability.

COW–CALF MANAGEMENT

Cow–calf producers are interested in managing their operations as economical units. The profitability of a commercial cow–calf operation can be assessed easily by analyzing the following criteria: (1) calf crop percentage weaned (e.g., number of calves produced per 100 cows in the breeding herd), (2) average weight of calves at weaning (5–9 months of age), and (3) annual cow cost (dollars required to keep a cow each year).

An example of the economic assessment of a commercial cow–calf producer who has an 85% calf crop weaned, 500-lb weaning weights, and a $300 annual cost would be as follows:

Calf crop % (0.85) × weaning weight (500 lb) =
425 lb of calf weaned per cow in the breeding herd

Annual cow cost ($300) ÷ lb of calf weaned (425 lb) = $70.59 per hundredweight

A break-even price of $70.59 per hundredweight means that the producer would have to receive more than 70¢ per pound, or $70.59 per 100 lb of calf sold, to cover the yearly cost of each cow in the breeding herd. Table 23.1 shows the break-even price for several levels of calf crop percentage, weaning weight, and annual cow cost. This information reflects different management levels in which the break-even price ranges from more than $1.00 per pound to less than $0.50 per pound. Profitability of the commercial cow–calf operation is determined by comparing the market price of the calves at the time they are sold with the break-even price.

Cow–calf operations are managed best by operators who know the factors that affect calf crop percentage (Fig. 23.1), weaning weight, and annual cow cost. The primary management objective should be to improve pounds of calf weaned per cow and reduce or control the annual cow costs.

Costs and Returns

Figure 23.2 shows the dollars returned over cash costs for average U.S. cow–calf operations. Note the wide deviation in net returns over the 20-year period. During 1986–93, higher prices for calves resulted in higher levels of profitability; however, returns for 1995–96 are negative.

Costs and returns are computed from an **enterprise budget** (sometimes called an ***enterprise analysis***). The enterprise budget is one of the best financial forms used to make management decisions.

The component parts of receipts and expenses for an average cow–calf enterprise are shown in Table 23.2. On a per-cow basis, the total receipts are $449.02 and the total expenses are $264.95. This means there is a return of $184.07; that is, a net return to capital, land, management, and risk.

MANAGEMENT FOR OPTIMUM CALF CROP PERCENTAGES

The primary management factors affecting calf crop percentages are as follows:

1. Heifers need to be fed adequate levels of a balanced ration to reach puberty at 15 months of age if they are to calve at the desired age of 2 years. Medium-frame-sized heifers of English breeds and crosses (e.g., Angus and Hereford) should weigh 650–750 lb at 15 months of age. Cattle of larger-frame-sized exotic breeds or crosses should weigh 100–150 lb more to ensure a high percentage of heifers cycling at breeding.

TABLE 23.1 Break-Even Price (per 100 lb) for Commercial Cow–Calf Operations with Varying Calf Crop Percentages, Annual Cow Costs, and Weaning Weights

Calf Crop Percent Weaned	Annual Cow Cost	Break-Even Prices for Calves of Different Weaning Weights		
		400 lb	500 lb	600 lb
95	$350	$ 92.10	$73.68	$61.40
95	300	78.95	63.16	52.63
95	250	65.79	52.63	43.85
85	350	102.94	82.35	68.63
85	300	88.24	70.59	58.82
85	250	73.53	58.25	49.02
75	350	116.67	93.33	77.78
75	300	100.00	80.00	66.67
75	250	83.33	66.67	55.56

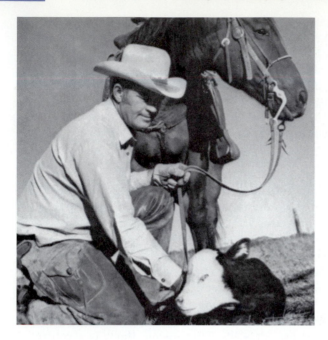

FIGURE 23.1 Percent calf crop, as measured by a live calf born and raised per cow, is the most economically important trait for the cow–calf producer. The bull, cow, calf, and producer each make a meaningful contribution to the level of productivity for this trait. Courtesy of Norden Laboratories, Inc.

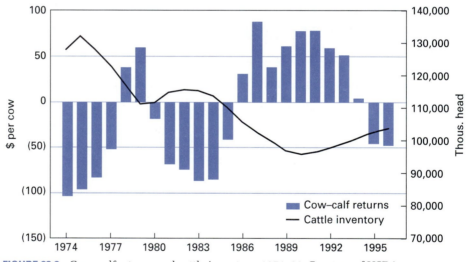

FIGURE 23.2 Cow–calf returns and cattle inventory, 1974–96. Courtesy of USDA.

2. Heifers should be bred to calve early in the calving season. Heifers calving early are more likely to be pregnant as 2- and 3-year-olds, whereas heifers calving late will likely not conceive during the next breeding season. Some producers save more heifers as potential replacements at weaning and as yearlings so selection for early pregnancy can be made at pregnancy test time.

3. Heifers typically have a longer postpartum interval than cows. This interval becomes shorter if first-calf heifers (heifers with their first calves) are separated from mature cows 60 days prior to calving and

TABLE 23.2   Cow–Calf Enterprise Budget (440 Head of Spring Calving Cows)

Production and Marketing Assumptions

Livestock Description	Unit	Market Wt	Market Month(s)	No. Head Marketed	% of Cow Herd
1. Steer calves	lb	500	Nov.	198	45.00
2. Heifer calves	lb	470	Nov.	172	39.09
3. Cull cows	lb	1,000		56	12.73
4. Cull bulls	lb	1,600		3	0.68

1.3% cow death loss; 13% replacement rate; 91% weaned calf crop

Operating Receipts—Livestock Sales

Description	Unit	Market Price	Production per Cow	Value per Cow	Total
1. Steer calves	lb	$0.9628	$225.00	$216.63	$95,317
2. Heifer calves	lb	0.8568	183.73	157.42	69,265
3. Cull cows	lb	0.5397	127.27	68.69	30,223
4. Cull bulls	lb	0.5757	10.91	6.28	2,764
5. Total Receipts				$449.02	$197,568

Direct Costs—Feed, Purchased and Raised

Description	Unit	Price	Quantity per Cow	Value per Cow	Total
6. Pasture rent	aum	$10.00	3.00	$ 30.00	$13,200
7. Aftermath	aum	9.00	1.00	9.00	3,960
8. Hay (alf, grass, tim)	ton	85.00	0.45	38.25	16,830
9. State lease	aum	4.00	3.00	12.00	5,280
10. Salt and mineral	lb	0.29	100.00	29.00	12,760
11. Total feed expenses				$118.25	$52,030

Other Operating Costs

12. Hired labor		$ 35.00	$15,400.00
13. Repairs: machinery, building, and fences		9.45	4,158.00
14. Machine hire/truck		1.24	547.75
15. Supplies		1.81	796.40
16. Veterinary medicine		7.50	3,300.00
17. Fuel, oil, lubricants		10.50	4,620.00
18. Breeding fees, pregnancy check, etc.		1.50	660.00
19. Miscellaneous		3.00	1,320.00
20. Utilities		1.85	814.00
21. Transportation		5.00	2,200.00
22. Marketing expenses		8.55	3,762.00
23. Herd bulls		8.52	3,748.80
24. Interest on operating capital 11.50%		12.20	5,368.02
25. Total other operating costs		$106.12	$46,694.97

Total Cost and Net Returns

Description	Value Per Cow	Total
26. Total cash operating expenses (line 11 + line 25)	$224.37	$ 98,725
Property and Ownership Costs:		
27. Depreciation	20.59	9,060
28. Taxes	6.26	2,754
29. Insurance	2.50	1,100
30. General overhead (5% of operating costs)	11.22	4,936
31. Total property and ownership costs	40.57	17,851
32. Total direct costs (line 28 + line 31)	264.95	116,576
33. Net receipts (line 5 – line 32)	$184.07	$ 80,992
(Returns to capital, land, management, and risk)		

Source: Gutierrez et al., 1990. Selected 1988–89 livestock enterprise budgets for Colorado. DARE Information Report IR: 90-2.

after calving. This separation allows the heifers to obtain their share of the feed essential for a rapid return to estrus.

4. Feeding programs are designed to have cows and heifers in a moderate body condition (visually estimated by the fat over the back and ribs) at calving time. Table 23.3 shows a body condition scoring system (BCS) currently being used. Thin cows at calving usually have a longer postpartum interval (Fig. 23.3). Cows that are too fat reflect a higher feed cost than is necessary for high and efficient production.

5. Cows, particularly first-calf heifers, should be observed every few hours at calving time. Some females will have difficulty calving (**dystocia**) and will need assistance in delivery of the calf. Calving difficulties should be kept to a minimum to prevent potential death of calves and cows. Calving difficulty is also undesirable because cows given assistance will usually have longer postpartum intervals.

6. Calving difficulty should be minimized but usually cannot be eliminated. A balance should be maintained between the number of calves born alive and the weight of the calves at weaning. Heavier calves at birth usually have heavier weaning weights. When calves are too heavy at birth, however, the death loss of the calves increases. Birth weight is the primary cause of calving difficulty; therefore, management decisions should be made to keep birth weights moderate. Bulls of the larger breeds or larger frame sizes should not be bred to heifers, and large, extremely growthy bulls, even in the breeds known for calving ease, should not be bred to heifers. Birth weight within a herd is influenced by genetics. Genetic differences are more important than certain environmental differences such as amount of feed during gestation. Bulls, to be used artificially, should have extensive progeny test records for birth weight and calving ease in addition to an individual birth-weight record.

FIGURE 23.3 The cow on the left is a body condition score (BCS) 4, whereas the cow on the right is a BCS 6. Cows with BCSs below 5 have a longer interval (days) between calving and pregnancy compared to cows having BCSs of 5 and higher. Courtesy of Oklahoma State University.

TABLE 23.3   System of Body Condition Scoring (BCS) for Beef Cattle

Group	BCS	Description
Thin condition	1	*Emaciated*—Cow is extremely emaciated with no palpable fat detectable over spinous processes, transverse processes, hip bones, or ribs. Tail-head and ribs project quite prominently.
	2	*Poor*—Cow still appears somewhat emaciated but tail-head and ribs are less prominent. Individual spinous processes are still rather sharp to the touch, but some tissue cover over dorsal portion of ribs.
	3	*Thin*—Ribs are still individually identifiable but not quite as sharp to the touch. There is obvious palpable fat along spine and over tail-head with some tissue cover over dorsal portion of ribs.
Borderline condition	4	*Borderline*—Individual ribs are no longer visually obvious. The spinous processes can be identified individually on palpation but feel rounded rather than sharp. Some fat cover over ribs, transverse processes, and hip bones.
Optimum moderate condition	5	*Moderate*—Cow has generally good overall appearance. On palpation, fat cover over ribs feel spongy and areas on either side of tail-head now have palpable fat cover.
	6	*High moderate*—Firm pressure now needs to be applied to feel spinous processes. A high degree of fat is palpable over ribs and around tail-head.
Fat condition	7	*Good*—Cow appears fleshy and obviously carries considerable fat. Very spongy fat cover over ribs and around tail-head. In fact, "rounds" or "pones" beginning to be obvious. Some fat around vulva and in crotch.
	8	*Fat*—Cow very fleshy and overconditioned. Spinous processes almost impossible to palpate. Cow has large fat deposits over ribs and around tail-head, and below vulva. "Rounds" or "pones" are obvious.
	9	*Extremely fat*—Cow obviously extremely wasty and patchy and looks blocky. Tail-head and hips buried in fatty tissue and "rounds" or "pones" of fat are protruding. Bone structure no longer visible and barely palpable. Animal's mobility might even be impaired by large fatty deposits.

Source: Richards et al., *J. Anim. Sci.* 62:300.

7. The bull's role in affecting pregnancy rate has a marked influence on calf crop percentage. Before breeding, bulls should be evaluated for breeding soundness by addressing physical conformation and skeletal soundness, palpating the genital organs, measuring scrotal circumference, and testing the semen for motility and morphology. **Libido** (sex drive) and mating capacity are additional important factors in how the bull affects pregnancy rate. These traits are not easily measured in individual bulls before breeding or even after breeding in multiple-sire herds. The typical cow-to-bull ratio is quoted by most cattle producers as 30 to 1. However, some bulls can settle more than 50 cows in a 60-day breeding season. In some large pastures with rough terrain, the cow-to-bull ratio might have to be less to assure a high calf crop percentage.

8. Crossbreeding affects calf crop percentage in several ways. Crossbred heifers usually cycle earlier and have higher conception rates than their straightbred counterparts. Crossbred calves are more vigorous and have a higher survival rate. An effective crossbreeding program can increase the calf crop by 8–12%.

9. The primary nutritional factor influencing calf crop percentage is adequate intake of energy, which can be expressed in pounds of total digestible nutrients (TDN). The quantity of TDN intake is important in helping initiate puberty, maintaining proper body condition at calving, and keeping the postpartum interval relatively short. Other nutrients of major importance are protein, calcium, and phosphorus. Additional vitamins and minerals are important only in areas where the soil or feed is deficient.

10. Calf losses during gestation are usually low (2–3%) unless certain diseases are present in the herd. Serious reproductive diseases such as brucellosis, leptospirosis, vibriosis, and infectious bovine rhinotracheitis (IBR) can cause abortions, which may markedly reduce the calf crop percentage. These diseases can be managed by blood-testing animals entering the herd or vaccinating for the diseases. Herd health programs vary for different operations, depending on the incidence of the diseases in the area. Details of these programs should be worked out with the local veterinarian.

11. Calf losses after 1–2 days following birth are usually small (2–3%) in most cow–calf operations. Severe weather problems, such as spring blizzards, can cause high calf losses where protection from the weather is limited. In certain areas and in certain years, health problems can also cause high death losses. Infectious calf scours and secondary pneumonia can occasionally reduce the potential calf crop 10–30%.

MANAGEMENT FOR OPTIMUM WEANING WEIGHTS

The primary management factors affecting calf weaning weights are as follows:

1. Calves born early in the calving season are heavier at weaning primarily because they are older. Calves are typically born over a several-week period but are weaned together on one specified day. Every time the cow cycles during the breeding season and fails to become pregnant, the weaning weight of her calf is reduced by 30–40 lb. Most commercial producers have a breeding season of 90 days or less so the calves are heavier at weaning and can be managed in uniform groups (Fig. 23.4).

2. The amount of forage available to the cow and the calf has a marked influence on weaning weights. The cow needs feed to produce milk for the calf. The calf, after about 3 months of age, will consume forage directly in addition to the milk it receives.

3. Growth stimulants, commonly given to nursing calves, will increase the weaning weight by 5–15% (Fig. 23.5). Ralgro (Zeranol), Synovex C, and Compudose, common growth stimulants, are implanted as pellets under

the skin of the ear. The pellets dissolve over a period of several weeks and supply the growth-stimulating substance that is absorbed into the bloodstream. This implant, however, should not be used on bulls and heifers to be used for breeding purposes. The implant sometimes interferes with the proper development and functioning of the reproductive organs.

4. Providing supplemental feed to the calves where it is inaccessible to the cows will increase the weaning weight of the calves. This practice of **creep feeding** should be used with caution because it is not always economical. It can be used on calves that have the ability to grow and not fatten. It helps calves make the transition of the weaning process and is a feasible practice under drought or marginal feed-supply conditions. Creep feeding of breeding heifer calves can impair the development of their milk secretory tissue and subsequently reduce milk production. This impairment is apparently caused by fat accumulating in the udder and crowding the secretory tissue.

5. Any diseases that affect the milk supply of the cow or growth of the calf will cause a reduction in the weaning weight of the calf (see Chapter 21).

6. Genetic selection for milk production and calf growth rate will increase calf weaning weight. Selection based on EPDs (weaning weight and milk) is the most effective way to genetically change weaning weights in a herd. Effective bull selection will account for 80–90% of the genetic improvement in weaning weight, although weaning-weight information can also be used in culling cows and in selecting replace-

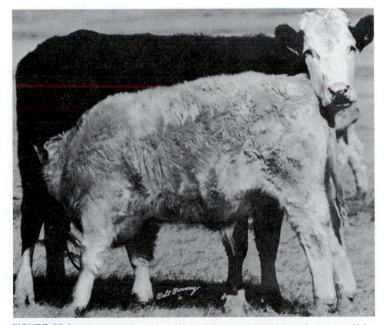

FIGURE 23.4 This calf is approximately 7 months old and soon will be weaned or separated from its dam. Commercial cow–calf producers manage their cows so that the calves will be heavy at weaning time. Courtesy of the *Charolais Banner.*

FIGURE 23.5   The squeeze chute is an essential piece of equipment for restraining cattle. The head can be restrained for inserting ear tags, treating eyes, implanting growth stimulants, or administering medication orally. Note the numbered ear tag, which is the most common form of individual animal identification. Courtesy of Dr. A. T. Ralston, Oregon State University.

ment heifers. It has been well demonstrated that effective selection can result in a 4–6-lb per-year increase in weaning weight on a per-calf basis.

7. Crossbreeding for the average cow–calf producer can result in a 10–30% (average 20%) increase in pounds of calf weaned per cow exposed in the breeding herd. Most of the increase occurs from improved reproductive performance; however, one-fourth to one-third of the 20% increase is due to the effect of heterosis on growth rate of the calf and increased milk production of the crossbred cow.

MANAGEMENT FOR LOW ANNUAL COW COSTS

In times of inflationary economic conditions, it may be difficult for a producer to lower annual cow costs. Some producers manage to lower costs, while other producers keep costs at a level similar to past years. Adequate income and expense records must be maintained so that cost areas can be carefully analyzed (Fig. 23.6). An enterprise budget (Table 23.2) is needed to assess cow costs and make management decisions to lower cow costs.

Table 23.4 shows cow productivity, annual cow costs, and calf break-even prices for average, low-cost, and high-cost producers in the NCA-IRM-SPA Database.

There is approximately a $200 per cow difference in net income when comparing low-cost producers ($61/cwt breakeven) with high-cost producers ($117/cwt breakeven). Table 23.5 shows that most of this $200 difference is due to financial cost per cow. While pounds weaned per cow is important, its con-

FIGURE 23.6 Cattle producers must keep accurate income and expense records to assess annual cow costs and the profitability of their operations. Courtesy of Duane Dailey, University of Missouri.

tribution to net income is far outshadowed by cost per cow. This difference shows producers where management decisions should first be focused.

Low-cost producers in the NCA-IRM-SPA Database were requested to identify the primary factors that determined their low break-even prices (Table 23.6).

Even though these "top five" are ranked as most important, the other, or miscellaneous costs, should not be overlooked. These smaller expense items (e.g., fuel, machinery, repair, supplies, utilities, and taxes) account for less than 20% of an average budget. Yet, they make up over 40% of the difference in cow cost between high-cost and low-cost producers. These smaller expense items should be compiled and evaluated on a regular basis.

The greatest consideration should be given to feed costs, as they constitute the largest part of annual cow costs, usually 50–70%. The period from the weaning of a calf to the last one-third of gestation in the next pregnancy is the time when cows can be maintained on comparatively small amounts of relatively cheap, low-quality feeds. Cow–calf operations having available crop aftermath feeds (e.g., stock-piled forage, corn stalks, grain stubble, and straw) usually have

TABLE 23.4   Cow–Calf Producer Profiles by Cost Group

	Cost Group		
	Average	Low One-Third	High One-Third
Percent calf crop	85%	83%	84%
Weaning weight	514 lb	527 lb	498 lb
Calf weight per cow exposed	438 lb	441 lb	418 lb
Cost per cow (financial)[a]	$377	$268	$490
Calf break-even price[a]	$0.86	$0.61	$1.17

[a] Before noncalf revenue adjustment.
Source: NCA-IRM-SPA Database (1995 NCA Cattlemen's College).

TABLE 23.5   Comparing Major Differences in Net Income

Item	High cost	Low cost	Difference
Financial cost per cow	$490	$268	$222
Pounds weaned per cow	441	418	18[a]

[a] Weaning weight valued at $0.80 per pound.
Source: NCA-IRM-SPA Database (1995 NCA Cattlemen's College).

TABLE 23.6   Top Five Ways Low-Cost Producers Reduce Costs

1. Reduce supplemental feed costs (40%*)
2. Rotational grazing (better pasture management) (30%*)
3. Right genetics (27%*)
4. Reduce labor costs (25%*)
5. Strong herd health program (19%*)

* Percent of respondents.
Source: NCA-IRM-SPA Database (1995 NCA Cattlemen's College).

the greatest opportunity to keep feed costs lower than other operations (Fig. 23.7). The most economical feed resource must be matched to the right biological type of cows—e.g., usually early puberty, calving ease, moderate milk, moderate mature weight, and fleshing ability (BCS) for early rebreeding. In many operations, the most economical feed resource comes from maximizing grazed forage with cows receiving minimal amounts of harvested and purchased feed.

Labor costs usually compose 15–20% of the annual cow costs. Labor costs are usually lower on a per-cow basis as herd size increases and in areas where moderate weather conditions prevail. It typically takes 15–20 hours of labor per cow unit per year. Operations that use labor inefficiently require twice as much labor per cow.

Interest charges on operating capital account for another 10–15% of the annual cow cost. Producers can reduce interest charges by carefully analyzing the costs of different credit sources.

Cows and heifers should be palpated at approximately 45 days after the end of the breeding season to determine if pregnancy has occurred. The producer should consider all marketing alternatives to maximize profits when selling open cows. Failing to check cows for pregnancy contributes to higher annual cow costs, lower calf crop percentages, and higher break-even costs of the calves produced.

STOCKER–YEARLING PRODUCTION

Several alternate stocker–yearling production programs are identified in Chapter 2, Fig. 2.4. Good producers understand the factors that affect the productivity and profitability of these various programs.

The primary factors affecting the costs and returns of stocker–yearling operations are marketing (both purchasing and selling the cattle), the gaining ability of the cattle, the amount and quality of available forage and roughage, and the health of the cattle. Table 23.7 shows that management decisions affecting purchase price and sale price have the greatest impact on profitability.

FIGURE 23.7 Corn-stalk fields are available to cattle after the grain has been harvested. There are millions of acres of stalk fields and other crop-aftermath feeds that can be grazed by cattle and other livestock. Courtesy of *BEEF.*

Stocker–yearling producers need to be aware of current market prices for the cattle they purchase or sell. They also need to understand the loss of weight of the cattle from the time of purchase to the time the cattle are delivered to their farm or ranch. This loss in weight, called *shrink,* can sometimes reflect the difference in the profit or loss of the stocker–yearling operation. It is common for calves and yearlings to shrink 3–12% from purchased weight to delivered weight. For example, yearlings purchased at 700 lb that shrink 8% will have a delivered weight of 644 lb. It typically takes 2–3 weeks to recover the weight loss.

The gaining ability of most stocker–yearling cattle is estimated visually. Cattle that are lightweight for their age, thin but healthy, with a relatively large skeletal frame size usually have a high gain potential. Cattle that are light for their age are typically most profitable for the stocker–yearling operator, whereas heavier cattle are most profitable for the cow–calf producer.

TABLE 23.7 Effect of a 10% Change on the Break-Even Price ($78/cwt) of a Stocker–Yearling Budget

Factor	Change (%)	Decrease in Break-Even Price ($/cwt)	Increase in Profit ($/head)
Sale price	+10%	$ 0.00	$ 56.01
Purchase price	−10	6.95	48.63
Average gain	+10	2.44	18.01
Pasture cost	−10	0.59	4.10
Interest cost	−10	0.33	2.29
Death loss	−10	0.20	1.42
All factors	10	10.02	130.59

Source: Cattle-Fax, Englewood, Colo.

Stocker–yearling cattle that are purchased and sold several times encounter stress situations of fatigue, hunger, thirst, and exposure to many disease organisms. The most common diseases include shipping fever complex and other respiratory diseases. These stress conditions make it necessary for stocker–yearling producers to have effective health programs for newly purchased cattle. Producers who have poor herd-health programs typically experience higher costs of gain and higher death losses.

The primary objective of the stocker–yearling operation is to obtain the most pounds of cattle gain within economic reason, while having assurance that high-quality forage yields can be obtained consistently each year. Forage management to obtain efficient production and consumption of nutritious feed is another essential ingredient of a successful stocker–yearling operation. Time of grazing and intensity of grazing (number of animals per acre) are important considerations if maximum forage production and utilization are to be maintained.

FEEDLOT CATTLE MANAGEMENT

The primary factors needed to analyze and properly manage a feedlot operation are the investment in facilities, cost of feeder cattle, feed cost per pound of gain, nonfeed costs per pound of gain, and marketing. A more detailed analysis of these factors is shown in Table 23.8.

Facilities Investment

The investment in facilities varies with type and location of feedlots. Larger commercial feedlots are quite similar regardless of where they are located in the United States. The general layout is an open lot of dirt pens with pen capacities varying from 100–500 head. The pens are sometimes mounded in the center to provide a dry resting area for cattle. The fences are pole, cable, or pipe. A feedmill to process grains and other feeds is usually a part of the feedlot. Special trucks distribute feed to fence-line feedbunks where cattle stand and eat inside the pens. Bunker trench silos hold corn silage and other roughages. Grains might be stored in these silos; however, they are more often stored in steel bins above the ground. The investment cost per head for this type of feedlot is approximately $150.

TABLE 23.8   The Major Component Parts of a Feedlot Business Analysis

Major Component, with Primary Factors Influencing Them				
Investment in Facilities	Cost of Feeder Cattle	Feed Cost Per Pound of Gain	Nonfeed Cost Per Pound of Gain	Total Dollars Received
Land	Grade	Ration	Death loss	Market choice
Pens	Weight	Rate of gain	Labor	Transportation
Equipment	Shrink	Feed efficiency	Taxes	Shrink
Feed mill	Transportation	Length of time	Insurance	Dressing percentage
Office	Gain potential		Utilities	Quality grade
			Veterinary expenses	Yield grade
			Repairs	Manure value

Feedlots for farmer-feeder operations vary from unpaved, wood-fenced pens to paved lots with windbreaks, sheds, or total confinement buildings. The latter might have manure collection pits located under the cattle, which stand on slotted floors. The feed might be stored in airtight structures. In most farmer-feeder operations, however, feeds are stored in upright silos and grain bins, particularly where rainfall is high. Feeds are typically processed on the farm and distributed with tractor-powered equipment to feedbunks located either inside or outside the pens. Investment costs for these feedlots vary from $200 to $500 per head.

Cost of Feeder Cattle

Before buying feeder cattle, the feedlot operator first estimates anticipated feed costs and the price the fed slaughter cattle will bring. These figures are then used to project the cost of feeder cattle or what the operator can afford to pay for them. Feeder cattle are priced according to weight, sex, **fill** (content of the digestive tract), skeletal size, thickness, and body condition. Most commercial feeders prefer to buy cattle with **compensatory gain.** These cattle are thin and relatively old for their weight. They have usually been grown out on a relatively low level of feed. When placed on feedlot rations, they gain rapidly and compensate for their previous lack of feed.

The feeder cattle buyer typically projects a high gain potential in cattle that have a large skeletal frame and little finish or body condition. However, not all cattle of this type will gain fast.

Heifers are usually priced a few cents a pound under steers of similar weight. The primary reason is that heifers gain more slowly, the cost per pound of gain is higher, and some feeder heifers are pregnant.

Feeder cattle of the same weight, sex, frame size, and body condition can vary several cents a pound in cost. This value difference is usually due to differences in fill. The differences in fill can amount to 10–40 lb in liveweight of feeder cattle. Feed and water consumed before weighing, distance and time of shipping, temperature, and the manner in which cattle are loaded and transported are some major factors affecting the amount of shrink. Shrink results primarily from loss of fill, but weight losses can occur in other parts of the body as well.

Feed Costs

Feed costs per pound of gain form the major costs of putting additional weight on feeder cattle. Typically, feed costs are 60–75% of the total costs of gain.

Feed costs per pound of gain are influenced by several factors, and the knowledge of these factors is important for proper management decision making. The choice of feed ingredients and how they are processed and fed are key decisions that affect feed costs.

Cattle that gain more rapidly and efficiently on the same feeding program have lower feed costs. Some of these differences are genetic and can sometimes be identified with the specific producer of the cattle. Most feeder cattle receive feed additives (e.g., Rumensin®) and ear implants (e.g., Ralgro® or Synovex®) that improve gain and efficiency and eventually the cost of gain.

Feed cost per pound of gain gets progressively higher as days on feed increase. Therefore, cattle feeders should avoid feeding cattle beyond their optimum combination of slaughter weight, quality grade, and yield grade.

Nonfeed Costs

Nonfeed costs per pound of gain is sometimes referred to as *yardage* cost. Yardage cost includes costs of gain other than feed. These costs can be expressed as either cost per pound of gain or cost per head per day. Obviously, cattle that gain faster will move in and out of feedlots sooner and accumulate fewer total dollar yardage costs.

Death loss and veterinary expenses caused by feeder cattle health problems can increase the nonfeed costs significantly. Most cattle feeders prefer to feed yearlings rather than calves because the death loss and health problems in yearlings are significantly lower than in calves.

Gross Receipts

The total dollar amount received for slaughter cattle emphasizes the need for the cattle feeder to be aware of marketing alternatives and the kinds of carcasses the cattle will produce. Most finished cattle from feedlots are sold directly to the packer, through a terminal market where professionals make the marketing transaction, or through an auction where the cattle are sold to the highest bidder. Nearly 80% of the fed cattle are sold directly to a packer. This marketing alternative requires the cattle feeder to be aware of current market prices for the weight and grade of cattle being sold.

Many slaughter cattle at large commercial feedlots are sold on a standard shrink (pencil shrink) of 4%, with the cattle being weighed at the feedlot without being fed the morning of weigh day. Feeders who ship their cattle some distance before a sale weight is taken should manage their cattle to minimize the shrink.

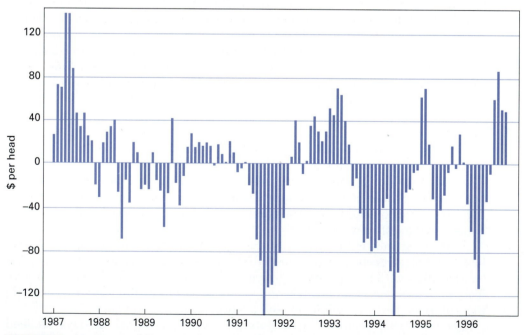

FIGURE 23.8 Average monthly returns ($ per steer) to cattle feeders in southern Great Plains, 1987–1996. Source: USDA.

TABLE 23.9   Fed Cattle Production Costs (all operations), 1975–90

Item	Dollars/cwt			
	1975	1980	1985	1990
Cash receipts				
Fed beef	$41.79	$66.11	$59.79	$65.61
Cash expenses				
Feeder cattle	18.41	48.54	39.75	39.85
Feed				
Silage	1.13	0.32	0.34	0.19
Dry grain	15.41	12.63	11.68	7.59
Protein supplements	1.76	1.98	1.96	2.17
Legume hay	0.68	0.92	1.13	1.64
Other roughages	0.22	0.38	0.68	0.96
Total feed costs	19.20	16.23	15.79	12.05
Other				
Veterinary and medicine	0.37	0.47	0.51	0.47
Fuel, lube, and electricity	0.33	0.47	0.38	0.46
Machinery and building repairs	0.40	0.26	0.30	0.29
Hired labor	0.37	0.60	0.81	0.85
Miscellaneous	0.45	1.03	1.25	1.49
Manure credit	−0.07	−0.10	−0.07	−0.07
Total variable expenses	39.71	67.50	58.72	55.39
Taxes and insurance	0.15	0.07	0.07	0.07
Hired management	0.08	0.15	0.14	0.13
Interest	2.42	3.76	3.92	4.72
Total fixed expenses	3.10	3.98	4.13	4.92
Total cash expenses	42.81	71.48	62.85	60.31
Receipts less cash expenses	−1.02	−5.37	−3.06	5.30

Source: USDA, ERS, *Economic Indicators of the Farm Sector—Livestock and Dairy,* 1975–90.

Most slaughter steers and heifers are sold on a liveweight basis with the buyer estimating the carcass weight, quality grade, and yield grade. Slaughter cattle of yield grades 4 and 5 usually have large price discounts. The price spread between the Select and Choice quality grades can vary considerably over time. Some marketing alternatives will not show a price differential between Select and Choice if the cattle have been well fed for a minimum number of days (120 days for yearling feeder cattle).

Cattle feeders who manage their cattle consistently for profitable returns know how to purchase high-performing cattle at reasonable prices. These cattle feeders formulate rations that will optimize cattle performance with feed costs. Also, their cattle have minimum health problems with a low death loss. The feeder feeds cattle the minimum number of days required to assure carcass acceptability and palatability and develops a marketing plan that will yield the maximum financial returns.

Some cattle are sold on a grade and yield basis (carcass weight and carcass grade), where total value is determined after the cattle have been slaughtered. This method of marketing is most useful to producers who know the carcass characteristics of their cattle.

COSTS AND RETURNS

The costs and returns for average fed cattle production in the United States are shown in Table 23.9. The enterprise budget is calculated on dollar-per-

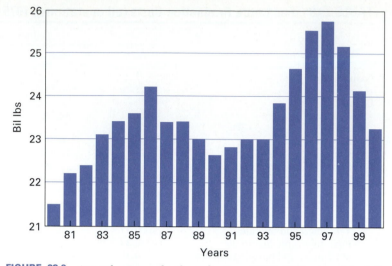

FIGURE 23.9 Annual carcass beef production (1994–2000 are projected). Source: National Cattlemen's Association and Cattle-Fax.

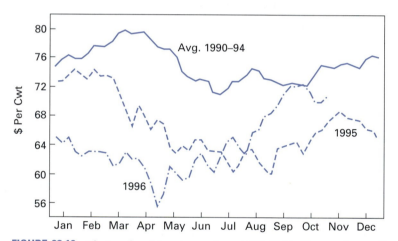

FIGURE 23.10 Choice slaughter steer prices, 1990–1996. These prices will reflect feeder cattle and carcass prices. Source: USDA.

hundredweight (cwt) basis. Note that the value of fed beef for 1990 was $65.61/cwt and total cash expenses were $60.31/cwt. The net returns of receipts over expenses was $5.30/cwt, or $63.60 for a 1,200-lb market steer. Figure 23.8 shows average monthly returns to cattle feeders in the southern Great Plains. Similar returns would be expected throughout the United States. Observe the contrasts in highly profitable versus highly unprofitable years.

PRODUCTION AND PRICES

Figures 23.9 and 23.10 show how carcass beef production and cattle prices are related. When production is low, prices are usually high. Producers save more replacement heifers and cull fewer cows during high prices. The increased number of breeding females eventually increase beef production beyond consumer demand. Then prices decrease. Low-cost producers with break-even prices below $65 per hundredweight will usually be profitable even when prices are low.

Chapter 36 examines in greater detail how producers can make effective management decisions.

CHAPTER SUMMARY

■ Well-managed commercial cow–calf producers keep cost-effective records to achieve low break-even prices (BE) by implementing the following formula:

$$BE = \frac{(\% \text{ calf crop} \times \text{weaning weight})}{\text{annual cow cost}}$$

■ Visual scores of body condition (BCS) in cows are used to minimize costs and obtain optimum levels of percent calf crop and weaning weights.

■ Low annual cow costs are achieved through extending the grazing during the year and reducing the consumption of supplemental and harvested feeds.

■ The primary factors affecting costs and returns of stocker–yearling and feedlot operations are (1) marketing (purchase price and sale price), (2) gaining ability of the cattle, (3) amount and quality of forage/feed, and (4) the health of the cattle.

REVIEW QUESTIONS

1. What are three criteria to assess profitability in commercial cow–calf operations?
2. At what age should beef heifers be first bred and first calved?
3. *True or False:* Heifers should be bred at the same time as cows.
4. *True or False:* Feeding programs should be designed to have cows and heifers in a moderate body condition.
5. What is the typical cow-to-bull ratio for pasture breeding?
6. How can crossbreeding improve calf crop weaned?
7. *True or False:* Calves born early in the calving season are heavier at weaning primarily because they are older.
8. How does the amount of forage available influence weaning weights of calves?
9. *True or False:* Growth stimulants given to nursing calves will increase weaning weights by 5–15%.
10. *True or False:* Creep feeding can increase weaning weights and is always economical.
11. How can genetic selection be used to increase weaning weight?
12. What constitutes the largest part of annual cow costs?

13. What are the primary factors affecting profitability in stocker–yearling operations?

14. What is the primary objective of stocker–yearling operations?

15. What are the primary factors affecting profitability of feedlot operations?

16. What are nonfeed costs per pound of gain referred to as?

17. How are most slaughter steers and heifers sold?

SELECTED REFERENCES

Publications

Albin, R. C., and Thompson, G. B. 1990. *Cattle Feeding: A Guide to Management.* Amarillo, TX: Trafton Printing, Inc.

Jarrige, R. and Beranger, C. (ed.). 1992. *Beef Cattle Production.* Amsterdam: Elsevier Science Publishers.

National Research Council. 1996. *Nutrient Requirements of Beef Cattle.* Washington, DC: National Academy Press.

Richards, M. W., Spitzer, J. C., and Warner, M. B. 1986. Effect of varying levels of nutrition and body condition at calving on subsequent reproductive performance. *J. Anim. Sci.* 62:300.

Ritchie, H. D. 1992. *Calving Difficulty in Beef Cattle.* Beef Improvement Federation, BIF-FS6a and FS6b.

Taylor, R. E. 1994. *Beef Production and Management Decisions.* New York: Macmillan Publishing Co.

Visuals

Beef Management Practices (sound filmstrips) covering: *Basic Beef Cattle Nutrition; Preventative Health Care; Handling Equipment and Facilities; Beef Cattle Castration; Dehorning Beef Cattle; Beef Cattle Identification;* and *Calving Management.* Vocational Education Productions, California Polytechnic State University, San Luis Obispo, CA 93407.

Beef Production Systems (sound filmstrips) covering: *Purebred Operations; Cow/Calf Production;* and *Feedlot Production.* Vocational Education Productions, California Polytechnic State University, San Luis Obispo, CA 93407.

Beef Selection Kit (sound filmstrips) covering: *Breeding Cattle Selection; Market and Feeder Cattle Selection;* and *Fitting and Showing Beef Cattle.* Vocational Education Productions, California Polytechnic State University, San Luis Obispo, CA 93407.

Conception to Feeder (videotape); *Energy Utilization in Feedlot Cattle* (videotape); *Cattle Handling and Transport* and *Beef Cattle Reproduction* (videotapes); *The Feedyard* (videotape); and *Cattle Production* (videotape). CEV, P.O. Box 65265, Lubbock, TX 79464-5265.

Dairy Cattle Breeds and Breeding

Production of milk per cow has been increased markedly in the past 50 years by improvements in breeding, feeding, sanitation, and management. The application of genetic selection, coupled with the extensive use of artificial insemination, has contributed to a successful dairy herd improvement program.

CHARACTERISTICS OF BREEDS

Six major breeds of dairy cattle—Holstein, Ayrshire, Brown Swiss, Guernsey, Jersey, and Red and White—are used for milk production in the United States. These breeds are shown in Color Plates Q and R (following p. 416) where production characteristics and other information about the breeds is also given.

It is important to recognize the variation for milk yield and percent fat that exists between breeds and within a breed. Figure 24.1 shows the variation in fat percentage for the Holstein and Jersey breeds. While the two breeds are distinctly different based on breed averages, it is possible to identify Holstein cows that have higher fat percentages than some Jersey cows.

Some Holstein cows produce extremely large amounts of milk—more than 100 lb/day at peak lactation. Thus, in Holsteins and other high-milk-producing cows, great stress is placed on udder ligaments, which can break down and no longer support the udder. If an udder breaks down, it is more susceptible to injury and disease, often necessitating culling the cow.

Ayrshires and Brown Swiss produce milk over a greater number of years than Holsteins, but their production levels are lower. Efficiency of milk production per 100 lb body weight is about the same for the major breeds of dairy cows.

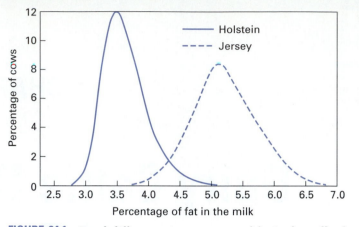

FIGURE 24.1   Breed differences in percentage of fat in the milk of Holstein and Jersey cows. Courtesy of G. E. Shook and W. H. Freeman and Company, © 1974.

Guernsey and Jersey cows produce milk having high percentages of milk fat and solids-not-fat, but the total amount of milk produced is relatively low. The efficiency of energy production of different breeds varies less among the breeds than do either total quantity of milk produced or percentage of fat in the milk.

Registration Numbers

Breed popularity can be estimated from registration numbers, as shown in Table 24.1. Most dairy cows are grade cows, which means they do not have pedigrees (with individual registration numbers) recorded in a breed association. However, producers with registered cattle are a source of breeding for the total dairy industry. Breeds with the largest registration numbers reflect the demand from the commercial dairy industry.

TABLE 24.1   **Registration Numbers for Major U.S. Breeds of Dairy Cattle**

Breed	Annual Registrations (in thousands)						Year Association Formed
	1995	1990	1985	1980	1975	1970	
Holstein	329.9	395.9	394.5	353.9	279.2	281.6	1871
Jersey	63.4	53.6	65.4	61.0	39.7	37.1	1868
Brown Swiss	10.8	11.7	11.9	12.9	14.0	16.4	1880
Guernsey	7.4	13.9	25.1	20.9	27.4	43.8	1877
Ayrshire	6.4	7.8	11.1	10.9	12.0	15.0	1875
Red and White	4.4	6.7	5.3	4.8	1.7	—	1964
Milking Shorthorn	3.2	2.3	3.4	4.9	4.6	5.4	1912

Source: Various dairy cattle breed associations.

DAIRY TYPE

Some descriptive terms describing ideal dairy type are *stature, angularity, level rump, long and lean neck, milk veins,* and *strong feet and legs.* In the past, dairy producers have placed great emphasis on dairy type, but research studies indicate that some components of dairy type may have little or no value in improving milk production and might even be deleterious if overstressed in selection. However, a properly attached udder and strong feet and legs are good indicators that a cow will remain a high producer for a long time.

Figure 24.2 shows the parts of a dairy cow and also gives a description of the component parts (general appearance, dairy character, body capacity, and udder) of preferred dairy type. The points on the scorecard show greatest emphasis for udder (40 points) and dairy character where both total 60 points out of a possible 100 points.

Some dairy breed associations have a classification program to evaluate the type traits. For example, the Holstein Association Linear Classification Program has 29 primary and secondary linear descriptive traits that measure functional conformation or type. A classifier, approved by the association, scores each cow or bull for the several traits and gives a final score if these scores are to be official records recognized by the association.

In the Holstein Association program a final score is calculated from rating the four major categories of classification traits: general appearance, dairy character, body capacity, and mammary system. The emphasis for each category for cows and heifers is shown in Table 24.2.

The final score represents the degree of physical perfection of any given animal. It is expressed in the following numbers and words:

Excellent (EX): 90–100   Good (G): 75–79

Very good (VG): 85–89   Fair (F): 65–74

Good plus (G+): 80–84   Poor (P): 50–64

The final score is used in computing PTAT (predicted transmitting ability for type). The PTAT identifies genetic differences in sires that can be considered in selection programs to improve dairy cattle type.

Type has value from a sales standpoint. Although some components of type score are negatively related to milk production, type is important as a measure of the likelihood that a cow will sustain a high level of production over several years.

IMPROVING MILK PRODUCTION

Great strides have been made over the past 50 years in improving milk production through improved management and breeding (Table 24.3). For example, the total amount of milk produced in the United States continues to increase, yet the number of dairy cows in 1995 was less than half the number in 1950. The average production per cow in 1995 was more than three times greater than the average production in 1940 (Table 24.3). In the past 10 years alone, annual milk production in the United States has increased by approximately 300 lb per cow per year.

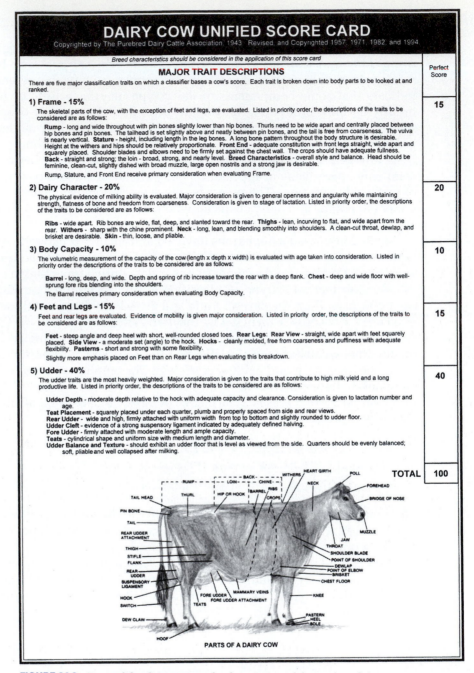

DAIRY COW UNIFIED SCORE CARD

Copyrighted by The Purebred Dairy Cattle Association, 1943. Revised, and Copyrighted 1957, 1971, 1982, and 1994.

Breed characteristics should be considered in the application of this score card

MAJOR TRAIT DESCRIPTIONS

	Perfect Score

There are five major classification traits on which a classifier bases a cow's score. Each trait is broken down into body parts to be looked at and ranked.

1) Frame - 15%

The skeletal parts of the cow, with the exception of feet and legs, are evaluated. Listed in priority order, the descriptions of the traits to be considered are as follows:

Rump - long and wide throughout with pin bones slightly lower than hip bones. Thurls need to be wide apart and centrally placed between hip bones and pin bones. The tailhead is set slightly above and neatly between pin bones, and the tail is free from coarseness. The vulva is nearly vertical. **Stature** - height, including length in the leg bones. A long bone pattern throughout the body structure is desirable. Height at the withers and hips should be relatively proportionate. **Front End** - adequate constitution with front legs straight, wide apart and squarely placed. Shoulder blades and elbows need to be firmly set against the chest wall. The crops should have adequate fullness. **Back** - straight and strong; the loin - broad, strong, and nearly level. **Breed Characteristics** - overall style and balance. Head should be feminine, clean-cut, slightly dished with broad muzzle, large open nostrils and a strong jaw is desirable.

Rump, Stature, and Front End receive primary consideration when evaluating Frame.

Perfect Score: 15

2) Dairy Character - 20%

The physical evidence of milking ability is evaluated. Major consideration is given to general openness and angularity while maintaining strength, flatness of bone and freedom from coarseness. Consideration is given to stage of lactation. Listed in priority order, the descriptions of the traits to be considered are as follows:

Ribs - wide apart. Rib bones are wide, flat, deep, and slanted toward the rear. **Thighs** - lean, incurving to flat, and wide apart from the rear. **Withers** - sharp with the chine prominent. **Neck** - long, lean, and blending smoothly into shoulders. A clean-cut throat, dewlap, and brisket are desirable. **Skin** - thin, loose, and pliable.

Perfect Score: 20

3) Body Capacity - 10%

The volumetric measurement of the capacity of the cow (length x depth x width) is evaluated with age taken into consideration. Listed in priority order the descriptions of the traits to be considered are as follows:

Barrel - long, deep, and wide. Depth and spring of rib increase toward the rear with a deep flank. **Chest** - deep and wide floor with well-sprung fore ribs blending into the shoulders.

The Barrel receives primary consideration when evaluating Body Capacity.

Perfect Score: 10

4) Feet and Legs - 15%

Feet and rear legs are evaluated. Evidence of mobility is given major consideration. Listed in priority order, the descriptions of the traits to be considered are as follows:

Feet - steep angle and deep heel with short, well-rounded closed toes. **Rear Legs: Rear View** - straight, wide apart with feet squarely placed. **Side View** - a moderate set (angle) to the hock. **Hocks** - cleanly molded, free from coarseness and puffiness with adequate flexibility. **Pasterns** - short and strong with some flexibility.

Slightly more emphasis placed on Feet than on Rear Legs when evaluating this breakdown.

Perfect Score: 15

5) Udder - 40%

The udder traits are the most heavily weighted. Major consideration is given to the traits that contribute to high milk yield and a long productive life. Listed in priority order, the descriptions of the traits to be considered are as follows:

Udder Depth - moderate depth relative to the hock with adequate capacity and clearance. Consideration is given to lactation number and age.
Teat Placement - squarely placed under each quarter, plumb and properly spaced from side and rear views.
Rear Udder - wide and high, firmly attached with uniform width from top to bottom and slightly rounded to udder floor.
Udder Cleft - evidence of a strong suspensory ligament indicated by adequately defined halving.
Fore Udder - firmly attached with moderate length and ample capacity.
Teats - cylindrical shape and uniform size with medium length and diameter.
Udder Balance and Texture - should exhibit an udder floor that is level as viewed from the side. Quarters should be evenly balanced; soft, pliable and well collapsed after milking.

Perfect Score: 40

TOTAL   100

PARTS OF A DAIRY COW

FIGURE 24.2 Parts of the dairy cow and a description of the preferred dairy type. Courtesy of the Purebred Dairy Cattle Association.

TABLE 24.2   Classification Traits Used for Final Scores in Holstein Cattle

Trait	Emphasis (%) Cows	Emphasis (%) Heifers
Frame	15	40
Dairy character	20	20
Body capacity	10	10
Feet and legs	15	30
Udder	40	—

Source: Holstein Association, Linear Classification Program.

SELECTION OF DAIRY COWS

The average productive life of a dairy cow is short (approximately 3–4 years). Many cows are culled primarily because of reproductive failure, low milk yield, udder breakdown, feet and leg weaknesses, and mastitis.

Heifers whose ancestral records indicate they will be high producers should be used to replace cows that are **culled** for low production. Such heifers should also be used to replace cows that leave the herd because of infertility, mastitis, or death, although the improvement gained thereby is generally modest.

A basis for evaluating a dairy cow is the quantity of milk (lb) and quality (total solids) that she produces. For a dairy operator to know which cows are good producers and which cows should be culled, a record of milk production is essential.

The **National Cooperative Dairy Herd Improvement Program** (**NCDHIP**) is a national industrywide dairy-production-testing and record-keeping program. It is often referred to as the **Dairy Herd Improvement (DHI) program.** In this program, USDA and Extension Service personnel work with dairy producers to help them improve milk production and dairy management practices. Records obtained are analyzed so dairy operators know how each cow compares with all cows in the herd and how the herd compares with other herds in the area. Producers can use several testing plans; some are official and others are unofficial, but any of them may be useful to a dairy operator interested in improving production in a herd. Seven important types of testing plans follow.

Dairy Herd Improvement Association (**DHIA**) is the most common official testing plan in the NCDHIP. All cows must be properly identified and all Dairy Herd Improvement (DHI) rules enforced. The results of official records might be published and are then used by the USDA to evaluate sires.

TABLE 24.3   Changes in Milk Production in the United States, 1940–95

Year	No. Cows (mil)	Average Milk Per Cow (lb)	Total Milk (bil lb)
1940	23.7	4,622	109.4
1950	21.9	5,314	116.6
1960	17.5	7,029	123.1
1970	12.0	9,751	117.0
1980	10.8	11,875	128.4
1990	10.0	14,645	148.3
1995	9.5	16,443	155.6

Source: USDA.

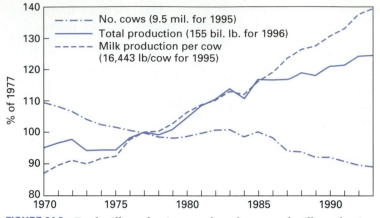

FIGURE 24.3   Total milk production, number of cows, and milk production per cow, 1970–1993. Courtesy of the USDA.

Dairy Herd Improvement Registry (DHIR) is another official testing program in NCDHIP. The same rules and procedures apply as in official DHIA records. The main difference is that records are sent to the offices of the breed associations and additional "surprise tests" may be conducted. DHIA and DHIR regulations are the same, and both are official testing programs. The DHIR program is for registered cattle. The following rules apply to both programs.

1. All cows in the herd must be entered into the official testing program.
2. All animals in the herd must be permanently identified.
3. Copies of pedigrees of all registered cows must be made available for DHIR.
4. Testing is done each month with not less than 15 days nor more than 45 days allowed between test periods.
5. Testing is conducted over a 24-hour period.
6. An independent supervisor must be present for supervising the weighing of the milk and the sampling of the milk for milk fat and other determinations.
7. Milk or milk fat records that are above values established by the breed association require retesting of the cow to ensure that an error was not made. An owner may request retesting if he or she feels the test does not properly reflect the production of the cows.
8. Surprise tests may be made if the supervisor suspects that they are needed to verify previous tests.
9. Any practice that is intended to or does create an inaccurate record of production is considered a fraudulent act and is not allowed.

Figure 24.4 shows an individual cow's record used in the DHIR program. Especially note the milk and milk fat recorded under the heading, "305 ME Season Production." Records are standardized to a lactation length of 305 days,

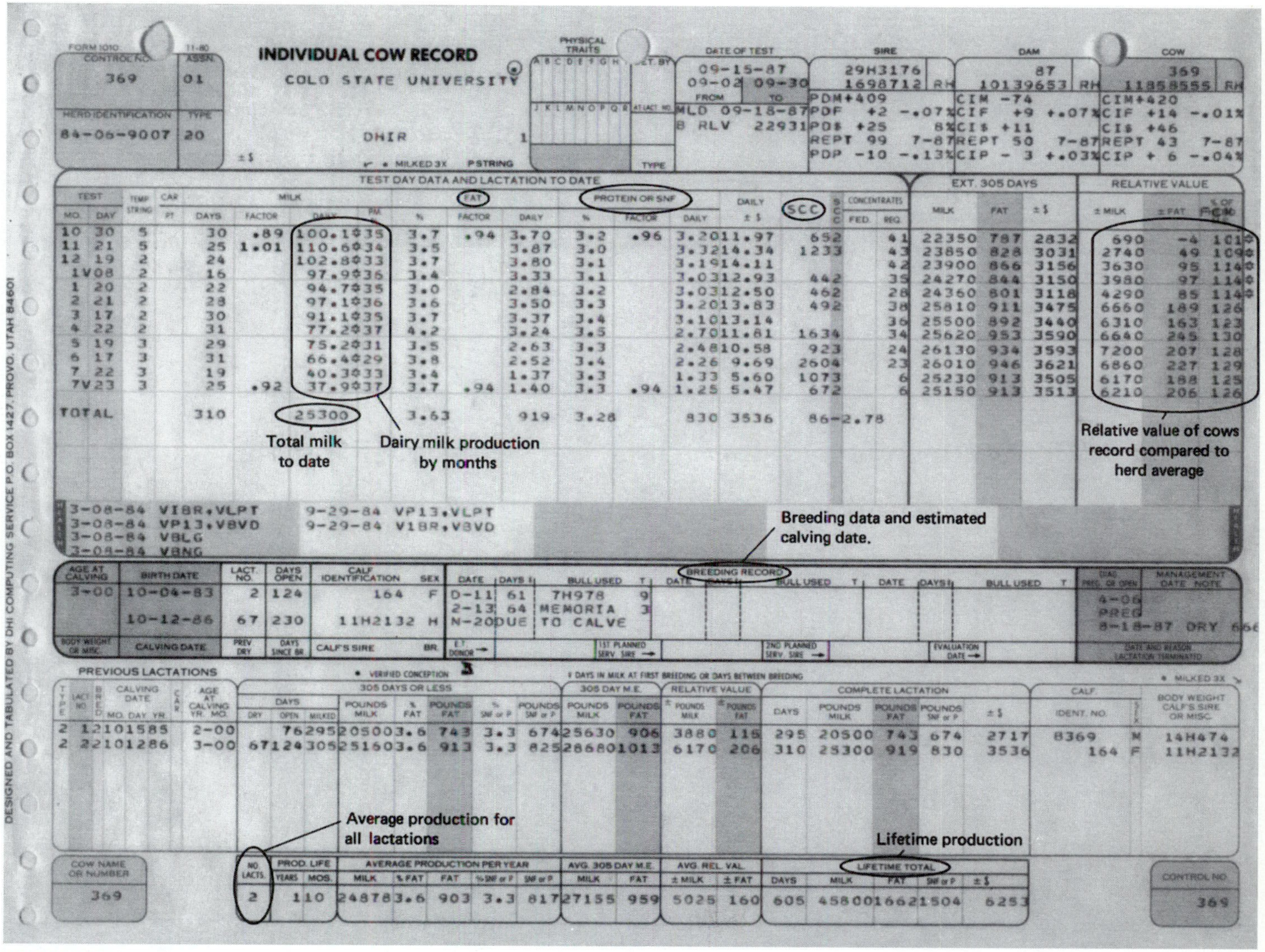

FIGURE 24.4 An individual cow record used in the DHIR program (some selected information is circled with a brief explanation given). Courtesy of Colorado State University.

two milkings per day (2×), and to a mature age of cow (ME = mature equivalent). Adjustments are made to records so they can be more accurately compared on a standardized basis.

Owner-sampler records from a program where the herd owner, rather than the DHIA supervisor, records the milk weights and takes the samples. The information recorded is the same as in official tests, but the records are for private use and are not published.

Tester-sampler records are similar to official records in that the DHIA supervisor samples and weighs the milk. The records are unofficial, however, and the enforcement of cow identification and other DHIA rules is less rigid than in official tests. These records are not for publication.

"**A.M.–P.M.**" records can be recognized as official DHIA records. In this plan the herd is tested each month throughout the year, but the supervisor takes only the morning (A.M.) milking for 1 month and the afternoon (P.M.) milking for the second month. Each daily milk weight is doubled to determine the daily milk weight for calculation by the test-interval method. The daily milk weights printed on the herd report are the average daily milk weights for the last two consecutive tests (one A.M. and one P.M.).

Milk Only records are unofficial records in which only milk weights are used; no tests are made. Each cow's record and the herd summary are based only on amount of milk produced. Unofficial records may also be combined in that owner-sampler records can be A.M.–P.M. or Milk Only. It is also possible to combine Milk Only and A.M.–P.M.

APCS is a popular new system where milk is recorded at two milkings. Samples are taken only A.M. or P.M.

A registry association for purebred animals exists for each dairy cattle breed in the United States. In addition, some breed associations honor cows with outstanding production performance and bulls with daughters that are outstanding in production. The Holstein selective registry, for example, has the Gold Medal Dam and Gold Medal Sire awards. A Gold Medal–winning cow must produce an average of 24,881 lb of milk and 500 lb of milk fat per year during her productive life; she must also have a minimum type score of 83 points based on 100 points. The Gold Medal cow should also have three daughters that meet the requirements. A Gold Medal–winning sire must have 10 or more daughters that meet high standards in milk and fat production and also in type score. Other breed associations have selective registries as a means of encouraging improvement in production.

BREEDING DAIRY CATTLE

The time of breeding is an important phase in dairy cattle management. Because they are milked each day, dairy cows are more closely observed than beef cows, allowing visual detection. When in estrus, dairy cows may show restlessness, enlarged vulvas, and a temporary decline in milk production. Also, when cows are in standing heat, they will permit other cows to mount them.

Technicians are available to artificially inseminate cattle, but well-trained dairy producers or employees are excellent inseminators. The use of semen from genetically proven sires is highly desirable, even though this semen may cost more than semen from an average bull. Considering the additional milk production that can be expected from heifers sired by a good bull, the extra investment can return high dividends. Semen costing $25–$150 per unit from genetically superior bulls may be a better investment than semen from less desirable bulls at $10–$20 per unit. Bulls should not run with the milking cows, because bulls are often dangerous. Bulls that provide semen for artificial insemination should be handled with caution as well.

The heritability of traits is indicative of the progress that can be made by selection. The susceptibilities to cystic ovaries, ketosis, mastitis, and milk fever are all lowly heritable (5–10%). The percentages of fat, protein, and solids-not-fat are all high (50%). Yearly milk (ME), protein, solids-not-fat, and fat yields are medium in hertability (25–30%). With the exception of teat locations and udder attachments, which are moderately heritable (25–30%), all other udder characteristics are lowly heritable (10–20%).

Strength of head and upstandingness are highly heritable (45 and 50%, respectively). Body weight, type score, levelness of rump, height of tail setting, depth of body, tightness of shoulders, and dairy characters are all moderately heritable (25–35%). Strength of pasterns, arch of back, heel depth, and straightness of hocks are lowly heritable (10–20%).

The genetic correlation between two traits is indicative of the amount of genetic change in trait A that might be expected from a certain amount of selection pressure applied to trait B. The most important genetic correlations are those that might be associated with milk yields during the first lactation. Fat, solids-not-fat, protein yield, lifetime milk yields, and length of productive life have high genetic correlations with first lactation milk yield (0.70–0.90). Overall type score, levelness of rump, udder texture, and strength of fore and rear udder are all negatively correlated with first lactation milk yield (–0.20 to –0.40). Dairy character and udder depth are positively correlated (0.35–0.40) with first lactation yield. Inbreeding tends to increase mortality rate and to reduce all production traits except fat percentage of milk and mature body weight.

There are several inherited abnormalities known in dairy cattle. This does not mean that dairy cattle have a higher number of inherited abnormalities than other farm animals, but that more is known about dairy cattle than about most other farm animals because dairy cattle are observed more closely. No breed of dairy cattle is free from all inherited abnormalities. Some of these abnormalities include achondroplasia (short bones), weavers, limber limbs, rectal-vaginal constriction, dumps, flexed pasterns (feet turned back), fused teats (teats on same side of udder are fused), hairless (almost no hair on calf), and syndactylism (only one toe on a foot). Many of these inherited abnormalities are lethal and most are recessive in their mode of inheritance.

Usually the occurrence of inherited abnormalities is infrequent. However, occasionally an outstanding sire might carry an abnormality or a specific dairy herd may have several genetically abnormal calves. As most inherited abnormalities result from recessive genes, both the sire and dam of genetically abnormal calves carry the undesirable gene. Breeding stock should not be kept from either parent.

One reason such rapid progress has been made in improving milk production is that genetically superior sires were identified and used widely in artificial insemination programs. Genetically superior bulls today might become the sires of 100,000 calves or more each in their productive lives by use of artificial insemination. Generally, the better sires will sire more calves than ordinary sires because dairy producers know the value of semen from outstanding bulls. Thus, selection is enhanced markedly for greater milk production by use of artificial insemination (Fig. 24.5).

Milk production, milk composition, efficiency of production, and characteristics that indicate that a cow will likely remain productive for several years are all highly important in selecting dairy cows. These traits are usually emphasized according to their relative heritability and economic importance. Milk and milk fat production are 20–30% heritable, and udder attachment is 30% heritable. Fertility is extremely important but low in heritability, so marked improvement in this trait is more likely to be accomplished environmentally through good nutrition and management rather than through selection.

Most dairy cattle in the United States are straightbred because crossing breeds has failed to make a significant improvement in milk production. No combination of breeds, for example, equals the straightbred Holstein in total milk production. Because milk production is controlled by many genes and the effect of each of the genes is unknown, it is impossible to manipulate genes that control milk production by the same method that one can use in the case of simple inheritance. Also, milk production is a sex-linked trait expressed only in the female. Furthermore, milk production is highly influenced by the environment. To improve milk production by genetic selection, the environment must be standardized among all animals present to ensure, insofar as possible, that differences between animals are due to inheritance rather than environment.

In addition to a high level of milk production, characteristics of longevity, regularity of breeding, ease of milking, and quiet disposition are important in dairy cows. Milk production, milk composition, efficiency of production, and characteristics that indicate that a cow will likely remain productive for several years are all highly important in selecting dairy cows.

Bulls are evaluated for their ability to transmit the characteristic of high-level milk production both by considering the production level of their ancestors (pedigree) and by considering the production level of their daughters (progeny testing). An index is used as a predictive evaluation of the bull's ability to transmit the characteristic of high-level milk production.

FIGURE 24.5 Cows being artificially inseminated with semen from a genetically superior bull. A well-managed AI program can make a significant contribution to improving milk production in dairy herds. Courtesy of Colorado State University.

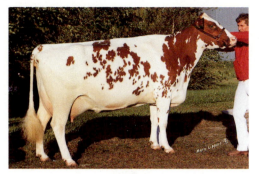

Ayrshire

Origin: Scotland
Average Weight:
   Bulls   1,850 lb
   Cows   1,200 lb
Color: Mahogany and white spotted, may have
          pigmented legs
Average milk yield (1995): 14,430 lb
Percentage of fat: 3.85%

Brown Swiss

Origin: Switzerland
Average Weight:
   Bulls   2,000 lb
   Cows   1,400 lb
Color: Solid blackish; hairs dark with light tips
Average milk yield (1995): 15,974 lb
Percentage of fat: 4.04%

Guernsey

Origin: Guernsey Island
Average Weight:
   Bulls   1,600 lb
   Cows   1,100 lb
Color: Light red and white; yellow skin
Average milk yield (1995): 13,398 lb
Percentage of fat: 4.48%

PLATE Q.
Major breeds of dairy cows with origin, identifying characteristics, and production listed. Photographs courtesy of Agri-Graphics.

Holstein

Origin: Holland
Average Weight:
  Bulls   2,200 lb
  Cows   1,500 lb
Color: Black and white
Average milk yield (1995): 19,091 lb
Percentage of fat: 3.66%

Jersey

Origin: Jersey Island
Average Weight:
  Bulls   1,500 lb
  Cows   1,000 lb
Color: Blackish hairs have white tips to give gray color
          or red tips to give fawn color; also can be solid
          black or white spotted
Average milk yield (1995): 13,396 lb
Percentage of fat: 4.70%

Red and White

Origin: Holland
Average Weight:
  Bulls   2,100 lb
  Cows   1,400 lb
Color: Red and white
Average milk yield (1995): 19,140 lb
Percentage of fat: 3.68%

PLATE R.
Major breeds of dairy cows with origin, identifying characteristics, and production listed. Photographs courtesy of Agri-Graphics.

The **Sire Index,** computed by the USDA, is based on comparing daughters of a given sire with their contemporary herd mates. Each sire is assigned a **predicted difference** based on the superiority or inferiority of his daughters to their herdmates. Many sires have daughters in 50–100 different herds, so the PTA (predicted transmitting ability) for milk, fat, and type is generally highly reliable and provides a sound basis for the selection of semen. The USDA publishes PTAs among sires semiannually.

SIRE SELECTION

The dairy industry (breed associations, National Association of Animal Breeders, and the USDA) use the **Best Linear Unbiased Prediction (BLUP)** method for estimating predicted transmitting ability (PTA) among sires. This method was developed primarily at Cornell University by Dr. C. R. Henderson. The BLUP procedure accounts for genetic competition among bulls within a herd, genetic progress of the breed over generations, pedigree information available on young bulls, and differing numbers of herdmate's sires, and partially accounts for the differential culling of daughters among sires. In the BLUP method, direct comparisons are made among bulls that have daughters in the same herd. Indirect comparisons are made by using bulls that have daughters in two or more herds. An example presented by the Holstein Association (1980) depicts the use of three different sires in two herds:

Herd 1	Herd 2
Daughters of sire A	Daughters of sire B
Daughters of sire C	Daughters of sire C

Direct comparisons can be made between sires B and C in herd 2. Indirect comparisons between sires A and B can be made by using the common sire C as a basis for comparison. PTAs are calculated for milk, protein, fat, type, and dollars returned.

The **Total Performance Index (TPI)** combines differences for milk production traits and difference for type traits into a single value. Figure 24.6 identifies a Holstein bull that is genetically superior in combination of type and milk traits. Figure 24.7 shows the Holstein cow that ranked first out of 700,000 cows for the best combination of milk production traits and type traits.

FIGURE 24.6 S-W-D Valiant is an outstanding Holstein bull. He ranks in the top percentage of all Holstein bulls in transmitting milk production traits and type to his offspring. Courtesy of American Breeders Service.

FIGURE 24.7 Beecher Arlinda Ellen is an outstanding Holstein cow. She has a 305-day milk record of 55,543 lb with a type score of Excellent-91. Courtesy of Agri-Graphics.

CHAPTER SUMMARY

- Based on registration numbers, the Holstein is the major breed of dairy cattle, followed by Jersey, Brown Swiss, Guernsey, Ayrshire, Red and White, and Milking Shorthorn.
- Selection for milk production has doubled the pounds of milk produced per cow from 1960 to 1990.
- The dairy industry has implemented effective production testing programs and record-keeping systems for a longer period of time compared to most other livestock.

REVIEW QUESTIONS

1. What are the six major breeds of dairy cattle in the United States?
2. Which breed produces the greatest amount of milk per day?
3. Which breed of dairy cattle is most popular?
4. *True or False:* Efficiency of milk production per 100 pounds of body weight is about the same for the major breeds of dairy cattle.
5. *True or False:* Most dairy cattle are registered.
6. What are the major reasons that many dairy cattle are culled?
7. What is Dairy Herd Improvement Association?
8. What is the most reliable sign that a cow is in estrus?
9. *True or False:* Expensive semen from genetically superior bulls is more desirable than less expensive semen from average bulls.
10. What traits are most important to select for in order to improve milk production in dairy cattle?
11. Why are most dairy cattle in the United States purebred rather than crossbred?
12. How are bulls evaluated for their ability to transmit the characteristic of high milk production?

SELECTED REFERENCES

Publications

Bath, D. L., Dickinson, F. N., Tucker, H. A., and Appleman, R. D. 1985. *Dairy Cattle: Principles, Practices, Problems, Profits.* Philadelphia: Lea & Febiger.

Holstein Friesian Association of America. 1987. *Linear Classification Program.* Brattleboro, VT.

Legates, J. E. 1990. *Breeding and Improvement of Farm Animals.* 4th ed. New York: McGraw-Hill.

Schmidt, G. H., et al. 1988. *Principles of Dairy Science.* Englewood Cliffs, NJ: Prentice-Hall.

Trimberger, G. W., Etgen, W. E., and Galton, D. M. 1987. *Dairy Cattle Judging Techniques.* Englewood Cliffs, NJ: Prentice-Hall.

Wiggins, G. R. 1991. National genetic improvement programs for dairy cattle in the United States. *J. Anim. Sci.* 69:3853.

Visuals

Animals Acquisition/Reproduction: *Introduction to Animal Acquisition—Breed Identification/Advantages and Disadvantages; Heat Detection in Dairy Cows; Artificial Insemination;* and *The Calving Process* (videotapes). Agricultural Products and Services, 2001 Killebrew Dr., Suite 333, Bloomington, MN 55420.

Cattle Breed Identification: Dairy (video). CEV, Inc., 5147-A 69th St., Lubbock, TX 79424.

Dairy Breed Selection; Selecting Dairy Females; and *Dairy Breeding Systems* (sound filmstrips). Prentice-Hall Media, 150 White Plains Rd., Tarrytown, NY 10591.

Fitting and Showing Dairy Heifers (sound filmstrip); *The Dairy Judging Kit* (slides, manual, and cassette). Vocational Education Production, California Polytechnic State University, San Luis Obispo, CA 93407.

Linear Evaluation of Dairy Cattle (video). CEV, Inc., 5147-A 69th St., Lubbock, TX 79424.

CHAPTER 25

Feeding and Managing Dairy Cattle

The U.S. dairy industry has changed greatly from the days of the family milk cow. Today, it is a highly specialized industry that produces, processes, and distributes milk. A large investment in cows, machinery, barn, and milking parlor is necessary. Dairy operators who produce their own feed need additional money for land on which to grow the feed. They also need machinery to produce, harvest, and process the crops.

Although the size of dairy operations varies from 30 milking cows or less to more than 5,000 milking cows, the average dairy has approximately 100 milking cows, 30 dry cows, and 100 replacement heifers. The average dairy producer farms 200–300 acres of land, raises much of the forage, and markets the milk through cooperatives of which he or she is a member. These producers sell about 3 tons of milk daily, or about 2.2 million lb annually, worth about $230,000. Their average total capital investment may exceed a half million dollars. The average producer has a partnership with a family member or another person to make management of time and resources easier.

The dairy operator must provide feed and other management inputs to keep animals healthy and at a high level of efficient milk production. Feed and other inputs must be provided at a relatively low cost compared with the price of milk if the dairy operation is to be profitable.

NUTRITION OF LACTATING COWS

The average milk production per cow in the United States for a lactation period of 305 days is approximately 16,400 lb. Some herd averages exceed 25,000 lb, and some top-producing cows yield more than 40,000 lb of milk per year. Thus, some lactating cows may produce more than 150 lb of milk, more than 5 lb of milk fat, and more than 4.5 lb of protein per day. A great need for energy and total amount of feed is created by lactation. For example, a cow weighing 1,400 lb that produces 40 lb of milk daily needs 1.25 times as much energy for lactation as it needs for maintenance. If she is producing 80 lb of milk daily, she needs 2.5 times the energy for milk production than is needed for maintenance.

Providing adequate nutrition to lactating dairy cows is challenging and complex. Some reasons for this complexity are shown in Fig. 25.1. The dairy cow's nutritional needs vary widely during the production cycle. Dairy producers must formulate rations to match each period of the production cycle so as to optimize milk yield and reproduction, prevent metabolic disorders, and increase longevity.

It is difficult during the first 2–4 months after calving to provide adequate nutrition, because milk yield is high and intake is limited. As the nutrient intake is less than the nutrient demand for milk, the cow uses her body fat and protein reserves to make up the difference. Thus, the cow is in a negative energy balance and usually loses body weight during the months of heavy milk production (Fig. 25.1). This same period of time can also be challenging to reproduction, as conception rates are usually lower when cows are losing weight.

Body condition scores (1 = thin, 3 = average, 5 = fat) can be used to monitor nutrition, reproduction, and health programs for dairy herds. Condition scores of low 3s and high 2s are acceptable during the first few weeks after calving, but body condition scores in the low 4s are necessary as the cows move into the dry period. Scores lower or higher than these signal potential management problems. Cows can be grouped by condition score and stage of lactation to provide adequate nutrition at effective costs.

At 2–3 months into the lactation period, daily milk production peaks and then starts to decline; feed intake is adequate or higher than milk production demands. This contributes to body-weight gain (Fig. 25.1). The energy content

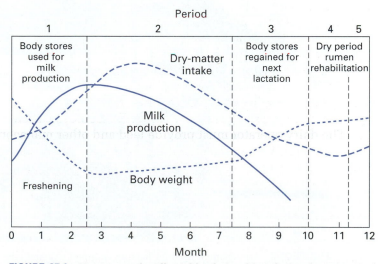

FIGURE 25.1 Nutrient and milk yield relationships during lactation and gestation. Courtesy of Hoffmann-LaRoche, Inc. *Nutritional needs of dairy cows.* (Growthlines). Fall 1989.

of the ration should be monitored after approximately 5 months of lactation to prevent the cow from becoming too fat.

Different types and amounts of feeds can be used to provide the needed energy, protein, vitamins, and minerals to lactating cows. The availability, palatability, and relative costs of different feeds are the primary factors influencing ration composition. Basic nutrition principles, presented in Chapters 15, 16, and 17, provide the foundation for developing dairy cattle feeding programs.

Most dairy cattle rations are based on roughages (hay and silages). Roughages are usually the cheapest source of nutrients; often they are produced by dairy farms, but increasing herd size, greater managerial demands, higher land-tax costs, and cheaper feeds are changing this traditional role. In some cases, roughages are fed free-choice and separate from the concentrates, which are fed in restricted amounts. In other cases, silages and concentrates are mixed together before being fed to the cows; this is known as a ***total mixed ration,*** which is becoming the preferred way of feeding dairy cows. Chopped hay and silage can be delivered to mangers on each side of an open alleyway (Fig. 25.2).

Lactating dairy cows cannot obtain an adequate supply of nutrients from an all-roughage ration. Concentrates are supplied in amounts consistent with the level of milk production, body weight of the cow, amount of nutrients in the roughage, and the nutrient content of the concentrates. Concentrates are usually provided to cows while they are being milked. However, high-producing cows requiring large amounts of concentrates do not have sufficient time in the milking parlor to consume all the necessary grain; therefore, additional concentrates are usually fed at another location.

The computer is increasingly used in dairy feeding programs today. Many dairy farms have feeding stations that use computers to determine how much

FIGURE 25.2 Arrangement used for feeding chopped hay and silage to dairy cows. The feed is delivered to the feeding area from a truck that proceeds along the open alleyway. Courtesy of Dr. Lloyd Swanson, Animal Science Department, Oregon State University.

concentrate a cow will be fed based on need for milk production and to control the rate (lb/hour) at which the concentrate is made available to the cow. These devices have two advantages: (1) they prevent cows from overconsuming concentrates, and (2) they prevent cows from eating concentrates too rapidly so as to prevent metabolic upsets.

There are economic advantages in feeding concentrates in relation to quantity of milk produced by each cow. However, this method tends to feed a cow that is declining in production too generously and to feed a cow that is increasing in production inadequately. It is a poor economic practice to allow cows in their last 2 months of lactation to have all the concentrates that they can consume. Heavy feeding at this stage of lactation does not result in increased milk production.

Young (2-year-old) cows that are genetically capable of high production should be fed large amounts of concentrates to provide the nutrition they need to grow as well as to produce milk. Without adequate nutrition, their subsequent breeding and lactation may be hindered.

Forages vary immensely in nutrient concentrations. Legumes (alfalfa) have high calcium, protein, and potassium relative to animal nutrient requirements, whereas other forages such as grasses and corn silage are considerably lower. Concentrates (grains) usually are low in calcium and high in phosphorus. The amount of protein supplement (soybean meal) and mineral supplement that must be added to a concentrate formulation obviously depends on the type and amount of forage being fed. Trace mineralized salt and vitamins A, D, and E are usually added to concentrates.

Table 25.1 shows examples of rations formulated to meet the nutrient requirements of dairy cows in various periods of production.

NUTRITION OF DRY COWS

How dry cows are fed and managed may influence their milk production level and health in the next lactation.

TABLE 25.1 Examples of Rations Needed to Meet the Nutrient Requirements of a 1,350-lb Dairy Cow with 3.8% Fat Test

Period[a]	Daily Milk Production (lb)	Daily Dry-Matter Intake (lb)	Ration (lb/day; as fed)
1	90	49	Alfalfa hay (28 lb)[b]; corn-oats grain mix (21 lb); soybean oil meal (5 lb); dical (0.5 lb); salt, vitamin, trace mineral mix (0.3 lb)
3	50	38	Alfalfa hay (27 lb)[b]; corn-oats grain mix (16 lb); soybean oil meal (4.5 lb); dical (0.3 lb); salt, vitamin, trace mineral mix (0.25 lb)
4	dry	24	Alfalfa hay (12 lb)[b]; corn silage (43 lb); monosodium phosphate (0.1 lb); salt, vitamin, trace mineral mix (0.1 lb)

[a] Refer to Fig. 25.1.
[b] Contains 20% crude protein.
Source: Hoffmann-LaRoche, Inc. *Nutritional needs of dairy cows.* Growthlines. Fall 1989.

A common practice of drying off lactating cows is abruptly to stop milking the cow; with high producers, however, this may be traumatic and dry-off may need to be more gradual (intermittent milking). The buildup of pressure in the mammary gland causes the secretory tissues to stop producing milk. At the last milking, the cow should be infused with a treatment for preventing mastitis.

Dairy producers plan for a 50–60-day dry period. Short dry periods usually reduce future milk yield because the cow has not adequately improved body condition and the mammary tissue has not properly regenerated. Long dry periods can lower milk yield by the cow becoming overly fat, and profitability may be less because feed costs are increased.

Dry cows should be separated from the lactating cows so they can be fed and managed consistent with their needs. Dry cows need less concentrates than lactating cows. If dry cows overeat, they will likely become fat. This excessive body fat may lower future milk yield and cause health problems, such as fatty liver, ketosis, mastitis, retained placenta, metritis, milk fever, or even death.

NUTRITION OF REPLACEMENT HEIFERS

Young heifers (5 months of age and older) can meet most of their nutritional needs from good-quality legume or grass-legume pasture in addition to 2–3 lb of grain. If the pasture quality is poor, they will need good-quality legume hay and 3–5 lb of grain. In winter, good-quality legume hay or silage and 2–3 lb of grain are needed. If the forage is grass (hay or silage), the heifers will need 3–5 lb of grain to keep them growing well.

Heifers should be large enough to breed at about 15 months of age and calve as 2-year-olds. Weight recommendations for heifers from different breeds are shown in Table 25.2. Weight is important because it affects when the heifer reaches puberty and can be bred. Also, heavier heifers at first calving produce more milk. For example, the optimum weight at calving for Holstein heifers is 1,200–1,250 lb. Lighter heifers produce less milk and heavier heifers are too costly for the increased production.

TABLE 25.2 Recommended Weights at Breeding for Replacement Heifers

Breed	Weight (lb)
Ayrshire	720–750
Brown Swiss	800–875
Guernsey	720–750
Holstein	800–875
Jersey	630–650

Source: Hoffmann-LaRoche, Inc. *Nutritional needs of dairy cows.* Growthlines. Fall 1989.

NUTRITION OF BULLS

It is generally recommended that dairy bulls not be kept on the farm and that all breeding be done AI. By not keeping bulls on the farm, producers (1) decrease the risk of a dangerous bull and (2) increase genetic progress by using superior AI bulls. Some commercial producers do keep bulls on their farms to breed cows naturally, primarily because bulls detect estrus more accurately and keep conception rates high.

CALVING OPERATIONS

Dairy cows that are close to calving should be separated from other cows, and each cow should be placed in a maternity stall that has been thoroughly cleaned and bedded with clean bedding. The cow in a maternity stall can be fed and watered there. A cow that delivers her calf without difficulty should not be disturbed; however, assistance may be necessary if the cow has not calved by 4–6 hours from the start of labor. Extreme difficulties in delivering a calf may require the services of a veterinarian, who should be called as early as possible.

As soon as a calf arrives, it should be wiped dry. Any membranes covering its mouth or nostrils should be removed, and its **navel** should be dipped in a tincture of iodine solution to deter infection. Producers should be sure newborn calves have an adequate amount of colostrum because colostrum contains antibodies to help the calf resist any invading microorganisms that might cause illness. The cow should be milked to stimulate her milk production.

Many commercial dairy operators dispose of bull calves shortly after the calves are born. Some bull calves are fed for veal, while others are castrated and fed for beef. Heifer calves are grown, bred, and milked for at least one lactation to evaluate their milk-producing ability.

Dairy calves rarely nurse their dams. Usually they are removed from their dam and fed milk or milk replacer for 6 to 8 weeks. Calves that are separated from one another usually have fewer health problems and thus a higher survival rate (Fig. 25.3).

Milk that is not salable or milk substitutes are fed to the young calves. Grain and leafy hay are provided to young calves to encourage them to start eating dry feeds and to stimulate rumen development. As soon as calves are able to consume dry feeds (45–60 days of age), milk or milk replacer is removed from the diet because of its high cost.

MILKING AND HOUSING FACILITIES FOR DAIRY COWS

Dairy farming is labor-intensive, so labor-saving machines are common parts of the facilities. Dairy cows usually are managed in groups of 20–50 head; they may be housed in free-stall (cows have access to stalls but are not tied) or loose housing systems. These facilities reduce cleaning, feeding, and handling time as well as labor. In tie-stall barns, cows are tied in a stanchion and remain there much of the year; feeding and milking are done individually in the stanchion. In this case, a pipeline milking system is used. Cows are milked in place and

FIGURE 25.3   An example of calf separation and isolation. These individual fiberglass calf huts are used for dairy calves from birth to weaning. The huts are 3.5 ft × 7.5 ft with an attached wire pen in front. Courtesy of Colorado State University.

milk is carried by a stainless-steel or glass pipe to the bulk tank, where it is cooled and stored. In loose housing systems, such as the loaf shed or free stall (Fig. 25.4), cows are brought to a specialized milking facility, called a *parlor,* for milking.

Many housing systems require daily or twice daily removal of manure and waste feed; this may be done mechanically (tractor-blade) or by flushing with water. Loose housing systems without free-stalls usually have cows on a manure pack, where the pack is typically removed once or twice a year.

FIGURE 25.4   A water flush-free stall barn with individual lock-in stanchions. These facilities provide for inside feeding of totally mixed rations. Courtesy of Colorado State University.

Manure, urine, and other wastes usually are applied directly to land and/or stored in a lagoon, pit, or storage tank for breakdown before being sprayed or pumped (irrigated) onto land.

MILKING OPERATIONS

An example of a modern milking parlor is a concrete platform raised about 30 inches above the floor (pit) of the parlor (Fig. 25.5). It is designed to speed the milking operation, reduce labor, and make milking easier for the operator. Cows enter the parlor in groups or individually, depending on the facility. The udder and teats are washed clean, dried, and massaged for about 20–30 seconds. Milking begins about 1 minute later and continues for about 6–8 minutes. After milking is completed, the milking unit is removed manually or by computer, and the teats are dipped in a weak iodine solution.

Modern milking machines (Figs. 25.6 and 25.7) are designed to milk cows gently, quickly, and comfortably. Some have a takeoff device to remove the milking cup once the rate of flow drops below a designated level. Removal of the milking unit as soon as milking is complete is important because prolonged exposure to the machine can cause teat and udder damage and lead to disease problems (mastitis).

The milking machine operates on a two-vacuum system: one vacuum is located inside the rubber liner and the other vacuum outside the rubber liner of the teat cup (Fig. 25.8). There is a constant vacuum on the teat to remove the milk and keep the unit on the teat. The intermittent vacuum in the pulsation chamber causes the rubber liner to collapse around the teat. This assists blood and lymph to flow out of the distal teat into the upper part of the teat and udder.

The milking parlor and its equipment must be kept sanitary. The pipes are cleaned and sanitized between milkings.

FIGURE 25.5 A trigon milking parlor equipped with computer units and automatic milker takeoffs. Two sides of the milking parlor will accommodate six cows each, while the third side handles four cows. Courtesy of Colorado State University.

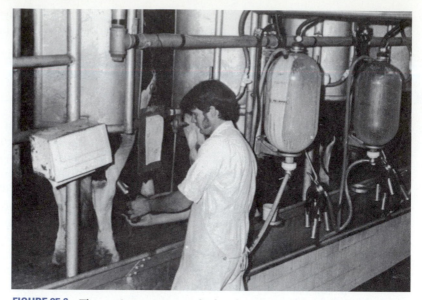

FIGURE 25.6   The suction cups are applied to the teats of the cow so that she can be milked by machine after she has come to her place in the milking parlor. Courtesy of Dr. Lloyd Swanson, Animal Science Department, Oregon State University.

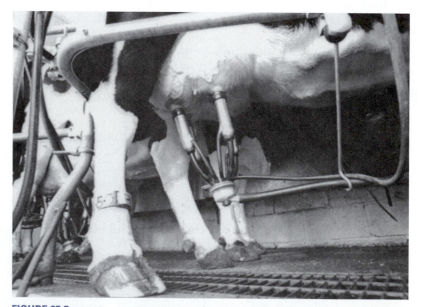

FIGURE 25.7   Modern milker on cow. The personnel in charge of milking observe closely to be certain that the suction cups are removed as soon as the cow has been milked (unless the machine can automatically reduce its suction at that time). Injury to the udder can result if suction continues for some time after milking has been completed. Courtesy of De Laval Agricultural Division, Alfa-Laval, Inc., Poughkeepsie, N.Y.

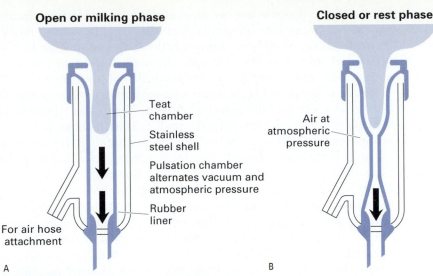

Open or milking phase

Closed or rest phase

Teat
chamber

Stainless
steel shell

Pulsation chamber
alternates vacuum and
atmospheric pressure

Rubber
liner

For air hose
attachment

Air at
atmospheric
pressure

A

B

FIGURE 25.8   Longitudinal section of the teat cup of a milking machine. (A) During the open or milking phase, the vacuum level is the same on both sides of the liner, permitting it to be in an open position. (B) During the closed or rest phase, atmospheric air is admitted into the pulsation chamber by the pulsator. The pressure differential between the inside of the liner and the outside causes the liner to collapse. Courtesy of De Laval Agricultural Division, Alfa-Laval, Inc., Poughkeepsie, N.Y.

Regular milking times are established with equal intervals between milkings. Most cows are milked twice per day; however, high-producing cows produce more milk if they are milked three times a day. The latter is labor-intensive and economically feasible only in herds with high levels of productivity.

CONTROLLING DISEASES

Dairy and beef cattle are afflicted by the same diseases but some diseases are more serious in dairy cattle than in beef cattle. Certain disease organisms can be transmitted through milk and can be a major problem for people who consume unpasteurized milk.

Several diseases, such as tuberculosis and brucellosis, can affect the health of dairy cattle and of humans consuming their milk. Requirements for the production of grade A milk include that the herd (1) must be checked regularly with the ring test for tuberculosis and (2) must be bled and tested for brucellosis. Further, milking and housing facilities must be clean and meet certain specifications. Grade A milk must have a low bacteria count and somatic cell count (low mastitis). Grade A milk sells at a higher price than lower grades of milk; therefore, most dairy producers strive to produce grade A milk.

Bang's disease (brucellosis) is caused by the *Brucella abortus* organism. It can markedly reduce fertility in cows and bulls and it can be contracted by humans as a disease known as *undulant fever*. Heifers should be calfhood-vaccinated to provide immunity from disease. Since most dairy cows are bred by artificial insemination, there is little danger of a clean herd becoming contaminated from breeding. Persons drinking unpasteurized milk are taking a serious risk unless they know that the milk comes from a clean herd.

Perhaps the most troublesome disease in dairy cattle is **mastitis**—inflammation and infection of the mammary gland. The disease destroys tissue, impedes milk production, and lowers milk quality. Mastitis costs the U.S. dairy industry more than $1.5 billion each year or approximately $200 per cow annually.

In its early stage of development, subclinical mastitis is undetectable by the human eye. This type of mastitis can be detected only with laboratory equipment (Somatic Cell Counter at the DHI testing laboratory) or with a cow-side test called the California Mastitis Test (CMT). In the latter test, a reagent is added with a paddle to small wells, then milk is squirted from each quarter into a specific well. If a reaction occurs, then the relative degree of subclinical mastitis infection can be determined (Fig. 25.9). As mastitis progresses, small white clots appear in the milk. At this stage it is called **clinical mastitis** and it is easily observed. One can also collect milk using a strip cup, which has a fine screen on a dark background such that white flakes, strings, or blood (from infected areas) can be seen. As mastitis progresses, the milk shows more clots, watery milk, and acute infection. At this stage of mastitis, the signs of mastitis in the cow are obvious: the udder is swollen, red, and hot and gives pain to the cow when touched. In the final stages of mastitis, the gel formation becomes dark and is present in a watery fluid. At this stage, the udder has been so badly damaged that it no longer functions properly.

Susceptibility to mastitis may be genetically related to a certain degree, but environmental factors such as bruises, improper milking, and unsanitary conditions are more prevalent causes of mastitis. High-producing animals may be more prone to stress and mastitis than low-producing animals. Also, cows with **pendulous** udders or odd-shaped teats are more likely to develop masti-

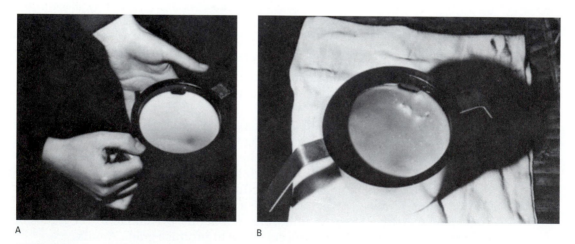

A   B

FIGURE 25.9   (A) Normal milk. (B) Ropey milk in strip cups. Courtesy of Dr. Lloyd Swanson, Animal Science Department, Oregon State University.

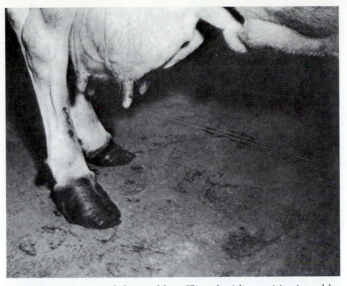

FIGURE 25.10   A pendulous udder afflicted with mastitis. An udder is pendulous when the udder floor is below the hock and the fore and rear attachments are broken down. Courtesy of Dr. H. P. Adams, Oregon State University.

tis (Fig. 25.10). Housing facilities and milking procedures are other important factors.

The best approach to controlling mastitis is using good management techniques to prevent its outbreak. These include using routine mastitis tests (DHI or CMT), treating infected animals, preventing infection from spreading, dipping teats after milking, using dry-cow therapy, and practicing good husbandry for cleanliness. Chronically infected cows may have to be sold.

When mastitis is diagnosed, treatment must begin immediately if the udder is to be saved. Most treatments involve infusing an antibiotic into the udder through the teat canal. Depending on the antibiotic used, milk from treated cows should not be sold for human use for 3–5 days after the final treatment is given.

Other diseases and health practices are discussed in Chapter 21.

COSTS AND RETURNS

Table 25.3 shows enterprise averages for 100 dairy farms. Comparisons for 26 profitable farms and 76 unprofitable farms are made for financial efficiency, total costs per cow, and cost per hundred pounds of milk. To be considered profitable, the dairy farm had to earn at least $30,000 of labor and management income per operator. Profitable farms averaged $47,111 per operator. Their return, compared to unprofitable farms was 13.8% versus −0.8% for return on equity and 11.2% versus 2.1% for return on assets.

Refer to Chapter 36 for a more detailed discussion of costs and returns.

TABLE 25.3   Comparison of Profitable and Unprofitable Dairy Farms[a]

Item	26 Profitable Farms	74 Unprofitable Farms
Financial Efficiency (percent of gross income)		
Operating expense	62.1%	67.4%
Interest expense	4.3	6.3
Depreciation expense	6.3	9.6
Net farm income	27.3	16.7
Total	100.0	100.0
Total Costs Per Cow		
Hired labor	$257	$226
Feed	620	663
Machinery	294	300
Livestock	391	420
Crops	211	246
Real estate	202	198
Other	170	190
Total	$2,145	$2,243
Cost per Hundred Weight of Milk		
Purchased feed and crop expenses	$4.17	$4.75
Breeding	0.19	0.21
Veterinarian and medicine	0.41	0.51
Milk marketing	0.42	0.43
Other expenses	0.69	0.76
Total	$5.88	$6.61

[a] Profitability not related to herd size or level of milk production—e.g., profitable farms' range in herd size was 34 to 270 cows, while milk production per cow ranged from 13,040 to 27,278 lb.
Source: University of Wisconsin (*1995 Dairy Profit Report*).

CHAPTER SUMMARY

■ Dairy farming is labor-intensive so facilities are important in saving labor, feeding cattle, and producing wholesome, clean milk.

■ Nutrition and health programs are extremely important to maintain high milk production in cows.

■ The most troublesome disease in many dairy operations is mastitis—the inflammation and infection of the mammary gland.

■ Forage and concentrates must be high in nutrient content and provided in large amounts to express the high genetic potential for milk production.

REVIEW QUESTIONS

1. What is the standard length of lactation period in dairy cattle in the United States?

2. *True or False:* It is often difficult to meet the nutrient requirements of high-producing dairy cattle during the first 2 to 4 months postpartum because nutrient demand is high but feed intake is limited.

3. A ration in which silages and roughages are mixed together and fed to dairy cattle is called a _____ .

4. Why are concentrates fed to lactating dairy cattle?
5. What nutrient requirement do young (two-year-old) cattle have in addition to those for maintenance and lactation?
6. Why are dry periods of 50 to 60 days necessary?
7. When should heifers be first bred?
8. What are two advantages of purchasing bull semen for use in artificial insemination?
9. Where is milking of dairy cattle most commonly performed?
10. Why is it important to remove the milking machine as soon as milking is complete?
11. How are dairy cows typically housed?
12. What is the most troublesome disease of dairy cattle?
13. What does the best approach to controlling mastitis include?

SELECTED REFERENCES

Publications

Babson Brothers Dairy Research Service. 1980. *The Way Cows Will Be Milked on Your Dairy Tomorrow.* 8th ed. Oak Brook, IL: Babson Brothers Dairy Research Service.

Bath, D. L., Dickinson, F. N., Tucker, H. A., and Appleman, R. D. 1985. *Dairy Cattle: Principles, Practices, Problems, Profits.* Philadelphia: Lea & Febiger.

The changing dairy farm. 1986. *New York's Food and Life Sciences Quarterly* 16.

Etgen, W. M. 1987. *Dairy Cattle Feeding and Management.* New York: Wiley.

Hoffmann-LaRoche, Inc. 1989–90. *Nutritional Needs of Dairy Cows.* Nutley, NJ: Hoffmann-LaRoche.

National Research Council. 1988. *Nutrient Requirements of Dairy Cattle.* Washington, DC: National Academy Press.

Oltenacu, P. A. 1994. Dairy Cattle Production. *Encyclopedia of Agricultural Science.* San Diego: Academic Press, Inc.

Schmidt, G. H., et al. 1988. *Principles of Dairy Science.* Englewood Cliffs, NJ: Prentice-Hall.

Visuals

Nutrition—*Common Feeds for Dairy Cattle; Nutrition Related Problems;* and *Storage and Handling of Feed.* Herd Health/Husbandry—*Distinguishing Normal and Abnormal Cattle; Normal Movements and Body Positions of Dairy Cattle; Animal Handling;* and *Young Stock Management.* Waste Management/Buildings and Equipment—*Waste Storage;* and *Buildings* (videotapes). Agricultural Products and Services, 2001 Killebrew Dr., Suite 333, Bloomington, MN 55420.

Dairy Management Practices (video). CEV, Inc., 5147-A 69th St., Lubbock, TX 79424.

CHAPTER 26

Swine Breeds and Breeding

Breeds of swine are genetic resources available to swine producers. Most market pigs in the United States today are crossbreds, but these crossbreds have their base and continued perpetuation from seedstock herds that maintain pure breeds of swine. A knowledge of the breeds, their productive characteristics, and their crossing abilities is an important part of profitable swine production.

CHARACTERISTICS OF SWINE BREEDS

In the early formation of swine breeds, breeders distinguished one breed from another primarily through visual characteristics. These visual, distinguishing characteristics for swine were, and continue to be, hair color and erect or drooping ears. Figure 26.1 shows the major breeds of swine and their primary distinguishing characteristics.

Relative importance of swine breeds is shown by registration numbers recorded by the different breed associations (Table 26.1). Although the registration numbers are for registered purebreds, the numbers do reflect the demand for breeding stock by commercial producers who produce market swine primarily by crossbreeding two or more breeds. Note how the popularity of breeds has changed with time. This change has resulted, in part, from using more objective measurements of productivity and the breeds' ability to respond to the demand and selection for higher productivity and profitability.

Competitors of purebred breeders are large corporate seedstock suppliers who offer boars and gilts of specialized bloodlines. These lines of breeding, usually called *hybrids,* originated from crossing two or more breeds, then applying some specialized selection programs. Farmers Hybrid Company was the first swine-breeding corporation; it formed in the 1940s. Since 1960, it has

Berkshire (black with white on legs, snout, and tail; erect ears)

Chester White (white; drooped ears)

Duroc (red; drooped ears)

Hampshire (black with white belt; erect ears)

Landrace (white; large, drooped ears)

Poland China (black with white on legs, snout, and tail; drooped ears)

Spotted (black and white spots; drooped ears)

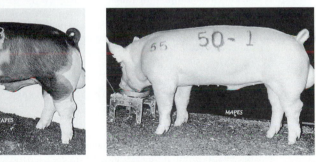

Yorkshire (white; erect ears)

FIGURE 26.1   The major breeds of swine in the United States. Courtesy of the National Swine Registry, other swine breed associations (Chester White Record, National Spotted Swine Record, and American Landrace Association) and Mapes Livestock Photos.

been joined by DeKalb, Kleen Leen, Land O'Lakes, and Pig Improvement Company (PIC).

TRAITS AND THEIR MEASUREMENTS

Table 26.2 gives an overview of the traits preferred in market barrows and gilts. These are specific targets for measurable traits.

TABLE 26.1   Number of Litters Recorded for Major U.S. Breeds of Swine

Breed	1995	1990	1985	Year Association Formed
Berkshire	1,934	2,071	2,158	1875
Chester White	3,077	5,554	5,158	1930
Duroc	13,709	22,179	21,852	1883
Hampshire	16,244	18,925	14,117	1893
Landrace	4,405	4,365	3,501	1950
Poland	1,600	1,848	2,318	1876
Spotted	3,251	6,443	9,310	1914
Yorkshire	19,007	23,861	24,335	1935
Total	63,227	85,236	82,749	—

Source: National Swine Registry.

TABLE 26.2   Symbol : A Standard of Performance

Symbol is the graphic representation of a standard of performance and carcass composition that provides a basis for the continuing improvement of the modern hog.

Symbol symbolizes the commitment of the entire pork industry to making lean pork the true meat of choice of the twenty-first century.

Symbol I   A Market Barrow—adopted in 1983
Was the target or goal for all pork producers. These figures probably represented the top 10% of all hogs marketed in 1983.

- 250 lb market barrow
- 180-lb, 32-inch-long carcass yields 105 lbs of lean pork
- Feed conversion efficiency of 2.5 from birth to a market age of 150 days
- Demonstrates a lean gain of ¾ lb per day of age
- Last rib fat depth measurement at slaughter is 0.7 of an inch
- Loin muscle area is 5.8 sq in
- Average of three backfat measurements is 1.0 in

Symbol II   A Market Gilt—adopted in 1996
Dramatic improvements in the nation's hogs have resulted in a need to update the 1983 figures which, in many cases, have been eclipsed by average hogs. The following represent new targets for the next century.

- 195 lb carcass
- Desirable muscle quality
- Minimum loin muscle of 6.5 (7.1)[a] sq. in. with appropriate
    —Color
    —Water-holding capacity
    —Ultimate pH
- Intramuscular fat level greater than or equal to 2.9% (2.5)
- High health production system
- Produced by an environmentally assured producer on PQA Level III
- Free of the stress gene
- Result of a terminal crossbreeding program
- From a maternal line capable of weaning 25 pigs
- Marketed at 156 (164) days of age
- Weighing 260 lbs
- Liveweight feed efficiency of 2.40 (2.40)
- Fat-free lean gain efficiency of 6.4 (5.90)
- Fat-free lean gain of 0.78 lbs per day
- All achieved on a corn/soy equivalent diet from 60 lbs
- Standard Reference backfat of 0.8 (0.6)
- Fat-Free Lean Index is 49.8 (52.2)

[a] Represents gilt numbers.
Source: National Pork Producers Council.

Sow Productivity

Sow productivity, which is of extremely high economic importance, is measured by litter size, number weaned per litter, 21-day litter weight, and number of litters per sow per year. A combination of pig number (born alive) and litter weight at 21 days best reflects sow productivity (Fig. 26.2). Litter weight at 21 days is used to measure milk production because the young pigs have consumed little supplemental feed before that time. Sow productivity traits are lowly heritable, which means they can be best improved through changing the environment (e.g., feed, health), rather than through selective breeding. Eliminating sows low in productivity while keeping highly productive sows is still a logical practice for economic reasons but not when making major genetic changes. It is important for commercial producers with effective crossbreeding programs to use breeds that rank high in sow and maternal productivity traits.

Growth

Growth rate is economically important to most swine enterprises and has a heritability of sufficient magnitude (35%) to be included in a selection program. Seedstock producers need a scale to measure weight in relation to age (Fig. 26.3). Growth rate can be expressed in several ways and typically it is adjusted to some constant basis, such as days required for swine to reach 230 lb.

FIGURE 26.2   This crossbred sow farrowed 17 pigs and raised 15 of them. The number of pigs raised per sow and the weight of the litter at 21 days of age effectively measure sow productivity. Courtesy of D. C. England, Oregon State University.

FIGURE 26.3 Using scales to measure the growth rate of individual boars is an essential part of a sound breeding system. Courtesy of Iowa State University.

Feed Efficiency

Feed efficiency measures the pounds of feed required per pound of gain. This trait is economically important because feed costs account for 60–70% of the total production costs for commercial producers. Obtaining feed efficiency records requires keeping individual or group feed records. However, feed efficiency can be improved without keeping feed records, such as by testing and selecting individual pigs for gain and backfat. This improvement occurs because fast-gaining lean pigs tend to be more efficient in their feed utilization. Feed efficiency records are obtained in some central boar-testing stations; however, costs and other factors need to be considered before similar records are taken on the farm.

Carcass Traits

Carcass traits are used to estimate the pounds or percentage of acceptable-quality lean pork in the carcass. The traits used to predict carcass composition and acceptable lean quality are (1) fat depth over the loin at the tenth rib, (2) loin muscle area, (3) loin muscle pH, (4) percent loin drip loss, and others. Fortunately, measures of fatness can be predicted reliably by backfat measurements taken on live pigs by ultrasonics (Fig. 26.4). Experienced persons can visually predict significant differences in lean-to-fat composition in live pigs with a fairly high degree of accuracy.

Structural Soundness

Structural soundness—the capacity of breeding and slaughter animals to withstand the rigors of confinement rearing and breeding—is vital to the swine industry today. Breeders feel that most unsoundness results from intensive swine production; actually, though, intensive rearing only makes unsoundness more noticeable. Some seedstock producers raise their breeding stock in modern facilities (similar to the manner in which most commercial producers raise their offspring) and then cull the unsound ones.

A

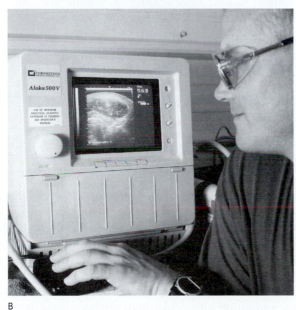

B

FIGURE 26.4 (a) Ultrasound measurements being taken on a pig; (b) screen showing ultrasound image of loin eye and fat (above the loin eye muscle) from the pig shown in (a). Courtesy of DEKALB Swine Breeders.

Recent studies report unsoundness, as measured by leg soundness score, to be 50% in heritability; therefore, improvement can be made through selection. Soundness can be improved through visual selection if breeders decide to cull restricted, too straight-legged, unsound boars lacking the proper flex at the hock, set of the shoulder, even size of toes, or proper curvature and cushion to the forearm and pastern.

Inherited defects and abnormalities generally occur with a low degree of frequency across the swine population. Most of the important genetic abnormalities are inherited as simple recessive genes. It is important to know these abnormalities exist and that certain herds may experience a relatively high incidence. Certain structural unsoundness characteristics might be categorized as inherited defects. **Cryptorchidism** (retention of one or both testicles in the abdomen), scrotal **hernias** (rupture), and **inverted nipples** (nipples or teats that do not protrude but are inverted into the mammary gland), and **PSE** carcasses (pale, soft, and exudative) are considered the most commonly occurring genetic problems. The PSE condition and the porcine stress syndrome (**PSS**) have occurred in recent years from selecting for extreme muscling.

The PSS gene in its homozygous mutant form (nn) results in either death from physical stress or PSE meat. However, the presence of the normal allele (N) nearly always prevents stress death or the rigidity response to testing, using the anesthetic halothane.

Research studies have shown the carrier animals to be 1 to 4% higher than normal pigs (NN) in lean percentage and to be equal, if not superior, in growth rate. All of these studies, however, have suggested the muscle quality of these pigs (Nn) to generally be undesirable and the frequency of PSE to be in the 30 to 50% range.

There is evidence that the PSS gene does not account for all of the pork industry's PSE problems. However, because the gene offers only a minimal increase in lean percentage which significantly increases the PSE frequency, the gene should be eliminated from the U.S. pig population.

If pork is to be elevated to the "meat of choice" as noted in industry literature, then carcass leanness and muscling cannot be the entire focal point. In addition to eliminating the PSS gene, selection methods to increase pork quality appear advisable. Consumer preference studies on pork suggest an increase in the lipid level of meat would make it more competitive with chicken and possibly other meats.

EFFECTIVE USE OF PERFORMANCE RECORDS

Commercial producers recognize that the performance level in their herd is influenced by two factors: genetics and environment (e.g., feed, health, housing). Commercial producers can effectively manage the genetics in their herd by identifying seedstock producers who have excellent genetic improvement programs. A sound genetic improvement program should include four features: (1) accurate, complete performance records including animal identification, consistent measurement of all boars and gilts (not on-again, off-again or limited, partial performance-testing), and ranking of animals within defined contemporary groups; (2) assessment of the genetic merit of economically important traits (growth rate, feed efficiency, carcass merit, and reproductive performance) based on the individual's performance relative to its contemporary group and incorporating the performance of relatives; (3) indexes weighting traits relative to their economic importance in commercial pork production (the indexes should correctly rank the individuals relative to their intended use in crossbreeding and marketing systems); and (4) selection of the highest-ranking boars and gilts based on selection indexes. Seedstock producers should utilize selection indexes as their primary selection criteria.

The rate of genetic improvement in a commercial pork producer's herd will parallel the rate of genetic progress made by the seedstock supplier the commercial producer chooses (Fig. 26.5). Therefore, the commercial producer

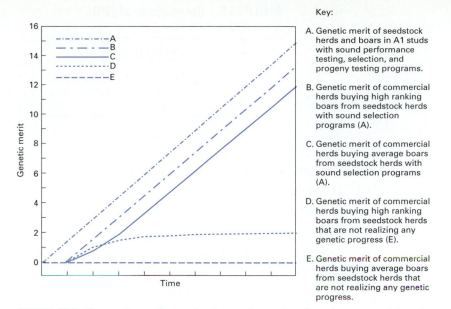

Key:

A. Genetic merit of seedstock herds and boars in A1 studs with sound performance testing, selection, and progeny testing programs.

B. Genetic merit of commercial herds buying high ranking boars from seedstock herds with sound selection programs (A).

C. Genetic merit of commercial herds buying average boars from seedstock herds with sound selection programs (A).

D. Genetic merit of commercial herds buying high ranking boars from seedstock herds that are not realizing any genetic progress (E).

E. Genetic merit of commercial herds buying average boars from seedstock herds that are not realizing any genetic progress.

FIGURE 26.5 Comparison of genetic change in seedstock and commercial swine herds. Source: *Pork Industry Handbook* (PIH-9).

should select seedstock producers whose improvement programs have the four features previously mentioned.

There are several excellent genetic evaluation programs available to seedstock breeders (e.g., SWINE-EBV, Nebraska SPF, and STAGES). STAGES (the Swine Testing and Genetic Evaluation System), which is a series of computer programs implemented by breed associations, will be highlighted here. STAGES computes genetic evaluations for several traits within a single herd. Postweaning traits include growth rate (days to 230 lb or average daily gain) and backfat thickness; reproduction and lactation traits include number born alive and litter weight on all sows.

The genetic evaluation which results from STAGES is expressed as an EPD (Expected Progeny Difference).

Several breeds each have an across-herd sire summary that is published on an annual basis. Sire summaries provide unbiased estimates of the genetic merit of boars that have a legitimate performance record or have progeny with performance records. These estimates are in the form of EPDs.

The EPD is a prediction of the progeny performance of an animal compared to the progeny of an average animal in the breed based on all information currently available. This allows for the direct comparison of all boars evaluated in the sire summary based on the predicted differences between their progeny's performance. These EPDs are expressed as plus or minus values with the average EPD for the breed being approximately zero. Table 26.3 compares two boars assuming the EPDs have equal accuracies.

Progeny of boar A are expected to reach 230 pounds 3.5 days sooner than progeny of boar B and have 0.10 inch less backfat. This assumes the boars are bred to sows of similar genetic merit.

Across-herd sire summaries for the Duroc, Hampshire, and Yorkshire breeds report EPDs for the following traits or combination of traits:

TABLE 26.3 Comparison of EPDs for Two Boars

| | Expected Progeny Difference | |
Boar ID	Days to 230 lb	Backfat
A	−3.5	−0.05
B	0	+0.05

NBA (Number Born Alive)

LWT (21-Day Litter Weight)

D/230 (Days to 230 Pounds)

BF (Backfat)

MLI (Maternal Line Index)—combines growth and maternal traits

SPI (Sow Productivity Index)—combines NBA and LWT

TSI (Terminal Sire Index)—combines D/230 and BF

An accuracy value is listed with each EPD. Accuracies range from 0–1.0 and give a confidence estimate of the EPD. Accuracies that approach 1.0 indicate the EPD is a good estimate of the boar's true genetic potential for that trait.

The results of the National Pork Producers Council Terminal Line National Genetic Evaluation Program (NGEP) were published in 1995. This was the largest, most comprehensive genetic evaluation program ever done in the swine industry.

A large representative sample of boars (795) from nine different sire lines (breeds and hybrids) were each mated to a similar genetic base of sows to produce 1990 litters. These progeny test pigs were from matings to give 100% heterosis.

More than 3,200 test pigs were evaluated for growth and carcass traits. Evaluations were also made for loin meat quality traits and loin eating-quality traits with more than 30 traits measured for these three major categories. Index 1 and Index 2 were calculated with each index combining several traits into an economic value expressed as dollars per pig. Index 1 establishes the value of a sire line to a producer selling hogs on a backfat-based, carcass merit buying system. Index 2 gives the value of a sire line if retail values can be realized in addition to the values achieved in Index 1. Thus, realizing these value differences per pig is dependent on the marketing alternative that is utilized. Table 26.4 gives a summary of several selected traits and Index 1 and Index 2.

The differences in sire lines in the NGEP are useful in planning effective selection programs. However, individual boar selection within a breed or line using EPDs and other performance data should receive the greater emphasis.

National consumer preference studies throughout the United States were made on samples of pork from the NGEP to not only evaluate the pork but also compare it to chicken breast. The major findings were as follows:

1. Price (58%) was the most important attribute when comparing chicken to pork. Next most important was lipid level (24%).

TABLE 26.4   **Selected Trait Summary (National Genetic Evaluation Program)**

				TRAIT					
Sire Line	Days to Reach 250 lb	Feed Conversion (lb feed/lb gain)	Backfat (in.)	Loin Area (sq. in.)	Loin Drip Loss (%)[a]	Loin pH[b]	Tenderness (lb)[c]	Index 1 ($/pig)[d]	Index 2 ($/pig)[e]
Berkshire	175	3.07	1.25	5.74	2.43	5.91	12.63	−$4.14	−$4.05
Danbred HD	176	2.88	0.98	6.75	3.34	5.75	12.78	2.12	1.52
Duroc	170	2.91	1.13	6.14	2.75	5.85	12.43	0.64	10.51
Hampshire	175	2.92	1.01	6.58	3.56	5.70	12.89	1.34	1.55
NGT Large White	173	2.94	1.17	5.62	2.92	5.84	13.40	−0.97	−8.87
Nebraska SPF Duroc	168	2.89	1.11	6.35	2.81	5.88	12.72	1.50	8.25
Newsham Hybrid	172	2.83	0.98	6.45	2.99	5.82	13.46	3.53	0.89
Spotted	174	3.14	1.24	5.83	2.88	5.83	13.02	−4.77	−8.20
Yorkshire	175	2.93	1.05	6.17	2.85	5.84	13.49	−0.74	−1.70

[a] Important because 1% drip loss equals a loss of 1 lb for every 100 lb meat.
[b] Higher pH associated with low drip loss, darker color, more firmness, increased tenderness—all positive attributes.
[c] Pounds of pressure to push probe into cooked loin (Instron machine).
[d] Index 1 combines days to 250 lb, feed conversion, and backfat at 10th rib.
[e] Index 2 combines intramuscular fat in loin, pH, tenderness, drip loss, and loin muscle area that were estimated using results of consumer preference study, then added to traits in Index 1.

2. Higher lipid levels and greater tenderness were preferred when taste-testing pork.

3. Increases in lipid level have more than twice the economic value to these consumers as increased tenderness or more preferred pH levels.

4. The most preferred visual characteristics must be balanced against the most preferred taste characteristics to identify the pork chop with the greatest long-term consumer appeal.

The EPD measures how an individual's offspring can be expected to perform compared to offspring of an average individual. For example, a boar with an EPD of −3.0 for days to 230 lb will be expected to sire offspring which reach market weight 3 days sooner than offspring of an average boar whose EPD is 0.

The EPDs for individual traits are combined to calculate economic indexes. The indexes are expressed as expected profit per litter relative to an average of 100. Four indexes are calculated, because different breeders may wish to select for different traits. The General Purpose Index puts equal emphasis on post-weaning and reproduction traits. It would be appropriate for selection of animals to be used in a rotational crossbreeding program or for general selection purposes where all traits are of importance. The Maternal Line Index puts twice as much weight on reproduction as on postweaning traits, while the Terminal Sire Index includes postweaning traits only. These two indexes can be used to select dam and sire lines, respectively, for use in a terminal crossbreeding program. Finally, a Sow Productivity Index is calculated, which involves reproduction traits only.

To compare animals within a group more objectively, indexes have been developed for use in testing stations and farm programs. The index is a single numerical value that is determined from a formula that combines the values for several traits. One such index, currently in use, combines the performance of daily gain, feed efficiency, and backfat into one numerical value for individual boars. In the formula that follows, ADG (average daily gain), F/G (feed per pound of gain), and BF (backfat) represent the averages for all the pigs in the group that are tested together. The index is as follows:

I (index) = 100 + 60 (averaging daily gain of individual—ADG)
−75 (feed efficiency of individual or group of litter mates—FG)
−70 (backfat probe or ultrasound of individual—BF)   .

SELECTING REPLACEMENT FEMALES

Sow productivity is the foundation of commercial pork production. The sow herd also contributes half of the genetic composition to the growing–finishing pigs. These two factors show the importance of replacement gilt selection so that highly productive gilts can be retained in the swine herd. Fast-growing, sound, moderately lean replacement gilts with good body capacity should be selected from litters where 8–14 excellent pigs were weaned (Fig. 26.6). Among sows that have farrowed and will rebreed, those that have physical problems, bad dispositions, extremely small litters (two pigs below the herd average), and poor mothering records should be culled.

A balance between sow culling and gilt selection needs to be established. Replacement gilts should be available in sufficient numbers to replace culled sows. Gilts, replacing sows, represent a major opportunity for genetic change in the sow herd. As sows generally produce larger litters of heavier pigs than gilts, replacing large numbers of sows with gilts may reduce production levels. This production differential and the low relationship among the performance of successive litters argue for low rates of culling based on sow performance to maintain high levels of production.

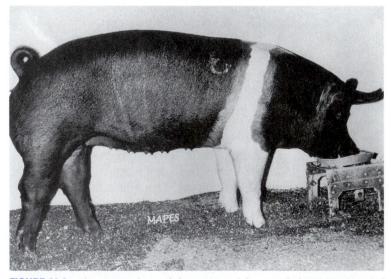

FIGURE 26.6 This Hampshire gilt has many of the visual characteristics and performance records identified in Table 26.2. Source: National Swine Registry.

The higher productive rate of sows over gilts must be balanced against the genetic change made possible by bringing gilts into production. A total gilt replacement level of 20–25% is suggested for each farrowing. Pork producers may find economic advantages in timing the culling of sows to take advantage of high prices for sows.

The gilt selection and sow-culling scheme suggested assumes that there are no major genetic antagonisms between maternal traits (litter size and milking ability) and rate of gain and low-backfat thickness. Some evidence indicates that the so-called meaty gilt does not make a good sow. There is, however, no documented evidence that selecting fast-growing, low-back fat gilts that are not too extreme in muscling will adversely affect sow performance. Guidelines for gilt selection by age and weight are given in Table 26.5.

BOAR SELECTION

Boar selection is extremely important in making genetic change in a swine herd. Over several generations, selected boars will contribute 80–90% of the genetic composition of the herd. This contribution does not diminish the importance of female selection. Boars should have dams that are highly productive sows. Productivity of replacement gilts is highly dependent on the level of sow productivity passed on by the boars.

TABLE 26.5   Gilt Selection Calendar

When	What
Birth	Identify gilts born in large litters. Hernias, cryptorchids, and other abnormalities should disqualify all gilts in a litter from which replacement gilts are to be selected.
	Record birth dates, litter size, identification.
	Equalize litter size by moving boar pigs from large litters to sows with small litters. Pigs should nurse before moving.
	Keep notes on sow behavior at time of farrowing and check on disposition, length of farrowing, any drugs (such as oxytocin) administered, condition of udder, and extended fever.
3–5 weeks	Take 21-day weight of litter. Wean litters. Feed balanced, well-fortified diets for excellent growth and development.
	Screen gilts identified at birth by examining underlines and reject those with fewer than 12 well-spaced teats. If possible, at this time select and identify about two to three times the number of gilts needed for replacement.
180–200 lb	Weigh gilts and obtain ultrasound measurement of backfat thickness. Evaluate for soundness.
	Select for replacements the fastest-growing, leanest gilts that are sound and from large litters. Save 25–30% more than needed for breeding.
	Remove selected gilts from market hogs. Place on restricted feed.
	Allow gilts to have exposure to a boar along the fence that separates them.
	Observe gilts for sexual maturity. If records are kept, give advantage to those gilts that have had several heat cycles prior to final selection.
Breeding time	Make final cull when the breeding season begins and keep sufficient extra gilts to offset the percentage nonconception in the herd.
	Make sure all sows and gilts are ear-marked, ear-tagged, or otherwise identified.

Source: Pork Industry Handbook, PIH-27, Guidelines for Choosing Replacement Females, 1992.

TABLE 26.6   Suggested Selection Standards for Replacement Boars

Trait	Standard
Litter size	10 or more pigs farrowed; 8 or more pigs weaned
Underline	12 or more fully developed, well-spaced teats
Feet and legs	Wide stance both front and rear; free in movement; good cushion to both front and rear feet; equal-sized toes
Age at 230 lb	155 days or less
Feed/gain, boar basis (60–230 lb)	240 lb of feed/100 lb (cwt) of gain or less
Daily gain (60–230 lb)	2.0 lb/day or more
Backfat ultrasound (adjusted to 230 lb)	0.8 in. or less

Source: Adapted from *Pork Industry Handbook* (PIH-9), Boar selection for commercial production.

Boars should be selected from the top 50% of the test group regardless of whether the selected boars come from a central test station or from the farm. Boars meeting the standards shown in Table 26.6 and having excellent EPDs should be considered as potential herd sires. An example of a genetically superior boar is shown in Fig. 26.7.

CROSSBREEDING FOR COMMERCIAL SWINE PRODUCERS

The primary function of seedstock breeders is to provide breeding stock for commercial producers. Breeding programs for seedstock producers are designed to improve genetically the economically important traits. Within a specific breed, rate of improvement will be dependent on heritability of the traits, how much selection is practiced, and how quickly generation turnover occurs. The most rapid genetic changes that have occurred over the past few decades are growth rate and leanness.

Commercial producers and seedstock producers select for the same economically important traits. Commercial producers have an added advantage over seedstock producers, in making genetic change, because crossing two or more swine breeds will increase productivity. This genetic phenomenon of heterosis or hybrid vigor is shown in Table 26.7. There is a 40% increase in total litter market weight of crossbreds, which is a cumulative effect of larger litters, high pig survival, and increased growth rate of individual pigs. This marked increase in productivity tells why more than 90% of the market hogs in commercial production are crossbreds. Crossbreeding allows genetic improvement in some traits with low heritabilities, such as sow productivity. There is little hybrid vigor for traits with high heritabilities, such as carcass traits.

An effective crossbreeding program takes advantage of using hybrid vigor and selecting genetically superior breeding animals from breeds that best complement one another. A producer needs to know the relative strengths and weaknesses of the available breeds shown in Table 26.8. These breed comparisons can change over time as a result of genetic changes made in the individual breeds. Thus, breed comparisons must be made on a regular basis. In

FIGURE 26.7 "Ulf 166", a Yorkshire boar who has demonstrated exceptional genetic superiority as evidenced by the data supplied.

- During a three year period (1991–1994), "Ulf" was either first-, second-, or third-ranking boar in the Yorkshire breed for NBA (Number Born Alive), LTWT (21-Day Litter Weight), BF (Backfat), and MLI (Maternal Line Index).
- "Ulf's" domination of the Yorkshire National Sire Summary was most evident in the January 1994 edition:
  —"Ulf" ranked No. 1 for MLI, while his sons ranked 2d, 3d, and 4th, 5th, 6th, 7th, 9th, and 10th.
  —"Ulf" ranked No. 3 for NBA, while his sons ranked 1st and 4th.
  —"Ulf" ranked No. 3 for LTWT, while his sons ranked 1st, 2d, 4th, 6th, and 8th.
  —"Ulf" ranked No. 2 for SPI (Sow Productivity Index), while his sons ranked 1st, 3d, 5th, 6th, 7th, and 8th.

Source: Swine Genetics International Ltd.

addition, the variation within a breed must be considered; individual pig selection could vary from the performance evaluations shown in Table 26.8.

The crossbred female is an integral and important part of an effective crossbreeding program. Even though crossbreeding provides an opportunity to reap the benefits of many genetic sources, an unplanned crossing program will not yield success or profit for the pork producer. A well-planned crossbreeding system capitalizes on heterosis, takes advantage of breed strengths, and fits the producer's management program.

The two basic types of crossbreeding systems are the **rotational cross** system and the **terminal cross** system. The rotational cross system combines two or more breeds, where a different breed of boar is mated to the replacement crossbred females produced by the previous generation (Fig. 26.8). In a simple terminal cross system, a two-breed single or rotational cross female is mated to a boar of the third breed. A more complex terminal crossing system, shown in Fig. 26.9, uses four breeds. In this example, crossbred boars are mated to crossbred females.

Figure 26.10 shows a **rotaterminal** crossbreeding system, which combines the three-breed rotational system and the terminal system of crossbreeding. Thus, all

TABLE 26.7   Heterosis Advantage for Production Traits

Item	First-Cross Purebred Sow	Multiple-Cross Crossbred Sow	Crossbred Boar
Percentage of Advantage over Purebred			
Reproduction			
Conception rate	0.0%	8.0%	10.0%
Pigs born alive	0.5	8.0	0.0
Litter size at 21 days	9.0	23.0	0.0
Litter size, weaned	10.0	24.0	0.0
Production			
21-day litter weight	10.0	27.0	0.0
Days to 220 lb	7.5	7.0	0.0
Feed/gain	2.0	1.0	0.0
Carcass composition			
Length	0.3	0.5	0.0
Backfat thickness	−2.0	−2.0	0.0
Loin muscle area	1.0	2.0	0.0
Marbling score	0.3	1.0	0.0

Source: *Pork Industry Handbook* (PIH-39), 1987.

the pigs produced by terminal cross boars are sold because the gilts would lack the superiority needed in maternal traits (reproduction and milk). The female stock can be purchased through a system primarily emphasizing reproductive performance. This latter system maximizes heterosis; however, purchasing replacement females increases the risk of introducing disease into the herd.

A two- or three-breed rotational crossbreeding system may better fit a producer's management program even though heterosis is less than with a terminal cross system. A two-breed rotation results in 67% of the heterosis of a terminal cross, whereas a three-breed rotation gives 86% of the heterosis of a terminal crossing system.

TABLE 26.8   Comparative Performance Evaluations for Swine Breeds

Trait	Berkshire	Chester White	Duroc	Hampshire	Landrace	Poland China	Spotted	Yorkshire
Conception rate (%)	+	+	A	A	−−	A	A	−
Litter size (no. pigs raised to weaning)	−	++	A	−	++	−−	−	++
21-day litter weight (lb)	−	−	A	A	++	−	−−	+
Rate of gain (lb/day)	−	−	++	A	A	−	−	A
Feed efficiency (lb feed/lb gain)	−	A	+	+	A	−	−	A
Backfat (in.)	−	A	A	+	A	A	−	A
Loin eye area (sq in.)	A	A	A	+	A	+	A	A

Evaluation system:
++ = substantially superior to average
+ = superior to average
A = average or near average
− = below average
−− = substantially below average
Sources: Adapted from the *Pork Industry Handbook* (PIH-39); and *National Hog Farmer* (June and Oct. 1990).

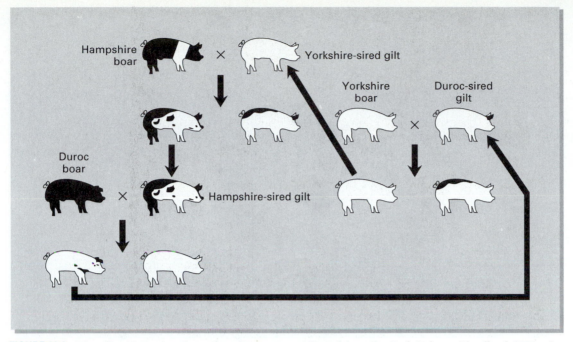

FIGURE 26.8   Three-breed rotational crossbreeding system. Adapted from the *Pork Industry Handbook* (PIH-39).

The three-breed rotational cross is a popular crossbreeding system, although many of the larger commercial operations are using terminal crossbreeding. The three-breed rotational cross combines the strong traits of a third breed not available in the other two breeds. Sires from three breeds are systematically rotated each generation and replacement crossbred females are selected each generation. These females are mated to the sire breed to which they are least

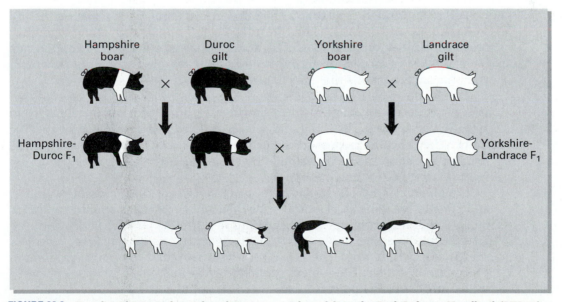

FIGURE 26.9   Four-breed terminal crossbreeding system. Adapted from the *Pork Industry Handbook* (PIH-39).

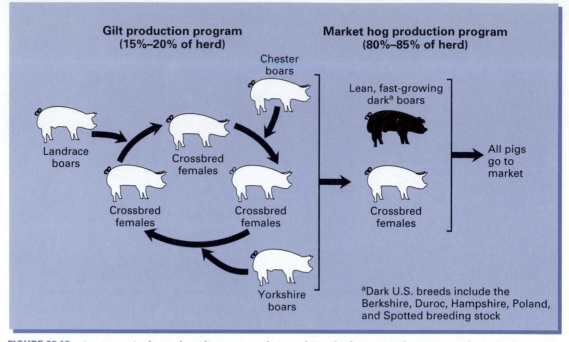

FIGURE 26.10   A rotaterminal crossbreeding system that combines both rotational crossing and terminal crossing. Adapted from the *Pork Industry Handbook* (PIH-106), Genetic principles and their applications.

related. Because reproductive performance is to be stressed in the initial two-breed cross, growth, feed efficiency, and superior carcass composition may be lower than desired. These traits can be emphasized in the individual boars selected from the third breed, although the reproduction and maternal traits should also receive attention.

Although limited comparative research information is available on the use of crossbred and hybrid boars, these boars are apparently more aggressive breeders, have fewer problems in leg soundness, and improve conception rate (Table 26.7) in comparison to purebred boars. Crossbred or hybrid boars can combine superior traits that may not be available in one purebred breed.

Hybrid boars sold by some commercial companies should not be confused with crossbred boars sold by private breeders. Hybrid boars are developed from specific line crosses. These lines have been selected and developed for specific traits. When specific crosses are made, the hybrid boar is used on specific cross females to obtain the recommended breed combination and to realize a high level of heterosis in their offspring.

Commercial pork producers have many selection tools, crossing systems, and genetic breeding-stock sources for their use. Producers should capitalize on heterosis and breed strengths, and require complete performance records on all selected breeding stock. Although there is no one best system, breed, or source of breeding stock, producers must evaluate their total pork production program and integrate the most profitable combination of factors associated with a crossbreeding program.

CHAPTER SUMMARY

■ The most important visual characteristics in identifying breeds of swine are hair color, color pattern, and erect or drooping ears.

■ Based on registration number, Yorkshire is the most numerous breed of swine, followed by Duroc and Hampshire.

■ Traits of high economic importance that are included in most selection programs are (1) sow productivity (number in litter and litter weight at 21 days), (2) growth rate, (3) feed efficiency, (4) carcass traits (pounds of acceptable-quality pork), and (5) skeletal soundness.

■ Approximately 50% of the boars are hybrids (several breeds combined) and 50% are from specific single breeds (purebreds).

■ Crossbreeding in swine is widely used whereby the outstanding traits of two or more breeds are combined and heterosis increases the pounds of market pigs sold per litter.

REVIEW QUESTIONS

1. What are the three most popular breeds of swine in the United States today?

2. What are measures of sow productivity?

3. What two measures combined are the best indication of sow productivity?

4. How can sow productivity best be improved?

5. *True or False:* Growth rate of swine is sufficiently heritable to include it in a selection program of genetic improvement.

6. Why is feed efficiency such an economically important trait for most commercial swine producers?

7. What are three quality measures of carcass traits?

8. Which carcass trait is most important?

9. Why are structural soundness traits of importance to commercial swine producers?

10. What is a major disadvantage of replacing large numbers of sows with gilts?

11. What is a major advantage of replacing large numbers of sows with gilts?

12. What are the four features which should be included in a sound program of genetic improvement?

13. What is the primary purpose of seedstock producers?

14. What are breeding programs of seedstock producers designed to do?

15. Why are more than 90% of all hogs in commercial production crossbred?

16. What are the two basic types of crossbreeding systems for swine?

17. A _____ crossbreeding system combines aspects of the three-breed rotational cross and terminal cross systems.

18. Which three breeds of swine are strongest in reproductive traits?

19. Which breed of swine is strongest in growth performance traits?

20. Which breed is superior in carcass traits?

SELECTED REFERENCES

Publications

Boar selection guidelines for commercial pork producers. 1988. *Pork Industry Handbook,* PIH-9. Swine Extension Specialists: various State Universities.

Crossbreeding systems for commercial pork production. 1988. *Pork Industry Handbook.* Swine Extension Specialists: various State Universities.

Genetic Evaluation. Terminal Line Program Results. 1995. Des Moines, IA: National Pork Producers Council.

Guidelines for choosing replacement females. 1992. *Pork Industry Handbook,* PIH-27. Swine Extension Specialists: various State Universities.

Legates, J. E. 1990. *Breeding and Improvement of Farm Animals.* 4th ed. New York: McGraw-Hill.

National Hog Farmer. 1999 Shepard Road, St. Paul, MN 55116.

Pond, W. G., Maner, J. H., and Harris, D. L. 1991. *Pork Production Systems.* Westport, CT: AVI Publishing Co.

Porcine Stress Syndrome. 1992. *Pork Industry Handbook,* PIH-26. Swine Extension Specialists: various State Universities.

Stewart, T. S., et al. 1991. Genetic improvement programs in livestock: Swine testing and genetic evaluation system (STAGES). *J. Anim. Sci.* 69:3882.

Seedstock Edge (periodical). West Lafayette, IN: National Swine Registry.

Visuals

Breeding Herd Selection and Management (sound filmstrip). Vocational Education Productions, California Polytechnic State University, San Luis Obispo, CA 93407.

Swine Breed I.D. (slide set). Vocational Education Productions, California Polytechnic State University, San Luis Obispo, CA 93407.

Swine Breed Identification; Female Selection in Swine; Boar Selection; and *Swine Breeding Systems* (sound filmstrips). Prentice-Hall Media, 150 White Plains Rd., Tarrytown, NY 10591.

The Swine Industry in the U.S.; Breeds of Swine; Types of Breeding Programs; Selection of Breeding Stock; Low Cost Swine Breeding Units; How to Handle Newly Purchased Breeding Stock; and *Breeding Management* (videotapes). Agricultural Products and Services, 2001 Killebrew Dr., Suite 333, Bloomington, MN 55420.

CHAPTER 27

Feeding and Managing Swine

Swine production in the United States is concentrated heavily in the nation's midsection known as the Corn Belt. This area produces most of the nation's corn, which is the principal feed used for swine.

TYPES OF SWINE OPERATIONS

There are four primary types of swine operations: (1) **feeder pig production,** in which the producer maintains a breeding herd and produces feeder pigs for sale at an average weight of approximately 40 lb; (2) **feeder pig finishing,** in which feeder pigs are purchased and then fed to slaughter weight; (3) **farrow-to-finish,** in which a breeding herd is maintained where pigs are produced and fed to slaughter weight on the same farm; and (4) **purebred** or **seedstock** operations, which are similar to farrow-to-finish except that their salable product is primarily breeding boars and gilts.

FARROW-TO-FINISH OPERATIONS

Since farrow-to-finish is the major type of swine operation, primary emphasis is given to it in this chapter. The information presented here is also pertinent to the other operations, as most of them represent a certain phase of a farrow-to-finish operation.

The selecting and breeding practices used to produce highly productive swine are presented in Chapter 26. The primary objective in the farrow-to-finish production phase is to feed and manage sows, gilts, and boars economically to ensure a large litter of healthy, vigorous pigs from each breeding female.

Boar Management

Boars should be purchased at least 60 days before the breeding season so they can adapt to their new environment. A boar should be purchased from a herd that has an excellent herd health program and then isolated from the new owner's swine for at least 30 days. During this time, the boar can be treated for internal and external parasites if the past owner has not recently done so. The boar should be revaccinated for erysipelas and leptospirosis. Boars not purchased from a validated brucellosis-free herd should have passed a negative brucellosis test within 30 days of the purchase date. An additional 30-day period of fence-line exposure to the sow herd is recommended to develop immunities before the boar's use.

Younger boars (8–12 months of age) should be fed 5–6 lb/day of a balanced ration containing 14% protein, whereas older boars should receive 4–5 lb/day. The amount of feed should be adjusted to keep the boar in a body condition that is neither fat nor thin.

Young, untried boars should be test-mated to a gilt or two before the breeding season, as records show that approximately 1 of 12 young boars are infertile or sterile. The breeding aggressiveness, mounting procedure, and ability to get the gilts pregnant should be evaluated.

Successful breeding of a group of females requires an assessment of the number of boars required and the breeding ability of each boar. Generally, a young boar can pen-breed 8–10 gilts during a 4-week period. A mature boar can breed 10–12 females. Handmating (producers individually mate the boar to each female) takes more labor but extends the breeding capacity of the boar. Young boars can be handmated once daily, and mature boars can be used twice daily or approximately 10–12 times a week. Handmating can be used to mate heavier, mature boars with the use of a breeding crate. The breeding crate takes most of the boar's weight off the female. Breeding by handmating should correspond with time of ovulation (approximately 40 hours from the beginning of estrus) or litter size will be reduced.

Boars should be considered dangerous at all times and handled with care. It is unsafe for a boar to have tusks, as he may inflict injury on the handler or on other pigs. All tusks should be removed prior to the breeding season and every six months thereafter.

Management of Breeding Females

Proper management of females in the breeding herd is necessary to achieve optimum reproductive efficiency. Gilts may start cycling as early as 5 months of age. It is recommended that breeding be avoided during their first heat cycle as the number of ova ovulated is usually less than in subsequent heat cycles. Mixing pens of confinement-reared gilts and regrouping them with direct boar contact initiates early puberty.

Hot weather is detrimental to a high reproductive rate in the breeding females. High temperatures (above 85°F) will delay or prevent the occurrence of estrus, reduce ovulation rate, and increase early embryonic deaths. The animals will suffer heat stress when they are sick and have an elevated body temperature as well as when the environmental temperature is high.

Besides reducing litter size, diseases such as leptospirosis, pseudorabies, and those associated with stillbirths, mummified fetuses, and embryonic deaths may also increase the problem of getting the females pregnant. Attention to a good herd health program will assure a higher reproductive performance over the years.

It is important that sows and gilts obtain adequate nutrients during gestation to assure optimum reproduction. Feeding in excess is not only wasteful and costly but is also likely to increase embryonic mortality. A limited feeding system using balanced, fortified diets is recommended (Fig. 27.1). It ensures that each sow gets her daily requirements of nutrients without consuming excess energy. As a rule of thumb, 4 lb of a balanced ration provides adequate protein and energy; however, during cold weather an additional pound of feed may prove beneficial, especially for bred gilts. With limited feeding, it is important that each sow gets her portion of feed and no more. The individual feeding stall for each sow is best because it prevents the "boss sows" from taking feed from slower-eating or timid sows. During gestation, gilts should be fed so they will gain about 70–80 lb, and sows should gain approximately 30–40 lb.

Table 27.1 shows several sow rations that utilize different grains. Note how the rations are balanced for protein, specific amino acids, minerals, and vitamins.

Management of the Sow During Farrowing and Lactation

Farrowing is the process of the sow or gilt giving birth to its pigs. On the average, there are approximately 10 pigs farrowed per litter, with approximately 8 pigs weaned per litter. Well-managed swine herds typically have three to four more pigs per litter than these average figures. Extremely large litters are usually undesirable because the pigs are small and weak at birth and death losses can be high.

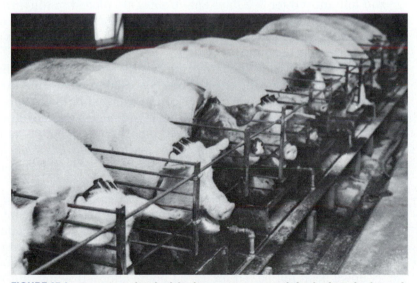

FIGURE 27.1 Sows in individual feeding units to control the feed intake for each sow. Courtesy of the University of Illinois.

TABLE 27.1   Selected Sow Rations (with corn, grain sorghum, barley, or wheat as the grain source)

Ingredient	Ration Number and lb/Ration					
	1	2	3	4	5	6
Corn, yellow	1,627	1,479	—	—	—	—
Grain sorghum	—	—	1,617	—	—	—
Barley	—	—	—	1,735	—	—
Wheat	—	—	—	—	1,724	1,565
Soybean meal, 44%	295	250	306	190	200	165
Dehydrated alfalfa meal, 17%	—	200	—	—	—	200
Calcium carbonate	19	11	20	22	20	12
Dicalcium phosphate	44	45	42	38	41	43
Salt	10	10	10	10	10	10
Vitamin–trace mineral mix	5	5	5	5	5	5
Total	2,000	2,000	2,000	2,000	2,000	2,000
Protein, %	13.40	13.50	13.90	14.20	14.90	14.90
Lysine, %	0.62	0.62	0.62	0.62	0.62	0.62
Tryptophan, %	0.17	0.18	0.17	0.19	0.21	0.22
Threonine, %	0.51	0.52	0.48	0.47	0.49	0.50
Methionine + cystine, %	0.54	0.50	0.42	0.43	0.55	0.54
Calcium, %	0.91	0.90	0.90	0.91	0.90	0.90
Phosphorus, %	0.70	0.70	0.70	0.70	0.70	0.71
Metabolizable energy, kcal/lb	1,476	1,406	1,419	1,335	1,417	1,352

Source: Pork Industry Handbook (PIH-23), Swine Diets, 1990.

Proper care of the sow during farrowing and lactation is necessary to ensure large litters of pigs at birth and at weaning. Sows should be dewormed about 2 weeks before being moved to the farrowing facility. The sow should be treated for external parasites, at least twice, a few days prior to farrowing time. The farrowing facility should be cleaned completely of organic matter, disinfected, and left unused for 5–7 days before sows are placed in the unit.

Before the sow is put in the farrowing pen or stall, her belly and teats should be washed with a mild soap and warm water. This helps eliminate bacteria that can cause diarrhea in nursing pigs.

Sows should be in the farrowing facility at the right time. Breeding records and projected farrowing dates should be used to determine the proper time. If farrowing is to occur in a crate or pen, the sow should be placed there no later than the 110th day of gestation. This gives some assurance the sow will be in the facility at farrowing time, as the normal gestation is 111–115 days in length. Farrowing time can be estimated by observing an enlarged abdominal area, swollen vulva, and filled teats. The presence of milk in the teats indicates that farrowing is likely to occur within 24 hours.

Before farrowing, while the sows are in the farrowing facilities, they can be fed a ration similar to the one fed during gestation. A laxative ration can be helpful at this time to prevent constipation. The ration can be made laxative by the addition of 10 lb/ton of epsom salts or potassium chloride, part of the protein as linseed meal, or the addition of 10% wheat bran, alfalfa meal, or beet pulp.

Sows need not be fed for 12–24 hours after farrowing, but water should be continuously available. Two or three pounds of a laxative feed may be fed at the first postfarrow feeding; the amount fed should be gradually increased until the optimum feed level is reached by 10 days after farrowing. Sows that are thin at farrowing may benefit from generous feeding early after farrowing. Sows nursing fewer than eight pigs may be fed a basic maintenance amount (6 lb/day) with an added amount, such as 0.5 lb for each pig being nursed. It is unnecessary to reduce feed intake before weaning. Regardless of level of feed intake, milk secretion in the udder will cease when pressure reaches a certain threshold level.

Baby Pig Management from Birth to Weaning

Most sows and gilts are farrowed in farrowing stalls or pens to protect baby pigs from the female lying on them (Fig. 27.2). Producers who give attention at farrowing will decrease the number of pigs that die during birth or within the first few hours after birth. Pigs can be freed from membranes, weak pigs can be revived, and other care can be given that will reduce baby pig death loss (Fig. 27.3). Manual assistance of the birth process should not be given unless obviously needed. Duration of labor ranges from 30 minutes to more than 5 hours, with pigs being born either head or feet first. The average interval between births of individual pigs is approximately 15 minutes, but this can vary from almost simultaneously to several hours in individual cases. An injection of oxytocin can be given to speed the rate of delivery after one or two pigs have been born.

Manual assistance, using a well-lubricated gloved hand, can be used to assist difficult deliveries. The hand and arm should be inserted into the reproductive tract as far as is needed to locate the pig; the pig should be grasped and gently but firmly pulled to assist delivery. Difficult births often associated with the occurrence of lactation failure where **mastitis** (inflammation of the udder), **metritis** (inflammation of the uterus), and **agalactia** (inadequate milk supply) may occur. Antibacterial solutions such as nitrofurazone are infused into the reproductive tract after farrowing, and sometimes intramuscular injections of antibiotics are used to decrease or prevent some of these infections.

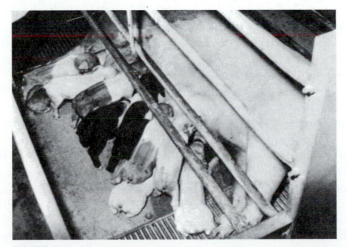

FIGURE 27.2 Sow and litter of baby pigs in a farrowing crate. The farrowing crate is a pen (approximately 5 × 7 ft) that confines the sow to a relatively small area. The panel, a few inches off the floor, separates an area for the baby pigs where the sow cannot lay on them. There is ample space for the pigs to nurse the sow. Courtesy of the University of Illinois.

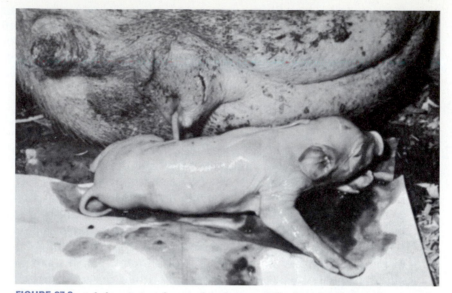

FIGURE 27.3   A baby pig just farrowed with its umbilical cord still extending into the vulva. Proper attention by the swine producer at farrowing time can assure a higher number of pigs weaned per litter. Courtesy of R. W. Henderson, Oregon State University.

It is important that the newborn pig receive colostrum to give immediate and temporary protection against common bacterial infections. Pigs from extremely large litters can be transferred to sows having smaller litters to equalize the size of all litters. Care should be given to ensure that baby pigs receive adequate colostrum before the transfer takes place.

Air temperature in the farrowing facility should be 70–75°F, with the creep area for baby pigs equipped with zone heat (heat lamps, gas brooders, or floor heat) having a temperature of 85–95°F. Otherwise the pigs will chill and die or become susceptible to serious health problems. Moisture should be controlled or removed from the farrowing facility without creating a draft on the pigs.

Management of environmental temperature from birth to 2 weeks of age is critical. At 2 weeks of age, the baby pig has developed the ability to regulate its own body temperature.

Soon after birth, the navel cord should be cut to 3–4 in. from the body and then treated with iodine. This disinfects an area that could easily allow bacteria to enter the body and cause joints to swell and abscesses to occur.

The producer should also clip the 8 sharp needle teeth of the baby pig. This prevents the sow's udder from being injured and facial lacerations from occurring when the baby pigs fight one another. Approximately one-half of each tooth can be removed by using side-cutting pliers or toenail clippers.

Good records are important to the producer interested in obtaining optimum production efficiency. The basis of good production records is pig identification. Ear notching pigs at 1–3 days of age with a system shown in Fig. 27.4 provides positive identification for the rest of the pig's life.

Baby pig management from 3 days to 3 weeks of age includes anemia and scour control, castration, and tail docking. Sow's milk is deficient in iron, so iron dextran shots are given intramuscularly at 3–4 days of age and again at 2 weeks of age (Fig. 27.5).

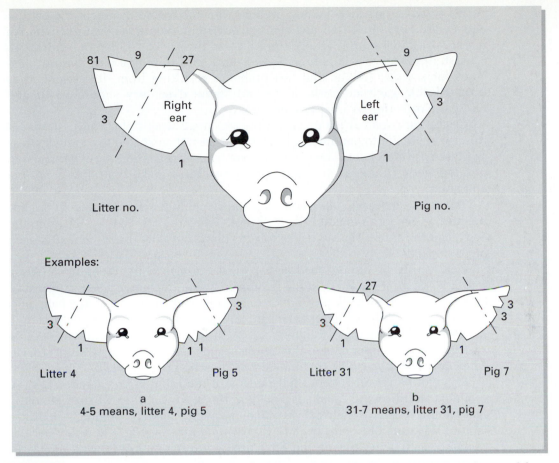

FIGURE 27.4 An ear-notching system for swine. Individual pig identification is necessary for meaningful production and management records. Drawing by Dennis Giddings.

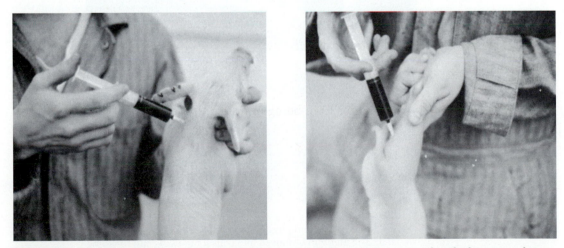

FIGURE 27.5 Iron shots being given to baby pigs. These injections can be given in either the ham or neck muscles. Courtesy of E. R. Miller, Michigan State University.

Baby pig scours are ongoing problems for the swine producer. Colostrum and a warm, dry, draft-free, and sanitary environment are the best preventative measures against scours. Orally administered drugs determined to be effective against specific bacterial strains have been found to be the best control measures for scours. Serious diarrhea problems resulting from diseases such as **transmissible gastroenteritis** (TGE) and swine **dysentery** should be treated under the direction of a local veterinarian.

Castration is usually done before the pigs are 2 weeks old. Pigs castrated at this age are easier to handle, heal faster, and are not subjected to as much stress as those castrated later. The use of a clean, sharp instrument, low incisions to promote good drainage, and antiseptic procedures make castration a simple operation.

Tail docking has become a common management practice to prevent tail biting during the confinement raising of pigs. Tails are removed ¼–½ in. from the body, and the stump should be disinfected along with the instrument after each pig is docked.

Pigs should be offered feed at 1–2 weeks of age in the creep area. These starter feeds can be placed on the floor or in a shallow pan. By 3–4 weeks after farrowing, the sow's milk production has likely peaked; however, the baby pigs should be eating the supplemental feed and growing rapidly (Fig. 27.6).

Rations for baby pigs (weighing 10–25 lb or 25–40 lb) are shown in Table 27.2. Note the components of the rations, especially that milk products (dried whey or dried skim milk) are included in the rations of 10–25-lb pigs.

Pigs can be weaned from their dams at a time that is consistent with the available facilities and the management of the producer. Pigs weaned at young ages should receive a higher level of management. General weaning guidelines include weaning only pigs over 12 lb, weaning pigs over a 2- to 3-day period, weaning the heavier pigs in the litter first, grouping pigs according to size in

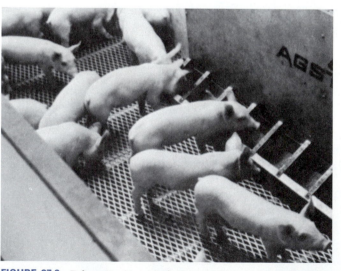

FIGURE 27.6 Baby pigs in a modern nursery unit. The pigs were weaned from the sow and placed into these pens when they weighed 15–20 lb. The pigs will be moved into pens for growing and finishing when they weigh approximately 40 lb. Courtesy of the University of Illinois.

TABLE 27.2   **Selected Baby Pig Rations**

| | Ration Number and lb/Ration | | | | |
| | Pigs 10–25 lb | | Pigs 25–49 lb | | |
Ingredient	1	2	1	2	3
Corn, yellow	1,060	955	1,396	1,159	625
Grain sorghum	—	—	—	—	625
Ground oats	—	—	—	200	—
Soybean meal, 44%	530	590	543	530	495
Dried whey	—	400	—	—	200
Dried skim milk	200	—	—	—	—
Sugar[a]	100	—	—	—	—
Fat	50	—	—	50	—
Calcium carbonate	15	15	15	15	13
Dicalcium phosphate	33	28	34	34	30
Salt	7	7	7	7	7
Vitamin–trace mineral mix	5	5	5	5	5
Total	2,000	2,000	2,000	2,000	2,000
Protein, %	19.50	19.60	17.90	17.80	17.60
Lysine, %	1.15	1.15	0.95	0.95	0.95
Tryptophan, %	0.26	0.26	0.23	0.23	0.23
Threonine, %	0.78	0.81	0.68	0.67	0.69
Methionine + cystine, %	0.64	0.64	0.60	0.58	0.56
Calcium, %	0.86	0.86	0.75	0.76	0.75
Phosphorus, %	0.70	0.70	0.65	0.65	0.66
Metabolizable energy, kcal/lb	1,529	1,451	1,478	1,495	1,495

[a] Dextrose or hydrolyzed cornstarch product.
Source: Pork Industry Handbook (PIH-23), Swine Diets, 1990.

pens of 30 pigs or less, providing one feeder hole for 2 pigs and one waterer for each 8–10 pigs, limiting feed for 48 hours, and using medicated water if scours develop.

Feeding and Management from Weaning to Market

A dependable and economical source of feed is the backbone of a profitable swine operation. Since approximately 60–70% of the total cost of pork production is feed, the swine producer should be keenly aware of all aspects of swine nutrition.

The pig is an efficient converter of feed to meat. With today's nutritional knowledge, modern meat-type hogs can be produced with a feed efficiency of less than 3.0 lb of feed per pound of gain from 40 lb to market. To obtain maximum feed use, it is necessary to feed well-balanced rations designed for specific purposes.

Swine rations for growing–finishing market pigs as well as rations for breeding stock are formulated around cereal grains, the largest component of swine rations. The most common cereal grains—those that provide the basic energy source—are corn, milo, barley, wheat, and their by-products. Because of its abundance and readily available energy, corn is used as the base cereal for comparing the nutritive value of other cereal grains. Milo (grain sorghum) is similar in nutritional content to corn and can completely replace corn in swine rations; however, milo needs more processing (grinding, rolling, steaming, etc.). Its energy value is about 95% the value of corn. Barley contains more protein and fiber than corn, but its relative feeding value is 85–95% of corn and it is less palatable than corn. Wheat is equal to corn in feeding value and is highly palatable. However, for best feeding results, wheat should be mixed half-and-half with some other grain. Which grain or combination of grains to use can best be determined by availability and relative cost.

Grinding the grains improves all grains for feeding, especially those high in fiber such as barley and oats. Pelleting the ration may increase gains and efficiency of gains from 5–10%. However, advantages of pelleting are usually offset by the higher cost of pelleting the ration.

Cereal grains usually contain lesser quantities of proteins, minerals, and vitamins than swine require; therefore, rations must be supplemented with other feeds to increase these nutrients to recommended levels. Soybean meal has been proven to be the best single source of protein for pigs when palatability, uniformity, and economics are considered. Other feeds rich in protein (meat and bonemeal or tankage) can compose as much as 25% of the supplemental protein if economics so dictate. Cottonseed meal is not recommended as a protein source unless the gossypol has been removed. **Gossypol** in relatively high levels is toxic to pigs. Table 27.3 shows recommended protein levels in corn–soybean meal diets for different weights of pigs being grown and finished for slaughter.

If economics dictate, a producer can supplement specific amino acids in place of some of the protein supplement. Typically, lysine may be supplemented separately because lysine is the first limiting amino acid in most swine diets.

Calcium and phosphorus are the minerals most likely to be deficient in swine diets, so care should be exercised to ensure adequate levels in the diet. The standard ingredients for supplying calcium and phosphorus in the swine diet are limestone and either dicalcium phosphate or defluorinated rock phosphate. If an imbalance of calcium and zinc exists, a skin disease called ***parakeratosis*** is likely to occur.

Swine producers may formulate their swine diets several ways: (1) by using grain and a complete swine supplement; (2) by using grain and soybean meal plus a complete base mix carrying all necessary vitamins, minerals, and antibiotics; or (3) by using grain, soybean meal, vitamin premix, trace-mineral pre-

TABLE 27.3   Percentage of Protein Required for Varying Weights of Pigs from Weaning to Slaughter

Period and Weight	Percentage of Protein in Ration
Starter	
10–40 lb	18–20
Grower	
40–100 lb	16
Finisher I	
100–160 lb	15
Finisher II	
160–240 lb	14

mix, salt, calcium and phosphorus sources, and an antibiotic premix. Which ration is used depends on the producer's expertise, the cost, and the availability of ingredients. The trace-mineral premix typically includes vitamins A, D, E, K, B_{12}, and other B vitamins (niacin, pantothenic acid, riboflavin, and choline).

Table 27.4 shows example diets for growing pigs (40–125 lb). Diets for finishing pigs (125 lb to market) are noted in Table 27.5. These diets are formulated using the common grains (corn, grain sorghum, barley, or wheat).

Feed additives are used by most swine producers because the additives have the demonstrated ability to increase growth rate, improve feed efficiency, and reduce death and sickness from infectious organisms. Feed additives available to swine producers fall into three classifications: (1) **antibiotics** (compounds coming from bacteria and molds that kill other microorganisms), (2) **chemotherapeutics** (compounds that are similar to antibiotics but are produced chemically rather than microbiologically), and (3) **anthelmintics** (dewormers). Antibiotics and chemotherapeutics should not be substituted for good management (e.g., sanitary conditions and proper health care).

Feed additive selection and the level to be used will vary with the existing farm environment, management conditions, and stage of the production cycle. Disease and parasite levels will vary from farm to farm. Also, the first few weeks of a pig's life are far more critical in terms of health protection. Immunity acquired from colostrum diminishes by 3 weeks of age, and the pig does not begin producing sufficient amounts of antibodies until 5–6 weeks of age. Also during this time, the pig is subjected to stress conditions of castration, vaccinations, and weaning. Therefore, at this age the additive level of the feed

TABLE 27.4   **Selected Growing Rations, 40–125 lb (with corn, barley, grain sorghum, or wheat as the grain source)**

Ingredient	Ration Number and lb/Ration				
	1	2	3	4	5
Corn, yellow	1,555	—	—	804	—
Barley	—	1,660	—	—	800
Grain sorghum	—	—	1,549	—	—
Wheat, hard winter	—	—	—	804	800
Soybean meal, 44%	395	293	400	342	350
Calcium carbonate	15	18	17	16	17
Dicalcium phosphate	25	19	24	24	23
Salt	7	7	7	7	7
Vitamin–trace mineral mix	3	3	3	3	3
Total	2,000	2,000	2,000	2,000	2,000
Protein, %	15.30	16.00	15.70	15.90	16.10
Lysine, %	0.75	0.75	0.75	0.75	0.75
Tryptophan, %	0.20	0.22	0.20	0.21	0.21
Threonine, %	0.58	0.55	0.55	0.57	0.55
Methionine + cystine, %	0.54	0.48	0.46	0.56	0.52
Calcium, %	0.65	0.65	0.66	0.65	0.66
Phosphorus, %	0.55	0.55	0.55	0.55	0.58
Metabolizable energy, kcal/lb	1,494	1,360	1,438	1,465	1,437

Source: Pork Industry Handbook (PIH-23), Swine Diets, 1990.

TABLE 27.5   Selected Finishing Rations, 125 lb to Market (with corn, barley, grain sorghum, or wheat as the grain source)

Ingredient	Ration Number and lb/Ration				
	1	2	3	4	5
Corn, yellow	1,662	—	—	—	—
Barley	—	1,770	—	—	—
Grain sorghum	—	—	1,649	—	—
Wheat, hard winter	—	—	—	1,754	—
Wheat, soft winter	—	—	—	—	1,703
Soybean meal, 44%	290	185	304	200	252
Meat and bone meal, 50%	—	—	—	—	—
Calcium carbonate	16	19	17	17	17
Dicalcium phosphate	22	16	20	19	18
Salt	7	7	7	7	7
Vitamin–trace mineral mix	3	3	3	3	3
Total	2,000	2,000	2,000	2,000	2,000
Protein, %	13.40	14.30	14.00	15.10	15.30
Lysine, %	0.62	0.62	0.62	0.62	0.62
Tryptophan, %	0.17	0.19	0.17	0.21	0.18
Threonine, %	0.51	0.48	0.48	0.49	0.49
Methionine + cystine, %	0.50	0.44	0.42	0.56	0.45
Calcium, %	0.61	0.61	0.61	0.61	0.60
Phosphorus, %	0.50	0.50	0.51	0.50	0.50
Metabolizable energy, kcal/lb	1,499	1,356	1,442	1,439	1,461

Source: Pork Industry Handbook (PIH-23), Swine Diets, 1990.

is much higher than when the pig is more than 75 lb and growing rapidly toward slaughter weight. Producers should be cautious because some additives have withdrawal times, which means the additive should no longer be included in the ration. This is necessary to prevent additives from occurring as residues in meat.

Market pigs that are fed properly prepared diets will likely have rapid and efficient liveweight gains. However, rapid and low-cost gains are also associated with excellent health programs and well-constructed facilities (Fig. 27.7).

MANAGEMENT OF PURCHASED FEEDER PIGS

Feeder pigs marketed from one producer to another producer are subjected to more stress than pigs produced in a farrow-to-finish operation. These stresses are fatigue, hunger, thirst, temperature changes, diet changes, different surroundings, and social problems. Almost every group has a shipping-fever reaction. Care of newly arrived pigs must be directed to relieving stresses and treating shipping fever correctly and promptly.

Management priorities for newly purchased feeder pigs should include (1) a dry, draft-free, well-bedded barn or shed with no more than 50 pigs per pen; (2) a specially formulated starter ration, having 12–14% protein, more fiber,

FIGURE 27.7   A group of pigs in a growing and finishing production unit. The building is totally enclosed and the floor is totally slatted. The manure from the pigs drops through the slats into a 36-in.-deep pit. The waste is removed from the pit through a water method. Courtesy of the University of Illinois.

and a higher level of vitamins and antibiotics than a typical starter ration; (3) an adequate intake of medicated water containing electrolytes or water-soluble antibiotics and electrolytes; and (4) prompt and correct treatment of any sick pigs.

MARKET MANAGEMENT DECISIONS

Slaughter weights of market hogs and which market is selected can have a marked effect on income and profitability of the swine enterprise. For many years the recommended weight to sell barrows and gilts was, in most instances, 200–220 lb. The primary reasons for selling these animals at a maximum weight of 220 lb were increased cost of gain and increased fat accumulations at heavier weights. Market discounts were common on pigs weighing over 220 lb because of the increased amount of fat. Today, swine producers can maximize income by marketing pigs weighing up to 250 lb. These heavier pigs are leaner, and they are not subject to the same market discounts of fatter pigs marketed several years ago. Decisions to market pigs at heavier weights should be based on feed costs and the gaining ability of the pigs.

Prices for slaughter pigs can vary among markets. Marketing costs, such as selling charges, transportation, and shrink (loss of liveweight) can also vary. If more than one market is available, producers should occasionally patronize different markets as a check against their usual marketing program. A single market is seldom the best market.

TABLE 27.6   Enterprise Records (costs and returns) for Farrow-to-Finish Swine Operations, 1992

Item	High Profit 112 farms	Low Profit 112 farms	Averages 336 farms
1. Return to capital, unpaid labor and management, $	$54,977	$18,182	$37,090
2. Net profit and return to management this period, $	$33,547	($7,277)	$13,555
3. Return per hour for all hours of labor & mgmt, $/hr	$22.86	$4.05	$13.68
4. Annual percent return on capital, %	39.38	0.39	19.88
5. Average price per cwt. of market hogs sold, $	$43.46	$42.69	$43.14
6. Average price per cwt. for all hogs sold, $	$42.84	$41.97	$42.51
7. Feed cost per cwt. of pork produced, $	$22.25	$27.06	$24.63
8. Other oper. costs (except hir. labor)/cwt. pork prod., $	$4.08	$5.99	$5.02
a. Utilities, fuel, elec., & telephone/cwt., $	$1.07	$1.61	$1.30
b. Veterinary services & medicine per cwt., $	$1.17	$1.51	$1.41
9. Depreciation, taxes, & ins. costs per cwt. of pork prod.	$1.95	$3.31	$2.62
10. Capital charge on fixed capital/cwt. of pork produced	$1.02	$1.83	$1.38
11. Capital charge on operating capital/cwt. of pork prod.	$1.19	$1.66	$1.41
12. Value of labor (all) per cwt. of pork produced, $	$4.06	$5.88	$4.86
13. Total cost per cwt. of pork produced, $	$34.54	$45.73	$39.91
14. Margin over all costs per cwt. of pork produced, not including inventory, $	$8.30	($3.76)	$2.60
15. Fixed costs per year per female maintained, $	$115.38	$165.69	$140.76
16. Fixed costs per year per crate maintained, $	$451.36	$654.28	$562.07
17. Fixed costs per pig weaned, $	$7.04	$11.53	$9.17
18. Net profit per year per female maintained, $	$308.00	($83)	$119.00
19. Net profit per year per crate maintained, $	$1,176	($312)	$470.00
20. Average no. of market hogs sold	1574	1311	1453
21. Average wt. of market hogs sold, lb	244	242	243
22. Pig death loss, birth to weaning (% of no. far. live)	12.29	15.33	13.66
23. Pig death loss, weaning to mkt. (% of no. weaned)	5.02	7.14	6.00
24. Breeding stock death loss (% of no. maintained)	3.69	4.93	4.22
25. Average female inventory, no. of head	109	104	107
26. No. of litters weaned per female per year	1.90	1.77	1.84
27. No. of pigs weaned per litter	8.54	8.10	8.41
28. No. of pigs weaned per female per year	16.25	14.46	15.51
29. No. of litters weaned per crate per year	7.26	7.06	7.30
30. No. of pigs weaned per crate per year	62	58	62
31. Total pounds of grain per cwt. of pork produced, lb	282	320	298
32. Total pounds of supplement per cwt. of pork prod, lb	69	80	75
33. Total pounds of feed per cwt. of pork produced, lb	351	401	373
34. Average cost of diets per cwt., $	$6.36	$6.81	$6.62
35. Hours of labor per cwt. of pork produced, hours	0.58	0.82	0.68
36. Cost of feed additives & drugs/cwt. of pork prod, $	$1.00	$1.67	$1.30
37. Average price of grain, $/bu	$2.21	$2.20	$2.21
38. Average price of supplement, $/cwt.	$14.51	$16.38	$15.55

Source: Stevermer, E. J. 1993. Iowa Livestock Enterprise Summaries. *Swine Enterprise Record.* EJS-206 (revised).

TABLE 27.7 Enterprise Records (costs and returns) for Feeding Feeder Pigs, 1992

Item	High Profit 14 farms	Low Profit 14 farms	Averages 41 farms
1. Return to capital, unpaid labor and management, $	$39,701	$8,464	$24,664
2. Net profit and return to management this period, $	$26,957	($7,908)	$11,588
3. Return per hour for all hours of labor & mgmt, $/hr	$44.69	$5.59	$23.50
4. Annual percent return on capital, %	40.27	8.29	28.43
5. Average price per cwt. of market hogs sold, $	$44.71	$42.53	$43.72
6. Average price per cwt. for all hogs sold, $	$44.87	$42.53	$43.78
7. Feed cost per cwt. of pork produced, $	$19.44	$22.66	$21.02
8. Other oper. costs (except hir. labor)/cwt. pork prod., $	$2.42	$3.25	$2.92
a. Utilities, fuel, elec., & telephone/cwt., $	$0.46	$0.78	$0.55
b. Veterinary services & medicine per cwt., $	$0.42	$0.89	$0.61
9. Depreciation, taxes, & ins. costs per cwt. of pork prod.	$1.26	$2.05	$1.51
10. Capital charge on fixed capital/cwt. of pork produced	$0.70	$1.23	$0.79
11. Capital charge on operating capital/cwt. of pork prod.	$1.28	$1.24	$1.24
12. Value of labor (all) per cwt. of pork produced, $	$1.81	$3.88	$2.66
13. Total cost per cwt. of pork produced, $	$26.90	$34.30	$30.15
14. Margin over all costs per cwt. of pork produced, not including inventory, $	$17.98	$8.23	$13.63
15. No. of feeder pigs purchased, no. head	1866	2603	1994
16. Average wt. of purchased feeder pigs, lb	50.17	49.94	50.14
17. Price per feeder pig purchased, $/head	$42.22	$39.30	$40.23
18. Price per cwt. paid for purchased feeder pigs, $	$85.59	$80.01	$81.38
19. Fixed cost per pig purchased, $	$3.50	$5.88	$4.13
20. Total no. of market hogs sold this period	1624	2530	1856
21. Average wt. of market hogs sold, lb	250	243	246
22. Pig death loss, percent of no. purchased (%)	3.04	3.14	3.05
23. Total pounds of grain per cwt. of pork produced, lb	260	283	275
24. Total pounds of supplement per cwt. of pork prod, lb	61	67	64
25. Total pounds of feed per cwt. of pork produced, lb	322	349	340
26. Average cost of diets per cwt., $	$6.08	$6.50	$6.20
27. Hours of labor per cwt. of pork produced, hours	0.24	0.45	0.36
28. Cost of feed additives & drugs/cwt. of pork prod, $	$1.01	$1.71	$1.25
29. Market price needed to break even, $/cwt.	$39.95	$44.14	$41.59
30. Average price of grain, $/bu	$2.18	$2.19	$2.19
31. Average price of supplement $/cwt.	$14.20	$15.27	$14.55

Source: Stevermer, E. J. 1993. Iowa Livestock Enterprise Summaries. *Swine Enterprise Record.* EJS-206 (revised).

COSTS AND RETURNS

Enterprise records for farrow-to-finish hog operations are shown in Table 27.6; similar information is given for finishing feeder pigs in Table 27.7. Compare the costs and returns for the high-profit versus the low-profit farms. Excellent management is reflected in lower costs and higher profits (compare items 2 and 13).

Costs and returns over several years are noted for farrow-to-finish operations (Fig. 27.8). Returns are significantly influenced by prices for barrows and gilts, which can fluctuate widely from year to year (Fig. 27.9). Excellent managers have larger positive returns and lower losses than shown in this figure. These top managers have low-cost production (especially low feed costs), while buying and selling pigs below and above average market prices.

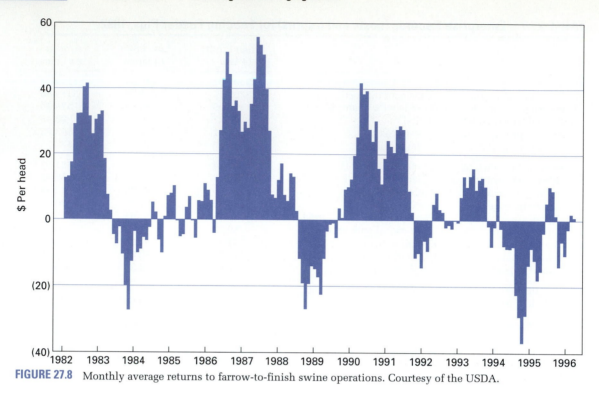

FIGURE 27.8 Monthly average returns to farrow-to-finish swine operations. Courtesy of the USDA.

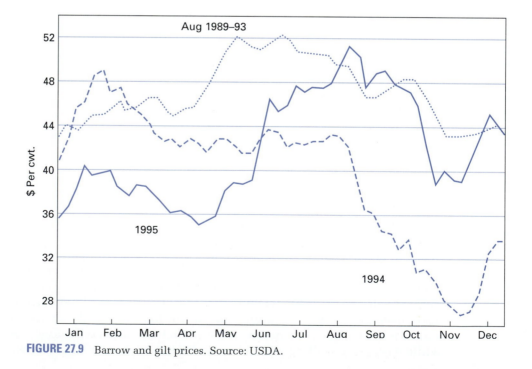

FIGURE 27.9 Barrow and gilt prices. Source: USDA.

Additional information on costs and returns is given in Chapter 36.

CHAPTER SUMMARY

- Swine operations are classified into (1) feeder pig production, (2) finishing purchased feeder pigs, (3) farrow-to-finish, and (4) purebred or seedstock.
- Management of the sow during farrowing and lactation and the baby pigs from birth to weaning are the most critical periods in a total swine management program.
- Sow rations and baby pig rations must have the proper combinations of amino acids, vitamins, and minerals (in addition to protein and energy) to assure adequate and cost-effective gains.
- Herd health programs must be well planned to prevent reproduction diseases, assure adequate lactation, and to prevent scours and other diseases in baby pigs.

REVIEW QUESTIONS

1. What are the four primary types of swine operations?
2. Which is the major type of swine operation?
3. Why should boars be purchased at least 60 days before the breeding season?
4. Why is it best to feed pregnant sows and gilts in individual feeding stalls?
5. What is the process of a sow or gilt giving birth called?
6. Why are gilts and sows farrowed in farrowing stalls?
7. Why is it important for newborn pigs to receive colostrum?
8. At what temperature should newborn pigs be maintained during the first two weeks?
9. Why is it important to maintain baby pigs at 85 to 95°F during the first two weeks after birth?
10. What is the basis of good production records?
11. What is the most common form of permanent identification used on pigs?
12. Why is clipping of needle teeth performed on baby pigs?
13. Why is docking of tails performed on baby pigs?
14. *True or False:* Cereal grains such as corn, milo, barley, and wheat form the basis of pig rations because they are important protein sources.
15. *True or False:* Swine rations can be supplemented with lysine because it is usually the first limiting amino acid in most swine diets.
16. What are the three classes of feed additives for swine rations?
17. What is the reason that heavy pigs exceeding market weight are often discounted by the packer?
18. What factors should be considered when deciding whether to market pigs at heavier weights?

SELECTED REFERENCES

Publications

Miller, E. R., Ullrey, D. E., and Lewis, A. J. 1992. *Swine Nutrition.* Boca Raton, FL: CRC Press, Inc.

National Hog Farmer (monthly periodical). P.O. Box 202162, Minneapolis, MN 55420-7162.

National Research Council. 1988. *Nutrient Requirements of Swine.* Washington, DC: National Academy Press.

Pond, W. G. 1994. Swine Production. *Encyclopedia of Agricultural Science.* San Diego: Academic Press, Inc.

Pond, W. G., Maner, J. H., and Harris, D. L. 1991. *Pork Production Systems.* Westport, CT: AVI Publishing Co.

Pork Industry Handbook (various years). Contains Fact Sheets (e.g., Management and Nutrition of the Newly Weaned Pig; Swine Diets; Swine Growing–Finishing Units; Pork Production Systems with Business Analyses). Cooperative Extension Services of several universities.

Stevermer, E. J. 1993. Iowa Livestock Enterprise Summaries. *Swine Enterprise Record.* EJS-206/Rev/ASB.

Visuals

Feeding Equipment Choices; Swine Manure Handling Systems; Environmental Control in Swine Buildings; Swine Buildings and Equipment: Material Alternatives; Swine Building Systems: Labor and Cost Comparisons; Gestation Management and Health; Sow Management During Farrowing; Baby Pig Management; Farrowing House Equipment; Farrowing Systems; Swine Nursery Facilities; Purchasing Feeding Pigs/Transporting Feeder Pigs; Facilities for a Feeder-to-Finish Swine Enterprise/Nutrition, Health; and *Management of a Feeder-to-Finish Swine Enterprise, Personal Capabilities of Feeder Pig Producers* (videotapes). Agricultural Products and Services, 2001 Killebrew Dr., Suite 333, Bloomington, MN 55420.

Swine Management Practices (two videotapes); *Swine Reproduction* (two videotapes); and *Swine Handling and Transport* (videotape). CEV, P.O. Box 65265, Lubbock, TX 79464-5265.

The Swine Management Series: *Introduction to Swine Management; Brood Sow and Litter; Swine Health Care; Swine Nutrition; Fitting and Showing Swine; Feeder Pig Selection;* and *Swine Nutrition* (sound filmstrips). Vocational Education Productions, California Polytechnic State University, San Luis Obispo, CA 93407.

Swine: Life-cycle Feeding (VHS videotape). VEP, California Polytechnic State University, San Luis Obispo, CA 93407.

CHAPTER **28**

Sheep Breeds and Breeding

Sheep apparently evolved in the dry mountainous areas of southwest and central Asia. Present-day domesticated sheep were derived from wild animals that existed in Asia some 8,000–10,000 years ago. It is believed that domestication occurred during the early civilizations of southwest Asia. The Romans introduced a white, wooled sheep into western Europe, and these sheep contributed heavily to the British breeds. Some 100–200 years ago, Europeans took the Merino and British breeds of sheep to South America, Australia, South Africa, and New Zealand. Additional information on the world and U.S. sheep industry is in Chapter 2.

MAJOR U.S. SHEEP BREEDS

Sheep have been bred for three major purposes: production of fine wool for making high-quality clothing; production of long wool for making heavy clothing, upholstering, and rugs; and production of mutton and lamb. In recent years, dual-purpose sheep breeds (which produce both wool and meat) have been developed by crossing fine-wool breeds with long-wool breeds and then selecting for both improved meat and wool production. Some breeds serve specific purposes; an example is the Karakul breed, which supplies pelts for clothing items such as caps and Persian coats.

Sheep breeds can be divided into three major categories:

1. **Ewe breeds** are usually white-faced and have fine or medium wool or long wool or crosses of these types. They are noted for reproductive efficiency, wool production, size, milking ability, and longevity. The Delaine

Merino, Rambouillet, Debouillet, Corriedale, Targhee, **Finnsheep,** and Border Leicester are breeds in the ewe breed category.

2. **Ram breeds** are meat-type breeds raised primarily to produce rams for crossing with ewes of the ewe breed category. They are noted for growth rate and carcass characteristics. The Suffolk, Hampshire, Shropshire, Oxford, Southdown, Montadale, and Cheviot are classed as ram breeds.

3. **Dual-purpose breeds** are used either as ewe breeds or as ram breeds. The Dorset and Columbia are good examples of breeds that are often used as ewe breeds to be crossed with a good ram breed. The Dorset, Lincoln, and Romney are often used as ram breeds to cross on other breeds of ewes to improve milking ability and fertility of the ewe flock. Several recently imported breeds such as the Texel offer a wide variety of genetics to choose from.

Characteristics

Breeds of sheep and their characteristics are listed in Table 28.1. Most breeds are distinguished by wool characteristics, color of face, size, and horned or polled condition. Inheritance of the horned or polled condition is sex-influenced and generally controlled by a single pair of genes. In the heterozygote, the genes are usually expressed as recessive in the ewe and dominant in the ram (Table 28.2).

The sheep breeds commonly used in the United States are shown in Figs. 28.1 and 28.2. The Merino and Rambouillet (fine-wool) breeds and the dual-purpose breeds that possess fine-wool characteristics have a herding instinct, so one sheepherder with dogs can handle a band of 1,000 ewes and their lambs on the summer range. Winter bands of 2,500–3,000 ewes are common. Ram breeds, by contrast, tend to scatter over the grazing area. They are highly adaptable to grazing fenced pastures where feed is abundant. Rams of the ram breeds (meat-type) are often used for breeding ewes of fine-wool breeding on the range for the production of market lambs. Crossbreeding in sheep has been used for more than a half century.

The Columbia and Targhee breeds were developed by the U.S. Sheep Experiment Station in Dubois, Idaho. Both breeds have proven useful on western ranges because they have the herding instinct and, when bred to a meat breed of ram, they raise better market lambs than do fine-wool ewes. The U.S. Sheep Experiment Station has made great strides in improving the Rambouillet, Columbia, and Targhee breeds.

The most important trait in meat breeds of sheep is weight at 90 days of age, with some emphasis also given to conformation and finish of the lambs. Weight at 90 days of age can be calculated for lambs that are weaned at younger or older ages than 90 days as follows:

$$\text{90-day weight} = \frac{\text{weaning weight} - \text{birth weight}}{\text{age at weaning}} \times 90 + \text{birth weight}$$

If a lamb weighs 10 lb at birth and 100 lb at 100 days, the 90-day weight is:

$$\frac{100 \text{ lb} - 10 \text{ lb}}{100 \text{ days}} \times 90 \text{ days} + 10 \text{ lb} = 91 \text{ lb}$$

If birth weight was not recorded, an assumed birth weight of 8 lb can be used.

TABLE 28.1 **Characteristics of Breeds of Sheep Within Each Type**

Breed	Size[a]	Carcass Conformation	Wool Fineness	Wool Length[b]	Fleece Weight[c]	Color	Horns or Polled	Other
Fine-wool breeds								
Merino	Small to medium	Poor	Very fine	Medium	Heavy	White	Rams horned, ewes polled	Good herding instinct; skin folds
Rambouillet	Large	Medium	Fine	Medium	Heavy	White	Rams horned, ewes polled	Good herding instinct
Long, coarse-wool breeds								
Romney	Medium to large	Medium	Coarse	Long	Heavy	White	Polled	Lambs do not fatten at small size
Lincoln	Large	Good	Coarse	Long	Heavy	White	Polled	Lambs do not fatten at small size
Ram (meat) breeds								
Suffolk	Large	Excellent	Medium	Short	Light	White with black, bare face and legs	Polled	Bare bellies; black fibers
Hampshire	Large	Excellent	Medium	Medium	Medium	White with black face	Polled	Wool blindness, good milkers, black fibers
Shropshire	Medium to large	Good	Medium	Medium	Medium	White, dark face and legs	Polled	Wool blindness, excellent milkers
Southdown	Very small	Excellent	Medium	Short	Very light	White with brown face and legs	Polled	Used in hothouse lamb production
Dorset	Medium	Good	Medium	Medium	Medium	White	Polled or horned	Highly fertile, good milkers
Cheviot	Small	Excellent	Medium	Medium	Medium	White	Polled	Very rugged
Oxford	Large	Excellent	Medium	Medium	Medium to heavy	White with brown face and legs	Polled	
Finnsheep	Small to medium	Poor	Medium	Medium	Medium	White	Polled	Very fertile
Dual-purpose breeds								
Corriedale	Medium	Good	Medium	Medium long	Heavy	White	Polled	Herding instinct
Columbia	Large	Good	Medium	Medium long	Heavy	White	Polled	Rugged; herding instinct
Targhee	Medium to large	Medium to good	Medium to fine	Medium	Heavy	White	Polled	Herding instinct
Miscellaneous breeds								
Navajo	Medium	Poor	Coarse	Long	Medium	Variable	Polled or horned	Wool for making Navajo rugs
Karakul	Large	Poor	Coarse	Long	Medium	Black or brown	Rams horned, ewes polled	Used for pelts

[a] Small—ewes weigh 120–160 lb and rams weigh 160–200 lb; medium—ewes weigh 140–180 lb and rams 180–250 lb; large—ewes weigh 150–200 lb and rams 225–350 lb.
[b] Short—2.0–3.5 in.; medium—2.5–4.5 in.; long—4–6 in. or longer.
[c] Light—5–8 lb; medium—6–10 lb; heavy—10–18 lb.

TABLE 28.2   Horn Inheritance in Sheep

Genotype	Phenotype of Ewes	Phenotype of Rams
HH	Horned	Horned
Hh	Polled	Horned
hh	Polled	Polled

Approximately 85–90% of the total income from sheep of the meat breeds is derived from the sale of lambs, with only 10–15% coming from the sale of wool. By contrast, the sale of wool accounts for 30–35% of the income derived from fine-wool and long-wool breeds. Even with wool breeds, the income from lambs produced constitutes the greater portion of total income.

The U.S. Sheep Experiment Station in Dubois, Idaho, has developed a new synthetic breed of sheep, called the *Polypay,* which has superiority in lamb production and in carcass quality. Four breeds (Dorset, Targhee, Rambouillet, and Finnsheep) provided the basic genetic material for the Polypay. Targhees were crossed with Dorsets and Rambouillets were crossed with Finnsheep, after which the two crossbred groups were crossed. The offspring produced by crossing the two crossbred groups were intermated and a rigid selection program was practiced. This population was closed to outside breeding. The Rambouillet and Targhee breeds contributed hardiness, herding instinct, size, a long breeding season, and wool of high quality. The Dorset contributed good carcass, high milking quality, and a long breeding season. The Finnsheep contributed early puberty, early postpartum fertility, and high lambing rate.

The Polypay breed has shown outstanding performance in conventional once-a-year lambing under range conditions and superior performance in twice-a-year lambing when compared with other breeds or breed crosses.

There are several U.S. breeds of sheep that are not listed in Table 28.1. Some of these breeds possess certain genes that may be useful for improving the most popular U.S. breeds either through systematic crossbreeding or in the establishment of new breeds.

Based on annual registration numbers, the Suffolk, Dorset, Rambouillet, and Hampshire are the most numerous breeds of sheep (Table 28.3). Suffolk and Hampshire are used extensively in crossbreeding programs to produce market lambs both in farm flocks and on the range. The most popular ewe breeds on the range are the Rambouillet, Columbia, and Corriedale. In a farm flock lamb-production system, the Dorset, Shropshire, Finnsheep, and Polypay are important ewe breeds.

BREEDING SHEEP

Reproduction

Sheep differ from many farm animals in having a breeding season that occurs mainly in the fall of the year. The length of the breeding season varies with breeds. Ewes of breeds with a long breeding season show heat cycles from mid- to late summer until midwinter. The breeds in this category are the Rambouil-

Dorset

Montadale

Polled Dorset

Cheviot

Hampshire

Suffolk

FIGURE 28.1   Some breeds of sheep commonly used either as straightbreds or for crossbreeding to produce market lambs. Photographs courtesy of Continental Dorset Club, Hudson, Iowa (Dorset and Polled Dorset); Montadale Sheep Breeders' Association, Indianapolis, Indiana (Montadale); American Cheviot Sheep Society, Lebanon, Virginia (Cheviot); American Hampshire Sheep Association, Columbia, Missouri (Hampshire); and National Suffolk Sheep Association, Logan, Utah (Suffolk).

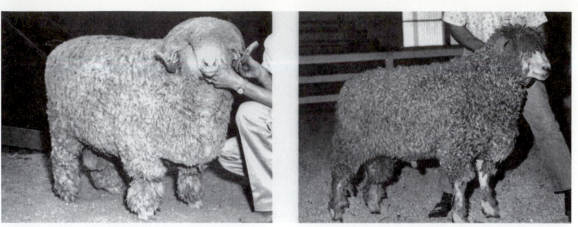

Rambouillet

Cotswold

Lincoln

Romney

Targhee

Finnsheep

FIGURE 28.2   Some breeds of sheep commonly used either as straightbreds or for crossbreeding to produce market lambs. Photographs courtesy of American Rambouillet Sheep Breeders Association, San Angelo, Texas (Rambouillet); American Cotswold Record Association, Rochester, New Hampshire (Cotswold); National Lincoln Sheep Breeders' Association, West Milton, Ohio (Lincoln); American Romney Breeders Association, Corvallis, Oregon (Romney); USDA, ARS, Western Region, U.S. Sheep Experiment Station, Dubois, Idaho (Targhee); and Animal Science Department, University of Minnesota (Finnsheep).

TABLE 28.3   **Major U.S. Sheep Breeds Based on Annual Registration Numbers (in thousands), 1975–95**

Breed	1995	1990	1985	1980	1975	Year U.S. Association Formed
			thousands			
Suffolk[a]	29.8	41.8	49.1	60.3	31.8	1935
Dorset	12.6	19.5	12.8	15.2	8.7	1898
Rambouillet	12.4	17.1	12.2	11.9	10.0	1889
Hampshire	11.2	16.4	16.9	21.4	17.2	1889
Columbia	5.0	7.8	8.0	10.0	6.0	1942
Southdown	5.8	5.9	4.8	4.4	3.6	1882
Corriedale	3.1	5.0	4.4	6.5	6.1	1816
Polypay	2.9	11.4	3.2	0	0	1979
Montadale	2.9	3.8	3.1	3.1	1.9	1945
Shropshire	2.6	2.9	3.4	4.4	3.8	1884
Cheviot	2.4	2.8	2.5	2.5	2.3	1891

Note: Only breeds with more than 2,400 annual registrations are listed.
[a] Registrations are combined from the National Suffolk Assoc. (20,002) and the American Suffolk Assoc. (9,850). An undetermined number of Suffolks are double-registered in both associations.
Source: Various breed associations; American Sheep Industry Assoc.

let, Merino, and Dorset. Breeds with an intermediate breeding season (Suffolk, Hampshire, Columbia, and Corriedale) start cycling in late August or early September and continue to cycle until early winter. Ewes of breeds having long or intermediate breeding seasons are more likely to fit into a program of three lamb crops in 2 years.

Ewes of breeds with short breeding seasons (Southdown, Cheviot, and Shropshire) do not start cycling until early fall and discontinue cycling at the end of the fall period.

Puberty is reached at 5–12 months of age and is influenced by breed, nutrition, and date of birth. The average length of **estrous cycles** is slightly over 16 days, and the length of estrus (when the ewe is receptive to the ram) is about 30 hours. The length of gestation (time from breeding until the lamb is born) averages 147 days but varies; medium-wooled and meat breeds have shorter gestation periods, whereas fine-wooled breeds have longer gestation periods.

Several factors affect **fertility** in sheep. Lambing rates vary both within and between breeds, with some ewes consistently producing only one lamb per year and others producing three or four lambs each year. To improve lambing rate, producers keep replacements from ewes that consistently produce two to four lambs each year.

Fertility of rams used in a breeding program should be checked by examining the semen. Semen can be collected by use of an artificial vagina and examined with a microscope. Ram fertility is evaluated by (1) checking of abnormal sperm (tailless, bent tails, no heads, and so on), (2) observing the percentage of live sperm (determined by a staining technique using fresh semen), and (3) checking the sperm motility and concentration in freshly collected semen. Two semen collections 3–4 days apart should be examined. When rams have not ejaculated for some time, there may be dead sperm in the **ejaculate.**

Other Factors Affecting Reproduction

Selection and Crossbreeding.   Crossbred ewe lambs, when adequately fed, are usually bred to lamb at 1 year of age. Generally, crossbred lambs are more likely to conceive as lambs than are purebred lambs.

Age.   Mature (3–7-year-old) ewes are more fertile and raise a higher percentage of lambs born than do younger or older ewes.

Light.   Light, temperature, and relative humidity affect reproduction in sheep. Sheep respond to decreased day lengths both by showing greater proportion of ewes in estrus and by higher **conception** rates.

Temperature.   Temperature has a marked effect on both ewes and rams. High temperatures cause heat sterility in rams because the testicles must be at a temperature below normal body temperature for viable sperm production. Embryo survival in the ewe is also influenced by temperature. When ambient temperature exceeds 90°F, embryo survival decreases during the first 8 days after breeding.

Health.   Environmental factors that affect the health and well-being of sheep—disease, **parasites,** lack of feed, or imbalance of the ration—reduce the number of lambs produced. Producers can increase their income and profit from a sheep operation by controlling diseases and parasites and by providing the sheep with an adequate supply of good feed. Sheep in moderate body condition are usually more productive than fat sheep. Ewes that are in moderate condition and that are gaining weight before and during the breeding season will have more and stronger lambs than ewes that are either overfat or **emaciated** at breeding.

Estrous Synchronization and AI.   In some intensively managed sheep operations, hormones can be used to synchronize estrus. Hormones can also be used to bring ewes into estrus at times other than during their normal breeding season.

Estrus can be synchronized if progesterone is given for a 12–14-day period in the feed, as an implant, via a *pessary* (intravaginal device), or by daily injections followed by injections of pregnant mare serum (PMS) the day progesterone is discontinued. Conception rates at this estrus are low; therefore, a second injection of pregnant mare serum 15–17 days later will bring the ewes into estrus and fertile matings will occur. There must be sufficient ram power to breed naturally a large number of ewes if they are synchronized or if artificial insemination can be used.

This system of synchronizing estrus can also be used to obtain pregnancies out of the normal breeding season and to obtain a normal lamb crop from breeding ewe lambs. Also, producers can use estrous synchronization for accelerated lambing if it is desired to raise two lamb crops per year.

If the ewes are to be artificially inseminated, high-quality fresh semen should be diluted with egg-yolk–citrate diluter shortly before insemination. Ram semen has been frozen but conception rates have been low.

Estrogen in Feeds.   Sometimes a flock of sheep may show very low fertility because of high estrogen content of the legume pasture or hay that the sheep are eating. Producers are advised to have the legume hay or legumes in the pasture checked for estrogen if they are experiencing fertility problems in their sheep.

The Breeding Season

Sheep may be hand mated or pasture mated. If they are **hand mated,** some **teaser rams** are needed to locate ewes that are in heat. Either a **vasectomized** ram (each vas deferens has been severed so sperm are prevented from moving from the testicles) can be used, or an apron can be put on the ram such that a strong cloth prevents copulation.

Tagging ewes (Fig. 28.3) before breeding will result in a larger percentage of lambs as well as a lamb flock of more uniform age.

Ewes should be checked twice daily for **heat.** Ewes normally stay in heat for 30 hours and ovulate near the end of heat. It is desirable to breed ewes the morning after they are found in heat in the afternoon. Ewes that are found in heat in the morning should be bred in the late afternoon.

If ewes are to be hand mated, the ram should breed them once. Recently bred ewes should be separated from other ewes for 1 or 2 days so that the teaser ram does not spend his energies mounting the same ewe repeatedly.

If ewes are to be **pasture mated,** they are sorted into groups according to the rams to which they are to be mated. It is best to have an empty pasture between breeding pastures so as not to entice rams to be with ewes in another breeding group.

FIGURE 28.3   Ewe that has been tagged prior to breeding. Wool has been shorn from the dock and vulva region. Courtesy of the *Sheep Production Handbook.*

Sheep can be identified by using numbered branding irons to apply scourable paint. The brisket of the breeding ram can be painted with scourable paint so that he marks the ewe when he mounts her. A light-colored paint should be used initially, then a dark color can be used about 14 days later to detect ewes that return in heat. An orange paint can be used initially, followed, successively, with green, red, and black. The paint color should be changed each 16–17 days. Because the ewes are numbered, they can be observed daily and the breeding date of each can be recorded. Also, a ram that is not settling his ewes can be detected and replaced. Occasionally, a ram may be low in fertility even though his semen was given a satisfactory evaluation before the breeding season.

A breeding season of 40 days results in a lamb crop of uniform age and identifies ewes for culling that have an inherent tendency for late lambing.

The Purebred Breeder

The goal of purebred breeders is to make genetic changes in the economically important traits. These breeders use selection as their method for genetic improvement, as crossbreeding is limited primarily to commercial producers. Purebred breeders with large operations may find it desirable to close their flocks and select ewe and ram replacements from within the closed flocks.

Normally, purebred breeders have selection programs to produce superior rams for commercial producers. At the same time, commercial producers prefer rams that contribute outstanding performance in a commercial crossbreeding program.

In general, great progress in sheep improvement can be made by selection within a breed for traits that are highly heritable and economically important. Crossbreeding can give significant genetic improvement in traits of low heritability such as fertility. Traits that are only moderately heritable can be improved by selecting genetically superior breeding animals and also by using crossbreeding.

Highly heritable traits (40% or higher) include mature body size, yearling type score, face cover, **skin folds,** clean fleece yield, yearling staple length, gestation length, loin eye area, fat weight, and retail cut weight. Lowly heritable traits (below 20%) include weaning type score, weaning condition score, multiple births, number of lambs weaned, fat thickness over loin, carcass weight per day of age, carcass grades, and dressing percentage. Moderately heritable traits (20–39%) include birth weight, 90-day weight, rate of gain, neck folds, grease fleece weight, fleece grade, lambing date, milk production, carcass length, and bone weight.

Principles applicable to breeding and improving sheep are presented in Chapters 12 (genetics), 13 (selection), and 14 (systems of mating).

Commercial Sheep Production

Lambs born in December or January and sold as small, finished lambs for Easter or early spring markets can be produced by breeding ewes of a highly fertile breed that produces large quantities of milk (such as Dorset) to a ram of a small breed that has excellent meat conformation (such as the Southdown). The lambs produced by such a mating receive ample milk to make rapid growth. Smaller size and excellent conformation contributed by the Southdown result in lambs having desirable conformation finishing at a relatively small size. They should be sufficiently finished for slaughter at 70–80 lb and make excellent small carcasses.

Lambs produced for slaughter directly from pasture at weaning time are usually produced in areas where good pastures are available. A producer may raise lambs of only one breed as straightbred animals; but more likely the producer with good pastures and an interest in marketing finished lambs at 90–100 days has some type of crossbreeding program. By crossbreeding, the producer can use ewes of a breed or breeds that are strong in traits like high fertility, wool that is fairly long and fine, high milking ability, and a long life, such as the Dorset, Rambouillet, Columbia, and Finnsheep. The producer can use rams of breeds that are rapidly growing, have desirable carcass conformation, and that finish well at 90–115 lb liveweight, such as the Suffolk or Hampshire.

Most lambs produced in farm flock operations are marketed at weaning (when they are approximately 100 days of age and weigh 90–120 lb). They are sufficiently finished by then to grade USDA Choice or Prime if they have good carcass conformation and have been adequately fed. A good method for producing such lambs is to use a three-breed rotational crossbreeding system. Producers can use breeds such as the Dorset, Suffolk, and Columbia in areas where relatively good grazing conditions exist.

Some range sheep operations have summer grazing areas that are sufficiently productive to supply the nutrients needed for lambs to grow and fatten on the range. Other range areas produce only enough forage for sheep to raise lambs that must be finished after weaning by placing them in feedlots and feeding grain and some hay for a period of 50–80 days. The use of good meat-type rams for breeding to ewes under this less desirable condition, as well as for breeding to ewes under highly desirable range conditions, is recommended. Under the more favorable summer range conditions, finished lambs at weaning can be produced. Under less desirable summer range conditions, **feeder lambs** can be produced when desirable meat-type rams are used. A range-sheep operator could use Rambouillet, Columbia, Panama, or Targhee ewes and breed them to Suffolk or Hampshire rams. All lambs would be marketed either as slaughter or feeder lambs. Replacement ewes could be purchased or raised by breeding the best range ewes to good rams of the same breed as that of the ewes.

Most market lambs are crossbreds (produced either from crossing breeds or from mating crossbred ewes with purebred rams), but the key to the success of **crossbreeding** is the improvement in production traits made by the purebred breeders in their selection programs. Crossbreeding can be used to combine meat conformation of the sire breeds with lambing ability and wool characteristics of the ewe breeds, and also to obtain the advantage of the fast growth rate of lambs that results from heterosis. The additive as well as the heterotic effects are evident in crossbred lambs. Maximum **heterosis** is obtained in three- or four-breed crosses, either rotational or terminal.

Crossing meat-type sheep with **Finnsheep** offers a means for rapidly increasing efficiency of lamb meat production. The Finnsheep breed was introduced into the United States in 1968, and its numbers are increasing rapidly in this country. Finnsheep (Fig. 28.4) are extremely prolific (two to six lambs per lambing), and research studies show that crossbred ewes of 25–50% Finnsheep breeding produce more lambs than straightbred or crossbred ewes of the meat breeds.

FIGURE 28.4   One of the first Finnsheep ewes imported into the United States with her first lambs, illustrating the high productivity of Finnsheep. Courtesy of Dwight and Mae Holaway.

Some research shows that two lamb crops can be raised per year or, if feed and other conditions do not support such intensive production, three lamb crops every 2 years can be produced. To achieve intense production of this type, producers need ewes that will breed throughout the year. At present, the Polypay seem to fit into such a program fairly well. Also, the Rambouillet and Dorset have a tendency to breed out of season.

To raise three lamb crops in 2 years, ewes should be bred in late summer, probably for a 30-day period (August), for lambs to arrive in January (lamb crop 1). The ewes and lambs need to be well fed so the lambs are sufficiently finished to go to market at 90 days of age (in April). The ewes are then bred in April (30-day breeding period) to lamb in August (lamb crop 2), and the ewes and lambs need good feed so the lambs are well finished and sufficiently large to be marketed at 90 days of age (November). The ewes are bred in November to lamb in April (third lamb crop) and the lambs are marketed at 90 days of age in July.

Three-Breed Terminal Crossbreeding

In some production schemes, crossbred ewes are bred to rams of a third breed as a terminal cross (all offspring are marketed). Some producers cross Columbias and Dorsets to develop large ewes that have long, fairly fine fleece, are hardy, and produce much milk. These crossbred ewes are bred to good Suffolk or Hampshire rams. A high percentage of lambs from these crosses are finished at weaning and weigh 90–100 lb at about 90–100 days of age. The entire lamb crop from this terminal mating is marketed. All the two-breed crossbred ewes that are productive are kept until they reach an age at which their production of lambs declines. Replacement ewes are produced by crossing Columbias and Dorsets. A producer carrying out such a program keeps a small flock of either straightbred Dorsets or Columbias to produce the two-breed crossbred ewes. Purebred Suffolk or Hampshire rams are purchased as needed and used for breeding as long as they are sound and inbreeding is not a problem.

INHERITED ABNORMALITIES

It is important to guard against certain genetic abnormalities when managing sheep. Although exceptions occur, sheep showing obvious genetic defects are usually culled. Some inherited abnormalities include the following.

Cryptorchidism is inherited as a simple recessive; therefore, a ram with only one testis descended into the scrotum should never be used for breeding. In addition, producers should cull rams that sire lambs having cryptorchidism, as well as the ewes that produced the lambs.

Dwarfism is inherited as a recessive and is lethal; therefore, ewes and rams producing dwarf offspring should be culled.

The mode of inheritance of **entropion** (turned-in eyelids) has not been determined, but it is known to be under genetic control. A record should be made of any lamb having entropion so that the lamb can be marketed.

Sheep have lower front teeth but lack upper front teeth. They graze by closing the lower teeth against the dental pad of the upper jaw. If the lower jaw is either too short (**overshot** or **parrot mouth**) or too long (**undershot**), the teeth cannot close against the dental pad and grazing is difficult. The mode of inheritance of these jaw abnormalities is unknown, but the conditions are under genetic control. Sheep having abnormal jaws should be culled.

Rectal prolapse is common in black-faced sheep. It is a serious defect, and both inheritance and the environment are influential in its occurrence. Lambs on heavy feeding or lush pastures are more likely to show rectal prolapse. If surgery is used to correct the condition, the animal should not be used as a breeding animal.

Skin folds, open-faced, closed-faced, and **wool blindness** are inherited and selection against these traits has been effective.

Spider syndrome is a severe skeletal deformity occurring in Suffolk sheep. The most striking feature is an outward bending of the forelimbs from the knees. Angular deformities of the hindlimbs are generally present. The genetic abnormality is due to a recessive gene and can be effectively selected against.

Many wool defects are known, including black fibers, black-tipped fibers, hairiness, fuzziness, and "high belly wool" in which wool that is typical of wool on the belly is present on the sides of sheep. Selection against all of these fleece defects should be practiced as a means of reducing the frequency of their occurrence.

CHAPTER SUMMARY

■ Breeds of sheep are classified primarily on wool production (e.g., Rambouillet), meat production (e.g., Suffolk), or a combination of the two (dual-purpose—e.g., Columbia).

■ Based on registration numbers, the most numerous sheep breed is Suffolk, followed by Dorset, Hampshire, Rambouillet, and Columbia.

■ Under most conditions, multiple births (i.e., twinning) are highly desirable, with other economically important traits being growth rate, wool production, and carcass merit (combination of quality grades and yield grades).

■ The breeding program for many large, range sheep flocks involves using wool breeds or dual-purpose breeds of ewes to mate to ram breeds excelling in growth rate and carcass characteristics.

REVIEW QUESTIONS

1. What are the three categories of sheep breeds?
2. What are the production characteristics of ewe breeds?
3. What characteristics are ram breeds noted for?
4. What is the most important trait of meat breeds?
5. What is the average length of the estrous cycle in ewes?
6. What is the average length of gestation in sheep?
7. At what age does puberty occur in sheep?
8. What factors influence the age at puberty in sheep?
9. How is ram fertility initially evaluated before the breeding season?
10. What four characteristics of ram semen are examined during fertility evaluation?
11. What factors affect reproduction in sheep?
12. What is tagging?
13. What is the primary goal of purebred sheep breeders?
14. What are some highly heritable traits of sheep which can be improved by genetic selection?
15. What is the key to success in a crossbreeding program?
16. What recently introduced breed of sheep is noted for multiple births?
17. Which three ewe breeds of sheep are notable for breeding throughout the year and fit well into a program of three lamb crops in two years?
18. What characteristic of ewe breeds makes them highly suitable for range production?

SELECTED REFERENCES

Publications

Botkin, M. P., Field, R. A., and Johnson, C. L. 1988. *Sheep and Wool: Science, Production, and Management.* Englewood Cliffs, NJ: Prentice-Hall.

Dickerson, G. E. 1978. *Crossbreeding Evaluation of Finnsheep and Some U.S. Breeds for Market Lamb Production.* North Central Regional Publication #246. ARS, USDA, and University of Nebraska.

Lasley, J. F. 1987. *Genetics of Livestock Improvement.* Englewood Cliffs, NJ: Prentice-Hall.

Maijala, K. 1991. *Genetic Resources of Pig, Sheep, and Goat.* New York: Elsevier Science Publishers.

Ross, C. V. 1989. *Sheep Production and Management.* Englewood Cliffs, NJ: Prentice-Hall.

The Sheepman's Production Handbook. 1996. American Sheep Industry Assoc., 6911 S. Yosemite, Englewood, CO.

Wilson, D. E., and Morrical, D. G. 1991. The national sheep improvement program: a review. *J. Anim. Sci.* 69:3872.

Visuals

Farm Flock (videotape; 10 min.). National Sheep Improvement program, Dept. of Animal Science, Iowa State University, Ames, IA 50010.

Farm Flock Production Systems; and *Personal Requirements/Personal Rewards* (videotapes). Agricultural Products and Services, 2001 Killebrew Dr., Suite 333, Bloomington, MN 55420.

An Introduction to Sheep Breed Identification (sound filmstrip). Prentice-Hall Media, 150 White Plains Rd., Tarrytown, NY 10591.

Range Flock (videotape; 10 min.). National Sheep Improvement Program, Dept. of Animal Science, Iowa State University, Ames, IA 50010.

Sheep Breed Identification (slide kit; 15 breeds of sheep). Vocational Education Productions, California Polytechnic State University, San Luis Obispo, CA 93407.

CHAPTER 29

Feeding and Managing Sheep

Sheep feeding and management are essential for the success of an operation. These areas must be integrated with a knowledge of how breeding and environmental factors affect sheep productivity and profitability. Several areas of feeding and management that affect efficient production are discussed in this chapter.

PRODUCTION REQUIREMENTS FOR FARM FLOCKS

Pastures

Good pastures are essential to the typical farm flock operator. Grass–legume mixtures, such as rye grass and clover or orchard grass and alfalfa, are ideal for sheep. Sheep can also graze on temporary pastures of such plants as Sudan grass or rape, which are often used to provide forage in the dry part of summer when permanent pastures may show no new growth. Crops of grain and grass seed may also provide pasture for sheep during autumn and early spring. Sheep do not trample wet soil as severely as do cattle; pasturing sheep on grass-seed and small-grain crops in winter and early spring when the soil is wet does not cause serious damage to the plants or soil.

Fencing

A woven-wire or electric fence is necessary to contain sheep in a pasture. Forage can be best utilized by rotating (moving) sheep from one pasture to another. This also assists in the control of internal parasites. Some sheep operators use temporary fencing (such as electric fencing) to divide pastures and for predator control. It is necessary to use two electrified wires, one located low enough to prevent sheep from going underneath the fence, and the other located high enough to prevent them from jumping over it.

Corrals and Chutes

It is occasionally necessary to put sheep in a small enclosure to sort them into different groups or to treat any ailing animals. Proper equipment is helpful in this task because sheep are often difficult to drive, particularly when ewes are being separated from their lambs. A **cutting chute** can be constructed to direct sheep into various small lots (Fig. 29.1). A well-designed cutting chute is sufficiently narrow to keep sheep from turning around and can be blocked so that sheep can be packed together closely for such purposes as treating diseases and reading ear tags. The chute can be constructed of lumber that is nailed to wooden posts set into the ground and properly spaced so that the correct width (14–16 in.) is provided when the boards are nailed on the inside. Pens used to enclose small groups of sheep can be constructed of a woven-wire fencing. A loading chute is used to place sheep onto a truck for hauling. A portable loading chute (Fig. 29.2) is ideal because it can be moved to different locations where loading and unloading sheep is necessary.

Shelters

Sheep do not normally suffer from cold because they have a heavy wool covering; therefore, open sheds are excellent for housing and feeding wintering ewe lambs, pregnant ewes, and rams. Although the ewes can be lambed in these sheds, the newborns need an enclosed and heated room when the weather is cold.

Lambing Equipment

Small pens, about 4 × 4 ft, can be constructed for holding ewes and their lambs until they are strong enough to be put with other ewes and lambs. Four-foot panels can be constructed from 1 × 4-in. lumber, and the two panels can be hinged together. The lambing pens (called *lambing jugs*) can be made along a wall by

FIGURE 29.1 Sheep being directed into a special pen by use of a cutting chute.

FIGURE 29.2   Sheep being unloaded directly onto pasture by use of a portable loading chute.

wiring these hinged panels. Heat lamps are extremely valuable for keeping new-born lambs warm. A heat lamp above each lambing jug can be located at a height that provides a temperature of 90°F at the lamb's level (Fig. 29.3).

It is advisable to identify each lamb and to record which lambs belong to which ewes. Often, a ewe and her lambs are branded with a scourable paint to

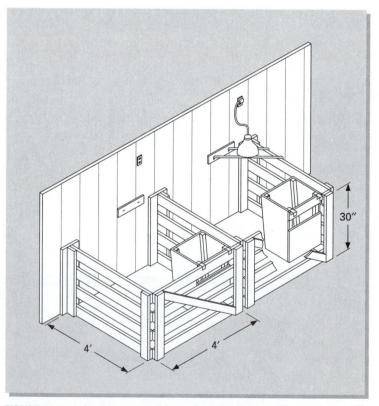

FIGURE 29.3   A lambing jug with heat lamp for ewes with newborn lambs. Courtesy of John H. Pedersen, Midwest Plan Service, Agricultural Engineering, Iowa State University.

identify them. Numbered ear tags can be applied to the lamb at birth. If new-born lambs are to be weighed, a dairy scale and a large bucket are needed. The lamb is placed in the bucket, which hangs from the scale, for weighing. It is advisable to immerse naval cords of newborn lambs in a tincture of iodine.

Feeding Equipment

Usually sheep are given hay in the winter feeding period. Feeding mangers (Fig. 29.4) should be provided because hay is wasted if it is fed on the ground. Some sheep may require limited amounts of concentrates. Concentrates can be fed in the same bunk as hay if the hay bunk is properly constructed, or separate feed troughs (Fig. 29.5) may be preferable.

Additional concentrates can be provided for lambs by placing the concentrate in a **creep** (Fig. 29.6), which is constructed with openings large enough to allow the lambs to enter but small enough to keep out the ewes. In addition to concentrates, it is advisable to provide good-quality legume hay, a mineral mix containing calcium, phosphorus, and salt. Examples of creep diets that are balanced for needed nutrients are shown in Table 29.1. Wheat or barley could replace up to half of the corn or sorghum (milo) in these diets. Lambs normally eat 1.5 lb of creep diet daily from 10–120 days of age (0.10 lb at 3 weeks and 3.0 lb at 120 days).

Water is essential for sheep at all times. It can be provided by tubs or automatic waterers. Tubs should be cleaned once each week. Large buckets are usually used for watering ewes in lambing jugs.

A separate pen equipped with a milk feeder may be needed for orphan lambs. Milk or a milk replacer can be provided free-choice if it is kept cold. Heat lamps kept some distance from the milk feeder can be provided.

Feed Storage

Areas should be provided for storage of hay and concentrates so that feeds can be purchased in quantity or so homegrown feeds can be stored in a dry place.

FIGURE 29.4   Sheep eating from a hay bunk.

FIGURE 29.5 A feed trough for providing concentrates for sheep. Note the iron rod that prevents the sheep from getting into the trough. Courtesy of the *Sheep Production Handbook,* 1988.

An open shed is satisfactory for hay storage. Some operators prefer a feed bin for concentrates; it is designed so that the feed can be put in at the top and removed from the bottom.

TYPES OF FARM FLOCK PRODUCERS

Some producers raise purebred sheep and sell **rams** for breeding. Commercial producers whose pastures are productive can produce slaughter lambs on pasture, whereas those whose pastures are less desirable produce feeder lambs. Lambs that are neither sufficiently fat nor large for slaughter at weaning time usually go to feedlot operators where they are fed to slaughter weight and condition. However, most feedlot lambs are obtained from producers of range sheep.

Purebred Breeder

Purebred sheep are usually given more feed than commercial sheep because the purebred breeder is interested in growing sheep that express their growth

FIGURE 29.6 Creep with feeder and heat lamp for lambs. A heat lamp is provided if the weather is cold. From R. A. Battaglia and V. B. Mayrose, *Handbook of Livestock Management Techniques* (New York: Macmillan, 1981), p. 381.

TABLE 29.1   Example Creep Diets for Lambs

Ingredient	Ration (percent)			
	1	2	3	4
Corn grain (yellow)	51.5	61.0	—	—
Sorghum grain	—	—	53.0	65.0
Soybean meal (44%)	22.0	12.5	—	—
Cottonseed meal (41%)	—	—	25.0	13.0
Alfalfa hay (midbloom)	20.0	20.0	15.0	15.0
Molasses (cane)	5.0	5.0	5.0	5.0
Trace mineral salt (sheep)	0.5	0.5	0.5	0.5
Calcium carbonate	1.0	1.0	1.5	1.5
Vitamin A	500 IU/lb			
Vitamin E	10 IU/lb			

Source: Sheep Production Handbook.

potential. Records are essential to indicate which ram is bred to each of the ewes and which ewe is the mother of each lamb. All ram lambs are usually kept together to compare and identify those most desirable for sale. The ram lambs are usually retained by the purebred breeder until they are a year of age, at which time they are offered for sale. Buyers usually seek to obtain rams from purebred breeders in early summer.

Purebred breeders have major responsibilities to the sheep industry. Because purebred breeders determine the genetic productivity of commercial sheep, they should rigidly select animals that are kept for breeding and should offer only those rams for sale that will contribute improved productivity for the commercial producer.

Commercial Slaughter Lamb Producers

The producer of slaughter lambs, whose feed and pasture conditions are favorable, strives to raise lambs that finish at 90–120 days of age weighing approximately 120 lb each. Any lamb that is small or lacks finish requires additional feeding. Usually, the price decline that occurs from June through September offsets the improvement that is made in value of lambs by feeding, so that feeding the culled farm flock lambs is generally unprofitable. Creep feeding of concentrates early in the life of lambs helps, but, because lambs may be disoriented when their creep is moved, they may not come to the creep after being put onto a new pasture. Therefore, creep feeding on pasture on which pasture rotation is practiced may not be beneficial. Keeping sheep healthy while providing adequate water on lush pasture will result in early market lambs.

Commercial producers can give some assurance to raising heavy, well-finished market lambs at weaning by using good ram selection and crossbreeding programs. Rams that are heavy at 90 days of age are more likely to sire lambs that are heavy at this age than are rams that are light in weight at 90 days of age. The breeds used in a three-breed rotation should include one that is noted for milk-producing ability, one that is noted for rapid growth, and one that is noted for ruggedness and adaptability. The Dorset could be considered for milk production, either the Hampshire or Suffolk for rapid growth, and the Cheviot for its ruggedness. All of these breeds make desirable carcasses.

Commercial producers castrate all ram lambs. Young ewes that are selected to replace old ewes should be large and growthy and should be daughters of ewes that produce relatively many lambs. Ewes that have started to decline in production and poorly productive ewes should be culled.

Commercial Feeder Lamb Producers

Some commercial sheep are produced under pasture conditions that are insufficient for growing the quantity or quality of feed needed for heavy slaughter lambs at weaning. Lambs produced under such conditions may be either fed out in the summer or carried through the summer on pasture with their dams and finished in the fall. Sudan grass or rape can be seeded so that good pasture is available in the summer, and lambs can be finished on pasture by giving them some concentrates. If good summer pasture cannot be made available, it is advisable to wait until autumn, at which time the lambs are put on full feed in a feedlot.

Commercial Feedlot Operator

Lambs that come to the feedlot for finishing are treated for internal **parasites** and for certain diseases, particularly **overeating disease.** They are provided water and hay initially, after which concentrate feeding is allowed and increased until they receive all the concentrates they will consume. Death losses may be high in unthrifty lambs.

The feedlot operator hopes to profit by efficiently increasing the lambs' weight. Lambs on feed should gain about 0.5–0.8 lb per head per day. Feeder buyers prefer feeder lambs weighing 70–80 lb over larger lambs because of the need to put 20–30 lb of additional weight on the lambs to finish them. A feeding period of 40–60 days should be sufficient for finishing thrifty lambs. Table 29.2 shows some example diets for growing–finishing lambs.

Feeding lambs is riskier than producing them because of death losses, price fluctuations of feed and sheep, and the necessity of making large investments.

TABLE 29.2 **Example Diets for Growing–Finishing Lambs**

Ingredient	Ration (pounds)			
	1	2	3	4
Corn grain (yellow)	620	1,465	—	—
Sorghum grain (milo)	—	—	390	1,475
Alfalfa hay (mature)	1,100	300	300	300
Cottonseed hulls	—	—	800	—
Soybean meal (44%)	140	100	—	—
Cottonseed meal (41%)	—	—	350	80
Molasses (cane)	120	100	120	100
Calcium carbonate	—	15	20	25
Trace mineral salt (sheep)	10	10	10	10
Ammonium chloride	10	10	10	10

Source: Sheep Production Handbook.

FEEDING EWES

Figure 29.7 shows the expected weight changes in a 160-lb ewe raising twin lambs. A ewe of similar weight, raising a single lamb, would have about two-thirds of the weight changes shown in Fig. 29.7. Economical feeding programs are implemented to correspond with these expected weight changes.

Mature, pregnant ewes usually need nothing more than lower-quality roughages (e.g., hay, wheat straw, or corn stover) during the first half of pregnancy, after which some concentrates and good-quality legume hay should be fed. Any grains such as corn, barley, oats, milo, and wheat are satisfactory feeds. Sheep usually chew these grains sufficiently so that grinding or rolling is not essential. Rolled or cracked grains may be digested slightly more efficiently and may be more palatable, but finely ground grains are undesirable for sheep unless the grains are pelleted.

Sheep need energy, protein, salt, iodine, phosphorus, and vitamins A, D, and E. In some areas selenium is deficient and must be supplied either in the feed or by injection. Mature, ruminating sheep have little need for quality protein or B vitamins, but immature sheep have variable requirements. Rumen microorganisms can synthesize protein from nonprotein nitrogenous substances in the ration.

Sheep will perform poorly or die quickly when the water supply is inadequate, so the importance of supplying water cannot be overstated. Water that is not clean will not be well accepted by sheep. Water intake is influenced by amount of food eaten, protein intake, environmental temperature, mineral intake, water temperature, pregnancy, water content of feed eaten (including rain and dew on pastures), and the odor and taste of the water.

Energy is perhaps the most common limiting nutrient for ewes. Underfeeding is the primary cause for a deficiency of energy because most feed materials are high in energy. Dry range grasses and mature forages such as grain straws

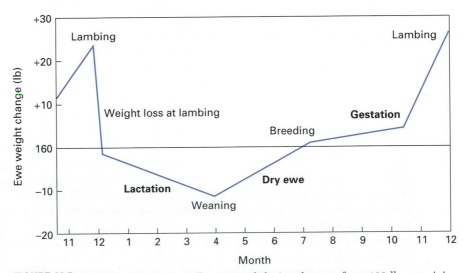

FIGURE 29.7 Weight changes normally expected during the year for a 160-lb ewe giving birth to and raising twin lambs. Courtesy of the *Sheep Production Handbook*.

may be high in gross energy but so low in digestibility that sheep cannot obtain all their energy needs from them.

The amount of protein in the ration for sheep is of greater importance than the quality of protein because sheep can make the essential amino acids by action of microorganisms in the rumen. The oil meals (soybean, linseed, peanut, and cottonseed) are all high in protein. Soybean meal is the most palatable and its use in a ration encourages a high feed intake. Nonprotein sources of nitrogen such as urea and **biuret** can be used to supply a portion, but not all, of the nitrogen needs of sheep. Not more than one-third of the nitrogen need should be supplied by urea or biuret, and these materials are not recommended for young lambs with developing rumens or for range sheep on low-energy diets. If protein intake is limited by mixing with salt, adequate water must be provided. The mineral needs for sheep are calcium, phosphorus, sulfur, potassium, sodium, chlorine, magnesium, iron, zinc, copper, manganese, cobalt, iodine, molybdenum, and selenium. All of these minerals are found in varying amounts in different tissues of the body. For example, 99% of the calcium, 80–95% of the phosphorus, and 70% of the magnesium occurs in the skeleton and more than 80% of the iodine is found in the thyroid gland.

The requirement for minerals is influenced by breed; age, sex, and growth rate of young animals; reproduction and lactation of the ewe; level and chemical form injected or fed; climate; and balance and adequacy of the ration.

Salt is important in sheep nutrition and it can become badly needed when sheep are grazing on lush pastures. Also, sheep may consume too much salt when they are forced to drink brackish water.

In some areas, iodine and selenium are deficient. The use of iodized salt in an area where iodine is deficient may supply the iodine needs. Selenium can be added to the concentrate mixture that is being fed or it can be given by injection.

The most critical periods of nutritional needs for the ewe are at breeding and just before, during, and shortly after lambing. A lesser but still important period for the ewe is during lactation. After the lamb is weaned, the ewe normally can perform satisfactorily on pasture or range without any additional feed. If accelerated lamb production is practiced, the ewes should be well fed after the lamb is weaned so the ewe can be bred again.

Example diets for ewes are shown in Table 29.3. Note how the amount of feed changes from maintenance to gestation to lactation. In both the gestation and lactation periods, the maintenance requirement must be met first.

Ewes are normally run on pasture or range where they obtain all their nutritional needs from grasses, browse, and forbs, except during the winter or dry periods when these plants are not growing. During these periods when there is no plant growth, the sheep must be given supplemental feed.

Occasionally lambs are raised on milk replacer or on cows' milk if the ewe dies during lambing or her udder becomes nonfunctional. During these two periods, frozen colostrum should be available for the lambs. If milk or milk replacer is used, it can be bottle-fed twice daily or it can be self-fed if it is kept cold. It is important to feed cold milk when it is self-fed to prevent the lamb from overconsumption. There should be heat lamps not far from where the lambs consume the cold milk so they can go to the heated area to become warm and to sleep.

Rams should be fed to keep them healthy but not fat. A small amount of grain along with good-quality hay satisfies their nutritional needs in winter.

TABLE 29.3   Example Diets (as fed) for 155-lb Ewes at Different Stages of Production

Stage of Production	Ration No.	Alfalfa Hay (midbloom; lb)	Corn Silage (mature; lb)	Corn Grain (lb)	Soybean Meal (44% CP; lb)	Salt–Trace-Mineral Mix (lb)[a]
Maintenance	1	3.0	—	—	—	0.05
	2	—	6.0	—	0.20	0.05
Gestation	1	3.5	—	—	—	0.05
(first 15 weeks)	2	—	6.0	—	0.25	0.05
Gestation (last 4 weeks)						
130–150% lambing	1	3.5	—	0.75	—	0.05
	2	—	6.0	0.75	0.40	0.05
180–225% lambing	1	3.5	—	1.25	—	0.05
	2	—	7.0	1.00	0.50	0.05
Lactation (first 6–8 weeks)						
Suckling single	1	4.0	—	2.00	—	0.05
	2	—	9.0	1.00	0.85	0.05

[a] Contains 50% trace-mineral salt (for sheep) and 50% dicalcium phosphate.
Source: *Sheep Production Handbook.*

Young bred ewes should be fed some grain along with all the legume hay they will consume because they are growing and also need some nutritional reserve for the subsequent lactation. Ewe lambs should be grown out but not fattened in the winter. Limited grain feeding along with legume hay satisfies their nutritional needs.

At lambing time, the grain allowance of ewes needs to be increased to assist them in producing a heavy flow of milk. Also, the lambs need to be fed a high-energy ration in the creep. Lambs obtain sufficient protein in the milk given by their mothers, but they need more energy. Rolled grains provided free-choice in the creep are palatable and provide the energy needed.

Lambs on full feed are allowed some hay of good quality and all the concentrates they will consume. Some feedlot operators pellet the hay and concentrates, whereas others feed loose hay and grains. After the grasses and legumes start to grow in the spring, all sheep generally obtain their nutritional needs from the pasture.

CARE AND MANAGEMENT OF FARM FLOCKS

The handling of sheep is extremely important. A sheep should never be caught by its wool because the skin is pulled away from the flesh, causing a bruise. When a group of sheep is crowded into a small enclosure, the sheep will face away from the person who enters the enclosure. When the sheep's rear flank is grasped with one hand, the sheep starts walking backward; this allows one to reach out with the other hand and grasp the skin under the sheep's chin. When a sheep is being held, grasping the skin under the chin with one hand and grasping the top of the head with the other hand enables the holder to pull the sheep forward so its brisket is against the holder's knee. If a sheep is to be moved forward, the skin under the chin can be grasped with one hand and the

dock (the place where the tail was removed) can be grasped with the other hand. Putting pressure on the dock makes the sheep move forward, while holding its chin with the other hand prevents it from escaping.

Lambing Operations

Before the time the ewes are due to lamb, wool should be clipped from the dock, udder, and vulva regions. This process is called **crutching** or tagging. If weather conditions permit, the ewes may be completely shorn. Also, all **dung tags** (small pieces of dung that stick to the wool) should be clipped from the rear and flank of pregnant ewes. Young lambs will try to locate the teat of the ewe and may try to nurse a dung tag if it is present.

Ewes should be checked periodically to locate those that have lambed. The ewe and newborn lamb should be placed in a lambing jug. The ewe that is about ready to lamb can be watched carefully but with no interference if delivery is proceeding normally. In a normal presentation, the head and front feet of the lamb emerge first. If the rear legs emerge first (**breech presentation**), assistance may be needed if delivery is slow because the lamb can suffocate if deprived of oxygen for too long. If the front feet are presented but not the head, the lamb should be pushed back enough to bring the head forward for presentation.

As soon as the lamb is born, membranes or mucus that may interfere with its breathing should be removed. When the weather is cold, it may be necessary to dry the lamb by rubbing it with a dry cloth. If a lamb becomes chilled, it can be immersed in warm water from the neck down to restore body temperature, after which it should be wiped dry. The lamb should be encouraged to nurse as soon as possible. A lamb that has nursed and is dry should survive without difficulty if a heat lamp is provided. The lamb can be identified by an ear tag, a tattoo, or both.

Some ewes may not want to claim their lambs. It may be necessary to tie a ewe with a rope halter so she cannot butt or trample her lamb.

Castrating and Docking

Because birth is a period of stress for lambs, it is best to wait 3–4 days before castrating and **docking** lambs. To use the elastrator method of castration and docking, a tight rubber band is placed around the scrotum above the testicles (for castration) and around the tail about an inch from the buttocks (for docking). Some death losses can occur when tetanus-causing bacteria invade the tissue where the elastrator was applied. Another castration and docking practice is to remove surgically the testicles and tail. The emasculator is also useful for docking; the skin of the tail is pulled toward the lamb, the emasculator is applied about an inch from the lamb's buttocks, and the tail is cut loose next to the emasculator. A fly repellent should be applied around any wound to lessen the possibility of **fly strike** (fly eggs are deposited during warm weather).

Occasionally, ewes develop a vaginal or uterine **prolapse** (the reproductive tract protrudes to the outside through the vulva). This condition is extremely serious and leads to death if corrective measures are not taken by trained personnel. The tissue should be pushed back in place, even if it is necessary to hoist the ewe up by her hind legs as a means of reducing pressure that the ewe is applying to push the tract out. After the tract is in place, the ewe can be harnessed so that external pressure is applied on both sides of the vulva. In some cases, it may be necessary to suture the tract to make certain that it stays in place. Once a ewe has prolapsed her reproductive tract, the tract shows a weakness that is likely to recur; therefore, a record should be kept so that the ewe can be culled after she weans her lamb.

Shearing

Sheep are usually shorn in the spring, before the hot weather months. Shearing is typically done by professional personnel, though shearing classes are available to teach the operator. The usual method of shearing involves clipping the fleece from the animal with power-driven shears, leaving sufficient wool covering to protect the sheep's skin. In the shearing operation, the sheep is set on its dock and cradled between the shearer's legs, which are used to maneuver the sheep into the positions needed to ease the shearing.

The wool that is properly removed during shearing remains in one large piece. It is spread with the clipped side out, rolled with the edges inside, and tied with paper twine. It is best to remove dung tags and coarse material that is clipped from the legs and put these items in a separate container. The tied fleece is put into a huge sack. When buyers examine the wool, they can obtain core samples from the sack. The core sample is taken by inserting a hollow tube that is sharp at the end into the sack of wool. The sample obtained is examined to evaluate the wool in the sack rather than having to remove the fleeces to examine them. If undesirable material is obtained in the core sample, the price offered will be much lower than if only good wool is found.

If the wool is kept for some time before it is sold, it should be stored in a dry place and on a wooden or concrete floor to avoid damage from moisture.

FACILITIES FOR PRODUCTION OF RANGE SHEEP

Sheep differ from cattle in that they graze weedy plants and brush as well as grasses and legumes. Because of their different grazing patterns, cattle and sheep can be effectively grazed together in some range areas. Total pounds of liveweight produced can be higher compared to grazing the separate species on the same range.

Range sheep are produced in large flocks primarily in arid and semiarid regions. Sheep of fine-wool breeding tend to stay together as they graze, which makes herding possible in large range areas. Range sheep are moved about either in trucks or by trailing so they can consume available forage at various elevations. Requirements for the production of range sheep are usually different from those for farm flock operations. One type of range sheep operation is described here, but it must be noted that variations exist.

Range sheep are usually bred to lamb later than sheep in farm flocks; therefore, they can be lambed on the range. Few provisions are needed for lambing when sheep are lambed on the range, but under some conditions a tent or a lambing shed may be used to give range sheep protection from severe weather at lambing time. Temporary corrals can be constructed using snow fences and steel posts when it is necessary to contain the sheep at the lambing camp or at lambing sheds.

Sheep on range are usually wintered at relatively low elevations in areas where little precipitation occurs. Wintering sheep are provided **feed bunks** if hay is to be fed, and windbreaks to give protection from cold winds. Some producers of range sheep provide pelleted feed to supplement the winter forage. Pellets are usually placed on the ground but are sometimes dispersed in grain troughs.

A sheep camp or sheep wagon is the mobile housing used by the sheep herder. The camp is moved by truck or horses because the sheep need to be moved over large grazing areas. A sheepherder usually has a horse and dogs to assist in herding the sheep. The sheep are brought together to a night bedding area each evening.

MANAGING RANGE SHEEP

Range sheep are grazed in three general areas: (1) the **winter headquarters,** which is a relatively low and dry area and which sometimes provides forage for winter grazing; (2) the **spring–fall range,** which is a somewhat higher elevation area and which receives more precipitation; and (3) the **summer grazing** area, which is at high, mountainous elevations and which receives considerable precipitation, resulting in lush feeds.

The Winter Headquarters

The forage on the winter range, where there is usually less than 10 inches of precipitation annually, is composed of sagebrush and grasses. The grasses are cured on the ground from the growth of the previous summer; consequently, winter forage is of lower quality than green forage because the plants in the winter forage are mature and because they have lost nutrients. The soil in these areas is often alkaline and the water is sometimes alkaline.

Because forage in the wintering area is of poor quality, supplemental feeding that provides needed protein, carotene, and minerals (such as copper, cobalt, iodine, and selenium) is usually necessary. A pelleted mixture made by mixing sun-cured alfalfa leaf meal, grain, solvent-extracted soybean or cottonseed meal, beet pulp, molasses, bone meal or dicalcium phosphate, and trace-mineralized salt is fed at the rate of 0.25–2.0 lb per head per day, depending on the condition of the sheep. The feed may be mixed with salt to regulate intake so that feed can be before the sheep at all times. If the intake of feed is to be regulated through the use of salt, trace-mineralized salt should be avoided. Adequate water must also be provided at all times, because heavy salt intake is quite harmful if sheep do not have water for long periods of time.

The ewes are brought to the winter headquarters about the first of November. Rams are turned in with the ewes for breeding in November if lambing is to take place in sheds or if a spring range that is not likely to experience severe weather conditions is available for lambing. Otherwise, the rams are put with the ewes in December for breeding.

January, February, and March are critical months because the sheep are then in the process of exhausting their body stores and because severe snowstorms can occur. If sheep become snowbound, they should each be given 2 lb of alfalfa hay plus 1 lb of pellets per day containing at least 12% protein. Adequate feeding of ewes while they are being bred and afterward results in at least a 30% increase in lambs produced and about a 10% increase in wool produced. In addition, death losses are markedly reduced. Sheep that are stressed by inadequate nutrition, either as a result of insufficient feed or a ration that is improperly balanced, are highly susceptible to pneumonia. Heavy death losses can result.

The Spring–Fall Range

Pregnant ewes are shorn at the winter headquarters (usually in April). They are then moved to the spring–fall range, where they are lambed. If they are lambed on the range, a protected area is necessary. An area having scrub oak or big sagebrush on the south slopes of foothills and ample feed and water is ideal for range lambing. Portable tents can be used if the weather is severe.

Ewes that have lambed are kept in the same area for about 3 days until the lambs become strong enough to travel. The ewes that have lambs are usually fed a pelleted ration that is high in protein and fortified with trace-mineralized salt and either bonemeal or dicalcium phosphate. Feeding at this time can help prevent sheep from eating poisonous plants.

Ewes with lambs are kept separate from those yet to lamb until all ewes have borne their lambs. In addition, ewes that are almost ready to lamb are separated from those that will not lamb for some time yet. Thus, after lambing gets under way, three separate groups of ewes are usually present until lambing is completed.

If the ewes are bred to lamb earlier than is usual for range lambing and a crested wheat-grass pasture is available, ewes may be lambed in open sheds. If good pasture is unavailable, the ewes may be confined in yards around the lambing sheds, starting a month before lambing. In this event, the ewes must be fed alfalfa or other legume hay and 0.50–0.75 lb of grain per head per day. The ewes should have access to a mixture of trace-mineralized salt and bone meal or dicalcium phosphate.

Although shed lambing is more expensive than range lambing, higher prices for lambs marketed earlier have made shed lambing advantageous. Fewer lambs are lost in shed lambing, and lambing can take place earlier in the year. The heavy market lambs that result produce enough income to offset the costs of shed lambing.

Summer Grazing

The Summer Range. Sheep are moved to summer range shortly after lambing is completed if weather conditions have been such that snow has melted and lush plant growth is occurring. The sheep are put into bands of about 1,000–1,200 ewes and their lambs. In some large operations, the general practice is to put ewes with single lambs in one band and ewes with twins in another. The ewes with twin lambs are given the best range area so the lambs will have added growth from the better forage supply. Sheep are herded on the summer range to assist them in finding the best available forage.

In October, prior to the winter storms, the lambs are weaned. Lambs that carry sufficient finish are sent to slaughter and other lambs are sold to lamb feeders. It is the general practice among producers of Rambouillet, Columbia, and Targhee sheep to breed some of the most productive and best-wooled ewes to rams of the same breed to raise replacement ewe lambs. Most of these ewe lambs are kept and grown out, and only the less desirable ones are culled. The remainder of the ewes are bred to meat-type rams, such as the Suffolk or Hampshire, and all their lambs are marketed for slaughter, as feeders or as stockers (animals used in the flock for breeding).

The Fall Range.   As soon as the lambs are weaned, the ewes are moved to the spring–fall range. Later they go to the winter headquarters for wintering.

The number of ewes that can be bred per ram during the breeding period of about 2 months is 15 for ram lambs, 30 for yearling rams, and 35 for mature, but not aged, rams. These numbers are general and depend greatly on the type of conditions existing on the range.

CONTROLLING DISEASES AND PARASITES

Sheep raised by most producers are confronted with a few serious diseases, several serious internal parasites, and some external parasites. Some common diseases of sheep include the following types.

Enterotoxemia (overeating disease) is often serious when sheep are in a high nutritional state (e.g., lambs in the feedlot), though it can also affect sheep that are on lush pastures. The disease can be prevented by administering type D toxoid. Usually, three treatments are given: two about 4 weeks apart and a booster treatment 6 months later. Losses from this disease among young lambs can be prevented by vaccinating pregnant ewes.

E. coli **complex**–*Clostridium E* is another disease common in sheep. At least three organisms are involved: (1) *E. coli,* (2) *Clostridium perfringens,* and (3) a virus called *rotavirus.* This disease affects lambs from birth to a few days of age. The disease can be prevented by vaccinating all pregnant ewes twice in later pregnancy with type C and D *Clostridium perfringens* toxoid. The lambing pen should be thoroughly cleaned. Broad-spectrum **antibiotics** given to afflicted lambs help. The ewe flock should be vaccinated with *Clostridium perfringens* type C and D toxoid. Enterotoxemia may be caused by *Clostridium perfringens* type D or C. The *Clostridium perfringens* type CD toxoid given to the ewe will give the lamb protection from both types of enterotoxemias.

Lamb dysentery, caused by *C. perfringens* type B, occurs very early in life and often during wet weather. It can be prevented by vaccinating ewes with type BCD vaccine.

Footrot is one of the most serious and common diseases affecting the sheep industry. The disease can be treated with systemic medication. It can be cured by severe trimming so that all affected parts are exposed, treating the diseased area with a solution of one part formalin solution to nine parts of water, and then turning the sheep onto a clean pasture so that reinfection does not occur. A tilting squeeze is useful for restraining sheep that need treatment (Fig. 29.8). Formaldehyde must be used with caution because the fumes are damaging to the respiratory system of both the sheep and the person applying the formaldehyde.

Once all sheep in the flock are free of footrot, it can best be prevented by making sure that it is not reintroduced into the flock. Rams introduced for breeding should be isolated for 30–60 days for observation. If footrot develops,

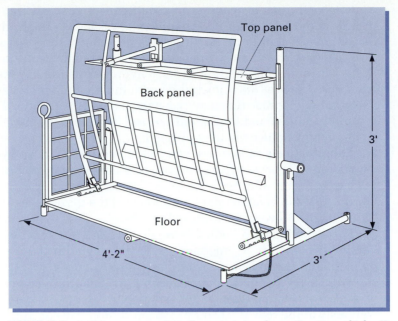

FIGURE 29.8 A tilting-squeeze chute for restraining sheep. Courtesy of John H. Pedersen, Midwest Plan Service, Agricultural Engineering, Iowa State University.

rams should continue in isolation until free of the disease. A vaccine is available for footrot.

Pneumonia usually occurs when animals have been stressed by other diseases, parasites, improper nutrition, or exposure to severe weather conditions. When pneumonia is recognized early, antibiotics are usually effective against it. The afflicted animal should be given special care and kept warm and dry.

Ram epididymitis is an infection of the epididymis that reduces fertility. No effective treatment is available.

Sore mouth usually affects lambs rather than adult sheep. It is caused by a virus and can be contracted by humans. It can be controlled by vaccination, and sheep should be vaccinated as a routine practice.

Sheep are subject to a nutritional disease known as **white muscle disease.** To prevent it, pregnant ewes should be given an injection of selenium during the last one-third of pregnancy and the lambs should be given an injection of selenium at birth.

Selenium can be added to the feed of pregnant ewes to prevent white muscle disease of their lambs. If trace-mineralized salt is provided, it should contain sufficient selenium to supply the needs of sheep.

Shipping fever is a highly infectious and contagious disease complex that usually affects lambs after stress of transportation. The most common type of shipping fever is *Pasteurellosis.* Bacterial agents associated with this disease are *Pasteurella hemolytica, Pasteurella multocida, Corynebacterium pyogenes,* and, to a lesser extent, *micrococci, streptococci,* and *pseudomonas.* Antibiotics and **sulfonamides** are usually effective as treatment. Care in transporting lambs helps prevent this disease.

Caseous lymphadenitis is a disease that occurs in greater frequency as sheep increase in age from lambs to old animals. It is due to a bacterium,

Corynebacterium pyogenes var pseudotuberculosis, which grows in lymph glands and causes a large development of caseous (a thick, cheeselike accumulation) material to form. It is a serious disease and one that is difficult to control. The abscesses can be opened and flushed with a solution of equal parts of 0.2% nitrofurazone solution and 3% hydrogen peroxide. One should not open an abscess and let the thick pus go onto the floor or soil where well sheep will be traveling because this may cause the disease to spread. All infected animals should be culled and they should be isolated from noninfected sheep immediately on appearance of being infected.

Salmonellosis causes dysentery in lambs and abortion by ewes. The organism is found in feces of animals and may contaminate water in stagnant pools. It usually occurs after lambs or other sheep have been stressed. There is no cure for this disease. If lambs show symptoms of the disease, they should be isolated from pregnant ewes and from well lambs.

Polyarthritis (stiff lamb disease usually associated with overeating) affects lambs. It responds to treatment with the oxytetracycline, chlortetracycline, tylosin, and penicillin.

Milk fever is due to hypocalcemia. Afflicted animals can be treated with injection of calcium salts. A 20% solution of calcium borogluconate given **intravenously** at the rate of 100 ml per sheep should have an afflicted animal up and in good condition within 2 hours.

Abortion may occur in late pregnancy due to vibriosis, enzootic abortion, or leptospirosis. Most of these are controlled by vaccination.

Sheath rot may occur in rams. There is no treatment that gives satisfactory results.

Urinary calculi (kidney stones) occurs when the salts in the body that are normally excreted in the urine are precipitated and form stones that may lodge in the kidneys, ureters, bladder, or urethra. Providing a constant supply of clean water helps immensely in preventing calculi formation. The ideal ratio of cal-

TABLE 29.4   **Summary of Compounds for Control of Internal Parasites of Sheep**

Internal Parasites	Drench	Remarks
Coccidia	Sulfonamides	Relatively effective; will form crystals
Coccidiosis	Sulfamethazine	in kidney of lamb, which can cause
	Sulfaguanidine	death
Liver fluke	Albendazole	A new drug released by the FDA for experimental use in Texas, Oregon, Washington, Idaho, and Louisiana
	Carbon tetrachloride	Toxic—use at recommended doses, effective only against adult flukes
Lungworm	Levamisol (Tramisol)	Effective only against adult lungworm
Stomach and round worms	Tramisol	Safe, effective; will kill arrested worms
	Phenothiazine	Control is much improved with the use of fine particle size
	Thiabendazole	Safe, effective
Tapeworm		
Broad	Lead arsenate	May be mixed with phenothiazine or
	Dipenthane 70 (teniatol, teniazine)	thiabendazole for control of roundworms and tapeworms with one drench
Fringe	No approved control	

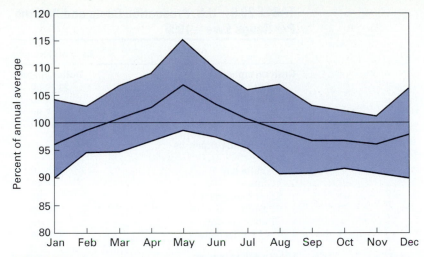

FIGURE 29.9 Seasonal lamb price index (1970–84). The dark line is the average monthly price compared with the average for all years (100). The shaded area shows the range in prices by years. Courtesy of the Western Livestock Marketing Information Project.

cium to phosphorus is 1.6 calcium to 1.0 phosphorus. Ammonium chloride can be included in the ration to help prevent urinary calculi.

Pregnancy disease (ketosis) is a metabolic disease that affects ewes in late pregnancy, particularly if they are carrying twins or triplets. The problem is that ewes carrying twins or triplets must break down body fat to provide their energy needs in later pregnancy. It is possible that use of fat breakdown may not provide the glucose needs, resulting in hypoglycemia. Pregnancy disease may be prevented by feeding some high-energy grain such as corn, barley, or milo, but it is aggravated by one or more large lambs reducing rumen space.

Grass tetany (or **grass staggers**) is due to a deficiency of magnesium at a particular time, most frequently in spring when lactating ewes are put onto lush pasture where there is insufficient magnesium available to the sheep. An injec-

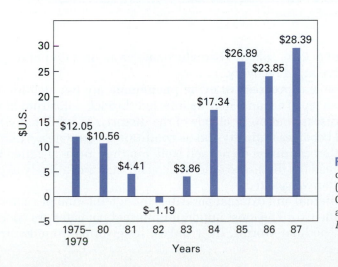

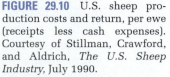

FIGURE 29.10 U.S. sheep production costs and return, per ewe (receipts less cash expenses). Courtesy of Stillman, Crawford, and Aldrich, *The U.S. Sheep Industry,* July 1990.

TABLE 29.5   **U.S. Production Costs and Returns Per Range Ewe—1996**

Item	Income/Expense
Cash receipts	**Dollars/head**
Lambs	$179.60
Cull ewes	4.90
Wool	7.00
Total cash receipts	$191.50
Cash operating expenses	
Feed	
Alfalfa hay	26.16
Grain	29.49
Salt & minerals	.89
Pasture	18.10
Public grazing	11.44
Total feed costs	$86.08
Other cash expenses	
Veterinary and medicine	2.25
Livestock hauling	2.30
Marketing	14.48
Shearing and tagging	2.75
Transportation	1.96
Repairs (machinery, buildings, fences)	2.00
Hired labor	18.75
Ram rental	3.00
Miscellaneous	2.67
Interest on operating capital	6.19
Total other cash expenses	$56.35
Total cash expenses	$142.43
Cash profit	$49.07
Property & ownership costs	
General farm overhead	4.36
Taxes	13.82
Insurance	.75
Depreciation	10.00
Total property & ownership costs	$28.93
Net profit	$20.14

Source: USDA.

tion of 50–100 ml of 20% calcium borogluconate or an injection of magnesium sulfate should give rapid recovery.

Johnes disease and **progressive ovine pneumonia** are two debilitating diseases that affect sheep. The lamb may be infected through colostrum of the ewe that has Johnes disease but no ill effects of the disease are exhibited until later when the animal becomes unthrifty and its condition deteriorates. Afflicted animals that are in thin condition do not sell well, but they must be culled because they can help spread the disease to other sheep and they are unlikely to produce lambs when they are emaciated.

Sheep have internal and external parasites, though internal parasites are the more serious of the two. The most important internal parasites are coccidiosis, stomach worms, nodular worms, liver flukes, lungworms, roundworms, and tapeworms.

TABLE 29.6 Colorado Lamb-Feeding Budget—1996

Item	Income/Expense
Cash receipts	**Dollars/head**
Slaughter lamb (130 lb)	123.50
Total cash receipts	$123.50
Cash expenses	
Feeder lamb (83 lb)	$80.51
Feed	
Corn (190 lb)	19.00
Ground hay (95 lb)	4.75
Protein supplement (36 lb)	3.20
Total feed costs	$26.95
Death loss (4%)	4.94
Yardage	3.20
Veterinary & medicine	.50
Machine hire	1.50
Miscellaneous	1.00
Interest on operating capital	1.31
Total cash expenses	$119.91
Net margin	$3.59

Source: USDA.

The treatments effective in controlling internal parasites of sheep are summarized in Table 29.4.

Common external parasites of sheep include blowfly maggots, **keds** (sheep ticks), lice, mites, screwworms, and sheep bots. External parasites can be controlled by making two applications of an effective insecticide that is not harmful to sheep. The two applications should be spaced so that eggs hatched after the first application will not result in egg-laying adults prior to the second application.

DETERMINING THE AGE OF SHEEP BY THEIR TEETH

The age of a sheep can be determined by its teeth. Lambs have four pairs of narrow lower incisors called *milk teeth* or *baby teeth*. At approximately a year of age the middle pair of milk teeth is replaced by a pair of larger, permanent teeth. At 2 years, a second is replaced. This process continues until, at 4 years of age, the sheep has all permanent incisors. The teeth start to spread apart and some are lost at about 6–7 years of age. When all the permanent incisors are lost, the sheep has difficulty grazing and should be marketed.

COSTS AND RETURNS

The objective of farm flock operators is the efficient production of slaughter lambs ready for market in April and May, when prices are usually highest (Fig. 29.9).

The items used to evaluate costs and returns for range ewes are shown in Table 29.5. Note that the total cash receipts per ewe ($191.50) and the total cash expenses per ewe ($142.43) represent a cash profit. The costs and returns for several years are shown in Fig. 29.10. Table 29.6 shows an enterprise budget for lamb feeding.

For a more detailed discussion of managing costs and returns, see Chapter 36.

CHAPTER SUMMARY

■ Sheep need good facilities for lambing and working them for sorting, branding, and treatment.

■ Lamb producers plan feeding, health, and management programs so that approximately 120-pound lambs can be weaned and harvested at 90–120 days of age from primarily forage feeding. Lambs weighing less than 100 pounds usually go to commercial feedlots where they receive relatively high-concentrate rations.

■ Sheep are usually shorn in the spring, prior to the hot weather months. Late spring storms sometimes challenge the sheep flock, especially under range conditions where there is limited protection.

■ Sheep are challenged by health problems when compared to most other livestock. Therefore, preventative herd health and treatment programs must be implemented in a timely manner.

REVIEW QUESTIONS

1. *True or False:* Although sheep normally do not suffer from cold weather because they have a heavy wool covering, newborn lambs should be sheltered in an enclosed and heated room when weather is cold.
2. Where are most feedlot lambs obtained?
3. Under what condition will farm flocks produce a larger proportion of feedlot lambs?
4. At what age and weight are commercial slaughter lambs typically finished?
5. What disease of growing lambs is of particular concern to commercial feedlot operators?
6. *True or False:* Mature, pregnant ewes usually need high-quality roughage and concentrate feeds during the first half of gestation.
7. What factors influence water intake of sheep?
8. _____ is the most common limiting factor in the diet of ewes.
9. When do the most critical periods of nutritional needs occur for ewes?
10. In what three general areas are range sheep grazed?
11. When does the breeding season of most sheep occur?
12. *True or False:* Because forage on winter range is of poor quality, supplemental feeding of sheep is usually needed to meet nutritional requirements.
13. What is the number of ewes that can be bred per ram during a 60-day breeding period?

14. What is the objective of farm flock operators?

15. Why is it important that lambs reach market weight and finish condition by mid- to late spring?

16. *True or False:* Ewes nursing twin or triplet lambs should be placed on the higher-quality pasture when available, than ewes nursing one lamb.

SELECTED REFERENCES

Publications

Battaglia, R. A., and Mayrose, V. B. 1981. *Handbook of Livestock Management Techniques.* New York: Macmillan.

Botkin, M. P., Field, R. A., and Johnson, C. L. 1988. *Sheep and Wool: Science, Production, and Management.* Englewood Cliffs, NJ: Prentice-Hall.

Chappell, G. L. M., Ely, D. G., and Aaron, D. K. 1992. Integrated Sheep Production: Sheep as a Business and Nutrition Programs. *Sheep Profit Day Proc.* Lexington, KY: Univ. of Kentucky.

Ensminger, M. E., and Parker, R. O. 1986. *Sheep and Goat Science.* Danville, IL: Interstate Printers and Publishers.

National Research Council. 1985. *Nutrient Requirements of Sheep.* Washington, DC: National Academy Press.

Sheep Housing and Equipment Handbook. 1982. Ames, IA: Midwest Plan Service.

The Sheep Production Handbook. 1996. American Sheep Industry Assoc., 6911 S. Yosemite, Englewood, CO.

Stillman, R., Crawford, T., and Aldrich, L. July 1990. *The U.S. Sheep Industry.* USDA/ERS, Staff Report no. AGES 9048.

Visuals

Production Systems for Sheep; Sheep Reproduction and Management Programs; and *Nutrition and Health in Sheep* (sound filmstrips). Prentice-Hall Media, 150 White Plains Rd., Tarrytown, NY 10591.

Sheep Management Practices (two videotapes). CEV, P.O. Box 65265, Lubbock, TX 79464-5265.

The Sheep Management Series (sound filmstrips): *Ewe and Lamb Management; Fitting and Showing Sheep; Docking Sheep: Controlling Internal Parasites of Sheep; Sheep Castration;* and *Basic Sheep Handling Skills.* Vocational Education Productions, California Polytechnic State University, San Luis Obispo, CA 93407.

Sheep Obstetrics; Assuring Baby Lamb Survival; Castration, Docking and Identification; Raising Orphan Lambs; Marketing Practices; Feeding the Farm Flock; Applying Health Care Practices; Feed and Water Delivery Systems; Sheep Handling; and *Using Equipment and Sheep Psychology* (videotapes). Agricultural Products and Services, 2001 Killebrew Dr., Suite 333, Bloomington, MN 55420.

CHAPTER 30

Horse Breeds and Breeding

BREEDS OF HORSES

Horses have been used for so many different purposes that many breeds have been developed to fill specific needs. The major breeds of horses and their primary uses are listed in Table 30.1. No attempt has been made in Table 30.1 to indicate all the ways each breed is used. For example, the Thoroughbred (Fig. 30.1) is classified primarily as a racehorse for long races, but it is used for several other functions as well.

The *light* breeds of horses provide pleasure for their owners through activities such as racing, riding, and exhibition in shows. Color Plates S and T (following p. 512) show examples of these breeds. Quarter Horses are excellent as cutting horses (horses that separate individual cattle out of a herd) and for running short races. Ponies, such as Shetland and Welsh, are selected for their friendliness and safety with children. The American Saddle Horse and the Tennessee Walking Horse have been selected for their comfortable gaits and responsive attitudes. The American Saddle Horse has regional showring popularity and is considered "the peacock of the showring." The Palomino, Appaloosa, and Paint horses are color breeds developed for showing and working livestock.

Beauty in color and markings is important in horses used for show and breeding purposes, and color breeds have been developed accordingly. The Appaloosa, for example, has color markings of three patterns: leopard, blanket, and roan. It appears that these color patterns are under different genetic controls. In Appaloosas, Paints, and Palominos, coloration can be affected or eliminated by certain other genes, such as the gene for roaning and the gene for graying. The gene for gray can eliminate both colors and markings, as shown by gray horses that turn white with age.

TABLE 30.1 Characteristics and Uses of Selected Breeds of Horses

Breed	Color	Height (in hands)[a]	Weight (lb)	Uses
Riding and harness horses				
American Quarter Horse	All colors except paint	14.2–15.2	1,000–1,250	Short racing, showing, stock work
American Saddle Horse	Chestnut, bay, brown, black	15.0–16.0	1,000–1,150	Showing, pleasure riding, 3 and 5 gaited
Arabian	Bay, chestnut, brown, gray, black	14.2–15.2	850–1,000	Pleasure riding, showing
Morgan	Bay, chestnut, brown, black	14.2–15.1	950–1,150	Pleasure riding, driving, showing
Standardbred	Bay, chestnut, roan, brown, black, gray	14.2–16.2	850–1,200	Harness racing
Tennessee Walking Horse	All colors	15.0–16.0	1,000–1,200	Pleasure riding, showing
Thoroughbred	Bay, brown, gray, chestnut, black, roan	15.2–17.0	1,000–1,300	Long racing
Ponies				
Hackney	Bay, chestnut, black, brown	11.2–14.2	450–850	Light harness, showing
Pony of America	Appaloosa	11.5–13.5	700–800	Riding by children, showing
Shetland	Bay, chestnut, brown, black, spotted, mouse	9.2–10.0	300–400	Riding by children, showing
Welsh	Bay, chestnut, black, roan, gray	11.0–13.0	350–850	Riding by children, showing
Draft[b]				
Belgian	Chestnut, roan usually	15.2–17.0	1,900–2,400	Heavy pulling
Clydesdale	Bay, brown, black, roan	15.2–17.0	1,700–2,000	Heavy pulling
Percheron	Black, gray usually	15.2–17.0	1,600–2,200	Heavy pulling
Shire	Bay, brown, usually black, gray	16.2–17.0	1,800–2,200	Heavy pulling
Suffolk	Chestnut	15.2–16.2	1,500–1,900	Heavy pulling
Color registries				
Appaloosa	Leopard, blanket, roan	14–16	900–1,250	Pleasure riding, showing, stock work
Buckskin	Buckskin, dun, grulla	14–16	900–1,250	Pleasure riding, showing, stock work
Paint	Tobiano, overo	14–16	900–1,250	Pleasure riding, showing, stock work
Palomino	Palomino	14–16	900–1,250	Pleasure riding, showing, stock work
Pinto	Pinto	14–16	900–1,250	Pleasure riding, showing, stock work
White and cremes	White and creme	14–16	900–1,250	Pleasure riding, showing, stock work

[a] Height is measured in inches but reported in *hands*. One hand equals 4 in.
[b] Draft horses are heavy horses used for pulling; other horses are called *light horses* and are used primarily as pleasure horses.

FIGURE 30.1   The Thoroughbred, Foolish Pleasure, in action. Thorough-
breds are great race horses. Note the extreme muscular development partially due to conditioning. Courtesy of New York Racing Association,
Louis Weintraub, Photo Communication Company, and the Jockey Club,
New York, New York.

Another group of horses is the draft horse (Color Plate U)—large and powerful animals that are used for heavy work. The Percheron, Shire, Clydesdale,
Belgian, and Suffolk are examples of draft horses (Table 30.1).

Popularity of Breeds

Registration numbers are a measure of breed popularity. Table 30.2 shows the
registration numbers for draft, light, and pony breeds. Based on these numbers,
the Quarter Horse is the most popular horse breed in the United States.

BREEDING PROGRAM

Reproduction

Mares of the light breeds reach sexual maturity at 12–18 months of age, whereas
draft mares are 18–24 months of age when sexual maturity is reached. Mares
come into heat every 21 days during the breeding season if they do not become
pregnant. Heat lasts for 5–7 days. Ovulation occurs toward the end of heat.
Because of the relatively long duration of heat and because ovulation occurs
toward the end of heat, horse owners often delay breeding the mare for two
days after she has first been observed in heat. Some breeders have the mare
bred every other day while she is in heat.

Although about 10% of all ovulations in mares are multiple ovulations,
twinning occurs in only about 0.5% of the pregnancies that carry to term. The
uterus of the mare apparently cannot support twin fetuses; consequently, most
twin conceptions result in the loss of both embryos. The length of gestation is
about 340 days (approximately 11 months) with usually only one foal being
born. Mares usually come into heat 5–12 days following foaling, and fertile
matings occur at this heat if the mare has recovered from the previous delivery.

TABLE 30.2   Major U.S. Draft, Light Horse, and Pony Breeds Based on Annual Registration Numbers (in thousands), 1980–95

Breed	1995	1990	1985	1980	Year Association Formed
Light horse					
Quarter Horse	110.0	110.6	157.4	137.1	1940
Thoroughbred	35.2	43.6	50.4	39.4	1894
Paint	34.0	22.4	12.7	9.7	1962
Arabian	13.0	17.7	30.0	19.7	1908
Standardbred	11.9	16.6	18.4	15.2	1938
Appaloosa	11.2	10.7	16.2	25.4	1938
Tennessee Walking	6.7	8.0	7.6	6.8	1935
Anglo and Half-Arab	4.2	4.3	10.1	14.3	1950
Pinto	3.8[a]	2.9	3.9	—	1956
Saddlebred	3.2	3.6	4.4	3.9	1891
Morgan	3.0	3.6	4.5	4.5	1907
Paso Fino	1.6[b]	1.6	1.3	.6	1973
Palomino	1.5	2.2	1.2	1.6	1936
National Show Horse	0.6[a]	0.9	0.9	0.6	1982
Draft					
Belgian	3.0	3.7	4.2	0.7	1887
Percheron	2.0	1.3	1.5	0.3	1876
Clydesdale	0.5	0.4	0.2	0.2	1879
Pony					
Miniature	11.0	11.2	—	—	1971
Shetland	—	0.7	0.6	1.0	1888
Welsh	0.7[a]	0.7	0.5	0.9	1906

[a] 1994 registrations.
[b] 1993 registrations.
Source: Adapted from the American Horse Council; *Horse Industry Directory,* 1996; and various breed associations.

Selection

Most breeders know that weaknesses exist either in their entire herd or in some individuals of the herd. Breeders generally attempt to find a stallion that is particularly strong in the trait or traits that need strengthening in the herd. Obviously, it is also necessary to be sure that no other weaknesses are brought into the herd, so it is better to use a stallion with no undesirable traits, even if he is only average in the trait that needs correcting in the herd. To emphasize one trait only for correcting a weakness while bringing another weakness into the herd will mean, after 25–30 years of breeding, little or no improvement.

An effective breeding program is a completely objective one. There is no place in a breeding program for sympathy toward an animal or for personal pets. Every attempt should be made to see that the environment is the same for all animals in a breeding program. A particularly appealing foal that is given special care and training may develop into a desirable animal. However, many foals less appealing in early life might also develop into desirable animals if given special care and training. A desirable animal that has had special care and training may transmit favorable hereditary traits no better than a less desirable animal that has not had this treatment. Selecting a horse for breeding that has had special care and training may in fact be selecting only for special care and training. Certainly, these environmentally produced differences in horses are not inherited and will not be transmitted. In fact, it is often wise to select animals that were developed under the type of environment in which they are expected to perform. If stock horses are being developed for herding cattle in rough, rugged country, selection under such conditions is more desirable than where conditions are less rigorous. Horses that possess inherited weaknesses tend to become unsound in a rugged environment and, as a result, are not used for breeding. Such animals might never show those inherited weaknesses in a less rugged environment.

An ideal environment for most horse-breeding programs has quality forage distributed over an area that requires horses to exercise as they graze. Where the land is limited, highly productive, and valuable, the animals may have to be kept in a small area and forced to exercise a great deal. Forced exercise tends to keep the animals from becoming too fat and gives strength to the feet and legs. Horses need regular, not sporadic, exercise. Regular exercise, even if quite strenuous, is healthy for genetically sound animals and may reveal the weaknesses of those that are not. Strenuous exercise can, however, be harmful to animals that have not previously exercised for a considerable period of time.

An environment should be provided that identifies horses having genetic superiority for their intended performance. For example, horses that are being bred for endurance in traveling should be made to travel long distances on a regular basis to determine if they can remain sound. Horses bred for jumping should be trained to jump as soon as they are physically mature so that those lacking the ability to jump or those that become unsound from jumping can be removed from the breeding program. Draft horses should be trained to pull heavy loads early in life (3 or 4 years of age) to determine their ability to remain sound and their willingness to pull. Performance should be measured before animals are used in a breeding program.

Any horse that is unsound should not be used for breeding regardless of the purposes for which the horse is being bred. Such abnormalities as toeing in or toeing out, sickle hocks, cow hocks, and contracted heels will likely lead to unsoundness and difficulties or lack of safety in traveling. Interference and forging actions are extremely objectionable because they can cause the horse to stumble or fall. Eye, mouth, and respiration defects should be selected against in all horses.

CONFORMATION OF THE HORSE

Body Parts

To understand the conformation of the horse, one should be familiar with the body parts (Fig. 30.2). A more detailed skeletal structure of the horse is shown in Chapter 18.

Feet and Legs

Major emphasis is placed on feet and legs in describing conformation in the horse—for identifying both correctness and conditions of unsoundness. The old saying, "no feet, no horse" is still considered valid by most horse producers. As feet and leg structure is important, a review of skeletal structure of the front leg and hind leg is given in Figs. 30.3 and 30.4.

Appaloosa

Arabian

Morgan

Paint

Miniature

PLATE S.

Breeds of light horses and a pony breed (Miniature). Courtesy of Appaloosa Horse Club, (Appaloosa); International Arabian Horse Association (Arabian); American Morgan Horse Association (Morgan); American Paint Horse Association (Paint); and Thomas Nebbia, © National Geographic Association (Miniature).

Quarter Horse

Standardbred

Thoroughbred

Tennessee Walking Horse

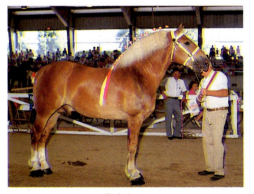

Belgian

PLATE T.
Breeds of light horses and a draft breed (Belgian). Courtesy of the American Quarter Horse Association (Quarter Horse); the U.S. Trotting Association (Standardbred); Gainesway Farm (Thoroughbred); Tennessee Walking Horse Breeders and Exhibitors Association (Tennessee Walking Horse); and the Belgian Draft Horse Corporation (Belgian).

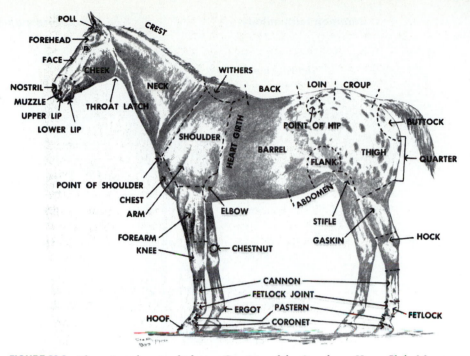

FIGURE 30.2   The external parts of a horse. Courtesy of the Appaloosa Horse Club, Moscow, Idaho.

Figure 30.5 shows the front legs from a front view. From this view, a vertical line from the point of the shoulder should fall in the centers of the knee, pastern, cannon, and foot. Each leg is divided into two equal halves. Deviations from this ideal position are shown with the common terminology associated with them.

The front legs from a side view are shown in Fig. 30.6. A vertical line from the shoulder should fall through the center of the elbow and the center of the foot. The angle of the pastern in the ideal position is 50 to 55°.

Figure 30.7 shows the correct hind leg position from a rear view; the less desirable positions of the feet and legs are also shown. In the ideal position, a vertical line from the point of the buttocks should fall through the centers of the hock, cannon, pastern, and foot.

The ideal position of the hind legs from a side view is shown in Fig. 30.8. Deviations from the ideal position are shown and described. The ideal position is shown where a vertical line from the point of the buttocks touches the rear edge of the cannon from the hock to the fetlock and meets the ground behind the heel.

The Hoof

Care of the horse's feet is essential to keep the horse sound and serviceable. Regular cleaning, trimming, and shoeing (depending on frequency of use) are needed. The hoof will, in mature horses, grow ¼–½ inch per month, so trimming is needed every 6–8 weeks.

The external parts of the hoof are shown in Fig. 30.9.

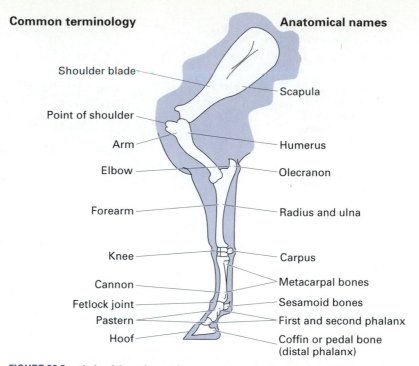

FIGURE 30.3   Skeletal front leg with common terminology and anatomical names. Courtesy of Colorado State University Extension Service Publication MOOOOOG.

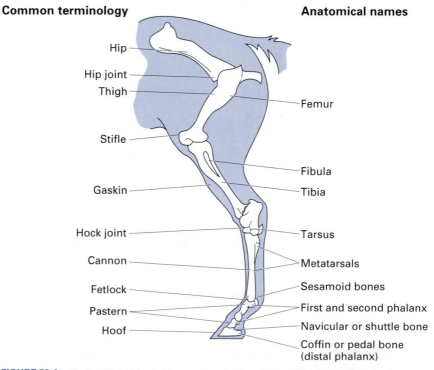

FIGURE 30.4   Skeletal hind leg with common terminology and anatomical names. Courtesy of Colorado State University Extension Service Publication MOOOOOG.

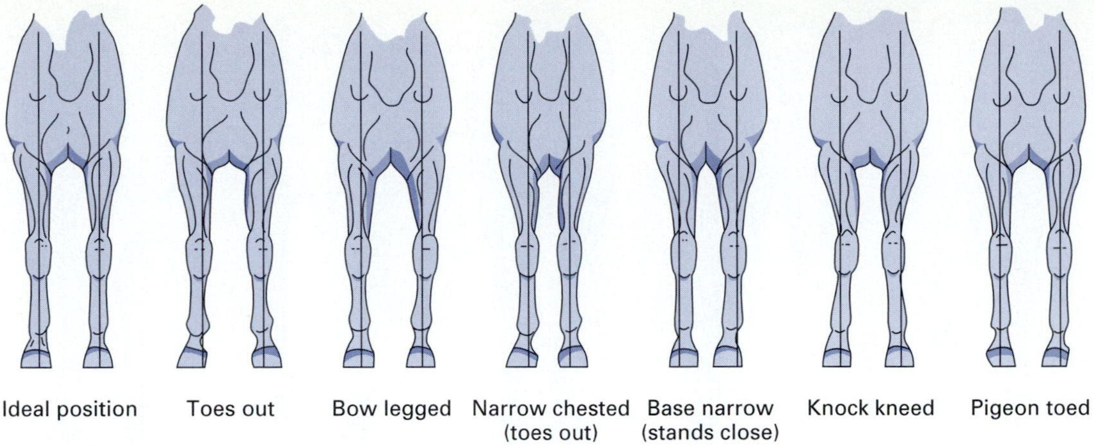

| Ideal position | Toes out | Bow legged | Narrow chested (toes out) | Base narrow (stands close) | Knock kneed | Pigeon toed |

FIGURE 30.5 Correct and faulty conformation of front feet and legs from a front view. Courtesy of Bill Culbertson.

UNSOUNDNESS AND BLEMISHES OF HORSES

Two terms, *unsoundness* and *blemish,* are used in denoting abnormal conditions in horses. An **unsoundness** is any defect that interferes with the usefulness of the horse. It may be caused by an injury or improper feeding, be inherited, or may develop as a result of inherited abnormalities in conformation. A **blemish** is a defect that detracts from the appearance of the horse but does not interfere with its usefulness. A healed wire cut or saddle sore, for instance, may leave a blemish without interfering with the usefulness of the horse.

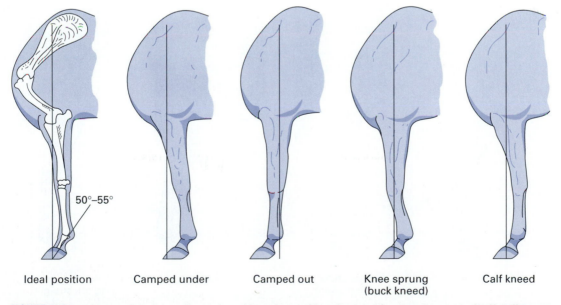

| Ideal position | Camped under | Camped out | Knee sprung (buck kneed) | Calf kneed |

FIGURE 30.6 Correct and faulty conformation of front feet and legs from a side view. Courtesy of Bill Culbertson.

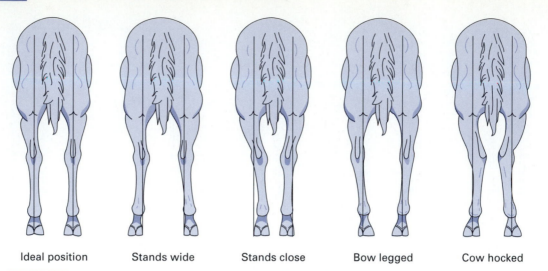

FIGURE 30.7 Correct and faulty conformation of hind feet and legs as shown in a rear view. Courtesy of Bill Culbertson.

Horses may have anatomical abnormalities that interfere with their usefulness. Many of these abnormalities are either inherited directly or develop because of an inherited condition. Abnormalities of the eyes, respiratory system, circulatory system, musculature, and conformation of the feet and legs are all important.

Some of the major unsoundnesses and blemishes include the following (several of which are identified by location on the body in Fig. 30.10).

Bog spavin is a soft swelling on the inner, anterior aspect of the hock. Although unsightly, the condition usually does not cause lameness. **Bone**

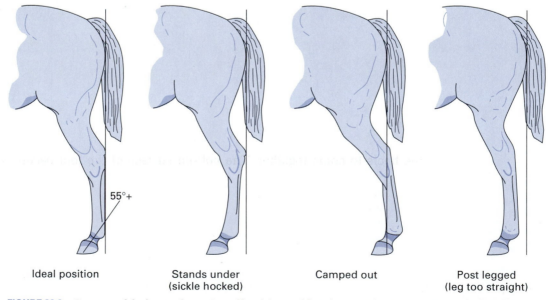

FIGURE 30.8 Correct and faulty conformation of hind feet and legs from a side view. Courtesy of Bill Culbertson.

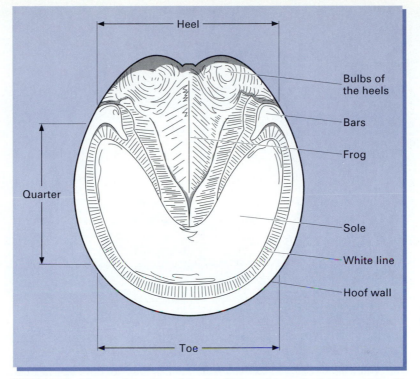

FIGURE 30.9 The conformation of the hoof (viewed from the bottom side) with several parts identified. Courtesy Colorado State University Extension Service Publication MOOOOOG.

spavin is a bony enlargement on the inner aspect of the hock. Both bog and bone spavin arise when stresses are applied to horses that have improperly constructed hocks. Lameness usually accompanies this condition, but most animals return to service after rest and, in some cases, surgery.

Capped hock is a thickening of the skin at the point of the hock. **Curb** is a hard swelling above the cannon bone (a hand's width below the hock). The plantar ligament becomes inflamed and swollen, usually owing to poor conformation or a direct blow.

Cataract is inherited as a dominant trait. It can be eliminated if the afflicted horse does not produce several foals before the cataract develops.

Contracted heels is commonly an inherited conformation defect. Contracted heels may result when the frog or cushion of the foot is damaged and shrinks, allowing the heels to come together. The bottom surface of the foot becomes smaller in circumference than at the coronet band.

The term *cow hocked* (Fig. 30.7) indicates that the points of the hocks turn inward. Such hocks are greatly stressed when the horse is pulling, running, or jumping.

Heaves is a respiratory disease in which the horse experiences difficulty in exhaling air. The horse can exhale a certain volume of air normally, after which an effort is exerted to complete exhalation. A horse with a mild case of heaves can continue with light work, but horses with moderate to severe heaves have a limited ability to work.

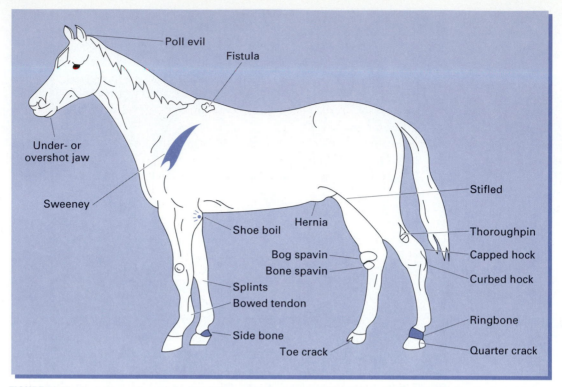

FIGURE 30.10 Locations of several potential unsoundnesses and blemishes. Courtesy of Colorado State University Extension Service Publication MOOOOOG.

Hyperkalemic periodic paralysis (HYPP) is a muscle disease that is genetically controlled. HYPP affects some lines of Quarter Horses, Paints, and Appaloosas. Affected horses experience tremors as a result of abnormal muscle fiber activity. A mutation in the gene responsible for sodium and potassium regulation is apparently the cause. Ideally, those horses who carry the trait should not be allowed to breed.

Laminitis, also called **founder,** is an inflammation of the laminae of the foot causing severe pain and lameness. The front feet are affected more frequently than the hind feet. Overeating of grain, consumption of large amounts of water by overheated horses, and overworking horses on hard surfaces are frequent causes of laminitis.

Moon blindness is periodic ophthalmia in which the horse is blind for a short time, regains its sight, and then again becomes blind for a time. Periods of blindness may initially be spaced as much as 6 months apart. The periods of blindness become progressively closer together until the horse is continuously blind. This condition received the name *moon blindness* because the trait is first noticed when the periods of blindness occur about a month apart. It was originally thought that the periods of blindness were associated with changes in the moon. Causes have been associated with infections, parasites, and riboflavin deficiency.

Navicular disease is an inflammation of the navicular bone in the front foot. The exact cause is unknown, but hard work, small feet, and trimming heels too low may contribute to the development of the disease. Corrective shoeing or surgery (cutting the palmar digital nerve) are recommended treatments, although the latter may not restore horses to their original usefulness.

Quittor is a deep sore that drains at the coronet. The infection is caused by puncture wounds, corns, and the like, and results in severe lameness.

Ring bone is a condition in which the cartilage around the pastern bone ossifies. Ring bone shows as a hard bony enlargement encircling the areas of the pastern joint and coronet.

Ruptured blood vessels is a defect of circulation in which the blood vessels in the lungs are fragile and may rupture when the horse is put under the stress of exercising. Some racehorses have died due to hemorrhage from these fragile blood vessels.

Shoe boil, or **capped elbow,** is a soft swelling on the elbow. Common causes are injury to the elbow while the horse is lying down or getting up or injury from a long heel on a front shoe.

Sickle hocked (Fig. 30.8) is the term used to describe the hock when it has too much set or bend. As a result, the hind feet are set too far forward. The strain of pulling, jumping, or running is much more severe on a horse with sickle hocks than on a horse whose hocks are of normal conformation.

Sidebones is an abnormality that occurs when the lateral cartilages in the foot ossify. During the ossification process, lameness can occur, but some horses regain soundness with proper rest and shoeing.

A **stifled** horse is one in which the patella (the kneecap in humans) has been displaced. Older horses seldom become sound once they are stifled, while younger horses usually recover. There is a surgical operation for this condition.

Stringhalt is an involuntary flexion of the hock during movement. It is considered a nerve disorder. Surgery can improve the condition.

Sweeny refers to atrophied muscles at any location, although many people use it to refer only to shoulder muscles. In the shoulder sweeny, the nerve crossing the shoulder blade has been injured.

A **thoroughpin** is a soft enlargement of the tendon sheath of the large tendon (tendon of Achilles) of the hock and the fleshy portion of the hind leg. Stress on the flexor tendon allows synovial fluid to collect in the depression of the hock. Lameness rarely occurs.

Toeing-in, or **pigeon-toed** (Fig. 30.5), refers to the turning in of the toes of the front feet. **Toeing-out** (Fig. 30.5) refers to the turning out of the toes of the front feet. These conditions influence the way in which the horse will move its feet when traveling. Toeing-out or moving the front feet inward is considered the more serious defect because it can lead to further interference and faults.

Tying-up, or **Equine rhabdomyolysis,** is a condition categorized as either early-form or late-form. Affecting the unconditioned or over-exerted horse, tying-up is characterized by a shuffling gait, heavy sweating, and overly contracted rump and thigh muscles. Severe forms may result in muscular trauma. Prevention involves feeding according to the horses workload and assuring horses are prepared for strenuous work.

Windgalls, sometimes referred to as **wind puffs** or *puffs,* occur when the joint capsules on tendon sheaths around the pastern or fetlock joints are enlarged. The disease is common, but not serious, in hardworking horses.

GAITS OF HORSES

The major gaits of horses, along with their modifications, are as follows:

1. **Walk** is a four-beat gait in which each of the four feet strikes the ground separately from the others.

2. **Trot** is a rapid, diagonal, two-beat gait in which the right front and left rear feet hit the ground in unison, and the left front and right rear feet hit the ground in unison. The horse travels straight without swaying sideways when trotting.

3. **Pace** is a lateral two-beat gait in which the right front and rear feet hit the ground in unison and the left front and rear feet hit the ground in unison. There is a swaying from right to left when the horse paces.

4. **Gallop** is the fastest gait with four beats.

5. **Canter** is a fast three-beat gait. Depending on the lead, the two diagonal legs hit the ground at the same time, with the other hind leg and foreleg hitting at different times.

6. **Rack** is a snappy four-beat gait in which the joints of the legs are highly flexed. The forelegs are lifted upward to produce a flashy effect. This is an artificial gait, whereas the walk, trot, pace, gallop, and canter are natural gaits. The rack is popular in the showring for speed and animation.

7. **Running walk** is the fast ground-covering walk of the Tennessee Walking Horse. It is an artificial gait that is faster than the normal walk. The horse moves with a gliding motion as the hind leg oversteps the forefoot print by 12–18 inches or more.

EASE OF RIDING AND WAY OF GOING

When a horse's foot strikes the ground, a large shock is created that would be objectionable to the rider if no shock absorption occurred. There are several shock-absorbing mechanisms existing in horses' feet and legs. The horse has lateral cartilages on all four feet that expand outward when the foot strikes the ground. This absorbs some of the shock. The pastern on each leg absorbs some of the shock when the foot strikes the ground by bending somewhat. A pastern that is too straight will not absorb much of the shock and one that is too long and weak will let the leg go to the ground. These kinds of pasterns will soon result in unsound horses. Thus, it is very important that a pastern has the proper slope so it can absorb the optimal amount of shock without the leg going to the ground or causing too much concussion on the joints and, ultimately, the rider.

The front legs each have two joints that allow movement and absorb shock: the joint between the ulna and the humerus, and the joint between the humerus and the scapula. Having movement at these joints results in some absorbing of shock. Also, the hind legs each have two joints that bend and, thus, absorb some shock. These joints are between the metatarsus and tibia and between the tibia and femur.

If the horse's feet and legs have proper conformation, a pleasant ride can be enjoyed. If there are abnormalities due to inheritance, injury, improper nutrition, or disease, the horse will give the rider a less pleasurable ride.

Abnormalities in Way of Going

A horse that toes out with its front feet tends to dish or swing its feet inward (wings in) when its legs are in action. Swinging the feet inward can cause the striding foot to strike the supporting leg so that **interference** to forward movement results. A horse that toes in (pigeon-toed) tends to swing its front feet outward, giving a **paddling** action.

Some horses overreach with the hind leg and catch the heel of the front foot with the toe of the hind foot. This action, called **overreaching,** can cause the horse to stumble or fall. **Forging** occurs when the hind foot hits the shoe on the front foot.

Figure 30.11 shows the way of going well as the horse moves straight and true, each foot moving in a straight line. The other illustrations show the path of flight of each foot when the structure of the foot and leg deviates from the desired norm. How the length and slope of the hoof affects way of going is shown in Fig. 30.12.

Since few horses move perfectly true, it is important to know which movements may be unsafe. A horse that wings in (interferes) is potentially more unsafe than a horse that wings out (paddles), as the former horse may trip itself.

DETERMINING THE AGE OF A HORSE BY ITS TEETH

The age of a horse can be estimated by its teeth (Figs. 30.13, 30.14, and 30.15). A foal at 6–10 months of age has 24 baby or milk teeth (12 incisors and 12 molars). The incisors include three pairs of upper and three pairs of lower incisors.

Chewing causes the incisors to become worn. The wearing starts with the middle pair and continues laterally. At 1 year of age, the center incisors show

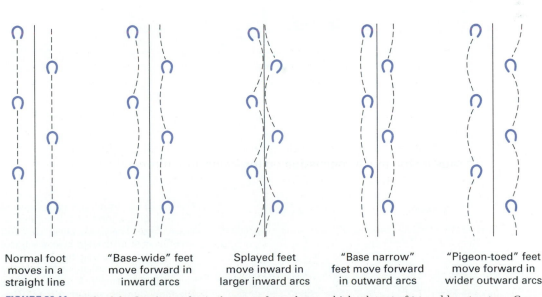

| Normal foot moves in a straight line | "Base-wide" feet move forward in inward arcs | Splayed feet move inward in larger inward arcs | "Base narrow" feet move forward in outward arcs | "Pigeon-toed" feet move forward in wider outward arcs |

FIGURE 30.11 Path of the feet (way of going) as seen from above, which relates to feet and leg structure. Courtesy of Colorado State University Extension Service Publication MOOOOOG.

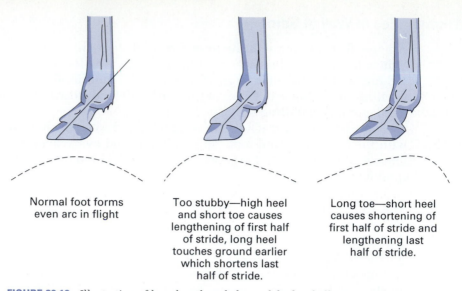

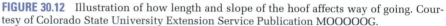

Normal foot forms
even arc in flight

Too stubby—high heel
and short toe causes
lengthening of first half
of stride, long heel
touches ground earlier
which shortens last
half of stride.

Long toe—short heel
causes shortening of
first half of stride and
lengthening last
half of stride.

FIGURE 30.12   Illustration of how length and slope of the hoof affects way of going. Courtesy of Colorado State University Extension Service Publication MOOOOOG.

wear; at 1.5 years, the intermediates show wear; and at 2 years, the outer, or lateral, incisors show wear. At 2.5 years, shedding of the baby teeth starts. The center incisors are shed first. Thus, at 2.5 years, the center incisors become permanent teeth; at 4 years, the intermediates are shed; at 5 years, the outer, or lateral, incisors are shed and replaced by permanent teeth.

6 days

1 year

6 weeks

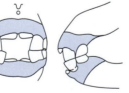

2½ years

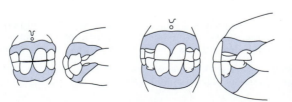

FIGURE 30.13   Front and side views of the teeth of the horse from 6 days to 3.5 years of age. The young horse has a full set of baby or milk teeth by 6 months of age. It starts shedding the baby teeth and developing permanent teeth at 2.5 years of age. From J. F. Bone, *Animal Anatomy and Physiology*, 5th ed. (Corvallis, OR: State University Book Stores, © 1988).

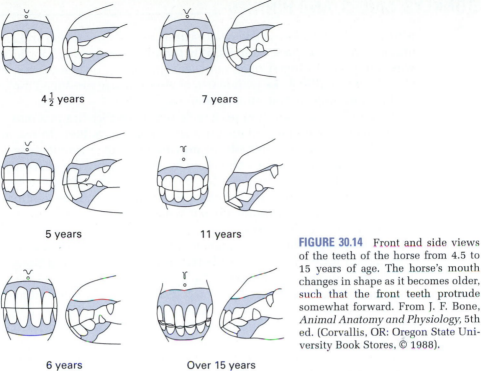

FIGURE 30.14 Front and side views of the teeth of the horse from 4.5 to 15 years of age. The horse's mouth changes in shape as it becomes older, such that the front teeth protrude somewhat forward. From J. F. Bone, *Animal Anatomy and Physiology,* 5th ed. (Corvallis, OR: Oregon State University Book Stores, © 1988).

A horse at 5 years of age is said to have a **full mouth,** because all the teeth are permanent. At 6 years, the center incisors show wear; at 7 years, the intermediates show wear; and at 8 years, the outer, or lateral, incisors show wear. Wearing is shown by a change from a deep groove to a rounded dental cup on the grinding surface of a tooth.

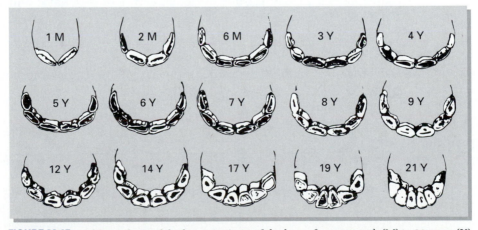

FIGURE 30.15 Table surfaces of the lower incisors of the horse from 1 month (M) to 21 years (Y) of age. From J. F. Bone, *Animal Anatomy and Physiology,* 5th ed. (Corvallis, OR: Oregon State University Book Stores, © 1988).

DONKEYS, MULES, AND HINNIES

Mules and hinnies have been used as draft animals in mining and farming operations and also as pack animals. When mules were needed for heavy loads, it was important to breed mammoth jacks to mares of one of the draft breeds. Crossing of smaller jacks with mares of medium size produced mules that were useful in mining and farming operations.

Donkeys, mules, and hinnies have some characteristics that make them more useful than horses for certain purposes. Their sure-footedness, for instance, makes them ideal pack animals for moving loads over rough areas. In addition, they are rugged, can endure strenuous work, and have the characteristic of taking care of themselves (e.g., mules do not gorge themselves when given free access to grain, so they do not normally founder from overeating). However, donkeys, mules, and hinnies do not respond to the wishes or commands of humans as well as do horses.

Recently, there has been a movement to use small donkeys as pets for children. Small mules from crosses between male donkeys and pony mares have also been popular as children's pets.

The primary reason trucks and tractors replaced horses and mules in farming operations was that work could be done more rapidly; one person could do much more with tractors and trucks than with horses and mules. The high cost of petroleum products (gasoline and diesel fuels) and the possible scarcity of these items may result in more use of horses and mules in farming, logging, and mining operations and a partial return to horsepower. It is unlikely that horses and mules will ever replace trucks and tractors, but certain operations on farms and in the timber industry may be done to advantage by horses and mules.

CHAPTER SUMMARY

- Horses are typically classified as light, draft, or pony breeds.
- Based on registration numbers, the Quarter Horse is the most popular horse breed.
- Mares have a 21-day estrous cycle, an estrus of 5–7 days, and a gestation length of 350 days.
- Unsoundness in horses leads to diminished usefulness and should be strictly selected against.
- Gaits of horses include the walk, trot, pace, gallop, canter, rack, and running walk.

REVIEW QUESTIONS

1. What is the most popular breed of horse in the United States today?

2. What is the length of the estrous cycle of mares?

3. What is the duration of gestation in horses?

4. *True or False:* Stallions for a breeding program should always be selected based on their ability to correct and improve upon a weakness in a herd, without consideration to new weaknesses which they may possess and introduce into the herd.

5. What is unsoundness?

6. What is a blemish?

7. What are the seven major gaits of horses?

8. How are a trot and pace similar? How do they differ?

9. *True or False:* The age of horses can readily be estimated by examining its teeth.

10. When is a horse said to have a full mouth?

11. What characteristics of donkeys, mules, and hinnies make them more desirable than horses for certain purposes?

12. What characteristics of donkeys, mules, and hinnies make them less desirable than horses?

13. What was the primary reason that mechanized vehicles replaced horses and mules in farming operations?

14. What event would likely spur the return to power from horses and mules for certain operations in the farming, timber, and mining industries?

SELECTED REFERENCES

Publications

Blakely, R. L. 1985. Miniature horses. *National Geographic* 167:384.

Bone, J. F. 1988. *Animal Anatomy and Physiology.* 5th ed. Corvallis, OR: Oregon State University Book Stores.

Bowling, A. 1996. *Horse Genetics.* Oxon, United Kingdom: CAB International.

Butler, D. 1985. *Principles of Horseshoeing II.* Grand Prairie, TX: Equine Research Inc.

Ensminger, M. E. 1990. *Horses and Horsemanship.* Danville, IL: Interstate Publishers, Inc.

Evans, J. W. 1989. *Horses: A Guide to Selection, Care and Enjoyment.* 4th ed. San Francisco: W. H. Freeman.

Evans, J. W. (ed.). 1992. *Horse Breeding and Management.* Amsterdam: Elsevier Science Publishers B.V.

Evans, J. W., Borton, A., Hintz, H. F., and Van Vleck, L. D. 1990. *The Horse.* 2d ed. San Francisco: W. H. Freeman.

Harris, S. 1993. *Horse Gaits, Balance and Movement.* Grand Prairie, TX: Equine Research Inc.

Horse Industry Directory. 1996. Washington, DC: American Horse Industry Council.

Jones, W. E. 1982. *Genetics and Horse Breeding.* Philadelphia: Lea & Febiger.

Lasley, J. F. 1987. *Genetics of Livestock Improvement.* Englewood Cliffs, NJ: Prentice-Hall.

Rich, G. A. 1981. *Horse Judging Guide.* Colorado State University Extension Service Publication MOOOOOG.

Stashak, T. S. (ed.). 1987. *Adam's Lameness in Horses.* 4th ed. Grand Prairie, TX: Equine Research Inc.

Visuals

Basic Horseshoeing Principles (55-min. video). Laporte, CO: Butler Publishing.

Beginner's Guide to Buying a Horse (sound filmstrip). Vocational Education Productions, California Polytechnic State University, San Luis Obispo, CA 93407.

Care of Your Horse's Feet (SL529—slide set). Educational Aids, National 4-H Council, 7100 Connecticut Ave., Chevy Chase, MD 20815.

Foaling Your Mare (VHS video; 50 min; 1988); *Foaling Fundamentals* (VHS video; 60 min; 1988); *Conformation* (VHS video); *Broodmare Management* (VHS video); *Stallion Management* (VHS video). Equine Research Inc., P.O. Box 535547, Grand Prairie, TX 75053.

Horse Breeds and Color Types (slides). Vocational Education Productions, California Polytechnic State University, San Luis Obispo, CA 93407.

Horse Judging (21 videotapes). CEV, P.O. Box 65265, Lubbock, TX 79464-5265.

Survival of the Fittest (two-part 16-mm film; 26 min. each). Covers Quarter Horse conformation—the relation of form to function. American Quarter Horse Association, Amarillo, TX 79168.

CHAPTER 31

Feeding and Managing Horses

Horses relate well to people and provide many forms of pleasure. Although many people enjoy horses, most own only a few; however, whether a person keeps only one horse for personal recreation or operates a large breeding farm, basic horse knowledge is vital. Information on managing horses, such as feeding, facilities, disease prevention, and parasite control, is important. Additional material on management is discussed in Chapter 36.

FEEDS AND FEEDING

Horses graze on grass and legume pastures and eat roughages such as hay and grains of all kinds. Concentrate mixtures containing grains, protein supplements, and vitamin and mineral additives are prepared and sold by commercial feed companies, especially to owners with only one or two horses. Wet molasses can be added to these concentrate mixtures to make sweet feed. The forage and concentrate mixtures may be mixed and pelleted to make a complete, higher-priced convenience feed for horse owners.

Although horses spend less time chewing than ruminants, the normal, healthy horse with a full set of teeth can grind grains such as oats, barley, and corn so that cracking or rolling these feeds is unnecessary. Wheat and milo, however, should be cracked to improve digestibility. The horse's stomach is relatively small, composing only 10% of the total digestive capacity. Only a small amount of digestion takes place in the stomach, and food moves rapidly to the small intestine. From 60–70% of the protein and soluble carbohydrates are digested in the small intestine, and about 80% of the fiber is digested in the cecum and colon. The large intestine has about 60% of the total digestive capacity, with the colon being the largest component. Bacteria that live in the cecum aid digestion there. Minerals, proteins as amino acids, lipids, and readily available carbohydrates such as glucose are absorbed in the small intestine.

Owners of pleasure horses are interested in having them make a desirable appearance, so many liberally feed their horses. Perhaps more horses are overfed than are underfed. Also, many people want to be kind to their animals, keeping them housed in a box stall when weather conditions are undesirable. This may not be best for the physiological state of the horse. Certainly, if any deficiency exists in the feed provided, such a deficiency is much more likely to affect horses that are not running on good pasture, where they can forage for themselves.

Young, growing foals should be fed well to allow for proper growth, but overfeeding and obesity are discouraged. Quality of protein and amounts of protein, minerals, and energy are important for young growing horses. Soybean meal or dried milk products in the concentrate mixture provide the amino acids and minerals that might otherwise be deficient or marginal in the weanling ration.

Good-quality pasture or hay supplemented with grain can provide the nutrition needed by young horses. An appropriate salt-mineral source should be provided at all times and clean water is essential (Fig. 31.1).

During the first 8 months of gestation, pregnant mares perform well on good pastures or on good-quality hay supplemented with a small amount of grain and an appropriate source of salt-mineral. As the fetus grows during the last trimester of pregnancy, the mare requires more concentrate and less fibrous, bulky hay. Oats, corn, or barley make an excellent feed grain for pregnant mares. Pregnant mares should be in good body condition but not obese, as with any horse. Animals used for riding or working, whether they are pregnant or not, need more energy than those not working. Horses being exercised

FIGURE 31.1 Horses need a readily available supply of fresh, clean water. Courtesy of Robert Henderson, Oregon Agricultural Experimental Station.

heavily should be fed ample amounts of concentrates to replace the energy used in the work.

Generally, lactating mares require more grain feeding than do geldings and pregnant or nonpregnant mares. Lactation is the most stressful nutritional period for a mare. If lactating mares are exercised, they must be fed additional grain and hay to meet the nutrient demand of the physical activity.

Stallions need to be fed as working horses during the breeding season and given a maintenance ration during the nonbreeding season. Feeding good-quality hay with limited amounts of grain is usually sufficient for the stallion.

Feed companies provide properly balanced rations for horse farms of all sizes. An owner who has only one or two horses may find it highly advantageous to use a prepared feed, since it is difficult and laborious to prepare balanced rations for only a few animals. Using commercially prepared feeds or custom-blended rations can prevent nutritional errors, save time, and, for larger farms, be more cost-effective.

Table 31.1 shows some examples of horse rations and describes when and how these rations should be fed.

MANAGING HORSES

Proper management of horses is essential at several critical times: the breeding season, foaling, weaning of foals, castration, and during strenuous work.

Breeding Season

Mares should be **teased** daily with a stallion to determine the stage of their estrous cycle (Fig. 31.2). When the mare is in standing heat (estrus), she is ready to be bred. When the stallion approaches the front of the mare, the mare reacts violently against the stallion if not in heat, but squats and urinates with a winking of the vulva if in heat. To ensure safety for the mare, stallion, and handlers, breeding hobbles (Fig. 31.3) should be fitted to the mare.

Cleanliness is paramount for both natural breedings and artificial insemination (AI) programs. Mares bred naturally should have the vulva washed and dried and the tail wrapped before being served by the stallion (Fig. 31.4). If the mare can be bred twice during heat without overusing the stallion, breeding 2 and 4 days after the mare is first noticed in heat is desirable. A mature stallion can serve twice daily over a short time and once per day over a period of 1 or 2 months. A young stallion should be used lightly at about three or fewer services per week. In an AI program, the stallion can be collected every other day to cover as many mares as possible, depending on the stallion's sperm numbers and motility. AI programs allow better management of the stallion.

Foaling Time

Mares normally give birth to foals in early spring, at which time the weather can be unpleasant. A clean box stall that is bedded with fresh straw should be made available. If the weather is pleasant, mares can foal on clean pastures. When a mare starts to foal, she should be observed carefully but not disturbed unless assistance is necessary. If the head and front feet of the foal are being

TABLE 31.1   Sample Rations for Horses of Different Ages and in Various Stages of Production

Creep feed for nursing foals. The grain should be fed at the rate of 0.5–0.75 lb of grain/100 lb body weight.

	Percentage in Grain Mix
Corn, rolled or flaked	34.0%
Oats, rolled or flaked	34.0
Soybean meal	22.0
Molasses	6.0
Dicalcium phosphate	2.0
Limestone	1.5
Trace-mineral salt	0.5

Grain mix for weanlings. The grain should be fed at the rate of 0.75–1.5 lb of grain/100 lb body weight. Select the grain according to the type of roughage fed. Allow free-choice consumption of either roughage type.

	Percentage in Grain Mix	
	Alfalfa Hay	Grass Hay
Corn, rolled or flaked	40.0%	34.0%
Oats, rolled or flaked	40.0	34.0
Soybean meal	12.0	23.0
Molasses	5.0	5.0
Dicalcium phosphate	2.5	3.0
Limestone	0	0.5
Trace-mineral salt	0.5	0.5

Grain mix for yearlings and mares. The grain should be fed at the rate of 0.5–1.0 of grain/100 lb of body weight. Select the grain according to the type of roughage fed. Allow free-choice consumption of either roughage.

Mares during late gestation and lactation can be fed the same grain mixes as yearlings. The grains should be fed at the rate of 0–0.5 lb/100 lb of body weight. Roughage consumption can vary from 1.5–2.5 lb/100 lb of body weight.

	Percentage in Grain Mix	
	Alfalfa Hay	Grass Hay
Corn, rolled or flaked	46.5%	38.0%
Oats, rolled or flaked	46.5	38.0
Soybean meal	0	15.5
Molasses	5.0	5.0
Dicalcium phosphate	1.5	2.0
Limestone	0	1.0
Trace-mineral salt	0.5	0.5

Grain mix for horses at maintenance and at work, dry mares during the first 8 months of gestation, or stallions. The grain should be fed as needed (to maintain body condition). Roughage consumption can vary from 1.5–2.5 lb/100 lb of body weight.

	Percentage in Grain Mix	
	Alfalfa Hay	Grass Hay
Corn	46.5%	46.5%
Oats	46.5	46.5
Molasses	5.0	5.0
Dicalcium phosphate	0	1.5
Monosodium phosphate	1.5	0
Trace-mineral salt	0.5	0.5

Source: Ginger Rich. *Horse Judging Guide,* Colorado State University Extension Service Publication, MOOOOOG.

FIGURE 31.2 Chute teasing. As the stallion is led along a chute holding several mares, the behavior of each mare determines whether she is in estrus. Courtesy of Colorado State University.

presented, it should be delivered without difficulty (Fig. 31.5). If necessary, however, a qualified person can assist by pulling the foal as the mare labors. Do not pull when the mare is not laboring and do not use a tackle to pull the foal. If the front feet are presented but not the head (Fig. 31.6), it may be necessary to push the foal back enough to get the head started along with the front feet. Breech presentations can endanger the foal if delivery is delayed; therefore,

FIGURE 31.3 Breeding hobbles. From R. A. Battaglia and V. B. Mayrose, *Handbook of Livestock Management Techniques* (New York: Macmillan, 1981).

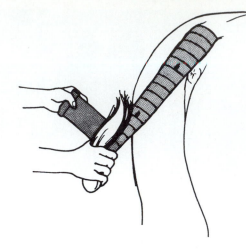

FIGURE 31.4 Tail of horse being wrapped prior to breeding. From R. A. Battaglia and V. B. Mayrose, *Handbook of Livestock Management Techniques* (New York: Macmillan, 1981).

assistance should be given to help the mare make a rapid delivery if breech presentation occurs. If it appears that difficulties are likely to occur, a veterinarian should be called as soon as possible.

As soon as the foal is delivered, its mouth and nostrils should be cleared of membranes and mucus so it can breathe. If the weather is cold, the foal should be wiped dry and assisted in nursing. As soon as the foal nurses, its metabolic rate increases and helps it stay warm. The **umbilical cord** should be dipped in a tincture of iodine solution to prevent harmful microorganisms from invading the body.

In general, it is highly desirable to exercise pregnant mares up to the time of foaling. Mares that are exercised properly while pregnant have better muscle tone and are likely to experience less difficulty when foaling than those that get little or no exercise while pregnant. During the horsepower era, many mares used for plowing and similar work foaled in the field without difficulty. The foal was usually left with the mare for a few days, after which the foal was left in a box stall while the mare was worked.

Weaning the Foal

When weaning time arrives, it is best to remove the mare and allow the foal to remain in the surroundings to which it is accustomed. Since the foal will make

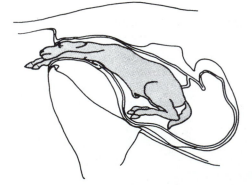

FIGURE 31.5 Correct presentation position of foal for delivery. From R. A. Battaglia and V. B. Mayrose, *Handbook of Livestock Management Techniques* (New York: Macmillan, 1981).

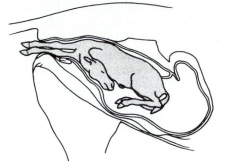

FIGURE 31.6 Malpresentation of foal for delivery; head and neck back. From R. A. Battaglia and V. B. Mayrose, *Handbook of Livestock Management Techniques* (New York: Macmillan, 1981).

every attempt to escape to find its mother, it should be left in a box stall or a secure and safe fenced lot. Fencing other than barbed wire should be used. Prior to weaning, the foal should become accustomed to eating and drinking on its own. High-quality hay or pasture and a balanced concentrate feed should be provided for the foal.

Castration

Colts that are not kept for breeding can be castrated any time after the testicles have descended; however, the stress of castration should not be imposed at weaning time. Some people prefer to delay castration until the colt has reached a year of age, whereas others prefer an earlier time. Genetic potential and nutrition, not time of castration, determine mature height and weight. Castration is performed because geldings can be handled more safely than stallions by owners and around other people.

The colt can be put onto a turntable or manually restrained with ropes for castration. Pain killers and muscle relaxants can be used. The scrotum is opened on each side by a qualified person and the membrane around each testicle is split to expose the testicle. The testicle is pulled from the body cavity far enough to expose the cord for clamping. The cord is crushed by the clamp and then severed. The crushing of the cord prevents excessive bleeding. It may be desirable to give the colt an injection of an antibiotic to prevent infection. If the scrotum was cleaned with a mild disinfectant before castration, the wound need not be washed with an antiseptic. Harsh disinfectants should not be applied to the wound. After the colt is castrated, he should be in a small, clean pasture for close observation. If fly infestation is a problem, a fly repellant should be applied about the scrotal area.

Horses can be permanently identified by tattooing them on the inside of the upper lip. This does not disfigure the animals and can be easily read by raising the upper lip. Other methods of identification include freeze branding, hot-iron branding, electronic implants, and chestnut implants.

Care of Hardworking Animals

Hardworking horses that are sweating and have elevated temperature, pulse, and respiration rates should be washed (cooled down) before receiving water and feed. The animals should be given only small amounts of water and walked until they have stopped sweating and their pulse and respiration are within

10% of resting levels. At that time, water, hay, and grain can be given. This handling process is important; otherwise, horses may become colicky or founder.

Hardworking horses do require extra energy, usually in the form of grain. However, care must be taken to avoid overfeeding grain to horses. Excess grain and, sometimes, lush green pasture can cause a horse to founder. Founder can cause death or severe lameness.

HOUSING AND EQUIPMENT

Barns should be located on a higher elevation than the surrounding area to assure good drainage. They should be accessible to utilities and vehicles and, preferably, have a southeast exposure.

Although many styles of barns exist, one constructed with an aisle between two rows of stalls provides easy access and efficient use of space. Feed can be placed in the stalls on either side as the feed cart goes down the aisle. If the stalls can be opened from the outside and cleaned with mechanical equipment, labor also is saved. The hay manger in stalls should be constructed at chest height to the horse. Hay racks placed above the horse's head force the horse to inhale hay dust when reaching for the hay, and health problems can result.

To alleviate mixing hay with grain, a separate grain feeder should be placed several feet from the hay rack. The waterer or water bucket should be located along the outside wall for drainage purposes. The water source should also be placed some distance from the hay and grain so feed will not drop in the water. Electric waterers that are float-controlled save labor but are prone to malfunction. Although water buckets require more labor, they allow water intake to be easily monitored.

Box stalls are used for foaling and for the mare and the foal when the weather is severe. The stall should be constructed to allow complete cleaning and proper drainage. Minimum dimensions for a foaling stall are 14 × 14 ft, whereas a regular box stall should be at least 10 × 10 ft.

Feed should be stored in an area connected with or adjacent to the barn. Truck access to the feed storage area is vital. The floor should be solid (e.g., concrete) and the rodent-proof bins should be easily cleaned. Overhead (loft) storage of hay requires a great deal of labor and expensive construction. Large amounts of hay stored in the barn where the horses are housed increases dust and danger of fire.

Lot **fences** should be constructed with wooden or steel posts and cable, or with wooden posts and 2-inch lumber. Barbed wire should not be used because it can cut a horse severely.

A dry, dust-free **tack room** is needed to house riding equipment, including saddles, blankets, bridles, and halters. In addition, it may be advisable to have a **washroom** where horses can be washed.

A **chute** is essential to care for horses that have been injured and need attention. If the horse is in a chute, it is unlikely that the animal can kick the person working around the horse. The chute can be used for AI, teasing, palpation, or treating health problems. The chute is also useful for injections or when blood samples are being taken. Often, horses strike with a front foot when a needle is inserted for vaccinations or blood samples; a properly constructed chute will protect people as well as the horse.

CONTROLLING DISEASES AND PARASITES

Sanitation is of vital importance in controlling diseases and parasites of horses. Horses should have clean stalls and be groomed regularly. A horse that is to be introduced into the herd should be isolated for a month to prevent exposing other horses to diseases and parasites. Horses may have illnesses caused by bacteria or viruses, internal or external parasites, poisonous plants, nitrates, monensin, moldy grains, etc.

The person who owns and cares for one horse or a dozen horses should engage the services of a trusted veterinarian. Veterinarians can help prevent diseases as well as treat animals that are diseased. Most veterinarians prefer to assist in preventing health problems rather than treating the animals after they are ill.

Horse manure is an excellent medium for microorganisms that cause **tetanus.** Horses should be given shots to prevent tetanus, which can develop if an injury allows tetanus-causing microorganisms to invade through the skin. Usually two shots are given to establish **immunity,** after which a booster shot is given each year. Because the same microorganisms that cause tetanus in horses also affect humans, those who work with horses should also have tetanus shots.

Strangles, also known as *distemper,* is a bacterial disease that affects the upper respiratory tract and associated lymph glands. High fever, nasal discharge, and swollen lymph glands are signs of strangles. The disease is spread by contamination of feed and water. Afflicted horses must be isolated and provided clean water and feed. A strangles bacterin is available; however, there are many strains of the disease.

Brood mares are subject to many infectious agents that invade the uterus and cause **abortion;** examples include *Salmonella* and *Streptococcus* bacteria and the viruses of **rhinopneumonitis** and **arteritis.** Should abortion occur, professional assistance should be obtained to determine the specific cause and to develop a preventative plan for the future.

Sleeping sickness, or **equine encephalomyelitis,** is caused by viral infections that affect the brain of the horse. Different types, such as the eastern, western, and Venezuelan, are known. Encephalomyelitis is transmitted by vectors such as mosquitoes. It can also be spread by horses rubbing noses together or sharing water and feed containers. Vaccination against the disease consists of two intradermal injections spaced a week to 10 days apart. These injections should be given in April on an annual basis.

Influenza is a common respiratory disease of horses. The virus that causes influenza is airborne, so frequent exposure may occur where horses congregate. The acute disease causes high fever and a severe cough when the horse is exercised. Rest and good nursing care for 3 weeks usually gives the horse an opportunity to recover. Severe aftereffects are rare when complete rest is provided. Horse owners who plan for shows should vaccinate for influenza each spring. Two injections are required the first year, with one annual booster thereafter.

Horses can become infested with internal and external parasites. Control of internal parasites consists of rotating horses from one pasture to another, spreading manure from stables on land that horses do not graze, and treating infested animals.

Pinworms develop in the colon and rectum from eggs that are swallowed as the horse consumes contaminated feed or water. These parasites irritate the anus, which causes the horse to rub the base of its tail against objects even to the point of wearing off hair and causing skin abrasions. Pinworms are controlled by oral administration of proper vermifuges.

Bots are the larvae stage of the bot fly. The female bot fly lays eggs on the hairs of the throat, front legs, and belly of the horse. The irritation of the bot fly causes the horse to lick itself. The eggs are then attached to the tongue and lips of the horse, where they hatch into larvae that burrow into the tissues. The larvae later migrate down the throat and attach to the lining of the stomach, where they remain for about 6 months and cause serious damage. A paste wormer is available to control bots.

Adult **strongyles** (bloodworms) firmly attach to the walls of the large intestine. The adult female lays eggs that pass out with the feces. After the eggs hatch, the larvae climb blades of grass where they are swallowed by grazing horses. The larvae migrate to various organs and arteries where severe damage results. Blood clots, which form where arteries are damaged, can break loose and plug an artery. Strongyles are treated with Ivermectin and other wormers.

Adult **ascaris worms** are located in the small intestine. The adult female produces large numbers of eggs that pass out with the feces. The eggs become infective if they are swallowed when the horse eats them while grazing. The eggs hatch in the stomach and small intestine, and the larvae migrate into the bloodstream and are carried to the liver and lungs. The small larvae are coughed up from the lungs and swallowed. When they reach the small intestine, they mature and produce eggs. These large worms may cause an intestinal blockage in young horses. The same chemicals used for the control of strongyles are effective in the control of ascarids.

Horses should be dewormed at least twice a year and up to four times a year. The type of wormer should be rotated every second time.

CHAPTER SUMMARY _____

■ The major digestive structures of the equine include the stomach, small intestine, cecum, and colon.

■ Nutrients should be provided to the horse in accordance with its requirements as determined by age, work schedule, lactation status, and gestation status.

■ Key management areas for horses include reproduction, weaning, housing, and health.

REVIEW QUESTIONS _____

1. Where is fiber digested in the gastrointestinal tract of a horse?
2. *True or False:* Most horses are overfed.
3. Horses being used for riding or working need more _____ in their diets than those not working.
4. At what critical times is proper management of horses essential?
5. How is the stage of the estrous cycle determined in mares?
6. When do mares normally give birth to foals?

7. *True or False:* Mares should not be worked or exercised during gestation to prevent injuring the fetus.

8. Why is castration of males performed on horses which are not to be used in a breeding program?

9. How can horses be permanently identified without disfiguring them?

10. Hardworking horses that are sweating and have elevated temperature, pulse, and respiration rates should be _____ before receiving water and feed.

11. What disorder may result from overfeeding grain or lush green pasture to horses?

12. _____ is of vital importance in controlling diseases and parasites of horses.

13. *True or False:* Because horse manure is an excellent medium for the microorganism that causes tetanus, and the tetanus-causing microorganism is the same for horses and humans, horses and those who work with horses should be immunized against tetanus.

14. How can internal and external parasites of horses be controlled?

15. What are four common internal parasites of horses which should be controlled?

SELECTED REFERENCES

Publications

Ambrosiano and Harcourt. 1989. *Complete Plans for Building Horse Barns Big and Small.* Grand Prairie, TX: Equine Research, Inc.

Battaglia, R. A., and Mayrose, V. B. 1981. *Handbook of Livestock Management Techniques.* New York: Macmillan.

Cunha, T. J. 1991. *Horse Feeding and Nutrition.* 2d ed. San Diego: Academic Press.

Ensminger, M. E. 1990. *Horses and Horsemanship.* Danville, IL: Interstate Publishers, Inc.

Ensminger, M. E., Oldfield, J. E., and Heinemann, W. W. 1990. *Feeds and Nutrition.* Clovis, CA: Ensminger Publishing Co.

Evans, J. W. 1989. *Horses: A Guide to Selection, Care, and Enjoyment.* San Francisco: W. H. Freeman.

Evans, J. W., Borton, A., Hintz, H. F., and Van Vleck, L. D. 1990. *The Horse.* 2d ed. San Francisco: W. H. Freeman.

Hickman, J. 1987. *Horse Management.* San Diego: Academic Press.

Loving, N. S. 1993. *Veterinary Manual for the Performance Horse.* Grand Prairie, TX: Equine Research, Inc.

National Research Council. 1989. *Nutrient Requirements of Horses.* Washington, DC: National Academy Press.

Shideler, R. K., and Voss, J. L. 1984. *Management of the Pregnant Mare and Newborn Foal.* Colorado State University Expt. Sta. Spec. Series 35.

Voss, J. L., and Pickett, B. W. *Reproductive Management of the Broodmare.* Colorado State University.

Visuals

Basic Horse Care (videotape). CEV, P.O. Box 65265, Lubbock, TX 79464-5265.

Breaking and Training the Western Horse (sound filmstrip). Vocational Education Productions, California Polytechnic State University, San Luis Obispo, CA 93407.

Feeding Horses (VHS video) and *Basic Horsemanship,* vol. II (VHS video). VEP, California Polytechnic State University, San Luis Obispo, CA 93407.

Tips for Safe Horse Handling (VHS video; 24 min.; 1990). Equine Research, Inc., P.O. Box 535547, Grand Prairie, TX 75053-9941.

CHAPTER 32

Poultry Breeding, Feeding, and Management

The term *poultry* applies to chickens, turkeys, geese, ducks, pigeons, peafowls, and guineas. The head and neck characteristics that distinguish several poultry types are shown in Fig. 32.1. Turkeys have some unusual identifying characteristics including a beard—a black lock of hair on the upper chest of the male turkey. They also have a **caruncle**—a red-pinkish fleshlike covering on the throat and neck, with the **snood** hanging over the beak.

BREEDS AND BREEDING

Characteristics of Breeds

Chickens. Chickens are classified according to class, breed, and variety. A *class* is a group of birds that has been developed in the same broad geographical area. The four major classes of chickens are American, Asiatic, English, and Mediterranean. A *breed* is a subdivision of a class composed of birds of similar size and shape. Some important breeds, strains, lines, and synthetics are shown in Fig. 32.2. The parts of a chicken are shown in Fig. 32.3. A *variety* is a subdivision of a breed composed of birds of the same feather color and type of comb.

Factors such as egg numbers, eggshell quality, egg size, efficiency of production, fertility, and hatchability are most important to commercial egg producers. Broiler producers consider such characteristics as white plumage color and picking quality, egg production, fertility, hatchability, growth rate, carcass quality, feed efficiency, livability, and egg production. Also, breeds, strains, and lines that cross well with each other are important to broiler breeders.

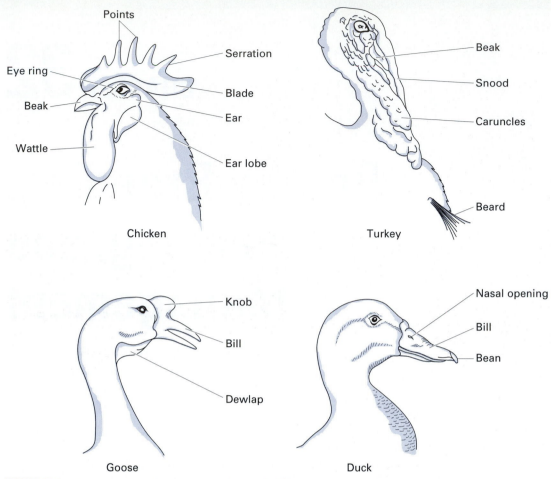

FIGURE 32.1   Distinguishing characteristics of several different types of poultry.

The breeds of chickens listed in Table 32.1 were developed many years ago. These specific breeds are not easily identified in the commercial poultry industry because the breeds have been crossed to produce different varieties and strains (see Fig. 32.2). Some breeds are exhibited at shows or propagated for specialty marketing and are more a novelty than part of today's poultry industry.

Turkeys.   There are eight varieties of turkeys, but most turkeys grown for meat are white. The eight varieties are (1) Bronze, (2) Narragansett, (3) White Holland, (4) Black, (5) Slate, (6) Bourbon Red, (7) Small Beltsville White, and (8) Royal Palm. Two types of turkeys are commercially important in the United States today: the small White and the large Broad White. Since the 1950s, the emphasis has been to produce more large, white turkeys of the type shown in Fig. 32.4. A corresponding decrease in the number of small whites has occurred.

Ducks.   Many breeds of domestic ducks are known, but only a few are of economic importance. The most popular breed in the United States is the White Pekin. It is most valuable for its meat, and produces excellent carcasses at 7–8

PURE BREEDS

Layers (White Eggs)

Leghorn

EXAMPLES OF CURRENT SYNTHETICS,
STRAINS OR LINES DEVELOPED FROM
ONE OR MORE OF THE PURE BREEDS

DEKALB XL·Link

H&N "Nick Chick"

Layers (Brown Eggs)

Rhode Island Red

Plymouth Rock

Hubbard Golden Comet

Meat Type (Broilers)

New Hampshire

Cornish

Indian River Broilers
Male and Female (Variety FS99)

FIGURE 32.2   Breeds, synthetics, lines, or strains of chickens. Courtesy of Dekalb Poultry Research, H&N International, Hubbard Farms, and Watt Publishing Company.

weeks of age. Another breed, the Khaki Campbell, is used commercially in some countries for egg production.

Geese.   Several breeds of geese are popular. The Embden, Toulouse, White Chinese, and Pilgrim are all satisfactory for meat production. The characteristics preferred by most commercial geese producers include a medium-sized carcass, good livability (low mortality), rapid growth, and a heavy coat of white or nearly white feathers. The Embden and White Chinese breeds meet these requirements. One variety of the Toulouse is gray; another variety is buff. The

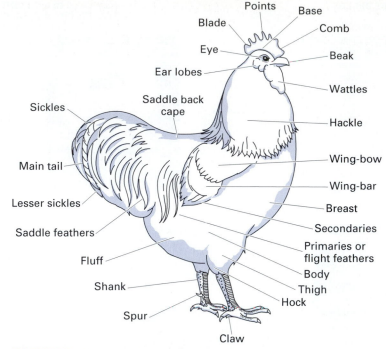

FIGURE 32.3 The external parts of a chicken.

FIGURE 32.4 Male and female turkeys that are nearing maturity. Courtesy of Hubbard Farms.

Pilgrim gander is white; the female is grayish. The White Chinese breed is pure white.

Breeding Poultry

Turkeys. There is a specific breeding cycle in the production of baby poults for distribution to growers who raise them to market weight. The pedigree flocks (generation 1) or parent stock are pure lines. The pure lines are crossed one with another to produce generation 2. Generation 2 lines are crossed one with another to produce generation 3. Eggs from these line crosses are selected into female or male lines and sent to hatcheries. Males from the male lines are

TABLE 32.1   Certain Breeds of Chickens and Their Main Characteristics

Breed	Purpose	Type of Comb	Color of Egg
American Breeds			
White Plymouth Rock	Eggs and meat	Single	Brown
Wyandotte[a]	Eggs	Rose	Brown
Rhode Island Red	Eggs	Single and rose	Brown
New Hampshire	Eggs and meat	Single	Brown
Asiatic Breeds			
Brahma[a]	Meat	Pea	Brown
Cochin[a]	Meat	Single	Brown
English Breeds			
Australorp	Eggs	Single	Brown
Cornish	Meat	Pea	Brown
Orpington[a]	Meat	Single	Brown
Mediterranean Breed			
Leghorn	Eggs	Single and rose	White

[a] These breeds are of minor importance to the U.S. poultry industry.

selected for such meat traits as thicker thighs, meatier drumsticks, plumper breasts, rate of growth, and feed efficiency. The females of the female line are selected for greater fertility, hatchability, egg size, and meat conformation. The male-line males are crossed with female-line females; the eggs produced are hatched and the poults are sold for the production of market turkeys.

Laying hens are approximately 30 weeks of age when they reach sexual maturity (Fig. 32.5). They are put under controlled lighting to stimulate the start of laying. The average hen usually lays for about 25 weeks and produces 88–93 eggs. The hen is considered "spent" at the end of the 25-week laying cycle. Most hens are marketed at that time for meat. Hens can be molted and stimulated to lay for another 25-week cycle. They require 90 days of molting and produce only 75–80 eggs during this later laying cycle; therefore, producers should compare the costs of growing new layers with that of recycling the

FIGURE 32.5   A flock of young, pullet turkey hens in the breeding program. Courtesy of Hubbard Farms.

old ones before deciding if it is economical to recycle old layers. The spent hens usually go into soup products.

Turkey eggs are not produced for human consumption because it costs much more to produce food from turkey eggs than from chicken eggs. It would cost about 50 cents each to produce turkey eggs for this purpose.

All turkey hens are artificially inseminated to obtain fertile eggs. As turkeys have been selected for such broad breasts and large mature size, it is difficult for toms to mate successfully with hens. Semen is collected from the toms and used for inseminating the hens. Usually 1 tom to 10 hens is sufficient for providing the semen needed.

The present-day turkey is much different from the turkeys of many years ago. Today's turkey grows more rapidly, converts feed into meat more efficiently, has more edible meat per unit of weight, and requires less time to reach market. Most turkeys grown today are white-feathered, so there are no dark pin feathers to discolor the carcass. A turkey requires only 2.8 lb of feed for every pound of gain (Fig. 32.6).

Chickens.   Sophisticated selection and breeding methods have been developed in the United States for increased productivity of chickens for both eggs and meat. The discussion here is directed primarily at improvement of chickens; however, methods described for chickens can also be applied to improve meat production in turkeys, geese, and ducks.

FIGURE 32.6   World's largest turkey feed conversion facility used to accurately measure individual feed efficiency of more than 3,000 female-line pedigree candidate toms per year. The company also tests individual feed conversion on approximately 7,400 other toms for potential use in their breeding program. Courtesy of Nicholas Turkey Breeding Farms.

Early poultry breeding and selection are concentrated on qualitative traits, which are, from a genetic standpoint, more predictable than quantitative traits. Qualitative traits, such as color, comb type, abnormalities, and sex-linked characteristics, are important; however, quantitative traits, such as egg production, egg characteristics, growth, fertility, and hatchability, are economically more important today.

Quantitative traits are more difficult to select for than are qualitative traits because the mode of inheritance is more complex and the role of the environment is greater. Quantitative traits differ greatly as to the amount of progress that can be attained through selection. For example, increase in body weight is much easier to attain than increase in egg production. Furthermore, a relationship usually exists between body size and egg size. Generally, if body size increases, a corresponding increase in egg size will occur. Most quantitative traits of chickens fall in the low to medium range of heritability, whereas qualitative traits fall in the high range.

Progress in selecting for egg production has been aided by the *trapnest,* a nest equipped with a door that allows a hen to enter but prevents her from leaving. The trapnest enables accurate determination of egg production of individual hens for any given period and helps to identify and eliminate undesirable egg traits and *broodiness* (the hen wanting to sit on eggs to hatch them). Furthermore, it allows the breeder to begin pedigree work within a flock.

The two most important types of selection applied primarily to chickens and, to a lesser extent, to other poultry today are mass selection and family selection. In **mass selection,** the older method, the best-performing males are mated to the best-performing females (Fig. 32.7). The program is effective in improving traits of high heritability. **Family selection** is a system whereby all offspring from a particular mating are designated as a family, and selection and culling within a population of birds are based on the performance level of the entire family. This type of selection is most adaptable to selection for traits of low heritability. This is not to say that progress for traits of high heritability cannot be made by using family selection—quite the opposite is true. However, mass selection is effective for traits of high heritability and certainly easier.

Progress through selection is rather slow for the reproductive traits because they are lowly heritable. Most breeding programs are geared toward the

FIGURE 32.7   A broiler breeding pen containing 1 male and 12 females. This is a typical breeding unit. Courtesy of Hubbard Farms.

improvement of more than one trait at a time. For example, to improve egg-laying lines, it might be necessary to attempt simultaneously to improve egg numbers, shell thickness, interior quality of the egg, and feed efficiency.

Outcrossing. Defined as mating unrelated breeds or strains, *outcrossing* is probably more adaptable to the modern broiler industry than to the egg-production industry, though it can be used to improve egg production. Numerous experiments have established fairly accurately which breeds or strains cross well with each other. Knowledge today is sophisticated to the extent that breeders know which strain or line to use as the male line and which to use as the female line.

Crossing Inbred Lines. Development of **inbred lines** for crossing is used largely for increasing egg production. *Inbreeding* is defined as the mating of related individuals or as the mating of individuals more closely related than the average of the flock from which they originated. The mating of brother with sister is the most common system of inbreeding for poultry; however, mating parents with offspring gives the same results. Both of these types of inbreeding increase the degree of inbreeding at the same rate per generation.

Hybrid chickens are produced by developing inbred lines and then crossing them to produce chickens that exhibit hybrid vigor. The technique employed in production of hybrid chickens is essentially the same as that used in development of hybrid plants. The breeder works with many egg-laying strains or breeds while producing hybrids. The strains or breeds are inbred (mostly brother or sister) for a number of generations, preferably five or more. In the inbreeding phase, many undesirable factors are culled out—selection is most rigid at this stage.

In all phases of hybrid production, the strains are completely tested for egg production and for important egg-quality traits. After inbreeding the birds for the necessary number of generations, the second step—crossing the remaining inbred lines in all possible combinations—is initiated. These crossbreds are tested for egg production and for egg-quality traits. Crossbreds that show improvement in these traits are bred in the third phase, in which the best test crosses are crossed into three-way and four-way crosses in all possible combinations. Most breeders prefer the four-way cross. The offspring that result from three-way and four-way crosses are known as *hybrid chickens.*

Strain Crossing. Strain crossing is more easily done than inbreeding and crossing inbred lines because it entails only the crossing of two strains that possess similar egg-production traits. It is definitely a type of crossbreeding if the two strains are not related to each other.

The computer is used in chicken breeding largely to help select breeding stock. An overall merit index is computed using data on heritability of each trait considered, the relative economic importance of each trait, and genetic correlations among the traits. Birds having the highest index are used for breeding.

In production of broiler or layer hybrid chicks, the computer is used to determine which lines or strains are most likely to give superior chickens. In broiler hybrids, hatchability, growth rate, economy of feed use, and carcass desirability are important. In layer hybrids, hatchability, egg production, egg size, egg and shell quality, and livability are important.

Practically all line-cross layers are now free from **broodiness** because attempts have been made to eliminate genes for broodiness, and records are available on which lines produce broody chicks when crossed. Lines that produce broody chicks in a line cross are no longer used for producing commercial layers.

FEEDING AND MANAGEMENT

The success of modern poultry operations depends on many factors. Hatchery operators must care for breeding stock properly so that eggs of good quality are available to the hatchery, and they must incubate eggs under environmental conditions that ensure the hatching of healthy and vigorous birds (Fig. 32.8). Feeding, health, and financial programs are essential to sound poultry management.

Incubation Management

Each type of poultry has an incubation period of definite length (Table 32.2), and incubation-management practices are geared to the needs of the eggs throughout that time. The discussion here of incubation management applies specifically to chickens. Although the principles of incubation management generally apply to other types of poultry as well, specific information should be obtained from other sources.

FIGURE 32.8   Removal of a batch of broiler chicks from the incubator. Courtesy of John Colwell from Grant Heilman.

TABLE 32.2 Incubation Times for Various Birds

Type	Incubation Period (days)
Chicken	21
Turkey	28
Duck	28
Muscovy duck	35–37
Goose	28–34
Pheasant	23–28
Bobwhite quail	23
Coturnix quail	17
Guineas	28

All modern commercial hatcheries have some form of forced-air incubation system. Today's commercial incubator has a forced-air (fan) system that creates a highly uniform environment inside the incubator. Forced-air incubators are available in many sizes. All sizes are equipped with sophisticated systems that control temperature and humidity, turn eggs, and bring about an adequate exchange of air between the inside and outside of the incubator. The egg-holding capacity of forced-air incubators ranges from several hundred to 100,000 or more eggs (Fig. 32.9).

Temperature, humidity, position of eggs, turning of eggs, oxygen content, carbon dioxide content, and sanitation must be regulated.

FIGURE 32.9 Large commercial room-type incubator. Courtesy of Chick Master Incubator Co., Medina, Ohio.

Temperature.   Proper temperature is probably the most critical requirement for successful incubation of chicken eggs. The usual beginning incubation temperature in the forced-air incubator is 99.5–100.0°F. The temperature should usually be lowered slightly (by 0.25–0.5°F) at the end of the fourth day and remain constant until the eggs are to be transferred to the hatching compartments (approximately 3 days before hatching). Three days before hatching, the temperature is again lowered, usually by about 1.0–1.5°F. Just before hatching, the chicks change from embryonic respiration to normal respiration and give off considerable heat, which results in a high incubator temperature.

With reference to temperature, there are two especially critical periods during incubation—(1) the first through the fourth day and (2) the last portion of incubation. Higher-than-optimum temperatures usually speed the embryonic process and result in embryonic mortality or deformed chicks at hatching. Lower-than-optimum incubation temperatures usually slow the embryonic process and also cause embryonic mortality or deformed chicks.

Humidity.   A relative humidity of 60–65% is needed for optimum **hatchability.** All modern commercial forced-air incubators are equipped with sensitive humidity controls. The relative humidity should usually be raised slightly in the last few days of incubation. It has been shown that when the relative humidity is close to 70% in the last few days of incubation, successful hatching is greater. Providing optimum relative humidity is essential in reducing evaporation from eggs during incubation.

Position of Eggs.   Eggs hatch best when incubated with the large end (the area of the space called the *air cell*) up; however, good hatchability can also be achieved when eggs are set in a horizontal position. Eggs should never be set with the small end up because a high percentage of the developing embryos will die before reaching the hatching stage. The head of the developing embryo should develop near the air cell. In the last few days of incubation, the eggs are transferred to a different type of tray—one in which the eggs are placed in a horizontal position.

Shortly before hatching, the beak of the chick penetrates the air cell. The chick is now able to receive an adequate supply of air for its normal respiratory processes. The horny point (egg tooth) of the beak eventually weakens the eggshell until a small hole is opened. The egg is now said to be *pipped.* The chick is typically out of the shell within a few hours. The hatching process varies considerably among different species.

Turning of Eggs.   Modern commercial incubators are equipped with time-controlled devices that permit eggs to be turned periodically. Eggs that are not turned enough during incubation have little or no chance of hatching because the embryo often becomes stuck to the shell membrane.

Commercial incubators are equipped with setting trays or compartments that allow eggs to be set in a vertical position. They also have mechanisms that rotate these trays so that the chicken eggs rotate 90° each time they are turned. The number of times the eggs are turned daily is most important. If the eggs are rotated once or twice daily, the percent of hatchability will be much lower than if the eggs are rotated five or more times daily.

Oxygen Content.   The air surrounding incubating eggs should be 21% oxygen by volume. At high altitudes, however, the available oxygen in the air may be too low to sustain the physiological needs of developing chick embryos, many of which therefore die. Good hatchability can often be attained at high altitudes if supplemental oxygen is provided. This is necessary for commercial turkey hatching, but hatchability can be improved by selecting breeding birds that hatch well in environments in which the supply of oxygen is limited.

Carbon Dioxide Content.   It is vital that the incubator be properly ventilated to prevent excessive accumulation of carbon dioxide (CO_2). The CO_2 content of the air in the incubator should never be allowed to exceed 0.5% by volume. Hatchability is lowered drastically if the CO_2 content of air in the incubator reaches 2.0%. Levels of more than 2.0% almost certainly reduce hatchability to near zero.

Sanitation.   The incubator must be kept as free of disease-causing microorganisms as possible. The setting and hatching compartments must be thoroughly washed or steam-cleaned and fumigated between settings. In many cases, they should be fumigated more than once for each setting. An excellent schedule is to fumigate eggs immediately after they have been set and again as soon as they have been transferred to the hatching compartment. A good procedure for fumigation is:

1. Prepare 1.5 ml of 40% formalin for each cubic foot of incubator space.
2. Prepare 1.0 g of potassium permanganate for each cubic foot of incubator space.
3. Set the temperature of the setting compartment or hatching compartment at the proper reading (99.5–100.0°F and 98.5–99.0°F, respectively).
4. Turn on the fan.
5. Close the vents.
6. Set the relative humidity at the proper reading (65%).
7. Place potassium permanganate in an open container on the compartment.
8. Pour formalin into the potassium permanganate container and close the incubator door for approximately 20 minutes.
9. Open the door for 5–10 minutes or turn on the exhaust system to allow formaldehyde gas to escape.
10. After the gas has escaped, remove the potassium permanganate container, close the door, and continue normal operating procedures.

Candling of Eggs.   In maintaining a healthy and germ-free environment, eggs must be *candled*—examined by shining a light through each egg to see if a chick embryo is developing—at least once during incubation so that infertile and *dead-germ eggs* (eggs containing dead embryos) can be identified. Many operators candle chicken eggs on the fourth or fifth day and turkey and waterfowl eggs on the seventh to tenth day. Some operators candle eggs a second time when transferring eggs to the hatching compartment.

Managing Young Poultry

The main objective in managing young poultry is to provide a clean and comfortable environment with sufficient feed and water (Fig. 32.10).

House Preparation. The brooder house should be thoroughly cleaned, **disinfected,** and dried several days before it receives young birds. All necessary brooding equipment should have been tested ahead of time and be in the proper place. One must make sure that the brooders are working properly and *check the thermostats and microswitches!* Such advance preparation is probably as essential to success as any management practice applied to young birds. It is imperative that the disinfectant have no effect on the meat or eggs produced by birds. Food and Drug Administration regulations that govern the use of disinfectants should be strictly observed.

Litter. Litter should be placed in the house when equipment is checked. Some commonly used substances are planer shavings, sawdust, wood chips, peat moss, ground corn cobs, peanut hulls, rice hulls, and sugarcane fiber. The entire floor should be covered by at least 2 inches of litter, which must be perfectly dry when the birds enter.

The primary purpose of litter is to absorb moisture. Litter that gets extremely damp should be replaced. Some operators add small amounts of litter as the birds grow.

Floor Space. The availability of too much or too little floor space can adversely affect growth and efficiency of production. If floor space is insufficient, young birds have difficulty finding adequate feed and water. This can lead to feather picking and actual cannibalism. Too much space can cause birds to become bored, which can lead to problems similar to those caused by overcrowding.

FIGURE 32.10 A commercial poultry house containing young broilers. Environmental conditions are well controlled for these many hundred birds. Courtesy of Hubbard Farms.

The amount of space required varies with the type and age of the birds. Regardless of type, each bird should have at least 7–10 square inches of space around and under the hover (a metal canopy that covers most brooder stoves). Chicks of the egg-producing type and birds of similar size fare well for 5–6 weeks on 70 square inches of floor space. When these birds are between 6 and 10 weeks old, the space per bird must be increased to nearly 145 square inches.

Broilers, turkeys, and waterfowl should have an area of 110–145 square inches for at least their first 8 weeks. In certain situations, such birds could be allowed as much as 145–220 square inches before they are 8 weeks old.

Feeder Space.   Chick-trough-feeders should be placed within the hover guard area or partially under the hover (0.8 inches/chick or 80 inches/100 chicks to 6 weeks of age). Producers should be sure the 10-watt attraction lights under the hover are working. One linear inch of feeder space per bird is sufficient for most species from the age of 1 day to 3 weeks; 2 inches for 3–6 weeks; and 3 inches beyond 6 weeks. Quail require less space, whereas turkeys, geese, and ducks require slightly more. Most feeders are designed so that birds can eat on either side of them.

The operator can determine how much feeder space is needed through observation. If relatively few birds are eating at one time, the feeding area is probably too large. If too many are eating at once, the area is probably too small. Young birds are usually fed automatically after the brooder stove enclosure is removed.

Water Requirements.   Three fountain-type waterers should be placed within the hover guard area (one fountain-type waterer/100 chicks). Generally, two 1-gallon-size water fountains are adequate for 100 birds that are 1 day old. More waterers can be added as necessary. Most commercial producers switch from water fountains to an automatic watering system when birds are put on automatic feeding. If trough-type waterers are used, allow at least 1 inch of water space per bird until 10–12 weeks of age. More space might be needed thereafter. Trough-type waterers designed so that birds can drink from both sides are available. A space of at least 0.5 inch per bird is needed if pan-type waterers are used.

Lighting Requirements.   The amount of light provided by the operator varies according to the type of bird being raised. No artificial lights are needed for chicks hatched during April through July. The time clock is set for 12 hours (5 A.M.–P.M.) so that lights are available when personnel are working in the area. A step-down lighting program is used for chicks hatched August through March. Many broiler producers use a 24-hour light regime, whereas others use systems such as 20 hours of light alternating with 4 hours of darkness. Growing pullets do not require as much light per day as broilers because the producer desires that future layers not reach sexual maturity too quickly. Generally, about 8–10 hours of light are sufficient for future layers until they are about 12–14 weeks old.

A system of lighting that takes into account the age of the bird, the time of year in which the bird was hatched (seasonal effect), and the type of housing (windowed or windowless) should be used. Pullets should be on a constant day-length regime or a decreasing-light regime at 10 weeks of age to prevent early sexual maturity. In most cases, some type of dim light is provided to prevent young chicks from piling up in periods of darkness. After birds are about 1 week old, some producers substitute light bulbs of relatively low intensity (15–25 watts) for a bulb of high intensity (40 watts is standard for the first week).

Other Management Factors. Young birds should have access to the proper feed as soon as they are placed in the brooder house. It is important that birds such as turkey poults start eating feed early. Some young poults die for lack of feed; although feed is readily available, they have not learned to eat it.

Young birds are commonly debeaked sometime between their first day of life and several weeks of age. **Debeaking** (also called *beak trimming*) is done with a machine by searing off approximately one-half of the upper mandible and removing a small portion of the lower one. Care should be taken to avoid searing the bird's tongue while debeaking. Debeaking is done to prevent birds from picking feathers from other birds; it also prevents cannibalism.

Birds should be vaccinated and given health care according to a prescribed schedule; a veterinarian should be consulted.

Managing 10- to 20-Week-Old Poultry

Management of poultry from approximately 10 to 20 weeks of age is quite different from managing younger birds. Improper practices in this most critical period could adversely affect subsequent production.

Confinement rearing is used by commercial producers of replacement birds—the birds that will be kept for egg production. In the three basic systems of confinement rearing, birds are raised on solid floors, slatted floors, or wire floors or cages.

Birds raised on floors in confinement should have from 1–2 square feet of floor space per individual. The amount of space required depends on the environmental conditions and on the condition of the house. Caged birds should be allowed 0.5 square feet to not more than 0.75 square feet.

Replacement chickens raised in confinement are generally fed a completely balanced growing, or developing, ration that is 15–18% protein. Most birds are kept on a full-feeding program, but in certain conditions some producers restrict the amount of feed provided. Restricted feeding can take different forms: total feed intake, protein intake, or energy intake may be limited. The main purposes of a restricted feeding program are to slow growth rate (thus delaying the onset of sexual maturity and reducing the number of small eggs produced) and to lower feeding cost. Feed intake must not be restricted whenever a restricted lighting system (which is also used to delay sexual maturity) is in effect.

Automatic feeding and watering devices are used in most confinement operations. Watering- and feeding-space requirements are practically the same as those for birds that are 6–10 weeks old. As long as birds are not crowding the feeders and waterers, there is no particular need to increase the feeding and watering space.

The lighting regime used in confinement rearing is very important. Birds raised in window-type housing receive the normal light of long-day periods unless the house is equipped with some type of light-check. In short-day periods, supplemental lighting can be used to meet the requirements of growing chickens. Replacement chickens should receive approximately 14 hours of light daily up to 12 weeks of age. To delay sexual maturity, the amount should then be reduced to about 8–9 hours daily until the birds have reached 20–22 weeks of age. The light is then either abruptly increased to 16 hours per day, or it is increased by 2–3 hours with weekly increments of 15–20 minutes added until 16 hours of light per day are reached.

Regardless of which system replacement pullets are reared under, they should be placed in the laying house when approximately 20 weeks of age so they can adjust to the house and its equipment before beginning to lay.

Management of Laying Hens

The requirements of laying hens for floor space vary from 1.5–2 square feet per bird for egg-production strains and from 2.5–3.5 square feet for dual-purpose and broiler strains. Turkey breeding hens require 4–6 square feet, but less area is needed if hens are housed in cages. If turkey hens have access to an outside yard, 4 square feet of floor space is best. Game birds such as quail and pheasants require less floor space than egg-production hens. In general, 3 square feet or more of floor space is adequate for ducks, whereas geese need approximately 5 square feet.

Breeder hens are housed on litter or on slatted floors, whereas birds used for commercial egg production are kept in cages (Fig. 32.11).

FIGURE 32.11 Composite pictures of an automated commercial cage house for laying hens. Feeding, watering, egg gathering (vertical elevator at end of row), and manure removal are done automatically.

Floor-type houses for breeder hens usually have 60% of the floor space covered with slats. Slatted floors are usually several feet above the base of the building and manure is often allowed to accumulate for a long time before being removed. Many slatted-floor houses are equipped with mechanical floor scrapers that remove the manure periodically. Frequent removal of manure lessens the chance that ammonia will accumulate.

Gathering of eggs in floor-type houses can be done automatically if some type of roll-away nesting equipment is used. In houses equipped with individual nests, the eggs are gathered manually. Some operators prefer colony nests (an open nest that accommodates several birds at a time).

Some type of litter or nesting material must be placed on the bottom of individual and colony nests if eggs are to be gathered manually. Birds should not be allowed to roost in the nests in darkness because dirty nests result.

Cage operations are used by most commercial egg-producing farms. Young chickens may be brooded in colony cages (Fig. 32.12), whereas laying hens may be housed in 4 to 6 tiers of decked laying cages. The individual cages range in size from 8–12 inches in width, 16–20 inches in depth (front to back), and 12–15 inches in height. The number of birds housed per unit varies with the operator. Density of the cage population is an important factor in production.

All cages are equipped with feed troughs that usually extend the entire length of the cage. Some troughs are filled with feed manually, others automatically. It is sometimes advantageous to dub (remove the combs and wattles) caged layers so they can obtain feed from automatic feeders by reaching the head through the openings.

Several types of watering devices (trough, nipple-type, or individual cup-type waterers) are available for use in individual and colony cages. The watering system should be equipped with a metering device that makes it possible to medicate the birds quickly when necessary by mixing the exact dosage of medicant required with the water.

The arrangement of cages within a house varies greatly. Some producers have a step-up arrangement, such as a double row of cages at a high position with a single row at a low position, on either side. Droppings from birds caged in the double row fall free of the birds in the single row. An aisle approximately 3 feet wide is usually between each group of cages. Rather than aisles, some systems have a movable ramp that can travel above the birds, from one end of

FIGURE 32.12   Rearing pullets in cages. Courtesy of J. F. Stephens.

the house to the other. In other systems, several double cages are stacked on top of each other.

In cage operations, manure is typically collected into pits below the cages and removed by mechanical pit scrapers or belts. Manure may also drop into pits that are slightly sloped from end to end and partially filled with water. The manure can be flushed out with the water into lagoons.

Housing Poultry

Factors such as temperature, moisture, ventilation, and insulation are given careful consideration in planning and managing poultry houses.

Temperature. Most poultry houses are built to prevent sudden changes in house temperature. A bird having an average body temperature of 106.5°F usually loses heat to its environment, except in extremely hot weather. Chickens perform well in temperatures between 35 and 85°F, but 55–75°F is optimal.

House temperature can be influenced by such factors as the prevailing ambient temperature, solar radiation, wind velocity, and heat production of the birds. A 4.5-lb laying hen can produce nearly 44 **British thermal units (Btu)** of heat per hour (1 Btu is the quantity of heat required to raise the temperature of 1 lb of water 1°F at or near 39°F). The amount of Btu produced varies with activity, egg-production rate, and the amount of feed consumed. Heat production by birds must be considered in designing poultry houses (about 40 of the 44 Btu produced per hour by a laying hen are available for heating). Although heat produced by birds can be of great benefit in severe cold, the house must be designed so that excess heat produced by birds in hot weather can be dissipated.

Moisture. Excess house moisture, especially in cold weather, can create an environment that is extremely uncomfortable, lead to a drop in production by laying birds, and, if allowed to continue for long, cause illness.

Much of the moisture in a poultry house comes from water spilled from waterers, water in inflowing air, and water vapor from the birds themselves and their droppings. Every effort should be made, of course, to minimize spillage from watering equipment. The amount of moisture in a poultry house can be reduced by increasing the air temperature or by increasing the rate at which air is removed. Because air holds more moisture at high temperatures than at low temperatures (Table 32.3), raising the air temperature causes moisture from the litter to enter the air and thus be dissipated. Air temperature in the house can be increased by retaining the heat produced by the birds themselves and by supplemental heat.

If incoming air is considerably colder than the air in the house, it must be warmed or it will fail to aid in moisture removal. In cold weather, most exhaust ventilation fans are run slower than normal, so most moisture removal is accomplished by increasing the air temperature in the house.

Ventilation. A properly designed ventilation system provides adequate fresh air, aids in removing excess moisture, and is essential in maintaining a proper temperature within the house. Type and amount of insulation and heat produced by the birds themselves must be considered in planning for ventilation.

TABLE 32.3   The Water-Holding Capacity of Air at Various Temperatures

Temperature (°F)	Pounds of Water per 1,000 lb of Dry Air
50	7.62
40	5.20
30	3.45
20	2.14
10	1.31
0	0.75
−10	0.45

Source: V. M. Meyer and P. Walther, *Ventilate Your Poultry House: For Clean Eggs, Healthy Hens, More Profit* (Ames: Iowa State University Cooperative Extension Service Pamphlet 292, 1963).

Ventilation is accomplished by positive pressure, through which air is forced into the house to create air turbulence, or by negative pressure, through which air is removed from the house by exhaust fans (Fig. 32.13). The positive pressure system is accomplished by having fans in the attic so that air is forced through holes. The negative pressure system is accomplished by locating exhaust fans near the ceiling. Some houses are ventilated so that the air is fairly warm and dry by having open walls. Also, when outside air is cold and dry, air can enter at low portions of the house and leave through vents near the ceiling as it warms. The heat created by the chickens warms the colder air, which then takes moisture from the house.

FIGURE 32.13   A battery of fans used to ventilate a large poultry house. Courtesy of Big Dutchman, a division of U.S. Industries, Atlanta, Georgia.

Two integral and necessary parts of the most common ventilation systems are the exhaust fan and the air-intake arrangement. The number of exhaust fans varies with size of the house, number of birds, and capacity of the fans (Fig. 32.13). Most fans are rated on the basis of their cubic feet per minute (cfm) capacity; that is, how many cubic feet of air they move per minute. Fans in laying houses should operate at 4–4.5 cfm per bird when the temperature is moderate. In summer, cfm per bird could be as high as 10.

Fans also have a static pressure (water pressure for low house temperatures and a high speed for warm house temperatures). Others operate at only one speed; the amount of air removed is controlled by shutters that open, so that more air can be removed when the temperature of the house exceeds the thermostatically fixed temperature. These shutters then close at lower temperatures.

An air-intake area must be provided for the ventilation system. It is usually a slotted area in the ceiling or near the top of the walls. There should be at least 100 square inches of inlet space for each 400 cfm of fan capacity.

Insulation.   Because energy needs are becoming ever more critical, the insulation of today's poultry houses must be superior to that of the past. Sudden temperature changes inside the house must be avoided, especially for young birds. Houses in cold and windy areas require better insulation than those in milder climates, but good insulation is also essential in areas with high outside temperatures. The insulation ability of any material is measured by its R value. This value is based on the material's ability to limit heat loss, as expressed in Btu. The better the insulating material, the higher its R value. For example, if the outside temperature is 11°F cooler than the inside, a material having an R value of 11 loses approximately 1 Btu/hour to the outside for each square foot of area when the outside temperature is 20°F cooler than the inside.

By knowing how much heat the birds generate, local variations in ambient temperature, expected wind velocities, and relative humidities experienced, the appropriate R value can be established. In many areas of the United States, especially in the cold regions, wall insulation in poultry houses should have an R value of 15; ceiling insulation, at least 20 (heat loss through the floor is negligible). In warmer areas, wall insulation should have an R value of 5–10; ceiling insulation, 10–15.

When computing the R value for a particular area of the house (e.g., walls), resistance of the outer surface, insulation, air space between the studding, inner-wall material, resistance of the inner surface, and, if windows are present, presence of glass must all be considered. Each of these factors has an R value, and the total of these values establishes the R value for that area of the house. A building-materials dealer can furnish the R-value ratings of these factors.

Because too much moisture may accumulate in the house, especially in cold weather, some type of vapor barrier should be present in the walls and ceiling. This barrier can be a part of the insulation itself (foil backing) or separate. It should be placed between the insulation and the inside wall. The foil-back portion of the insulation should be placed next to the inside wall.

Feeds and Feeding

Rations fed to poultry today are complex mixtures that should include all ingredients, in a balanced proportion, that have been found to be necessary for body maintenance, maximum production of eggs and meat, and optimum reproduction (**fertility** and hatchability). Poultry nutritionists are constantly searching for and testing new combinations of feedstuffs that, when incorporated into diets, permit better efficiency of production. Feeding can be expensive, because an adequate intake of energy, protein, minerals, and vitamins is essential.

The computer is used in formulating least-cost rations that meet the nutritional requirements of poultry. Data input that must be provided includes constraints that depend on age of birds and their productivity, cost of available ingredients, ingredients desired in the ration, composition of the nutritional materials, and requirements of the chickens to be fed.

The constraints set minimum and maximum percentages. For example, an upper limit on the amount of fiber that could be in the ration must be established because birds cannot use fiber effectively. Likewise, a low or minimal level of soybean meal is important to provide the needed amino acids.

Large computers are expensive, so only large poultry operations can afford them. However, lower-cost microcomputers are now being used to compute least-cost rations. Most poultry producers can afford a microcomputer.

Energy requirements are supplied mainly by cereal grains, grain by-products, and fats. Some important grains are yellow corn, wheat, sorghum grains (milo), barley, and oats. Most rations contain high amounts of grain (60% or higher, depending on the type of ration). A ration containing a combination of grains is generally better than a ration having only one type. Animal fats and vegetable oils are excellent sources of energy. They are usually incorporated into broiler rations or any high-energy rations.

Protein is so highly essential that most commercial poultry feeds are sold on the basis of their protein content. The types of amino acids present determine the nutritional value of protein (see Chapter 15). Excellent protein can be derived from both plants and animals. Most rations contain both plant and animal protein so that each source can supply amino acids that the other source lacks. The most common sources of plant protein are soybean meal, cottonseed meal, peanut meal, alfalfa meal, and corn gluten meal. Cereal grains contain insufficient protein to meet the needs of birds. The best sources of animal protein are fish meal, milk by-products, meat by-products, tankage, blood meal, and feather meal. The protein requirements of birds vary according to the species, age, and purpose for which they are being raised. The protein requirements of certain birds are shown in Table 32.4. Where variable values are listed, the highest level is to be fed to the youngest individuals.

Whatever ration is adequate for turkeys is generally suitable for game birds. Some game bird producers feed a turkey ration at all times. Others feed a complete game bird ration. The actual nutrient requirements of game birds are not as well known as are the requirement of chickens and turkeys.

A considerable number of minerals are essential, especially calcium, phosphorus, magnesium, manganese, iron, copper, zinc, and iodine. Calcium and phosphorus, along with vitamin D, are essential for proper bone formation. A deficiency of either of these elements can lead to a bone condition known as *rickets.* Calcium is also essential for proper eggshell formation.

The amount of calcium required varies somewhat with age, rate of egg production, and temperature. Chickens and turkeys up to 8 weeks of age require 0.8–1.2% of calcium in their diets. Calcium can be reduced to 0.6–0.8% in 8- to

TABLE 32.4   Protein Requirements of Poultry

Type	Age (weeks)	Percentage of Protein Required in Diet
Chickens		
Broilers	0–6	20.0–23.0%
Replacement pullets	0–14	15.0–18.0
Replacement pullets	14–20	12.0
Laying and breeding hens		14.5
Turkeys		
Starting	0–8	26.0–28.0
Growing	8–24	14.0–22.0
Breeders		14.0
Pheasant and quail		
Starting and growing		16.0–30.0
Ducks		
Starting and growing		16.0–22.0

Source: National Research Council, *Nutrient Requirements of Poultry,* 8th ed. (Washington, DC: National Academy of Sciences, 1984).

16-week-old birds. Laying hens require at least 3.4% calcium. The amounts of phosphorus required for chickens of different ages are as follows: 0–6 weeks, 0.4%; 6–14 weeks, 0.35%; and mature, 0.32%. Turkeys require from 0.3–0.6% phosphorus; game birds, approximately 0.5%. Requirements for other minerals vary greatly depending on the stage of growth and production.

The vitamins that are most important to poultry are A, D, K, and E (fat soluble), and thiamin, riboflavin, pantothenic acid, niacin, vitamin B_6, choline, biotin, folacin, and vitamin B_{12} (water soluble).

COSTS AND RETURNS

Table 32.5 shows the 1995 production costs for eggs, broilers, and turkeys. Note that feed costs comprise 60–70% of the total costs. Management practices that monitor and lower feed costs usually have a positive effect on net returns.

Figure 32.14 shows how wholesale broiler prices and production costs have decreased over a 25 year period of time. This vividly demonstrates how broiler

TABLE 32.5   Poultry Meat and Egg Production Costs and Returns, 1995

Item	Production Cost (cents per doz/lb)		Net Return (cents per doz/lb)
	Feed	Total	
Market eggs	28.8	47.0	8.9
Broilers	16.0	26.4	7.6
Turkeys	21.9	35.6	5.4

Source: USDA.

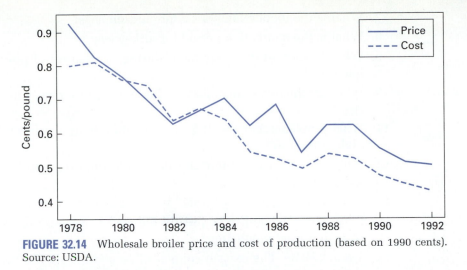

FIGURE 32.14 Wholesale broiler price and cost of production (based on 1990 cents). Source: USDA.

prices are so competitive with other meats, yet a profit can be maintained by lowering production costs. From the wholesale product, value-added poultry products are created which can increase consumer demand and yield even higher profits. The latter is true if low-cost production continues to be emphasized. These same principles need greater emphasis in other meat-producing animals.

Additional information affecting costs and returns is discussed in Chapter 36.

CHAPTER SUMMARY

■ The most important poultry in the United States are chickens, turkeys, ducks, and geese.

■ Breeds of chickens are relatively unimportant as the commercial broiler and egg producers utilize hybrid chickens resulting from sophisticated breeding methods.

■ The production of broilers and eggs is very intensively managed in an integrated system involving breeding, hatching, feeding, processing, and marketing name-brand products to the consumer.

REVIEW QUESTIONS

1. How are breeds of chickens classified?
2. What are the four major classes of chickens?
3. What characteristics are most important to commercial egg producers?
4. What Mediterranean breed of chicken is of major importance to the U.S. poultry industry?
5. What three American breeds of chickens are important to the U.S. poultry industry?
6. What English breed of chicken is important to the U.S. broiler industry?

7. What is the incubation period for chicken eggs?

8. What is the incubation period for turkey eggs?

9. What factors must be regulated during incubation of poultry eggs for hatching?

10. What are the two critical periods for temperature regulation of hatching eggs?

11. What problems may result from incubation temperatures which are too high or too low?

12. Why is maintaining humidity near 60% important in hatching incubators?

13. How should eggs be positioned in an incubator for good hatchability?

14. Why will hatchability be greatly reduced if eggs are not turned often enough during incubation?

15. *True or False:* The oxygen content of incubators should be maintained at 21%.

16. *True or False:* It is vital to maintain the carbon dioxide content of incubators at or below 0.5% to maintain good hatchability of eggs.

17. What is the primary purpose of litter to be placed on the floor of brooder houses?

18. What is debeaking or beak trimming, and why is it performed?

29. Replacement laying hens should be managed to reach puberty and begin laying eggs at _____ of age.

20. What factors are carefully considered when planning and managing poultry houses?

21. What are the two integral parts of a ventilation system for poultry?

22. What mineral is especially important in the diets of laying hens?

23. What accounts for 60 to 70% of total costs of poultry production?

SELECTED REFERENCES

Publications

American Poultry History. 1996. Mount Morris, IL: Watt Publishing Co.

Broiler Industry (monthly publication). Mount Morris, IL: Watt Publishing Co.

Egg Industry (monthly periodical). Mount Morris, IL: Watt Publishing Co.

Livestock and Poultry. Washington, D.C.: USDA (Economic Research Service).

Moreng, R. E., and Avens, J. S. 1985. *Poultry Science and Production.* Reston, VA: Reston Publishing Co.

National Research Council. 1994. *Nutrient Requirements of Poultry.* 9th ed. Washington, DC: National Academy Press.

North, M. O., and Bell, D. D. 1990. *Commercial Chicken Production Manual.* Westport, CT: AVI Publishing Co.

Turkey World (monthly publication). Mount Morris, IL: Watt Publishing Co.

Watt Poultry Yearbook. 1996. Mount Morris, IL: Watt Publishing Co.

Visuals

The Business of Laying Eggs (videotape; 15 min.). American Farm Bureau Federation, 225 Touhy Ave., Park Ridge, IL 60068.

Commercial Chicken Production (videotape; 57 min.). Dept. of Animal Science, University of Delaware, Newark, DE 19711.

Hatchery Operations (80 slides and audiotape; 18 min.). Poultry Science Department, Ohio State University, 674 W. Lane Ave., Columbus, OH 43210.

How to Do a Poultry Autopsy (silent filmstrip); and *Small Flock Brooding Techniques* (sound filmstrip). Vocational Education Productions, California Polytechnic State University, San Luis Obispo, CA 93407.

CHAPTER 33

Goat Breeding, Feeding, and Management

Geological records reveal that goats existed five million years ago and bones of goats closely associated with people have been found dating back 12,000 years. It appears that people at that time existed in small groups, moving to where food and water could be found for themselves and the goats. The goats were particularly important to people because they provided food (milk and meat), skins for making certain items of clothing, and mohair for many uses.

Goats thrive on **browse** (brushy plants) and **forbs** (broad leaf plants), but they also eat grass extensively, depending on forage species and stage of maturity. Choice of plants and plant parts is important to goats because they are more likely to select the more nutritious parts and, because of their browsing over a wide territory, they are less likely to suffer from internal parasites.

Goats have a large impact on the economy and food supply for people of the tropical world and also several Mediterranean countries. In many countries, people consume more goats' milk than cows' milk, and goats are an important source of meat. In the United States, the consumption of both goat milk and goat meat is increasing, and the production of mohair is an important industry in Texas.

There are five major kinds of domesticated goats: (1) the **dairy goat** is used largely for the production of milk and, to a lesser extent, for meat; (2) the **Angora goat** is used mainly for the production of mohair and, to a lesser extent, for brush clearance and the production of meat; (3) the **meat goat** is used for brush clearance, for the production of meat, and, in some parts of the world, for skins and fine leather; (4) the **Cashmere goat** is noted for the soft cashmere fibers used in producing high-quality clothing; and (5) the **pygmy goat** is used as a laboratory ruminant animal and pet in the United States but is an important disease-resistant meat and milk producer in West Africa (where it is known as the West African Dwarf) and other countries.

In some parts of the world, goat meat production is extremely important and, in fact, meat goats outnumber dairy goats worldwide. Goat numbers in the United States are modest compared to cattle and sheep because of the highly specialized and effective dairy industry built around the dairy cow, as well as the specialized beef cattle and sheep industries. Yet in some parts of the world, including India, Iran, Pakistan, and Sudan, large quantities of dairy products are produced from goats.

Milk produced by dairy goats differs from cows' milk in that all carotene has been converted into vitamin A in goat milk. The type of curd formed from goat milk is different from the curd from cow milk because of differences in the major caseins; milk fat in goat milk is in smaller globules than in cow milk, does not rise or coalesce as readily, and contains much more short-chain fatty acids. Goat milk is more readily digested and assimilated by people and animals because of these differences. The dairy goat is a desirable animal for providing milk for the family because it is small and less expensive to feed than the cow. Goats can consume large quantities of browse, which is not very palatable to cattle.

IMPORTANCE OF GOATS

The number of goats in the different countries of the world is presented in Table 33.1. Most goats are found in developing countries, where they contribute greatly to the needs of the people there. There are many families that have a few goats for the production of milk, meat, and hides for family use.

TABLE 33.1   Number of Goats in the World, in Different Areas of the World, and in Developed and Developing Nations

Area or Country	Number of Goats (mil)
Africa	175
Nigeria	24
Ethiopia	17
Asia	401
China	123
India	119
Pakistan	43
Bangladesh	30
Europe	16
Greece	6
Spain	3
North and Central America	16
Mexico	10
United States	2
South America	25
Brazil	12
Argentina	4
World	639

Source: FAO Production Yearbook, 1995.

TABLE 33.2   Leading Countries in Goat Meat and Fresh Goatskin Production

Country	Goatmeat (mil lb)	Country	Goatskins (mil lb)
China	1,830	China	337
India	1,043	India	269
Pakistan	946	Pakistan	257
Bangladesh	231	Bangladesh	74
Nigeria	286	Nigeria	45
World total	7,176	**World total**	1,469

Source: FAO Production Yearbook, 1995.

One major by-product of goats is skins. In areas with extensive goat numbers for milk and meat production, hides become important. The leading countries in the production of goat meat and goatskins are given in Table 33.2.

PHYSIOLOGICAL CHARACTERISTICS

Goats have many characteristics that are similar to sheep. Some of the production traits of goats are summarized in Table 33.3.

CHARACTERISTICS OF BREEDS OF DAIRY GOATS

The external parts of a goat are shown in Fig. 33.1. The dairy goat and dairy cow are about equal in efficiency of converting feed into milk, even though the dairy goat produces much more milk in relation to its size and weight than does the dairy cow. Feed needed for maintenance is higher per unit of body weight for goats than for dairy cows.

TABLE 33.3   Production Characteristics of Goats

Trait	Time/lb
Gestation length	144–155 days (average 150 days)
Length of estrous cycle	15–18 days (average 16 days)
Length of estrus	1–3 days (average 2 days)
Age at puberty	120 days to over a year[a]
Normal breeding season	Late summer, early fall, or late winter[b] (breeds near the equator are less seasonal)
Size of kids at birth	1.5–11.0 lb[c] (average 5.5 lb)
Adult size	
Does	40–190 lb (average 130 lb)
Bucks	50–300 lb[d] (average 160 lb)

[a] Age of puberty depends on the breed and nutrition. Pygmy goats tend to reach puberty at an early age, some kid at nine months of age.
[b] Normal breeding season depends on the breed. Pygmy does are capable of breeding over a long breeding season but are more fertile when bred in the fall or early winter. Angora goats usually have a highly restrictive breeding season during the fall.
[c] Pygmy and Angora goats are smaller at birth than are dairy goat kids.
[d] Pygmy goats are the smallest in size, Angora goats are intermediate, and dairy goats are the largest. Bucks of all breeds are larger than corresponding does.

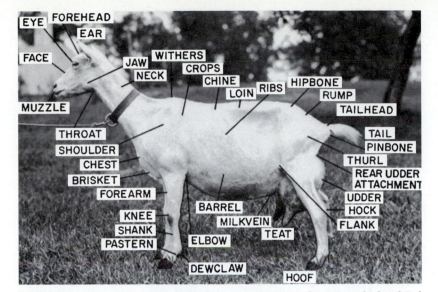

FIGURE 33.1 The external parts of the dairy goat. Courtesy of R. F. Crawford and Ted Edwards, *Emerald Dairy Goat Association Newsletter.* Photograph by Ole Hoskinson.

Gall (1981) compared the efficiency of Holstein dairy cows with Alpine Beetel dairy goats. The cows weighed 1,325 lb and produced 13,260 lb of milk, whereas the dairy goats weighed 108 lb and produced 1,083 lb of milk per lactation. The dry-matter intakes per lactation and per quart of milk for the Holstein cows were 13,300 lb and 2.3 lb, and for the goats they were 1,173 lb and 2.4 lb per quart of milk, respectively.

The six major dairy goat breeds registered in the United States—Toggenburg, Saanen, Alpine, Nubian, LaMancha, and Oberhasli—are all capable of high milk productivity. These breeds are shown in Figs. 33.2–33.7.

The Toggenburg breed is medium in size, is vigorous, and has high milk production (Fig. 33.2). The highest record of milk production for one lactation by a U.S. Toggenburg female is 5,750 lb. The Saanen is a large, all-white breed capable of producing the most milk (Fig. 33.3). A world-record milk production for one lactation by a Saanen female is 7,713 lb. The Alpine breed is a large, somewhat rangy goat (Fig. 33.4). Alpines also produce much milk; the record for one lactation of a female is 5,729 lb.

The Nubian is a tall, proud-looking goat breed that differs from the other five breeds in being a better meat producer and having long, wide, pendulous ears and a Roman nose (Fig. 33.5). The U.S. record milk production of a Nubian female is 4,420 lb, which is lower than for the three Swiss breeds previously described. However, Nubian milk is distinctive for its higher milk fat content (7.4% versus 3.5% for Swiss breeds). It is also considered a dual-purpose breed.

The LaMancha goat breed of California origin (Fig. 33.6) is smaller and differs from all other breeds in having almost no external ears ("gopher" ears) or extremely short ears ("elf" or "cookie" ears). Milk production is comparatively lower. Records for one lactation of up to 4,510 lb have been recorded. Fewer females of this breed and the Oberhasli breed have been officially tested for milk production. Therefore, these breeds' true capacities have not yet been demonstrated.

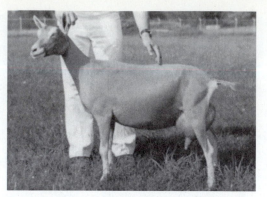

FIGURE 33.2 A Toggenburg doe with typical "badger" face white stripes. Note the fore attachment of the udder and the well-placed teats. Courtesy of the *Dairy Goat Journal.*

FIGURE 33.3 A Saanen doe. Courtesy of Nancy Lee Owen.

FIGURE 33.4 A French Alpine doe. Courtesy of Eva Rappaport.

The Oberhasli (Fig. 33.7) is a medium-sized dairy goat breed, also of Swiss origin, with usually solid red or black colors. Production records of up to 3,300 lb milk have been reported.

Gall (1981) reported the following annual average milk production records from the major dairy-goat breeds in the United States (only goats having more than five lactations during 276 to 305 days were used): Toggenburg, 1,940 lb; Saanen, 2,035 lb; Alpine, 2,024 lb; Nubian, 1,662 lb; and LaMancha, 1,719 lb.

FIGURE 33.5 A Nubian doe. Note the ear size and shape and the long legs. Courtesy of Cindy Schneider.

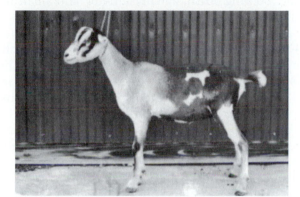

FIGURE 33.6 A LaMancha doe. Note the external earless condition. This individual is homozygous for earlessness. Courtesy of Cindy Schneider.

FIGURE 33.7 Oberhasli does are usually solid red or blackish red. Courtesy of Mary Slabach.

CARE AND MANAGEMENT

Simple housing for goats is adequate in areas where the weather is mild because goats do best when they are outside on pasture or in an area where they can exercise freely. If the weather is wet at times, an open shed with a good roof is necessary. Lactating goats can be fed in an open shed in stormy weather and taken to the milking parlor for milking. If the weather is severely cold, an enclosed barn with ample space is needed. At least 20 square feet per goat is needed if goats are to be housed in a barn, but 16 square feet per goat may suffice in open sheds.

The use of several small pastures permits goats to be periodically moved, thus increasing grazing efficiency and reducing the risk of parasite infestation. Fences must be properly constructed with woven wire approximately 6 feet high and with 6-inch stays. Fencing for goats is different from fencing for cows; a woven-wire fence is best for containing goats because they can climb a rail fence. In addition, a special type of bracing at corners of the fence is necessary because goats can walk up a brace pole and jump over the fence. With electric fences, two electric wires should be used—one located low enough to keep goats from going underneath and one high enough to keep them from jumping over. However, an electric fence may not contain bucks of any breed; some may go through the fence despite the shock. Good fences are needed to keep goats away from neighbors, trees, and gardens.

Several pieces of equipment are needed for normal care: a tattoo set, hoof shears and hoof knife, a grooming brush, an emasculator, and a balling gun for administering boluses (large pills for dosing animals).

All breeds of dairy goats are available in polled strains. The polled condition has advantages because a polled goat is less likely to catch its head in a woven-wire fence than is a horned goat. Horned animals should be disbudded during the first week following birth by use of an electric dehorning iron. Goat breeders were plagued by the genetic linkage between hermaphroditism and polledness. However, polled goats with good fertility can be found.

Goats' feet should be kept properly trimmed to prevent deformities and footrot. A good pruning shear is ideal for leveling and shaping the hoof, but final trimming can be done with a hoof knife. Footrot should be treated with formaldehyde, copper sulfate, or iodine. Affected goats are best isolated from the others and placed on clean, dry ground after treatment for footrot.

All kids should be identified by a tattoo in the ear, except for LaMancha and Pygmies, where identification is in the tail-web. Ear tags are also useful and are easy to read. The tattoo serves as permanent identification in case an ear tag is lost. All registered goats must be tattooed because the goat registry associations do not accept ear-tag identification.

Male kids not acceptable for breeding should be castrated. This can be done by constricting the blood circulation to the testicles by use of a rubber band elastrator or by surgically removing the testicles. Some breeders nonsurgically crush the cords to the testicles with an emasculator. After the blood supply to the testicles is discontinued, they usually atrophy but exceptions occur. After using the emasculator, the blood circulation to the testicles may continue and although the male is sterile, he continues to produce testosterone and has normal sex drive. A problem with the elastrator is that tetanus can occur when the tissue dies below the elastrator band. Surgical removal of the testicles creates a wound that attracts flies in the warm season. A fly repellant should be applied around the wound. It is recommended that a veterinarian castrate older goats. Whatever method is used, caution must be taken to prevent infection, gangrene, and other complications.

It is important that the udder of dairy goats has strong fore and rear attachments and that the two teats be well spaced. Goats are milked by hand (Fig. 33.8) or by milking machines. The milking stanchion should be elevated to place the goat at a convenient level for the person doing the milking (Fig. 33.9). Clean, sanitary conditions of the milking stanchion, the milking machine, the goat, and her udder and teats are very important prior to milking. After milking, the teats should be dipped in a weak iodine solution.

Considerations of special importance in managing dairy goats are care at time of breeding, kidding, and feeding.

Time of Breeding

The female goat comes into heat at intervals of 15–18 days until she becomes pregnant. Young does can be bred first when they weigh 85–95 lb at about 6–10 months of age, which means that does in good general condition may be bred to have kids at 1 year of age. Most goats are seasonal breeders; the normal breeding season occurs in September, October, and November. If no effort is made to breed does at other times, most of the young will be born in February, March, or April. Normal goat lactations are 7–10 months in duration. A period of at least 2 months when no goats are lactating might occur, but with staggered breeding, the kidding and milking seasons can be extended. Housing goats in the dark for several hours each day in the spring and summer months (to simulate the onset of shorter days) causes some to come into estrus earlier than usual. Conversely, artificial additional light in the goat barn may delay estrus in the autumn.

One service is all that is needed to obtain pregnancy, but it is generally wise to delay breeding for a day after the goat first shows signs of heat. The doe stays in heat from 1–3 days, but the optimum period of standing heat may last only a few hours at the end of estrus. A female in heat is often noisy and restless, and her milk production may decrease sharply. She may disturb other goats in the herd, so it is advisable to keep her in a separate stall until she goes out of heat.

Male goats during the breeding season usually have a rank odor, especially when confined to a pen. Housing the males downwind apart from the does will reduce some of the nuisance of male odors. Cleanliness of the male's long beard and shaggy hair helps reduce "bucky" odors.

Careful records should be kept showing breeding dates and the bucks used. From breeding records, it is possible to calculate when does should deliver kids, as the gestation period is about 150 days. Detailed breeding records are essential for knowing ancestors, which in turn helps in the selection of superior males to produce especially desirable offspring.

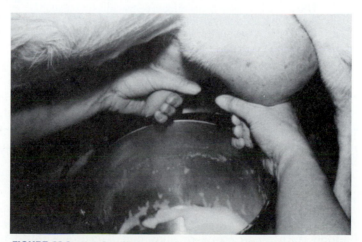

FIGURE 33.8 Hand-milking the goat. Courtesy of G. F. W. Haenlein, University of Delaware.

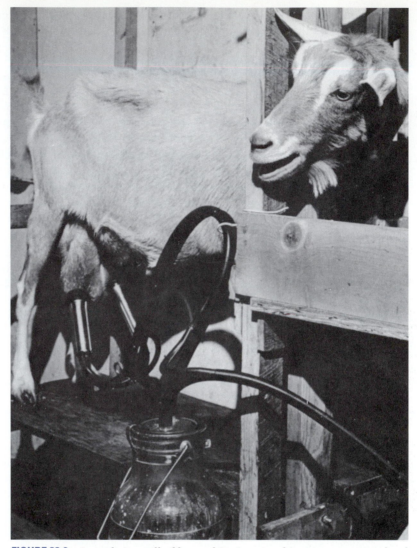

FIGURE 33.9   A goat being milked by machine in a stanchion at a convenient level for milking. Courtesy of Eva Rappaport.

Delivery of twins is common and delivery of triplets occurs on occasion, particularly among mature does. More males are conceived and born than are females. The ratio is approximately 115:100. Mature does average 2 kids, whereas younger does average 1.5 kids. Kids at birth weigh 5–8 lb, but singles may be heavier than twins or triplets, and males are usually heavier than females.

Time of Kidding

A doe almost ready to deliver young should be placed in a clean pen that is well bedded with clean straw or shavings. The doe should have plenty of clean water, some laxative feed (such as wheat bran), and fresh, soft legume hay. She should be carefully observed but not disturbed unless assistance is necessary, as indicated by excessive straining for 3 hours or more. If the kid presents the front feet and head, delivery should be easy without assistance. If only the front feet, but not the head, are presented, the kid should be pushed back enough to bring the head forward in line with the front feet. Breech presentations are not uncommon, but these deliveries should be rapid to prevent the kid from suffocating. If assistance is needed, a qualified attendant with small well-lubricated hands should pull when contractions occur. Gentleness is essential; harsh or ill-timed pulling can cause severe internal damage.

As soon as the kid arrives, its mouth and nostrils should be wiped clean of membranes and mucus. In cold weather, it is necessary to take the newborn to a heated room for drying. A chilled kid must be helped to regain its body temperature with a heat lamp or even by immersing its body up to the chin in water that is as hot as can be tolerated when the attendant's elbow is immersed in it for 2 minutes. The navel should be kept out of the water. Kids should be encouraged to nurse as soon as they are dry. Nursing helps the newborn kid to keep warm.

Difficulty in kidding may be caused by a low selenium supply, diseases, and internal parasites. It is advisable to use disposable rubber or plastic gloves when assisting with delivery and to wash and disinfect everything diligently. Some diseases (including brucellosis) are transmissible to people; however, brucellosis has not been recently reported in U.S. goats.

Feeding

Because goats are ruminants, they can digest roughage effectively. However, the types and proportions of feed should be related to the functions of the goats. For example, dry (nonmilking) does and bucks that are not actively breeding perform satisfactorily on ample browse, good pasture, and good-quality grass and legume hay. If grass is short and hay is of poor quality, and goats are milking well, feeding of supplemental concentrates is necessary. Overfeeding of supplemental concentrates can cause diarrhea or obesity, which interferes with reproduction and subsequent lactation.

Heavily lactating does and young does that must grow while lactating should be given good-quality hay (ad libitum) and concentrates liberally. Good-quality pasture should supplement hay well, but even then lactating goats require some roughage in the form of hay or other source of long fiber to prevent scouring.

Pregnant does should be fed to gain weight to ensure adequate nutrition of the kids. A doe should be in good flesh, but not fat, when she kids because she draws from her body reserves for milk production.

Grains such as corn, oats, barley, and milo may be fed whole because goats crack grains by chewing. If protein supplements are mixed with the grain or if the feed mix is being pelleted, the grain may be rolled, cracked, or coarsely ground. Some people prefer to mix the hay and grain and prepare the ration in a total pellet form. There is added expense in pelleting, but much less feed is wasted.

In some areas, deficiencies in minerals such as phosphorus, selenium, and iodine may exist. The use of iodized or trace-mineralized salt along with dicalcium phosphate or steamed bonemeal usually provides enough minerals if legume hay or good pastures are available. Calcium is usually present in sufficient quantity in legume hay and phosphorus is adequately provided in grain feeds. To be safe, it is best to provide a mixture of trace-mineralized salt and bonemeal free-choice.

Young, growing goats and lactating goats need more protein than do bucks or dry does. A ration containing 12–15% protein is desirable for bucks and dry does, but 15–20% protein may be better for young goats and for does that are producing much milk.

Generally, kids must be allowed to nurse their dams to obtain the first milk—colostrum. After 3 days, kids may be removed from their mothers and given milk replacer by means of a lamb feeder or hand-held bottle until they are large enough to eat hay and concentrates. It is advisable to encourage young kids to eat solid feed at an early age (2–3 weeks) by having leafy legume hay and palatable fresh concentrates such as rolled grain available at all times. Solid feeds are less expensive than milk replacers, and when the kids can do well on solid feeds, milk replacers should not be fed. Small kids need concentrates until their rumens are sufficiently developed to digest enough roughage to meet all their nutritional needs. At 5–6 months of age, young goats can do well on good pasture or good-quality legume hay alone.

CONTROLLING DISEASES AND PARASITES

Major diseases affecting goats are Johnes disease, caseous lymphadenitis, caprine pleuropneumonia, ecthyma, enterotoxemia, goat pox, herpesvirus, Pasteurella hemolytica, tetanus, and viral leukoencephalomyelitis—each are caused by a virus or a bacteria. As goats are highly resistant to **bluetongue,** there is no reason for alarm even if sheep having bluetongue are grazing in the same pasture.

Johnes disease also affects sheep, and it is a serious health problem in goats. Affected goats become unthrifty, emaciated, and unproductive.

A troublesome disease, **caseous lymphadenitis,** is characterized by nodules in the lymph area (throat). When one of the infected nodes is opened, a thick, caseous material is exposed. The disease is difficult to control. One should check animals in a herd from which breeding stock is considered for purchase and then not buy animals from a herd that has infected animals. Also, an isolation program is desirable for purchased animals to see if they develop the disease. Infected animals should be isolated; then infected nodes are opened and flushed with H_2O_2.

Caprine Arthritis Encephalitis (CAE) affects the central nervous system of young kids where head tremors, loss of coordination and partial paralysis can be observed. The arthritis form of the disease occurs in mature goats and affects the feet and other limb joints. Other symptoms are loss of condition and "hard udder" (a firm, swollen udder in freshly kidded goats). There is no effective treatment for the disease. Preventative methods include reducing contact with infected animals (e.g., at shows and sales) and blood testing suspects and removing infected animals from the herd.

Caprine pleuropneumonia results in high mortality. The services of a veterinarian should be obtained to help control this disease.

Contagious ecthyma, also called **soremouth,** is contagious to sheep and humans. Animals that recover from it are immune for several years. A live vaccine can be used on kids on premises that are infected.

Enterotoxemia is caused by *Clostridium perfringens* types C and D. Animals of all ages are susceptible. *C. perfringens* antitoxin given intravenously or subcutaneously will give dramatic response. Toxoid vaccination given to month-old kids followed by a second dose in 2 weeks and booster doses each year will control the disease. This disease, sometimes called *overeating* disease, can be prevented by avoiding feeding excess milk and grain.

Goat pox causes lesions on mucous membranes and skin. It is prevalent in North Africa, the Middle East, Australia, and Scandinavia but not in the United States.

Herpesvirus is serious in neonatal kids.

Young kids are highly susceptible to **Pasteurella hemolytica.** They can be treated with penicillin with successful results.

Tetanus is caused by *Clostridium tetani,* which infects wounds and often causes death. The causative organism is prevalent in soil contaminated with horse feces. Two doses of toxoid to goats over 1 month of age and then an annual booster injection will control the disease.

Mastitis is an inflammation of the udder predisposed by bruising, lack of proper sanitation, and improper milking. The milk becomes curdled and stringy. It is advisable to use a milk-culture sensitivity test and engage the services of a veterinarian for treating severe cases. An udder may be treated with a suitable antibiotic by sliding a special dull plastic needle up the teat canal into the udder cistern and depositing the antibiotic, or by systemic antibiotic treatment intramuscularly or intravenously. Hot packs may help reduce the edema and, in severe cases, frequent milking is necessary and most helpful. Dry goat treatments can effectively prevent mastitis.

Brucellosis is transmissible to humans. All goats should be tested for goat brucellosis, and reactor animals or suspects must be slaughtered. Goats do not get the same brucellosis that affects cows.

Footrot occurs when goats are kept on wet land. It is contagious, but the bacterial organism causing it does not live long in the soil. Any animal to be introduced into a herd should be isolated for several days to see if footrot is present. Severe hoof trimming of infected feet, followed by treating with formaldehyde (as described in Chapter 29), cures this disease. Generally, dairy goats dislike wet land and footrot is not a common problem in U.S. dairy goats.

Ketosis, or pregnancy disease, rarely occurs in goats. Affected goats are in pain and cannot walk. Exercising goats that are pregnant, reducing the amount of feed in the latter part of the pregnancy to avoid fat conditions, and being certain that they have clean, fresh water at all times helps prevent ketosis.

Milk fever can occur, but rarely does, in goats that are lactating heavily. Some heavy-milking goats may deplete their calcium stores. They cannot stand and may die of progressive paralysis. An intravenous injection of calcium gluconate results in rapid recovery. The goat may be up and normal in less than an hour after a calcium gluconate injection.

Goats are subject to many external and internal parasites. The external parasites include **lice** and **mites.** The animals may be sprayed, dipped, or dusted with appropriate insecticides similar to those used on dairy cows. Dairy feed stores are usually good sources of materials and advice for treatment.

Internal parasites of goats include **stomach worms** and **coccidia.** The same treatment for stomach worms that is effective in sheep or horses can be used. **Coccidiosis,** caused by a protozoan organism in the intestinal tract, is found mostly in young goats. **Drenching** the animal with sulfa drugs or certain antibiotics helps control coccidiosis. Routine treatment with recommended anthelmintics helps control internal parasites. Professional help in administering these drugs is advisable.

THE ANGORA GOAT

Angora goats (Fig. 33.10) are raised chiefly to produce mohair, which is made into fine clothing. It is estimated that world mohair production is approximately 15,000 tons per year. These goats are produced in Turkey, South Africa, and, in the United States, the Southwest (Texas has the most Angora goats of any state), the Ozarks of Missouri and Arkansas, and the Pacific Northwest. They do best on browse, so they are most commonly kept on rough, brushy areas. They are not as well adapted for intensive grazing pastures because they prefer broadleaf plants over some grasses, and because they are more likely to become parasitized when grazing near the soil level in overgrazed pastures.

Some Angora goats have a tendency to shed their mohair in the spring, but selection against this trait has been successful. Some Angora goats are not shorn for 2–3 years, which allows the mohair to grow to lengths of 1–2 feet. This special mohair brings a high price and is used for making wigs and doll hair. Shearing is usually done in spring or twice a year. The clip from does

FIGURE 33.10   An Angora goat showing a full fleece of mohair. Angora goats are useful for meat production, brush clearing, and mohair production. Courtesy of Mr. and Mrs. Don F. Kessi.

weighs 4–6 lb, whereas bucks and wethers may shear 5–8 lb. Highly improved goats often shear 10–15 lb of mohair per year. Three classes of mohair are produced: the **tight lock** with ringlets the full length of the fibers; the **flat lock,** which is wavy; and the **fluffy,** or **open fleece.** The tight-lock fleeces are fine in texture but low in yield. The flat lock lacks the fineness of the tight lock but is satisfactory for making cloth, and its yield is high. The fluffy fleece is of low quality, often quite coarse, and is easily caught in brush and thereby lost.

Fleeces should be sorted at shearing time. Coarse fleeces and those with burs should be kept separate from good fleeces. The clip from each fleece should be tied separately so that it can be properly graded.

Fertility can be low in Angora goats. Losses of kids on the open range or in scattered timber areas of the West Coast states are often high, especially because of predators such as coyotes, bobcats, and wild dogs. Generally, the kid crop that is raised to weaning is about 80% of the number conceived, as estimated on the basis of the number of does bred. Losses of 20% or more may occur when proper care is lacking at the time kids are born.

Angora goats usually have horns that are curved sideways. Swiss (dairy) goats have straight sickle horns. When they are in good condition on the range, mature bucks and wethers may weigh 150–200 lb. Mature does weigh 90–110 lb, but well-fed show animals usually weigh more. Kids are usually weaned at 5 months of age, which allows the doe to gain in weight at breeding time. Breeding over a period of days is usually done in late September and October. Since the gestation is 147–152 days, this time of breeding results in the kids arriving in the spring after the weather has moderated. Kidding may take place in open sheds when the weather is severe, or outside when the weather is favorable. Young does are usually first bred to kid at the age of 2 years but, if they are properly developed, they may be bred to kid when they are yearlings.

Angora goats need shelter in rainy periods or wet snows, especially after shearing; an open shed is sufficient. The long mohair coats of Angoras keep them warm during cold weather, but they may nevertheless crowd together and smother in severe snowstorms if no shelter is provided.

Angora goats are subject to the same parasites and diseases as dairy goats, and the same treatments are effective for both. In most other respects, though, the care given Angora goats resembles the care given sheep rather than that given dairy goats.

Angora, Spanish meat goats, or cull dairy goats are used in some areas to kill brush. One method is to stock the area heavily with goats and give them little feed as long as their health is not impaired. Keeping the goats hungry for green feed causes all brush to be destroyed in 2 years because the goats nip off the buds of new growth, thus starving the plant. It usually requires 2 years to starve hardy plants, but some of the less hardy ones will succumb the first year. Another method is to mix goats with cattle for 6–10 years. The goats tend to eat leaves of brushy plants while the cattle consume the grasses.

CASHMERE GOATS

Cashmere goats originated in Central Asia; they are white in color, have erect ears, and long, twisted horns. In recent years, there has been an increase in Cashmere goats and producers in the United States.

Cashmere goats produce the cashmere fiber which is extremely fine and highly valued. The average yield per animal is ½ to 2 pounds of unscoured fiber with 60% yield as clean fiber. The estimated world cashmere production is 3,000–4,000 tons per year. The fine undercoat is combed out with scrupulous care, the process taking 8 to 10 days. The majority of the world's production of cashmere is processed into the most luxurious, expensive cloth and knitwear in Scotland.

THE MEAT GOAT

Meat goats are found mainly in the southwestern and western United States, though they are raised primarily in Africa and the Middle East, where their main uses are for meat and skins. In West Africa there has been a preference for a small, well-muscled, agile goat—the West African Dwarf. These goats can obtain most of their feed from brushy plants and low trees. As many of the areas away from cities and towns are lacking in refrigeration, small goats that are slaughtered can be consumed by the family before the meat spoils.

Goats of any kind are usually run with sheep outside of the United States; this practice may be advantageous in the United States as well. Goats normally eat more browse and forbs than sheep.

Rams should never be run with doe goats, and buck goats should never be run with ewes. These animals will mate when the females come in heat (estrus); if they conceive, the pregnancy will usually terminate at about 3 or 4 months. There are reports of rare cases of live sheep–goat crosses delivered alive at term.

THE PYGMY GOAT

Pygmy goats were developed in Africa. Those brought to the United States came from the Cameroon area to the Caribbean and were then imported to the United States. Early Pygmy goats were brought to the United States as exotic zoo animals. Zoos sold Pygmy animals not only to other zoos but also to research scientists. A colony of Pygmy goats was established in Oregon and several other research herds were established, including herds at the University of California. Because Pygmies are small in size, the cost of keeping them is minimal. Pygmy goats are also raised for meat and milk.

Breeders of Pygmy goats enjoy showing their goats just as do dairy goat and Angora goat breeders. There are over 50 Pygmy goat shows held annually in the United States, where several hundred goats are shown.

Pygmy goats vary in color from black to white with a dorsal stripe down the back and dark on the legs. On all Pygmy goats except the black ones, the muzzle, forehead, eyes, and ears are accented in tones lighter than the dark portion of the body.

The Pygmy goat is also becoming popular as an animal for 4-H projects (Fig. 33.11). The goats are small and do not require a lot of space or feed. All goats are very friendly animals, which makes them ideal for young people to handle; they are also easily trained.

Refer to Chapter 36 for a more detailed discussion of management principles and their applications.

FIGURE 33.11 An Agouti Pygmy goat can be a useful 4-H project animal. Courtesy of Deana Mobley.

CHAPTER SUMMARY

■ Goats have a large impact on the economy and food supply of people in several countries of the world. The consumption of both goat milk and goat meat is increasing in the United States.

■ The major kinds of domesticated goats are (1) dairy goat, (2) meat goat, (3) Angora goat, (4) Cashmere goat, and (5) pygmy goat.

■ Besides the Angora goat, the most numerous goats are the leading breeds of dairy goats: Toggenburg, Saanen, Alpine, Nubian, LaMancha, and Oberhasli.

■ Goat management involves understanding the primary principles and production practices affecting reproduction, feeding, and health.

REVIEW QUESTIONS

1. What are the five kinds of goats and what are they used for?
2. *True or False:* Meat goats outnumber dairy goats worldwide.
3. What is the length of the estrous cycle in goats?
4. What is the duration of gestation in goats?
5. *True or False:* Dairy goats are about equal to dairy cows in efficiency of converting feed into milk.
6. *True or False:* Dairy goat milk is more easily digested by humans than is cows' milk.
7. What are the six major breeds of dairy goats in the United States?
8. Why is woven wire preferred to rail for fencing of goats?
9. How are goats permanently identified?
10. When is the normal breeding season for goats?
11. *True or False:* It is acceptable to run does with rams and bucks with ewes because goats and sheep will not interbreed.

SELECTED REFERENCES

Publications

Dairy Goat J. (monthly). Schroeder Publications, 6041 Monona Dr., Monona, WI 53716-3989.

Ensminger, M. E., and Parker, R. O. 1986. *Sheep and Goat Science.* Danville, IL: Interstate Printers.

FAO Production Yearbook, 1995. Rome, Italy: Food and Agriculture Organizations of the United Nations.

Haenlein, G. F. W. 1993. Producing quality goat milk. *International Journal of Animal Science* 8:79.

Haenlein, G. F. W., and Ace, D. L. (eds.). 1984. *Extension Goat Handbook.* Washington, DC: USDA Extension Service.

International symposium: Dairy goats. 1980. *J. Dairy Sci.* 63:1591–1781.

Mackenzie, D. and Goodwin, R. 1993. *Goat Husbandry.* 5th ed. Boston: Faber and Faber.

Maclean, J. T. 1987. *Goat Husbandry in the U.S.,* 1976–1986. Beltsville, MD: USDA.

National Research Council. 1981. *Nutrient Requirements of Goats.* Washington, DC: National Academy Press.

National Symposium on Goat Meat and Marketing. 1991. Langston, OK: Caprine Development Foundation.

Neimann-Sorenson, A. and Tube, D. E. (eds.-in-chief). World Animal Science, Series C: Production System Approach. 1982. I. E. Coop (ed.). *Sheep and Goat Production.* New York: Elsevier.

Proceedings of the Fifth International Conference on Goat Production and Diseases. 1992. New Delhi, India.

Wilkinson, J. M. and Stark, B. A. 1987. *Commercial Goat Production.* Boston: BSP Professional Books.

Visuals

Dairy Goat Kit: *Dairy Goat Management;* and *Fitting and Showing Dairy Goats* (sound filmstrips). Vocational Education Productions, California Polytechnic State University, San Luis Obispo, CA 93407.

Introduction to Dairy Goat Production, Facilities, Breeding/Prekidding Management, Kidding Management; Care of Kids: Birth to Weaning; Maintaining Herd Health/Milking Management; Marketing/Obtaining Financing; and *Dual-Purposes Goat* (videotapes). Agricultural Products and Services, 2001 Killebrew Dr., Suite 333, Bloomington, MN 55420.

Livestock Judging (slide set no. 7; major breeds of goats illustrated). Vocational Education Productions, California Polytechnic State University, San Luis Obispo, CA 93407.

CHAPTER 34

Animal Behavior

A knowledge of animal behavior is essential to understanding the whole animal and its ability to adapt to various management systems utilized by livestock and poultry producers. For example, the value and performance of farm animals can be increased when managers apply their knowledge of animal behavior.

Animal behavior is a complex process involving the interaction of inherited abilities and learned experiences to which the animal is subjected. Behavioral changes enable animals to adjust to changing conditions, improve their chances of survival, and serve humans. Producers who understand patterns of behavior can manage and train animals more effectively and efficiently.

Basically, there are two major fields of animal behavior, one is psychology and the other is **ethology.** Historically, psychology has been directed toward studying learning in humans and applying insights gained from nonhuman animal studies to understanding human behavior. There are many different schools within the broad field of psychology. Ethology, however, originated with naturalists (going back to Aristotle) who originally emphasized instinctive behavior, but who also studied learned behavior.

Instinct (reflexes and behavioral patterns) is inherently present at birth. All mammals, at birth, have the instinct to nurse even though they must first learn the location of the teat. Shortly after hatching, chicks begin pecking to obtain feed.

Habituation is lack of response to a repeated stimulus such as a low-flying aircraft. When animals first see and hear the airplane, the novelty of it may frighten them. However, after they are repeatedly exposed to this experience, they become habituated and are no longer frightened.

Conditioning is the process whereby an animal makes an association between a previously neutral stimulus (e.g., a bell) or behavioral response (e.g., lifting its foot) and a previously significant stimulus, such as a shock or food. There are two types of conditioning:

1. **Classical conditioning**—e.g., Pavlov's noted study showed an association formed between an unconditioned stimulus (sight of food, which caused salivation) and a neutral stimulus (sound of a bell). The animal initially salivated at the sight of food; later, merely the sound of the bell produced salivation because of the previous association between the two stimuli.

2. **Operant conditioning** is learning to respond in a particular way to a stimulus as a result of **reinforcement** when the proper response is made. Reinforcement is a punishment or reward for making the proper response. Animals avoiding an electric fence and cattle coming to the feedbunk when they see a feed truck are examples of operant conditioning. In the first example, the animal is negatively reinforced by the shock. Cattle are positively reinforced, in the second example, with feed from the feed truck when they arrive at the feedbunk.

Trial and error is trying different responses to a stimulus until the correct response is performed, at which time the animal receives a reward. For example, newborn mammals soon become hungry and want to nurse. They search for someplace to nurse on any part of the mother's body until they find the teat. This is trial and error until the teat is located; then, when the young mammals nurse, they receive milk as a reward. Soon they learn where the teat is located and find it without having to go through trial and error. Thus, the young become conditioned in nursing behavior through reinforcement.

Reasoning is the ability to respond correctly to a stimulus the first time that a new situation is presented. **Intelligence** is the ability to learn to adjust successfully to certain situations. Both short-term and long-term memory are part of intelligence.

Imprinting covers those processes where the helpless young bond to their caretakers—usually their dam. The way imprinting occurs varies between species. Odors and the dam licking the fluids from the newborn leads to bonding and rapid recognition in cattle and sheep. Creating a nest aids in the bonding of a sow with her young pigs.

SYSTEMS OF ANIMAL BEHAVIOR

Farm animals exhibit several major systems or patterns of behavior: (1) sexual, (2) caregiving, (3) care-soliciting, (4) agonistic, (5) ingestive, (6) eliminative, (7) shelter-seeking, (8) investigative, and (9) allelomimetic. Some of these behavior systems are interrelated, though they are discussed separately in this chapter. It is not the intent here to describe, in detail, the different behavior patterns for all farm animals. The major focus is to identify the behavioral activities that most significantly affect animal well-being, productivity, and profitability. By understanding animal behavior, producers can plan and implement more effective management systems for their animals.

SEXUAL BEHAVIOR

Observations on sexual behavior of female farm animals are useful in implementing breeding programs. Cows that are in heat, for example, allow themselves to be mounted by others. Producers observe this condition of "standing heat," or estrus, to identify those cows to be **hand-mated** or bred artificially. Ewes in heat are not mounted by other ewes, but vasectomized rams can identify them.

Males and females of certain species produce **pheromones,** chemical substances that attract the opposite sex. Cows, ewes, and mares may have pheromones present in vaginal secretions and in their urine when they are in heat. Bulls, rams, and stallions will smell the vagina and the urine using a nasal organ that can detect pheromones. A common behavioral response in this process is called **flehmen,** during which the male animal lifts its head and curls its upper lip.

It appears that in a sexually active group of cows, the bull is attracted to a cow in heat most often by visual means (observing cow-to-cow mounting) rather than by olfactory clues. The bull follows a cow that is coming into heat, smells and licks her external genitalia, and puts his chin on her rump. When the cow is in standing heat, she stands still when the bull chins her rump. When she reaches full heat, she allows the bull to mount.

When females are sexually receptive, they usually seek out a male if mating has not previously occurred. Females are receptive for varying lengths of time; cows are usually in heat for approximately 12 hours, whereas mares show heat for 5–7 days with ovulation occurring during the last 24 hours of estrus.

Vigorous bulls breed females several times a day. If more than one cow is in heat at the same time, bulls tend to mate with one cow once or a few times and then go to others. Other bulls may become attached to one female and ignore the others that are also in heat.

The ram chases a ewe that is coming into heat. The ram champs and licks, puts his head on the side of the ewe, and strikes with his foot. When the ewe reaches standing heat, she stands when approached by the ram.

The buck goat snorts when he detects a doe in heat. The doe shows unrest and may be fought by other does. Mating in goats is similar to that in sheep.

The boar does not seem to detect a sow that is in heat by smelling or seeing. If introduced into a group of sows, a boar chases any sow in the group. The sow that is in heat seeks out the boar for mating, and when the boar is located she stands still and flicks her ears. Boars produce pheromones in the saliva and preputial pouch, which attracts sows and gilts in estrus to the boars. Ejaculation requires several minutes for boars in contrast to an instantaneous ejaculation by rams and bulls.

The sequence of events in estrous detection in horses appears to be similar to swine (previously described). The stallion approaches a mare from the front and a mare not in heat runs and kicks at the stallion. When the mare is in standing heat she stands, squats somewhat, and urinates as he approaches. Her vulva exhibits **winking** (opens and closes) when she is in heat.

In chickens and turkeys, a courtship sequence between the male and female usually takes place. If either individual does not respond to the other's previous signal, the courtship does not proceed further. After the courtship has developed properly, some females run from the rooster, which chases them until they stop and squat for mating. The male chicken or turkey stands on a squatting female and ejaculates semen as his rear descends toward the female's cloaca. Semen is ejaculated at the cloaca and the female draws it into her reproductive tract while the male is mounting her.

Male chickens and turkeys show a preference for certain females and may even refuse to mate with other females. Likewise, female chickens and turkeys may refuse to mate with certain males. This is a serious problem when pen matings of one male and 10–15 females are practiced. The eggs of some females may be infertile. This preferential mating is a greater problem in chickens, as AI is the common breeding practice in turkeys.

Little relationship appears to exist between sex drive and fertility in male farm animals. In fact, some males that show extreme sex drive have reduced fertility because of frequent ejaculations that result in semen with reduced sperm numbers.

Research studies show that many individual bulls have sufficient sex drive and mating ability to fertilize more females than are commonly allotted to them. An excessive number of males, however, are used in multiple-sire herds to offset the few that are poor breeders and to cover for the social dominance that exists among several bulls running with the same herd of cows. The bull may guard a female that he has determined is approaching estrus. His success in guarding the female or actually mating with her is dependent on his rank of dominance in a multiple-sire herd. If low fertility exists in the dominant bull or bulls, then calf crop percentages will be seriously affected even in multiple-sire herds.

Tests have been developed to measure libido and mating ability differences in young bulls. While behavioral differences are evident between different bulls, these tests, based on pregnancy rates, have not proven accurate for use by the beef industry.

Bulls being raised with other bulls commonly mount one another, have a penal erection, and occasionally ejaculate. Individual bulls can be observed arching their back, thrusting their penis toward their front legs, and ejaculating.

Bulls can be easily trained to mount objects that provide the stimulus for them to experience ejaculation. AI studs commonly use restrained steers for collection of semen. Bulls soon respond to the artificial vagina when mounting steers, which provides them with a sensual reward.

Mating behavior has an apparent genetic base, as there is evidence of more frequent mountings in hybrid or crossbred animals.

Some profound behavior patterns are associated with sex of the animal and changes resulting from castration. This verifies the importance of hormonal-directed expression of behavior. Intact males have more aggressive behavior, whereas castrates are more docile after losing their source of male hormone.

CARE-GIVING BEHAVIOR

Care-giving behavior can originate from the sire or dam; however, most care-giving is maternally oriented.

When the young of cattle, sheep, goats, and horses are born, the mothers clean the young by licking them. This stimulates blood circulation and encourages the young to stand and to nurse. Sows do not clean their newborn but encourage them to nurse by lying down and moving their feet as the young approach the udder region. Thus, they help the young to the teats.

Most animal mothers tend to fight intruders, especially if the young squeal or bawl. Often cows, sows, and mares become aggressive in protecting their young shortly after parturition. Serious injury can occur to producers who do not use caution with these animals.

Strong attachments exist between mother and newborn young, particularly between ewe and lamb and cow and calf. Beef cows diminish their output of milk about 100–120 days after birth of young, and ewes do the same after 60–75 days. This reduction in milk encourages the young to search for forage, the consumption of which stimulates rumen development. It is at this time that care-giving by the mother declines.

If young pigs have a high-energy feed available at all times, they nurse less frequently. Without the strong stimulus of nursing, sows reduce their output of milk. Some sows may wean their pigs early and show little concern for them a few days after they are weaned. Pigs are usually weaned by producers at 21–35 days of age.

There is evidence that more cows calve during periods of darkness than during daylight hours. The calving pattern, however, can be changed by when cows are fed. Cows that are fed during late evening have a higher percentage of their calves during daylight hours.

CARE-SOLICITING BEHAVIOR

Young animals cry for help when disturbed, distressed, or hungry (Fig. 34.1). Lambs bleat, calves bawl, pigs squeal, and chicks chirp. Even adult animals call for help when under stress. The female and her offspring may recognize each other's vocal sound; however, it appears that the most effective way the dam recognizes her offspring is by smell. The young usually nurses with its rear end toward the female's head. This allows a dam to smell her offspring and decide to accept or reject it. A rejected young animal is usually bunted with the head of the dam and kicked with the rear legs when it attempts to nurse. The young animals are less discriminate in their nursing behavior than are their dams.

AGONISTIC BEHAVIOR

Agonistic behavior includes behavior activities of fight and flight and those of aggressive and passive behavior when an animal is in contact (physically) with another animal or with livestock and poultry producers.

Interaction with Other Animals

Unless castrated when young, the males of all farm animals fight when they meet other unfamiliar males of the same species. This behavior has great practical implications for management of farm animals. Male farm animals are often run singly with a group of females in breeding season, but it is often necessary to keep males together in a group at times other than the breeding

FIGURE 34.1   A calf separated from its dam has a distinct bawl, which communicates distress or dissatisfaction. Courtesy of the American Hereford Association.

season. The typical producer simply cannot afford to provide a separate lot for each male.

Bulls and other males may exert themselves during prolonged physical activity when fighting, and thus generate much heat. Therefore, bulls and other potential fighting males should be put together either early in the morning or late in the evening when the environmental temperature is lower than at mid-day. Often, fighting can be reduced when male animals are mixed and put into a new environment.

If a mature bull is put with young bulls, fighting usually occurs until the younger ones concede to the mature one. After the fighting is over, bulls may start mounting one another (one form of homosexuality).

Cows, sows, and mares usually develop a peck order, but fight less intensely than males. Sows that are strangers to each other sometimes fight. Ewes seldom, if ever, fight, so ewes that are strangers can be grouped together without harm.

Some cows withdraw from the group to find a secluded spot just before calving. Almost all animals withdraw from the group if they are sick.

Early and continuous association of calves is associated with greater social tolerance, delayed onset of aggressive behavior, and relatively slow formation of social hierarchies.

Status and social rank typically exist in a herd of cows, with certain individuals dominating the more submissive ones. The presence or absence of horns is important in determining social rank, especially when strange cows are mixed together. Also, horned cows usually outrank polled or dehorned cows where close contact is encountered, such as at feedbunks or on the feed ground.

Large differences in age, size, strength, genetic background, and previous experience have powerful effects in determining social rank. Once the rank is established in a herd, it tends to be consistent from one year to the next. There is evidence that genetic differences exist for social rank.

Animals fed together consume more feed than when they are fed individually. This competitive environment evidently is a stimulus for greater feed consumption. Dairy calves, separated from their dams at birth, appear to gain equally well whether fed milk in a group or kept separate. There is, however, evidence that they learn to eat grain earlier when group-fed compared with being individually fed. Cattle individually fed in metabolism stalls consume only 50–60% of the amount of feed they will eat if the animals are group-fed.

When fed in a group of older cows, 2-year-old heifers have difficulty getting their share of supplemental feed (Table 34.1). Two-year-old heifers and 3-year-old cows can be fed together without any significant age effect in competition for supplemental feed. These behavior differences no doubt explain some of the nutrition, weight gains, and postpartum interval relationships that are age-related when cows of all ages compete for the same supplemental feed.

Dominant cows raised in a confinement operation usually consume more feed and wean heavier calves. More submissive cows wean lighter calves (25%), and fewer of them are pregnant compared with more aggressive cows. Figure 34.2 shows a 1-hour feeding pattern for two cows of different social rankings.

Interactions with Humans

Producers rank the disposition or temperament of animals from docile to wild or *high-strung*. This evaluation is usually made when the animals are being handled through various types of corrals, pens, chutes, and other working facilities. The typical behavior exhibited by animals with poor dispositions is one of fear or of aggressive fighting or kicking.

There is evidence that farm animals develop good or poor dispositions from the way they have been treated or handled, though there is evidence that disposition has an inherited basis as well. A few heritability estimates for disposition are in the medium-to-high category, indicating the trait would respond to

TABLE 34.1 Weight Changes of 2-Year-Old Heifers Fed Separately As a Group or Together with Older Cows

Treatment	Weight Change
Pastured and fed with older cows	25-lb loss
Pastured and fed separately	46-lb gain

Source: Wagnon, 1965.

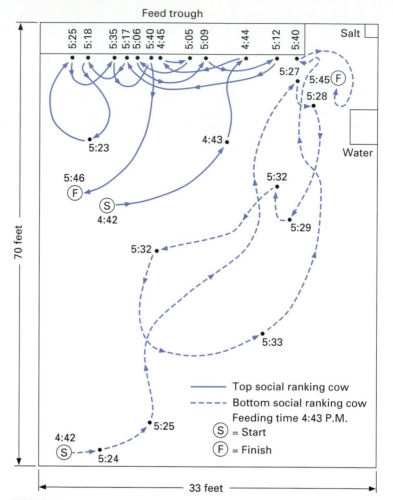

FIGURE 34.2 One-hour feeding pattern for two cows of different social rankings. Adapted from Schake and Riggs, 1972.

selection. Some producers cull or eliminate animals with poor dispositions from their herds and flocks because of the potential for personal injury and economic loss (broken fences and facilities), as well as to reduce the excitability of other animals.

Recent research has shown that cattle with a nervous, excitable temperament (e.g., being highly agitated when restrained) had lower weight gains in the feedlot and a higher incidence of dark cutters.

Behavior During Handling and Restraint

Most animals are handled and restrained several times during their lifetime. Ease of handling depends largely on the animal's temperament, size, previous handling experience, and the design of the handling facilities. Animals remember positive and negative experiences. Livestock with previous experience of calm, quiet handling will be less stressed and easier to handle in the future than animals that have had previous experiences with rough handling.

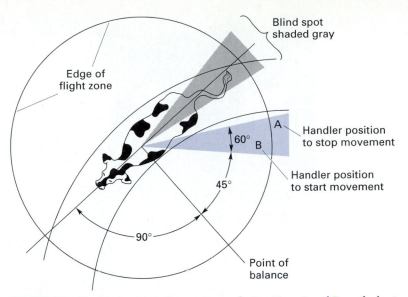

Blind spot shaded gray

Edge of flight zone

A

60°
B

Handler position to stop movement

45°

Handler position to start movement

90°

Point of balance

FIGURE 34.3 Handler positions for moving cattle. Positions A and B are the best places for the handler to stand. The flight zone is penetrated to cause the animal to move forward. Retreating outside the flight zone causes the animal to stop moving. Handlers should avoid standing directly behind the animal because they will be in the animal's blind spot. If the handler gets in front of the line extending from the animal's shoulder, the animal will back up. This is the point of balance. The solid curved lines indicate the location of the curved, single-file chute. Courtesy of Temple Grandin.

Understanding animal behavior can assist in preventing injury, undue stress, and physical exertion for both animals and producers. An example is knowing how to approach animals so they will respond to how the producer prefers the animals to move (Fig. 34.3). Most animals have a *flight zone*. When a person is outside this zone, the animal usually exhibits an inquisitive behavior. When a person moves inside the flight zone, the animal usually moves away. The size of the flight zone depends on the tameness or wildness of the animal.

Blood odor appears to be offensive to some animals; therefore, the reduction or elimination of such odors may encourage animals to move through handling facilities with greater ease. Animals are easily disturbed by loud or unusual noises such as motors, pumps, and compressed air.

With their 310–360° vision, cattle are sensitive to shadows and unusual movements observed at the end of the chute or outside the chutes (Fig. 34.4). For these reasons, cattle will move with more ease through curved chutes with solid sides (Fig. 34.5).

Round pens, having an absence of square corners, handle cattle that are more excitable with less injury occurring to them.

Some breeders claim that AI facilities should permit beef cows to be handled quietly and carefully, and that using facilities where cows have previously felt pain should be avoided. They also state that pregnancy rates will thereby increase. This may sound logical, but research has not substantiated these claims.

FIGURE 34.4 Shadows that fall across a chute can disrupt the handling of animals. The lead animal often balks and refuses to cross the shadows. Courtesy of Temple Grandin.

INGESTIVE BEHAVIOR

Ingestive behavior is exhibited by farm animals when they eat and drink. Rather than initially chewing their feed thoroughly, ruminants swallow it as soon as it is well lubricated with saliva. After the animals have consumed a certain amount, they ruminate (regurgitate the feed for chewing). Cattle graze for 4–9 hours per day and sheep and goats graze for 9–11 hours per day. Grazing is usually done in periods, followed by rest and **rumination.** Sheep rest and ruminate more frequently when grazing than do cattle—cattle ruminate 4–9 hours per day; sheep, 7–10 hours per day. A cow may regurgitate and chew between 300 and 400 **boluses** of feed per day; sheep between 400 and 600 boluses per day.

Under range conditions, cattle usually do not go more than 3 miles away from water, whereas sheep may travel as much as 8 miles a day. When cattle and sheep are on a large range, they tend to overgraze near the water area and to avoid grazing in areas far removed from water. Development of water areas, fencing, placing of salt away from water, and herding the animals are man-

FIGURE 34.5 Animals move more easily through curved chutes with solid sides because the solid sides prevent them from seeing people and other distractions through the fence. Courtesy of Temple Grandin.

agement practices intended to assure a more uniform utilization of the range forage.

Cattle, horses, and sheep have palatability preferences for certain plants and many have difficulty changing from one type or several types of plants to other types. Most animals prefer to graze lower areas, especially if they are near water. These grazing behaviors tend to cause overgrazing in certain areas of the pasture and to reduce weight gains.

The age of the cows and weather affect the typical behavior of cows grazing native range during the winter (Table 34.2). At the Range Research Station in Miles City, Montana, cows grazed less as temperatures dropped below 20°F, and at a −30°F, 3-year-olds grazed approximately 2 hours less than 6-year-olds. Also, with colder temperatures, cows wait longer before starting to graze in the morning. At 30°F, cows started grazing between 6:30 and 7:00 A.M., but at −30°F, they waited until about 10 A.M. to begin grazing.

TABLE 34.2   Activities of Cows Grazing on Winter Range

Activity	Hours
Grazing	9.45
Ruminating	
Standing	0.63
Lying	8.30
Idle	
Standing	1.11
Lying	3.93
Traveling	0.58
Total	24.00

ELIMINATIVE BEHAVIOR

Cattle, sheep, goats, and chickens void their feces and urine indiscriminately. Hogs, by contrast, defecate in definite areas of the pasture or pen. Ease of cleaning swine pens can be planned by knowing defecation patterns of the pigs. Horses tend to void their feces on scent piles of other horses and genders.

Cattle, sheep, goats, and swine usually defecate while standing or walking. All these animals urinate while standing, but not usually when walking. Cattle defecate 12–18 times a day; horses, 5–12 times. Cattle and horses urinate 7–11 times per day. Animals on lush pasture drink less water than when they consume dry feeds; therefore, the amount of urine voided may not differ greatly under these two types of feed conditions.

All farm animals urinate and defecate more frequently and void more excreta than normal when stressed or excited. They often lose a minimum of 3% of their liveweight when transported to and from marketing points. Much of the **shrink** in transit occurs in the first hour, so considerable weight loss occurs even when animals are transported only short distances. Weight loss can be reduced by handling animals carefully and quietly and by avoiding any excessive stress or excitement of the animals.

SHELTER-SEEKING BEHAVIOR

Animal species vary greatly in the degree to which they seek shelter. Cattle and sheep seek a shady area for rest and rumination if the weather is hot, and pigs try to find a wet area. When the weather is cold, pigs crowd against one another when they are lying down to keep each other warm. In snow and cold winds, animals often crowd together. In extreme situations, they pile up to the extent that some of them smother. Unless the weather is cold and windy, cattle and horses often seek the shelter of trees when it is raining. This may be hazardous where strong electrical storms occur because animals under a tree are more likely to be killed by lightning than those in the open.

INVESTIGATIVE BEHAVIOR

Pigs, horses, and dairy goats are highly curious, investigating any nonthreatening strange object. They usually approach carefully and slowly, sniffing and looking as they approach. Cattle also do a certain amount of investigating (Fig. 34.6). Sheep are less curious and more timid than some other farm animals. They may notice a strange object, become excited, and run away from it. An object such as a paper cup on the ground can be either threatening or attractive to the animals. A novel object may attract the animals when they are on pasture. However, this same object may cause bolting and balking if the handler attempts to force the animals to walk over it. In one situation, the object triggers a fear reaction and in the other situation an investigative response.

ALLELOMIMETIC BEHAVIOR

Animals of a species tend to do the same thing at the same time. Cattle and sheep tend to graze at the same time and rest and ruminate at the same time.

FIGURE 34.6 Cattle expressing investigative behavior as a producer is approaching them. Courtesy of R. W. Henderson.

Range cattle gather at the watering place at about the same time each day because one follows another. This behavior is of practical importance because the producer can then observe the herd or flock with little difficulty, notice anything that is wrong with a particular animal, and have that animal brought in for treatment. If one is artificially inseminating beef cattle, the best time to locate range cows in heat is when they gather at the watering place. This type of behavior is also useful in driving groups of animals from one place to another.

OTHER BEHAVIORS

Communication and maladaptive are two other behaviors that are common to the nine behavior systems previously presented. Some highlights of these two behaviors are given.

Communication

Communication exists when some type of information is exchanged between individual animals. This may occur with the transfer of information through any of the senses.

Females more easily adopt the young of others through transfer of the odor of one young animal to another. Cows have fostered several calves if their own calves are removed at birth and the foster calves are smeared with amniotic fluid previously collected from the second water bag. This is an example of imprinting.

Many farm animals learn to respond to the vocal calls or whistles of the producer who wants the animals to come to feed. The animals soon learn that the stimulus of the sound is related to being fed. This is an example of operant conditioning.

The bull vocally communicates his aggressive behavior to other bulls and intruders into his area through a deep bellow. This bellow and aggressive behavior are under the control of the male hormone (testosterone), as the castrated male seldom exhibits similar behavior. This communication behavior is part of the agonistic behavior system.

The bull also issues calls to cows and heifers, especially when he is separated from, but still within sight of, them. This type of communication could be included in the system of sexual behavior.

Horses have at least four vocal and three nonvocal sounds: (1) *squeal,* made during threats and encounters between individuals—high pitched; (2) *nickers,* made by the stallion during mating and between the mare and foal prior to feeding—low pitched; (3) *whinnies,* begin as a squeal and end in a nicker—occurs between horses that are distressed or seek social contact; (4) *groans,* occur during discomfort or anguish; (5) *blows,* nonvocal sounds as air passes through the nostrils during prefeeding or when in alarm; (6) *snorts,* produced during nasal irritation, conflict, or relief; and (7) *snores,* produced by inhaling air which has little part in communication.

Cattle are especially perceptive in their sight, as they have 310–360° vision. This affects their behavior in many ways—for example, when they are approached from different angles and when they are handled through various types of facilities.

Maladaptive or Abnormal Behavior

Animals that cannot adapt to their environment may exhibit inappropriate or unusual behavior. Some animals under intensive management systems, such as poultry and swine, are often kept in continuous housing to reduce costs of land and facilities. Frequently, both chickens and swine resort to cannibalism, which may lead to death if preventive measures are not taken. Some swine producers remove the tails of baby pigs to prevent tail chewing. Tail chewing can cause bleeding, and whenever bleeding occurs, the pigs are likely to become cannibalistic.

Some uncastrated male animals raised with other males masturbate and demonstrate homosexuality. In the latter situation, males mount other males, attempting to breed them. Some of the more submissive males may have to be physically separated from the more aggressive males to prevent injury or death.

The **buller-steer syndrome** is exhibited in steers that have been castrated before puberty. This demonstrates a masculine behavior of other than testosterone origin. Certain steers (bullers) are more sexually attractive for other steers to mount. As one steer mounts a buller, other steers are attracted to do the same. Thus, the activity associated with the buller-steer syndrome can cause physical injury, a reduction in feedlot gains, and additional labor and equipment expense as bullers are usually sorted into separate feedlot pens. Some feedlots experience 1–3% of their steers as bullers.

Growth implants and reduced pen space have been cited as reasons for an increased incidence of buller-steers. Some behaviorists cite evidence of similar homosexual behavior of males in free-ranging natural environments.

CHAPTER SUMMARY

- Animal behavior is a response to instincts and various stimuli to which the animals are exposed. It is a complex process involving the interaction of inherited abilities and learned experiences.

- Behavioral changes enable animals to adjust to changing conditions, improve their chance of survival, and serve humans by adapting to various management systems.

- Farm animals exhibit several major systems or patterns of behavior: (1) sexual, (2) care-giving, (3) care-soliciting, (4) agonistic, (5) ingestive, (6) eliminative, (7) shelter-seeking, (8) investigative, and (9) allelomimetic.

REVIEW QUESTIONS

1. _____ is the scientific study of an animals' behavior in response to its natural environment.

2. _____, or reflexes and responses, is inherently present at birth, whereas _____ is a lack of response to a repeated stimuli.

3. What are the two types of conditioning?

4. What is imprinting?

5. What are the nine major systems or patterns of behavior exhibited by farm animals?

6. An example of sexual behavior displayed by female farm animals in which they allow a male, or sometimes other females, to mount them and which permits producers to detect their estrus is _____.

7. A mare in estrus will display standing heat and _____ of the vulva when a stallion approaches.

8. *True or False:* Care-giving behavior can originate from either the sire or the dam, but is most often maternally oriented.

9. Behavioral activities of fight and flight are examples of what type of behavior?

10. Interactions with other animals and interactions with humans are types of _____ behavior.

11. Producers can take advantage of the knowledge of agonistic animal behavior to handle animals in a low-stress manner by moving in and out of an animal's _____ zone.

12. What are reasons for culling animals with poor dispositions from the herd or flock?

13. Rumination or *chewing the cud* in ruminants is an example of what type of behavior?

14. Imitative behavior, when animals of the same species do the same thing at the same time, such as grazing or herding, is an example of what type of behavior?

15. What are two behaviors that are common to the nine systems of behavior?

16. Bellowing and aggressive behavior of bulls is under control of what hormone?

17. Animals under intensive management systems which cannot adapt to their environment and which exhibit inappropriate or unusual behavior are displaying what type of behavior?

18. What is an example of maladaptive behavior exhibited by chickens and swine under confinement housing?

SELECTED REFERENCES

Publications

Fraser, A. F. and Broom, D. M. 1990. *Farm Animal Behavior and Welfare.* London: Bailliere Tindall.

Grandin, T. 1993. Teaching principles of behavior and equipment design for handling livestock. *J. Anim. Sci.* 71:1065.

Grandin, T. (ed.). 1993. *Livestock Handling and Transport.* Wallingford, Oxon, United Kingdom: CAB International.

Hafez, E. S. E. (ed.). 1975. *The Behaviour of Domestic Animals.* 3d edition. London: Bailliere Tindall.

Hart, B. L. 1985. *The Behavior of Domestic Animals.* New York: W. H. Freeman.

Houpt, K. A. 1991. *Domestic Animal Behavior for Veterinarians and Animal Scientists.* Ames, IA: Iowa State University Press.

Hurnik, J. F., Webster, A. B., and Siegel, P. B. 1995. *Dictionary of Farm Animal Behavior.* Guelph, Canada: Office of Educational Practice.

Lynch, J. J., Hinch, G. N., and Adams, D. B. 1992. *The Behavior of Sheep.* Australia: CSIRO Publications.

McGlone, J. J. 1994. Animal behavior (ethology). *Encyclopedia of Agricultural Science.* San Diego: Academic Press, Inc.

Monahan, P., and Wood-Gush, D. 1990. *Managing the Behavior of Animals.* New York: Chapman and Hall.

Visuals

Safety in Handling Livestock (sound filmstrip). Vocational Education Productions. California Polytechnic State University, San Luis Obispo, CA 93407.

Videos on handling beef cattle, dairy cattle, and swine. Livestock Conservation Institute, 1910 Lyda Drive, Bowling Green, KY 42104.

CHAPTER 35

Issues in Animal Sciences

The industries that comprise animal agriculture face an increasingly diverse set of issues which, left unresolved, may have serious negative consequences. As the major political parties have lost their position as the sole method of participation in policy formation, the specialized interest group has become a significant force in determination of public policy at all levels of government.

The issues which currently face animal industries can be categorized as environmental, diet-health, animal rights, socioeconomic, and food safety (Table 35.1). However, before these issues can be adequately described it is critical to understand the underlying changes in policy formation that impact how agriculturalists and their critics interact to influence decision makers. Special interest or pressure groups are smaller versions of political parties with correspondingly narrower agendas and concerns. The ability of a pressure group to advance its interests depends on a bias for an exclusive interest. Membership in these groups tends to be skewed toward upper-middle-class membership and are unlikely to serve a balanced set of political or social interests. Correspondingly, the communications received by decision makers from these groups tend to carry a strong upper-class bias.

Conflicts over particular issues arise from the efforts of a special interest group to bring an agenda into the public arena while an opposing group(s) works to neutralize or avoid the confrontation. Federal and state agencies contribute to political conflict either by direct participation or by providing the forum in which conflict can occur. Regulatory agency participation is heightened by the increasing independence from legislative or executive control and the likelihood that each agency operates with a relatively narrow focus of societal need.

Policy is created by the ability of interest groups to define an issue, call attention to the group's position, and then pressure government to take a

TABLE 35.1   Major Issues in the Animal Industries (ranked by species)

Issue	Species/Rank					
	Beef	Dairy	Horses	Poultry	Sheep	Swine
Animal welfare	12	3	1	2	7	9
Biotechnology	11	6	5	—	—	8
Consumer						
Diet and health	5	9	—	—	—	3
Food safety	1	7	—	1	—	2
Product perception	5	4	—	3	1	3
Quality assurance	6	—	—	—	5	4
Environment						
Air quality	10	2	—	—	—	5
Endangered species	4	—	—	—	—	—
Global warming	13	—	—	—	—	—
Water quantity and quality	9	2	3	—	—	5
Wildlife	13	—	—	—	—	—
Government policies						
Milk marketing	—	1	—	—	—	—
Predator control	—	—	—	—	4	—
Wool act	—	—	—	—	2	—
Marketing						
Integration and concentration	2	—	—	—	—	1
Global markets	7	5	4	—	6	6
Value-based marketing	3	—	—	—	—	7
Public lands	8	—	—	—	3	—

Source: Smith, 1993; *J. Dairy Sci.* 76:3254; representatives from animal industry organizations.

desired action (Fig. 35.1). Therefore, preventing the movement of an issue into the public arena is equally as powerful as creating awareness.

Animal, consumer, and environmental groups tend to be activist in nature, well funded, have a strong commitment to a narrow mission or cause, and are growing both in number of organizations and memberships. However, as the number of issue-oriented pressure groups continues to increase there are signs of coalition formation designed to consolidate both political and economic power.

As the set of issues or concerns addressed by an organization broadens, the income-generating ability generally increases. Crucial to the fund raising aspects of special interest group management is the need to maintain high media visibility even to the extent of staging events or utilizing scant or misleading information.

Given this scenario, the animal agriculture industries must carefully monitor issues to determine appropriate responses, while promoting caring stewardship, wholesome food production, and careful husbandry practices. Furthermore, agricultural organizations need to establish appropriate working coalitions with moderate special interest groups. Finally, organizations must improve their ability to motivate membership; establish proactive positions with the media, government, and public; and increase funding.

ASSESSING RISK

Assessing risk becomes important when evaluating most of the issues outlined in this chapter. Technology, shifts in societal attitudes, consumer perceptions, and

FIGURE 35.1 Animal rights groups protest in Washington, D.C., and in other areas where they can gather support. Courtesy of William E. Carnahan and *BEEF*.

the consequences of excessive regulations requires the development of better systems of determining short- and long-term risk associated with applications of technology or the distribution of resources. Improvements in the precision of analytical measurement place additional pressure on the need for risk analysis.

ANIMAL WELFARE

One of the common mistakes in evaluating the philosophical position of animal welfare and animal rights groups is to categorize them as all having the same agenda. It is much more logical to consider these groups along a spectrum which ranges from animal exploitation to animal liberation (Table 35.2).

Animal exploitation would be characterized as events or uses which are conducted without concern for the animals involved. Examples might include cock fighting or dog fighting. These activities typically occur in violation of existing laws. Animal use groups typically function under the premise that animals are for human use but that people have a responsibility to provide good animal care and to utilize management practices that minimize animal suffering. Examples of such groups might include livestock producers, breed associations, zoos, and equine sport enthusiasts. Animal control groups include regulatory agencies or organizations at all levels of government that deal with domestic or wild animal control. These groups typically focus on enforcement of existing laws.

TABLE 35.2   Animal Welfare/Animal Rights Groups: A Summary of the Views

Category/Group	Viewpoint
Animal exploitation	These groups argue that animals exist for human use or even abuse; animals are human property. These groups advocate or conduct activities that are illegal in most states (e.g., dog fighting, cock fighting, live pigeon shooting). Most of their activities involve pain or death of the animals primarily for entertainment.
Animal use	These groups believe that animals exist primarily for human use (e.g., livestock production, hunting, fishing, trapping, rodeos, zoos). These organizations have guidelines for the responsible care of the animals. They believe that harvesting of animals for food should be as painless as possible.
Animal control	These organizations enforce the laws, ordinances, and regulations affecting animals. Animals may be supplied for research; surplus animals are destroyed. Many advocate spaying or neutering animals.
Animal welfare	These are national groups, humane societies, and welfare agencies that support humane treatment of animals. They work within existing laws to accomplish goals. They publicize and document animal abuses to get laws changed. They do not provide animals for research. They require spaying and neutering, and are willing to euthanize surplus pets rather than let them suffer.
Animal rights	These national and local groups believe animals have intrinsic rights that should be guaranteed like human rights. These rights include not being killed, eaten, used for sport or research, or abused in any way. Some hold that pets have the right to breed. Most require spaying or neutering.
Animal liberation	These groups believe that animals should not be forced to work or produce for human benefit. They call for animal liberation.

Source: Adapted from K. B. Morgan, *An Overview of Animal-Related Organizations with Some Guidelines for Recognizing Patterns* (Kansas City, MO: Community Animal Control, 1989).

Animal welfare groups, such as the traditional humane societies, go a step further to extend care for animals that are suffering or homeless due to neglect. Historically, these groups have focused on pets or companion animals. Animal rights groups advocate that animals have rights which include not being consumed or used in sport or research or, in some cases, even used as pets or companions. These groups almost universally promote vegetarianism, are opposed to the taking of any life and may even believe that spaying/neutering are violations of an animal's right to breed. Animal liberationists are extreme groups that extend the animal rights agenda in violent ways and believe that their cause is so just that any means used to advance that cause are therefore acceptable.

The debate over animal rights has generated discussion both external to and within U.S. animal agriculture sectors. This discussion has yielded a variety of viewpoints. For example, the Colorado Cattlemen's Association has adopted an animal welfare code of ethics that articulates the values of its membership (Table 35.3). This ethic might well represent the traditional perspective of animal husbandry. Tom Regan (1985) argues that "the rights view will not be satisfied with anything less than the total dissolution of the animal industry as we know it today." Philosopher Bernard Rollin (1993) articulates a reasoned perspective that "the ethic which has emerged in mainstream society does not say we should not use animals or animal products. It does say that the animals we use should live happy lives where they can meet the fundamental set of needs dictated by their natures and where they do not suffer at our hands."

A common assumption is that the animal issue can be conveniently categorized into two notions—animal welfare and animal rights. However, as Rollin (1995) points out this assumption is flawed in that it: (1) attempts to place the

TABLE 35.3   Colorado Cattlemen's Association Animal Welfare Code of Ethics

Item	Policy
Statement of position	The multibillion dollar livestock industry in Colorado is dependent upon the welfare of the animals under its stewardship. It is the policy of the Colorado Cattlemen's Association to promote among its members good stewardship toward animals under their care. It is further the policy of the CCA to cooperate with the Colorado Department of Agriculture, the Colorado Federation of Animal Welfare Agencies, and other organizations, agencies, and individuals that share legitimate concerns about the humane treatment of animals.
General considerations in livestock raising	A. Livestock should be raised in conditions that meet their basic physical and behavioral needs B. Handling facilities. Properly designed, well-kept facilities allow humane, efficient cattle movement. Facilities should be constantly evaluated to see if they can be modified to allow better and more humane animal handling. C. People with a good knowledge of working cattle and cattle behavior allow the best use of these facilities. Staff should be monitored to make sure they understand the best way to work cattle. Training should be available for those who need additional instruction in handling livestock. This applies especially to those who have not previously handled livestock. D. Inducements. Inducements of any sort (hot shot, whips, etc.) should be used as little as possible and should be used only to the extent that it is necessary to facilitate animal movement. They should never be used in a punitive or angry manner. E. Livestock should have access to professional veterinary care as required both to prevent and treat injuries and disease. Use of pharmaceuticals should be used based on an evaluation of the animal's need, not simply out of habit.
Transport of animals	Density of the loading of livestock should be based upon careful consideration of the class of livestock and the planned duration of the trip. Under no circumstances should the animals be crowded to the point of causing undue stress during the transport. Length of time in the vehicle should be based upon the class and condition of the livestock. In no case should the animals be in the vehicle long enough to cause them inordinate amounts of stress.
Livestock auctions	A. Terminally sick or injured animals should be destroyed on the ranch and not be subjected to the additional stress of being shipped to auction (e.g., nonambulatory cattle, severe cases of bovine ocular neoplasia). B. It is essential that auction management continually monitor their facilities and staff to make certain that conditions that may foster animal abuse do not exist.
Statement of duty	It is a livestock producer's duty to oppose inhumane treatment of livestock at any stage of the animal's life. Persons who willfully mistreat animals will not be tolerated in our business. We will provide any assistance necessary to proper officials during the investigation of prosecution of individuals who abuse livestock under their care.

discussion into an "us versus them" context, and (2) it fails to account for the truth that animals have rights which dictate the need for husbandry practices that assure their welfare. However, there are clear differences between the perspectives of livestock producers and the Animal Liberation Front or the People for the Ethical Treatment of Animals.

Welfare concerns due to modern agricultural practices can be classified into three basic categories according to Rollin (1995): (1) production diseases such as liver abscesses in feedlot cattle as a result of diets high in concentrates and low in roughages, (2) scale effects from large animal units which yield less individual animal attention than traditional sized livestock enterprises, and (3) physical and psychological deprivation due to prolonged confinement. Specific concerns relative to each species are listed in Table 35.4 with some examples shown in Fig. 35.2.

TABLE 35.4 **Animal Welfare Concerns Relative to Animal Agriculture**

Beef	Swine	Dairy	Veal	Poultry	Horses
Branding	Confinement	Calf management	White veal production	Housing/ cages	Training techniques
Castration	Farrowing crates	Calf housing	Behavioral deprivation	Behavioral problems	Housing and stalling
Dehorning	Flooring systems	Stall systems	Flooring systems	Forced molting	Racing as two-year-olds
Cancer eye	Tail docking	Castration	Diet	Debeaking	Transport
Downers	Teeth clipping	Dehorning	Group housing	Toe trimming	Slaughter
Slaughter	Castration	Branding		Exercise	Wild horses
Gomer bulls	Early weaning	Tail docking		Nesting	Sports injuries
Feedlots	Halothene gene	Mastitis/ lameness		Dust bathing	Insurance fraud
Handling/ transport	Handling/ transport	Downers		Boredom	Performance drugs

Source: Rollin, 1995. Farm Animal Welfare.

FIGURE 35.2 Animal rights groups target the confinement practices of the livestock and poultry industries. Producers are concerned with comfort and well-being of their animals. Courtesy of John Colwell and Larry LeFever from Grant Heilman.

Animal protection groups have also expressed concern relative to the use of animals in research. Rollin (1995) has proposed the following model for assuring the well being of animals in research settings:

1. Animals should not suffer pain or distress unless the pain or distress was the experimental objective, or unless control of pain/distress would invalidate the results.

2. Animal should not be used repeatedly for invasive experiments.

3. Drugs which cause paralysis while leaving the animal conscious should not be used without anesthesia.

4. Husbandry and housing should fit the nature of the animal.

5. Oversight of these principles should be provided by local committees of nonscientists and scientists via protocol review and facility inspection.

The animal welfare and rights issue is not likely going to go away. The livestock industry will be best served by assuring that producers maintain a strong record of sound husbandry practices, continuing to educate consumers and policymakers, and continuing to research opportunities to improve facilities, management techniques, and animal handling.

BIOTECHNOLOGY

The term *biotechnology* creates mixed reactions from both consumers and producers. Biotechnology has been defined as the application of physical, chemical, and engineering principles to biological systems. Therefore, biotechnology ranges from the use of vaccines to the manipulation of the genome. Biotechnology offers potential improvements to the lives of people via enhancement of health, food safety, and production of a sustainable food supply. Biotechnical applications will help reduce dependence on agrichemicals by producing pest/disease-resistant plants and animals. Biotechnology will improve the efficacy of vaccines or even produce milk that boosts human immunity against bacterial infection.

The use of biotechnology offers excellent benefits to human medicine via mass production of proteins such as blood factor VIII, tissue plasminogen activator, and antithrombin III which serve to control bleeding common to hemophilia, break up blood clots during cardiac arrest, and as a blood thinner, respectively.

In the case of genetic engineering, the first thought or image of some people ranges from "progress" (3.9%) to "mutant/monster" (7.1%) as noted in Table 35.5. Almost 20 percent of those surveyed had negative perceptions about genetic engineering. Even management practices such as crossbreeding which have been in effect for years created some degree of disfavor amongst consumers (Table 35.6).

Clearly consumers have contradictory notions about genetic engineering. These techniques are viewed more favorably when applied to plants than animals and when they are believed to have the potential to improve life. However, a majority of consumers believe their use requires strict regulations and labeling restrictions. Furthermore, consumers are very unsure as to who to trust as a reliable source of information about genetic engineering. Several studies

TABLE 35.5   First Thought or Image When Thinking About Genetic Engineering

Genetic Engineering Category	Percent
Science/technology	11.8%
Test-Tube Baby	9.7
Plant/Animal/People	8.2
Negative/Frightened	7.5
Monster/Mutant	7.1
DNA/Chromosomes	5.8
Medicine	4.1
God/Creation	4.0
Progress	3.9
Crossbreeding	3.5
Neutral	2.6
Nazi/Hitler	2.3
Artificial/Tampering	1.8
Other	1.6
Don't know	26.1

Source: W. K. Hallman and J. Metcalfe, Rutgers University.

have found that scientists are not always seen as reliable sources of information about biotechnology.

Bovine somatotropin (BST) provides an excellent case study in terms of a biotechnical application being approved for agricultural use. In 1994, Monsanto's Posilac (trademarked) became the first synthetic or recombinant BST product approved by the Food and Drug Administration (FDA) for commercial marketing. The use of BST is projected to increase milk production in dairy cows 5–15 pounds per day accompanied by an increase in feed efficiency of 2 to 10 percent. BST is to be administered every 14 days over the last two-thirds of the lactation cycle. However, BST is not a substitute for good management (Figure 35.3).

BST is a naturally occurring complex protein which can be produced using recombinant techniques. All milk, whether from treated or nontreated cows, contains BST and milk composition is not different between treated and untreated cows. Because BST is a protein, it is broken into amino acids by the digestive process. Furthermore, BST is not functional in humans as it will not bind to human cells. Extensive research (20,000 treated cows in hundreds of

TABLE 35.6   Consumer Responses to Genetic Engineering Questions

Genetic Engineering Category	Percent Response
Do not approve of producing plants by cross-fertilization	20
Do not approve of producing hybrid animals by crossbreeding	62
Approve of bioengineering to produce plants	61
Approve of bioengineering to produce animals	28
Fruits/vegetables produced should be labeled	84
Genetic engineering would improve quality of life for people	69
Disagreed that "scientists know what they are doing, so only moderate regulations on genetic engineering are probably necessary"	63

Source: W. K. Hallman and J. Metcalfe, Rutgers University.

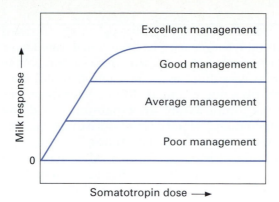

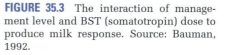

FIGURE 35.3 The interaction of management level and BST (somatotropin) dose to produce milk response. Source: Bauman, 1992.

projects) has verified the safety of milk and meat from dairy cows supplemented with BST. Cows treated with BST exhibit reproductive and health performance similar to that of high-producing cows.

It has been suggested that BST is the most thoroughly studied substance in the history of food-related research (Table 35.7). BST is a technology that allows dairy producers the opportunity to provide a safe and wholesome product while utilizing fewer feed resources and fewer cows in the process.

It is important to recognize that biotechnology does not always offer simple, cost-effective solutions. Furthermore, the speed of breakthroughs arising from genetic engineering is limited by the volume of genetic information carried in one cell—it is estimated that there are nearly 3 billion pieces of genetic information in bovine cells. Additionally, evidence is accumulating that genetic markers may affect different traits in different breeds. Biotechnology brings with it a new set of moral and ethical dilemmas which will have to be resolved by society. Issues such as patenting of genetically engineered animals, regulation of the release of biotechnically altered species into the environment, and the use of biotechnology in human medicine are only a few that will face consumers and producers alike.

TABLE 35.7 History of Bovine Somatotropin

Year	Event
1950s	Scientists conclude that BST is inactive in humans.
1982	Genetic splicing techniques used to mass produce BST.
1985	FDA determines that milk from BST-treated cows is safe.
1987	FDA-ordered studies conclude that milk from BST-treated cows contains no more BST than other milk.
1989	FDA officials call BST one of the safest products ever administered.
1990	Monsanto receives approval for BST sales in Mexico, Brazil, Bulgaria, Nambia, and Zimbabwe.
1990	A NIH panel concludes BST is safe for human consumption.
1993	Government approved with a 90-day moratorium on commercial utilization.
1994	BST utilized on a commercial basis in the United States.

ENVIRONMENTAL ISSUES

Societal approaches to natural resource use has been transformed over time. In the early history of European settlement of North America, the prevailing attitude assumed that resources were limitless, and environmental management focused almost exclusively on using resources from a short-term perspective without much regard for the future. Westward expansion was driven by the notion of "a land of milk and honey." Horace Greeley's exhortation to "go west young man" typified U.S. history prior to the late 1800s.

Gifford Pinchot, father of the U.S. Forest Service, advocated wise use with concerned planning for the future. Pinchot and other early federal lands managers believed that public resources should be managed by technically proficient professionals and the goal of management should be to create the "greatest good for the greatest number." The resource agencies of the federal government were established based upon this philosophy.

Henry David Thoreau and other late nineteenth century writers advanced the belief that wilderness has unique value and that certain resources or geographic areas ought to be protected from human influence. The National Wildlife Federation and the Nature Conservancy are organizations representative of these ideals.

Environmentalism initially emerged as a philosophy opposed to private use or control of public resources and typically viewed regulatory agencies as captured by commercial interests. Environmentalism has biosphere preservation as its first priority with human needs secondary. Environmental philosophy runs a broad range of actual belief systems including those who believe in recycling as a way to conserve finite resources to activists who strive to develop very restrictive regulations to the use of natural resources.

Some have adopted a philosophy of retreatism which is based on a belief that "humans are a cancer on the planet." Sacrificing standard of living and human needs are deemed necessary to protect the environment. Some followers of retreatism would go so far as to embrace terrorism as a legitimate means to advancing their cause.

Decisions regarding resource utilization and allocation will continue to result in conflict. Animal agriculture will likely face the greatest challenges in the areas of waste management, water utilization and quality, federal grazing lands management, deforestation and land degradation, and wildlife habitat.

Waste Management

Waste management issues are of particular concern to intensive agricultural systems such as farrow to finish confinement facilities, dairies, and feedlots. Specific issues often involve manure management and odor control. Industry has shown a strong willingness to overcome these challenges. The efforts of U.S. pork producers to improve environmental practices and management are outlined in Table 35.8. The stewardship beliefs of the National Cattlemen's Beef Association are outlined in Table 35.9.

Manure disposal is often accomplished by applying wastewater and solids to land resources as a means of fertilization. Animal waste is rich in organic materials, nitrogen, phosphorus, and potassium. The use of animal waste as fertilizer is a classic example of creating both economic and environmental benefits by linking the animal and crop/forage system.

TABLE 35.8   Changes in or Development of Programs Due to Concerns of Regulations About Environmental Quality 1990 to 1995

Practice	All Operations	% Farms with Marketings >10,000
Manure management	20.9	52.6
Dust control in buildings	8.7	36.0
Surface water monitoring	5.7	18.8
Groundwater monitoring	5.2	14.7
Employee training	4.6	43.2
Air quality monitoring	2.9	9.8

Source: USDA, APHIS—January 1996.

TABLE 35.9   Stewardship Beliefs of the National Cattlemen's Beef Association[a]

WHEREAS, productive natural resources are vital for the well-being, not only of the individual farmer, rancher, or feeder, but also for the local, state and national economy and society as a whole. Healthy natural resources provide a healthy watershed and a renewable source of feed for domestic animals and wildlife. Farming and ranching sustains open spaces and aesthetic features which contribute to recreational opportunities.

THEREFORE BE IT RESOLVED, that the National Cattlemen's Association promotes the prudent use of natural resources and offers the following Resource Stewardship recommendations. NCA further recognizes the value and benefit of periodic input and revision to keep the commitment to resource stewardship alive.

BE IT FURTHER RESOLVED, that NCA shall not be compelled to defend anyone in the beef cattle industry who has clearly acted to abuse grazing, water or air resources. To achieve these goals, the following environmental stewardship code is recognized by the industry:

1. Recognize the environment for its varying and distinct properties.
2. Manage for the whole resource, including climate, soil, topography, plant and animal communities.
3. Realize that natural resources are ever-changing and management must adapt.
4. Recognize and appreciate the interdependence of ecosystems.
5. Recognize that management practices should be site- and situation-specific, and must be locally designed and applied.
6. Recognize that successful management is an ongoing, long-term process and commit to stewardship, economic success and business continuity.
7. Strive to develop a management framework that involves family, employees, and business associates so that the entire team is committed to common goals.
8. Monitor and document for effective practices.
9. Never knowingly cause or permit abuses that result in permanent damage on public or private land.
10. Develop ways to communicate and share the vast practical experience of other resource stewards.
11. Become involved in organizations that provide an effective way to educate and support individuals.
12. Solicit input from a variety of sources on a regular basis as a means to improve the art and science resource management.
13. Help develop public and private research projects to enhance the current body of knowledge.
14. Recognize that individual improvement is the basis for any change.
15. Communicate with diverse interests to resolve resource management issues.

[a] Adopted by the membership of the National Cattlemen's Association, January 1995.

Water Utilization and Quality

Water quality issues are often at the forefront of environmental debates. Specific concerns include nonpoint source pollution, direct ground or surface water contamination, and the amount of water utilized by various agricultural practices. Water is a critical natural resource for agriculturalists because of its vital importance to cultivation of crops and forages, to maintenance of animal and human life, and its role in maintaining ecosystems. Particularly in the semiarid West and Southwest, conflicts over water rights and appropriation are frequent between agricultural interests, municipalities, water recreationists, and other users.

Several major water-quality issues that relate to animal production and agriculture in general are as follows:

1. The use of commercial fertilizers and manure increases the availability of nutrients needed by plants. However, excessive amounts of nitrogen and phosphorus from commercial fertilizers and manure or pathogenic microorganisms from manure may cause water-quality problems.

   Based on survey data from the Centers for Disease Control, it has been estimated that waterborne infection caused by bacteria, viruses, or protozoan pathogens affect 994,000 people and are responsible for 900 deaths every year in the United States. Sources of waterborne microorganisms included home septic systems; cropland irrigated with sewage effluent; sanitary landfills; land application of sludge, sewage effluent, and manure.

   Commercial fertilizers and manure can increase nitrate/nitrite levels in water that have health risks to both humans and farm animals. Proper fertilization practices, nitrification inhibitors, water-nitrogen management, and legume cropping rotations can be used to improve nitrogen utilization and reduce runoff and groundwater infiltration.

   Transport of phosphorus into water is an issue when erosion moves soil materials into surface water. Erosion can be prevented by using appropriate cropping systems and well-managed grazing programs with animals.

2. Proper planning of animal facilities and drainage systems is needed to prevent animal wastes from contaminating water sources.

3. Pesticides include a number of chemicals which if utilized inappropriately may lead to water pollution. Following pesticide storage, mixing, and application protocols in conjunction with integrated pest management programs which minimize synthetic pesticides can be effective preventative measures to water contamination. As is the case with all environmental issues, risks to human health, environment, and economic systems must be evaluated and integrated into a balanced policy.

Federal Lands

The Louisiana Purchase of 1803 initiated nearly 200 years of federal land acquisitions, redistributions, and management policies. Various Homestead Acts, cash sales, and land grants to support schools and agricultural colleges disposed of some federal lands before the turn of the century. The first national parks were created in 1972; the Federal Reserve Act of 1891 and the Weeks Act of 1911 allowed expansion of the lands managed as national forests. Table 35.10 outlines the percentage of the 11 contiguous western states comprised of federal lands.

TABLE 35.10   Federal Lands in the West as a Percentage of State Land Mass

State	Percent
Nevada	86.0
Utah	64.1
Idaho	63.1
Oregon	52.2
Wyoming	47.6
California	44.6
Arizona	42.7
Colorado	35.9
New Mexico	33.1
Montana	29.4
Washington	28.8

Source: Public Lands Statistics, 1976, U.S. Bureau of Land Management.

The conclusion of the Civil War was followed by the establishment of the open range livestock industry in the western states and territories. Stimulated by the expansion of the railroad system complete with refrigerated cars, increasing demand for beef both domestically and abroad, and the ready availability of cattle in Texas to be driven northward, the open range livestock industry boomed in the 1870s and into the 1880s. The legends of the cattle barons were created by foreign investment, primarily from Scotland and England, which allowed for the development of large ranches such as the Swan Cattle Company that ran over 120,000 cattle in its zenith. Large tracts of land were controlled in the West by homesteading on or near water sources. A common thread in the history of the western states is that the control of water allowed for the domination of other resources as well.

Unfortunately, the cattle boom associated with the open range livestock industry was speculative and thus many of the management decisions of the time were short-sighted and resulted in damage to the rangelands. A devastating drought in 1886 and the blizzards of the winter of 1887 combined to cause cattle death losses as high as 60 percent across the Great Plains region. Just as quickly as it had boomed, the open range livestock industry went "bust." It gave way to the forces of inclement weather, failed speculative ventures, and the continuous process of westward expansion and settlement. But significant damage to the western ecosystems had already been done. Restoration of the range resource yielded the first practitioners of range ecosystem management.

The 1905 grazing regulations were the first in a long line of policies that focused on improving and protecting the western rangelands. The Taylor Grazing Act helped establish the framework for the development of the Bureau of Land Management. Significant acts of legislation that affect livestock grazing on federal lands are summarized in Table 35.11. Each of these legislative actions have yielded a degree of conflict and confrontation between groups which place differing values on the federal lands. The federal lands are host to a multitude of uses including livestock and wildlife grazing, timber production and harvest, mining, a variety of recreational activities ranging from backpacking to skiing, preservation, and watersheds.

TABLE 35.11   History of Legislative Actions Affecting Public Land Grazing

Homestead Acts of 1862, 1909, 1916
Weeks Act of 1911
Taylor Grazing Act of 1934
Multiple Use-Sustained Yield Act of 1960
Wilderness Act of 1964
Classification and Multiple Use Act of 1964
National Environmental Policy Act of 1969
Forest and Rangeland Renewable Resources Planning Act of 1974
National Forest Management Act of 1976
Federal Land Management and Policy Act of 1976
Public Rangeland Improvement Act of 1978
Rangeland Reform Proposal (Department of Interior) 1993–94

Most recently, Secretary of the Interior Bruce Babbit proposed significant changes to rangeland policy in 1994. In general, the primary areas impacted by the proposal include: (1) public participation in rangeland management; (2) administrative practices; (3) range improvement and water rights ownership; (4) resource management criteria; and (4) grazing fees and related incentive programs. These policies are not yet implemented and remain mired in controversy.

Multiple use of federal lands will likely continue as will disagreements relative to the management of the resources. However, consensus building models of decision making coupled with judicious management of these lands will continue to assure their productivity and sustainability.

Air Quality

Odor from manure can be a particular problem for livestock producers who manage animals in intensive systems (e.g., large feedlots, dairies, poultry, and swine operations). Odor and noise regulations are currently handled by state and local regulations. As urban sprawl continues to move into traditional agricultural areas, odor and air quality issues must be carefully monitored by animal agriculturalists.

Federal regulations as defined in the Air Pollution Control Act affect animal agriculture primarily in regards to methane, reactive organic components, ammonia, and particulate matter. Managing these four pollutants is of public interest as they can contribute to acid rain, ozone depletion, and excessive particulates in the atmosphere.

Endangered Species

The Endangered Species Act of 1973 was passed to protect threatened and endangered species and their habitats from detrimental human activity. This legislation was reviewed in 1996. Once a plant or animal is placed on the endangered list, the Fish and Wildlife Service or National Marine Fisheries Service is delegated the responsibility of drafting a recovery plan. More than 7,000 species have been listed with 325 drafted recovery plans covering 411 species. Government species-recovery efforts have only allowed 5 delistings in the past two decades.

Meanwhile, it is estimated that better than three-quarters of the U.S. wildlife population depends on private land for at least part of the year. Conflict results when government mandates the use of privately held resources because of the discovery of an endangered plant or animal. Some groups contend that domestic livestock and other agricultural enterprises compete for scarce resources and thus cause habitat loss for endangered species.

Examples of specific species recovery plans that have raised issue with livestock producers include the desert tortoise, the sockeye salmon, and the timber wolf. Reintroduction of the wolf into Yellowstone National Park has stirred debate between those who believe the resurgence of predators better balances ecosystems and those whose livestock may fall victim to predatory species.

These issues are difficult to resolve because they oftentimes become highly emotional and frequently end up in divisive positions instead of consensus solutions. As resources are demanded for divergent uses and as the human population multiplies, the need for balancing economic interests versus endangered species' needs will be critical. The process of reaching consensus solutions is a difficult task. However, it is a responsibility that can no longer be ignored.

Global Warming

Global warming (sometimes called the *greenhouse effect*) becomes an issue with farm animals that are ruminants, because they emit methane (CH_4) into the atmosphere. When solar energy is radiated from the earth into the atmosphere it is absorbed by certain gases—for example, methane, carbon dioxide, nitrous oxide, and chlorofluorocarbons. Increased amounts of these gases absorb more infrared radiation, thus increasing the temperature of the atmosphere and causing global climate changes.

There is not complete agreement as to whether global warming is occurring and, if so, at what speed. Yet there are some scientific studies that show the need for concern and more careful assessment. A 1.8-mile Greenland Ice Core Project showed in 1994 that there is an alarming volatility in the world climate—even in decades without changes brought about by humans such as the greenhouse effect.

Methane is second in importance to carbon dioxide as a greenhouse gas. Methane is also produced from wetlands, bogs, and rice paddies as a result of anaerobic fermentation. Ruminant animals produce methane by digesting plant material, which releases some of the carbon in the gas. This carbon is originally a result of photosynthesis; thus, it is not new gas but recycled carbon.

Accusations have been made that cattle and other ruminants are the major sources of greenhouse gases that are destroying the protective ozone layer. However, realistic calculations show that methane accounts for only 18% of the world's greenhouse gases, and only 7% of the world's methane is produced by all ruminants—domestic and wild (e.g., deer). Beef cattle in the United States produce 0.5% of the world methane which is equivalent to 0.1% of the total greenhouse gases. Carbon dioxide emitted from automobiles contributes much more greenhouse gas than methane from ruminants. In fact, driving a few miles to purchase a hamburger produces more greenhouse gas than the methane required to produce the hamburger.

Additional charges against cattle production have been focused on the destruction of rainforests in Central America to increase forage availability for cattle then imported as beef to the United States. Consumption of hamburger in the United States has been blamed for the loss of rainforests in Central America. In reality, the United States imports less than 1% of its beef from Central

America. The reason for the loss of the rainforests for cattle production results primarily from social, economic, and political pressures in the countries where rainforests are located.

Conversion of Agricultural Land

The conversion of agricultural land to nonagricultural uses is becoming an increasingly significant challenge to preserving wildlife habitat and assuring the security of food supplies for future generations. Since 1978, Colorado has lost an average of 90,000 acres of agricultural land per year. This loss equates to an area 140 miles long and 1 mile wide.

Urban dwellers desire small acreage homesites, cities and communities grow, and growing recreational demand breaks up existing blocks of land. As large areas of land that once were managed as an ecosystem are divided into smaller parcels and then crisscrossed with infrastructure improvements such as roads and power lines, both wildlife populations and agricultural production capacity are negatively affected.

As the popularity of 5- to 35-acre homesite tracts grows, more pressure will be placed on traditional agricultural enterprises as they interface with urbanization.

CONSUMER ISSUES

Diet and Health

The relationship between diet and human health is a controversial and complex topic. The consumption of red meats, dairy products, and eggs have been linked to two of the most dreaded human diseases—coronary heart disease (CHD) and cancer. Consumer perceptions have been influenced to accept these alleged relationships, and consumers have reduced their consumption of some animal products accordingly.

Cholesterol and saturated fat in animal products have been the dietary components most frequently associated with CHD and cancer. One extreme view of this diet and health issue is that most animal products are high in fat and cholesterol and thus consumption of these products should be eliminated or drastically reduced. This view states that taking such action will reduce the incidence of CHD and cancer, and that life expectancy will be increased. Other people with a totally opposite viewpoint argue that many individuals who have high consumption levels of animal products live to an old age and are free of these serious diseases.

Sound, well-based research work often appears to be lost in an emotionally charged issue. Many organizations and other groups have occasionally based judgments and decisions on emotion rather than on the best accumulated research facts. Some individuals feel that consumer perceptions are directed by some self-proclaimed "diet and health experts" and a communications system that builds part of its readership on sensationalism.

The major known risk factors associated with coronary heart disease are genetics (a family history of CHD), high blood cholesterol, smoking, hypertension, physical inactivity, and obesity. Obesity caused by excess caloric intake is a major nutritional problem in the United States. It is generally accepted that consumption of animal fat by humans causes an increase in the level of cholesterol in the blood, while consumption of vegetable oils (polyunsaturated fats) causes a decrease in blood cholesterol concentration. These relationships led to the theory that there is a relationship between consumption of animal fats and the incidence of atherosclerosis ("plugging" of the arteries with fatty tissue), which in turn results in an increased likelihood of death from coronary heart disease.

Evidence supporting the proposed blood cholesterol–heart disease relationship is still theoretical. Studies in which dietary fat intake has been modified, either in kind or amount, did not show significantly reduced mortality rates. Changing diets from animal fats to vegetable fats has not improved the heart disease record. There are data that suggest that poor health conditions can also result from diets high in polyunsaturated fats.

Most consumers do not know the difference between saturated and polyunsaturated fats. Through many margarine and vegetable oil commercials, however, they have been informed that saturated is "bad" and unsaturated is "good." In a review of the diet and heart disease relationship, some medical doctors argue that the dietary-heart hypothesis became popular because a combination of the urgent needs of health agencies, oil-food companies, and ambitious scientists had transformed that fragile hypothesis into treatment dogma. Diet-heart enthusiasts then began using a show of hands or mail polls to influence national food policy.

Recent research evidence shows that when polyunsaturated fats are substituted for saturated fats in the diet, they do not always lower the blood cholesterol level. In most cases, consumption of polyunsaturated fat increases the concentration of cholesterol in tissue. Thus, animal fats may be necessary to keep down the accumulation of total body cholesterol.

Cholesterol is a naturally occurring substance in the human body. Every cell manufactures cholesterol on a daily basis. The average human *turns over* (uses and replenishes) 2,000 mg of cholesterol daily. Average dietary consumption of cholesterol is approximately 600 mg daily. Therefore, the body makes 1,400 mg each day to meet its needs.

Cholesterol cannot be used by the body unless it is joined with a water-soluble protein, creating complexes known as *lipoproteins.* There are several different types of lipoproteins. Research workers have identified two of these lipoproteins: HDL (high-density lipoprotein) and LDL (low-density lipoprotein). High blood levels of the LDLs have been generally associated with increased cardiovascular problems, while some research data show that higher blood levels of HDLs may reduce heart attacks by 20%. Additional research with laboratory animals has shown that those fed beef had HDL levels 33% higher than animals fed soybean diets.

There is evidence of genetic differences in the proportion of HDLs and LDLs in individuals. People who are overweight, nonexercisers, and cigarette smokers have higher proportions of LDLs than those who are lean, exercisers, and nonsmokers.

Although the exact roles of dietary cholesterol and blood levels in the development of coronary heart disease are not known, it would appear logical from the existing information to use prudence in implementing drastic changes in dietary habits. Certainly, those individuals with high health risks primarily due to genetic background should take the greatest precautions.

Some individuals argue that consumers eat too much red meat and that excessive red meat consumption causes cancer. Evidence to support this statement is questionable, and excessive consumption must be defined. Previous data have shown that the average daily per-capita beef consumption is approximately 2 oz. Per-capita red meat consumption is approximately 3 oz daily; this provides approximately 25 g of protein, which is less than half the average recommended daily allowance (RDA) for protein. The American Heart Association recommends 3.5 oz of cooked meat per person on a daily basis. Based on this recommendation, the average U.S. per-capita consumption of beef and other red meat is not excessive.

Figure 35.4 shows that on an average per-capita basis, red meat and poultry supply less than 16 grams of fat per day. This level of fat (15.6 g) is less than 24% of the 67 grams of fat per day recommended by the American Heart Association for a 2,000-calorie-per-day diet. The 15.6 grams of fat represent 30% of the calories from fat where each gram of fat contains 9 calories.

The U.S. government has released several reports that outline dietary guidelines and goals for American citizens. Some argue that it is the government's responsibility to provide people with information about diet; others support scientifically based guidelines, but believe Americans should have freedom of choice.

The *Dietary Guidelines* published by the USDA and Department of Human Health Services (HHS) were broad recommendations and a reasonable place to initiate a sound nutritional program. The revised 1990 *Guidelines,* which draw heavily on two diet and health reports by the National Academy of Sciences and the U.S. Surgeon General, recommended that not more than 30% of daily calories come from fat and that less than 10% come from saturated fat. Some of the earlier versions of the *Guidelines* implied that red meat was a health risk by recommending that Americans eat less meat. The revised guidelines, however, convey a more positive recommendation for lean red meat consumption. In 1992, the USDA used the Food Guide Pyramid (Fig. 35.5) in conjunction with

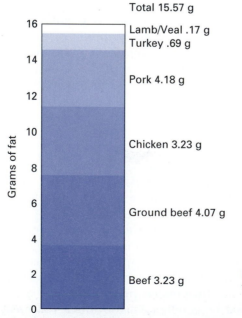

Total 15.57 g

Lamb/Veal .17 g
Turkey .69 g

Pork 4.18 g

Chicken 3.23 g

Ground beef 4.07 g

Beef 3.23 g

FIGURE 35.4   Average daily per-capita fat consumption from cooked meat. Courtesy of the USDA.

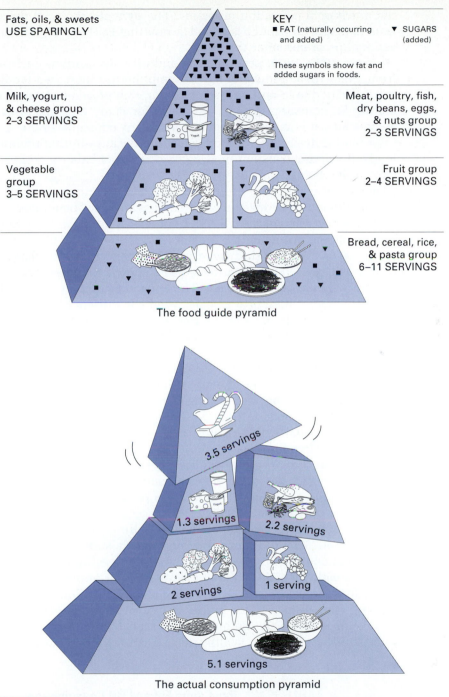

Fats, oils, & sweets
USE SPARINGLY

KEY
■ FAT (naturally occurring and added)   ▼ SUGARS (added)

These symbols show fat and added sugars in foods.

Milk, yogurt, & cheese group
2–3 SERVINGS

Meat, poultry, fish, dry beans, eggs, & nuts group
2–3 SERVINGS

Vegetable group
3–5 SERVINGS

Fruit group
2–4 SERVINGS

Bread, cereal, rice, & pasta group
6–11 SERVINGS

The food guide pyramid

3.5 servings

1.3 servings

2.2 servings

2 servings

1 serving

5.1 servings

The actual consumption pyramid

FIGURE 35.5 A comparison between the food guide pyramid and actual consumption. Source: USDA and National Live Stock and Meat Board.

the nutritional education programs. The pyramid replaced the wheel graphic used to display the four basic food groups that has been used in nutritional educational programs since the 1950s.

From 1985–1990, per-capita fat and oil consumption decreased; however, from 1990–1995, per-capita fat consumption has increased (Table 35.12). This is primarily due to an increased consumption of salad and cooking oils. Moderating fat consumption is a challenge for many consumers because taste is associated with the fat content of foods. Many consumers rank taste as the most important criteria in their food selection. Because of this preference for fat in foods, the actual consumption pyramid is out of balance compared to the recommended Food Guide Pyramid (Fig. 35.5).

In recent years the beef industry has met consumer demands by providing beef trimmed to a fat level of ⅛ inch or less and has encouraged the publication of the 1990 version of the USDA's *Handbook 8-13*. It contains nutritional information on beef trimmed to ¼ inch of fat.

In 1994, a research study involving 85,000 nurses noted that women who ate the highest amount of margarine were far more likely to suffer heart disease than women who ate less margarine. The apparent problem is caused by *transfatty* acids that occur naturally in dairy products but are also formed during margarine manufacture. One of the Harvard researchers said that consumers now need to rethink the earlier recommendation to substitute margarine for butter. Some margarine manufacturers are considering ways to eliminate transfatty acids from their products.

In many aspects, the dietary guidelines, consumption goals, and nutritional information have not changed all that much over the past 20 years. Eating a variety of foods to provide the necessary nutrients, consuming food in moderation to maintain a healthy weight, and getting adequate exercise are the basic approaches for most people. Lean, palatable red meat and the moderate consumption of dairy products and eggs not only fits into a healthy diet but also satisfies the taste preferences of many people. The red meat industry has become more consumer-driven and has reduced fat content in meat products by 25–80% during 1985–1995.

In the 1990s, people are showing less willingness to exercise and eat healthy. Interest is increasing in enjoying food and leisure activities. However, many consumers continue to have concerns about specific health issues. Their concerns are eating a balanced diet, maintaining the appropriate body weight, and eating foods lower in fat and cholesterol. Many nutritionists say most people eat primarily for taste, not for nutritional considerations—and what makes food taste good is primarily the ingredient of fat. These nutritionists also state that some fat is necessary in a healthy diet, so eating limited amounts of fat is accept-

TABLE 35.12   Per-Capita Fat and Oil Consumption

	1975	1980	1985	1990	1995
	Lb per capita				
Fats and oils	52	57	64	61	67
Vegetable	42	45	51	51	55
Animal	10	12	13	10	12

Source: USDA.

able. Foods should not be classified as "good versus bad." More important is selecting a variety of foods to fit into a balanced diet. Properly selected and prepared, animal products fit healthy diets.

Food Safety

Although the food industry does not have a perfect record in regards to food safety, U.S. consumers enjoy a plentiful food supply that can arguably be called the safest in the world. Again, relative risk must be carefully evaluated when considering food safety issues (Table 35.13).

Food safety is a challenge for both consumers, producers, processors, retailers, and food service outlets. Historically the food inspection system has been expected to assure food wholesomeness. The premise of total quality management is that quality cannot be inspected into a product. Instead, systems must be developed to produce quality and safety at every key step in the process.

Because food plays such an integral role in society, food safety concerns are often manifested as emotional responses to half-truths and inaccurate information. The Alar incident where incorrect claims about the safety of Alar (a preservative) use cost the U.S. apple industry millions of dollars is a classic example.

Food scientists rank food hazards as: (1) microbial contamination; (2) naturally occurring toxicants; (3) environmental contaminants (e.g., heavy metals); (4) pesticide residues, and (5) food additives.

Examples of food safety issues for the livestock industry include infection by *E. coli* 0157:H7 and concerns relative to bovine spongiform encephalopathy (BSE). Concerns about *E. coli* 0157:H7 bacterial infection outbreaks peaked in 1993 when 1,000 cases were reported in the United States. Over three-quarters of the 1993 incidents were attributed to improperly prepared and cooked ground beef, most of which were traced to a single food-service supplier.

Centers for Disease Control data in 1995 reported only 455 cases of *E. coli* 0157:H7 infection with fewer than one-quarter traced to beef. While the goal must be to eliminate food-borne illness outbreaks, these data show that large industries can make progress in food safety.

In fact microbial contamination has resulted in a variety of new processing technologies designed to improve food safety. Some of the innovations include steam vacuuming, hot water washing, weak acid rinses, and steam pasteurization. Most processors are working to implement a multiple set of hurdles or blockades to microbial growth. Such systems have tremendous potential to enhance food safety.

TABLE 35.13 Relative Risk of Death from a Variety of Circumstances

Relative Risk of Death (per Million)	Circumstance
220	Auto accident
38	Drowning
29	Fire-related
0.6	Lightning
0.2	Venom
0.02	Botulism
0.01	Salmonellosis

Source: Mossell, 1988.

Another proposed food safety technology is radiation pasteurization. Unlike the previously mentioned technologies, the use of radiation has been surrounded by a degree of controversy. The use of low doses of gamma rays, X rays, and electrons can safely control pathogenic microbes.

The use of this technology doesn't make foods radioactive and essentially has little effect on food because the cells are no longer active. Small changes to vitamins, such as B_1 and C, do occur but are not dangerous to human health. While many studies show that informed consumers are accepting of foods that have been radiation pasteurized, there continues to be opposition from some consumer groups.

Microorganisms

Microorganisms, such as molds, viruses, bacteria, yeasts, and parasites, are literally universally present. These microorganisms play potentially beneficial, harmful, or neutral roles in regard to food. Molds and yeasts are important to baking and cheese making, while some bacteria act as inhibitors to harmful bacteria. Harmful microorganisms either cause spoilage or disease. Disease causing microorganisms are referred to as *pathogens.*

Food spoilage microbes are usually visually apparent, while food-borne pathogens may be difficult to detect. The Centers for Disease Control reports that 77 percent of food-borne illness outbreaks originate due to improper handling and cooking in food-service establishments, 20 percent due to improper food handling and cooking in the home, and 3 percent due to food manufacturing defects or failure.

Sanitation and temperature control are critical elements in controlling potentially harmful bacteria. Processors, retailers, and food service operators can minimize contamination via proper equipment sanitation, disciplined personnel hygiene practices, proper handling and storage, cooking to appropriate temperatures, and systematically assessing food safety efforts.

Consumers should never purchase outdated goods or those in which containers have been broken or dented. Furthermore, refrigerator temperatures should be maintained at or below 40°F and freezer temperatures at or below 0°F. Consumers are advised to follow label directions for preparation and serving; avoid cross-contamination of foods; and to wash, rinse, and sanitize equipment, countertops, and cutting surfaces. Foods should never be thawed at room temperature; food should be cooked to appropriate temperatures (140°F or greater, typically, 155°F for ground meats, and 165°F for leftovers); and, during storage or preserving, foods should be kept outside the range of 40 to 140°F.

Bovine Spongiform Encephalopathy

Bovine spongiform encephalopathy (BSE) is an infectious fatal disease of the bovine central nervous system. Symptoms typically do not develop until cattle reach ages greater than 18 months. The infectious agent belongs to a group of proteins referred to as *prions.* It is theorized, although not definitely proven, that BSE originated in cattle in Great Britain as a result of feeding rendered sheep protein to bovines.

While there is no definitive scientific evidence to support or refute the possibility of transmission of BSE to humans, the USDA has taken precautions to assure that BSE does not infect the U.S. cattle herd. Since 1989, the U.S. has banned importation of live ruminant animals and ruminant-derived products from countries where BSE is present. It is important to note that BSE has not been diagnosed in the United States.

Residues

Much of the U.S. public feels that chemical residues from pesticides, hormones, and food additives are the greatest food safety threats. There is a perception that the presence of these residues in any amount is extremely dangerous. In reality, large enough doses of many substances (including water) can be harmful and even fatal. What is critical to assess is the level of the residue in the food, the amount consumed, and the risks associated with that level of consumption.

Consumers may perceive that giving farm animals anabolic steroids/hormones will produce meat with residues that are harmful to their health. An example of a growth promotant would be the use of an estrogenic compound administered via a slow release implant placed subcutaneously in the animal's ear. These products are utilized to increase growth and repartition feed energy to produce less fat and more lean. As a result of this process, the average amount of estrogen in a 3-ounce serving of beef is increased from 1.2 to 1.9 nanograms (one nanogram is a billionth of a gram). Similar-size servings of potatoes, ice cream, and soybean oil contain more nanograms of estrogen ($\times200$, $\times500$, and $\times1$ million times, respectively) compared to beef from implanted cattle. Considering that the average nonpregnant human female produces 480,000 nanograms of estrogen each day by normal physiological processes, the increased estrogen consumed in beef is of no physiological or medical consequence; that is, there are no problems associated with ingestion of meat from an animal.

In 1994, a jury awarded $42 million to 11 women whose mothers took DES (diethylstilbestrol—a synthetic estrogen) between 1947 and 1971 to prevent miscarriages during early pregnancy. This was the first time a trial was held claiming the hormone was related to reproductive problems and not to cancer. These recent awards raise additional perceptions that hormones are extremely dangerous health risks and should be avoided. However, in these situations, women were given massive therapeutic doses of DES, which is grossly different than the extremely small amounts of hormones found in meat and other foods.

There is a perception that so-called natural or organic animal products contain far fewer residues of antibiotics, hormones, drugs, and pesticides. Scientific tests of organic versus conventional beef have shown no evidence in reality of this perception. Also, the Food Safety and Inspection Service (USDA) ran tests on 40,000 random samples of meat and poultry in 1992 and found extremely low (0.29%) illegal residues in the product. There were no pesticide violations.

Product Perceptions

Consumer perceptions of animal products are covered, in part, in other parts of this chapter—for example, diet and health (perceptions that red meat, dairy products, and eggs are too high in fat and cholesterol); products from animals fed hormones (e.g., BST) or exposed to other chemicals are not safe because of the residues they contain; and other perceptions. These perceptions and others are discussed in more detail in other parts of this chapter.

Quality Assurance

Animal production began a new era in the 1990s as production practices were evaluated in light of their impact on the process of assuring that consumers received wholesome products of high perceived value. Quality assurance programs have been instituted in food and fiber industries as a means to coordinate genetic inputs, animal health care practices, and management protocols.

Quality assurance programs in the food animal industries have focused on (1) superior record keeping, (2) safe use of pharmaceutical products with a special emphasis on product administration in accordance with label restrictions, (3) minimizing intramuscular injections, (4) assuring quality of animal feeds, and (5) avoiding residues.

Most quality assurance efforts are designed in accordance with Hazard Analysis Critical Control Point (HACCP) protocols. HACCP programs utilize a seven-step process to assure the product quality is attained throughout the production process (Table 35.14).

Quality assurance efforts require an effective process of communicating critical information throughout the production and processing chain. Improved communication can go a long way toward adding value and strengthening positive consumer perceptions about animal products.

For example, contamination of wool with polypropylene products such as hay baling twine can significantly reduce the value of the clip. Contamination typically occurs when sheep bed down on areas where baling twine has been littered as a result of feeding hay. Wool producers are solving this problem by more carefully disposing of twine, avoiding the use of polypropylene feed tags, and communicating a price incentive for wool free from contamination.

MARKETING ISSUES

The various methods of selling and merchandising livestock and livestock products continue to undergo change as economic and policy conditions shift. In many cases, these changes have created controversial issues in regards to industry concentration, pricing methods, international trade policies, and government involvement in economic decision making.

TABLE 35.14   Seven Steps for Utilizing Hazard Analysis Critical Control Points

Number	Steps
1	Conduct a hazard analysis
2	Identify critical control points in the process
3	Establish preventative measure limits for each control point
4	Establish monitoring process for critical control points
5	Take corrective action
6	Institute effective record keeping system
7	Monitor system on continuous basis and assure it works

Concentration and Integration

Concentration occurs when fewer enterprises control an increasing large percentage of the total animal inventory in a particular industry. This control can be accomplished via ownership or contractual arrangement.

Vertical integration occurs when two or more stages of production are controlled by the same owner (e.g., cow–calf herds and stocker operations owned by the same firm). *Horizontal integration* exists when supportive or competitive goods and services are controlled or owned by the same company (e.g., broiler producers who own or control the company that supplies the feeds).

Concentration and integration were discussed in Chapters 2 and 3. The poultry industry has the greatest degree of integration, and the swine industry is moving rapidly toward greater vertical integration. While the other animal industries have less concentration and integration, there is evidence that suggests a growing trend of fewer owners consolidating into enterprises with larger numbers of animals per production unit.

The major issues associated with concentration and integration include the following:

1. Potential loss of competitiveness in cash markets due to contractual agreements that lock in a feeder or packer's supply and thus bypass traditional market outlets.

2. Concentration of animals into fewer and larger production units. This poses additional concerns for animal welfare and water quality, in particular.

3. The loss of traditional small to midsized farms. The fact that larger enterprises typically have significant cost advantages due to economics of scale may yield a loss of competitiveness for more traditional-sized farms. The benefits of the social structure in family farms and rural communities must be considered. However, traditional midsized farms can remain competitive by utilizing cost-saving practices and accessing marketing power and purchasing flexibility by forming networks and other cooperative ventures with other producers of similar size.

Global Markets

As domestic markets mature, livestock producers will look to export opportunities as a means to enhance profitability. Agricultural exports will likely play an important role in helping the United States reduce the trade deficit. For example, the U.S. beef cattle industry accounted for more than a $2 billion trade surplus for beef and beef byproducts in 1994.

The North American Free Trade Agreement (NAFTA) was implemented on January 1, 1994, and created the world's largest free trade area—360 million people annually producing $6.2 trillion worth of goods and services and exporting/ importing more than $1 trillion worth of goods. NAFTA phases out 90 percent of all tariffs between the United States, Mexico, and Canada over 10 years and eliminates remaining tariffs over a 15-year period.

NAFTA is not without its critics. However, there is strong evidence that U.S. animal agriculture will benefit over the long term. In the short term, fed beef cattle exports from Canada have risen due to the capacity of Canadian feedlots exceeding their packing plant capacity. However, with the elimination of tariffs, the United States has significantly increased its beef exports into Canadian markets.

NAFTA might have escaped criticism had not the Mexican peso devalued along with the emergence of severe drought conditions in Mexico's leading cattle regions at approximately the same time as the implementation of the agreement. If Mexico can strengthen its economy, the United States should experience dramatic growth in meat exports across the southern border.

A seven-year effort to rewrite the rules of world trade culminated in 117 nations approving GATT (General Agreement on Tariffs and Trade) in December 1993. The agreement was implemented in 1995 and became one of the most far-reaching pacts ever made. It cuts tariffs (taxes charged by a country on imported goods) by an average of 40%. It has the potential to increase trade and investments, create more jobs, and increase per-capita incomes in many countries.

However, many issues have surfaced with GATT, and others will continue to surface. Farmers in Europe, India, and Japan protested the GATT accord with signs and slogans of "American Imperialism" and "GATTASTROPHE." They fear the agreement will ruin millions of farmers (subsidies would be removed or reduced) and uproot centuries-old traditions. Some U.S. agricultural organizations expressed fear that there will be limited access for American products in return for allowing European farmers to continue to receive large subsidies for a long period of time.

Animal industries in the United States and elsewhere become involved in issues that affect per-capita income and purchasing power in other countries. As standards of living increase, people prefer larger amounts of animal protein in their diets (Chapter 1, Fig. 1.6). This increase in demand for animal products influences the U.S. export market positively. The only way to ensure that our (U.S.) standard of living continues to grow is to make ourselves more competitive, not to protect our economy (CAST, 1993).

The United States will face strong market competition from Germany, Japan, Australia, the Pacific Rim, and South America. Agricultural research efforts should be directed toward making U.S. producers more competitive in commodities where there is direct competition. The extension services should increase their efforts to assist animal industries experiencing serious competitive pressures from the lowering of trade barriers. This brings to the surface another issue as U.S. funding for agricultural research and the extension service continues to decrease.

Value-Based Marketing

During the 1980s and early 1990s, the red meat industries became more consumer-focused. There was a clear message that *waste fat* (fat on the external surfaces and fat between muscles) was very undesirable. Yet, *taste fat* (marbling or intramuscular fat), as it contributed to highly palatable products, was still preferred. A research study identified an optimum range of intramuscular fat in beef—now called the *window of acceptability*—where 3% intramuscular fat is a minimum (equivalent to "minimum slight" marbling, which is the bottom of the U.S. Select grade) and 7% intramuscular fat is the maximum (equivalent to "minimum moderate" marbling, which is the bottom of high Choice) because higher levels of intramuscular fat would not meet diet/health guidelines. To meet consumer demand for less external fat, packers and processors trimmed fat to ¼ inch, then ⅛ inch, with some cuts trimmed of all excess external fat.

There has been a serious challenge in marketing red meat animals because they and their products have been marketed as a commodity where, in general, price is established based on the average performance of a group of animals. Value-based marketing could replace the average-price buying with a system that recognizes and pays for value differences of individual animals/products within the same commodity. Value-based marketing would encourage and reward—at every stage of production and distribution—products with the desired quantity of intramuscular fat but with less trimmable fat. Other value-determining characteristics could be added into a grid system of pricing.

The major impediments to developing value-based marketing are as follows:

1. Using dressing percentage in the pricing of live, red meat animals encourages the production of excessively fat animals because fatter animals typically have higher dressing percentages. The average external fat specification of 1.00 inch in most box beef programs favors overfat cattle with high dressing percentages. In cattle, yield grades 2 and 3 are usually priced together with discounts given to yield grades 4 and 5. Some individuals have recommended a lower maximum external fat thickness for yield grade 3s (e.g., from 0.80 to 0.60 inch).

   In 1990, packers reported that 25% of the market lambs were U.S. Yield grade 4 or higher. Emphasis on dressing percentage is the major reason for the higher numerical yield grades.

2. Distrust between packers and feeders prevents more on-the-rail pricing and trading.

3. The price spread between Choice and Select quality grades in cattle encourages overfattening.

4. Labor union restrictions discourage economics of retailers purchasing trimmed, saw-ready, subprimal cuts.

5. The retailer is reluctant to send the correct monetary signals of value differences to the packer.

Breeding animals, feeding practices, and total management programs are currently available to produce animals/products that can meet the desired carcass and retail targets with appropriate combinations of muscle, external fat, seam fat, marbling, and bone. What is needed is a value-based marketing system that monetarily rewards superiority while identifying and discounting inferiority—this is opposed to the current system of average pricing which encourages mediocrity.

ISSUES AND OPPORTUNITIES

With challenging issues come opportunities. Jobs are created, recognition is given, and monetary rewards are available to successful problem solvers and decision makers. These successful individuals face the problems, meet the challenges, and bring reasonable resolution to the issues. The challenge is to separate the myths, perceptions, and emotions from the scientific truths and realities, then to effectively communicate these truths to the opinion influencers and eventually to all concerned. Consumers in the United States can be provided highly palatable animal products that are the cheapest and safest in the world. With the creativity of people, the environment can be sustained for generations for the economic, physical, and emotional well-being of the U.S. and world population—even with significant population increases.

CHAPTER SUMMARY

- Issues facing the livestock industry can be categorized as animal welfare, biotechnology, environment, diet and health, food safety, and marketing.
- Management of these issues in terms of public perception is important to the stability of the livestock industry.
- Application of the principles of sound stewardship and husbandry practices is critical to the future of livestock enterprises.

REVIEW QUESTIONS

1. *True or False:* Special interest groups represent the broad spectrum of societal opinion.

2. Animal agriculturalists would most likely fit into which three "animal groups"?

3. Rollin categorizes animal welfare concerns into three areas. What are they?

4. What is biotechnology?

5. What is BST?

6. *True or False:* Consumers universally understand and support application of genetic principles to agriculture.

7. To what important diseases of humans have the consumption of red meats, dairy products, and eggs been putatively linked?

8. *True or False:* Consumption of red meats, dairy products, and eggs is a major risk factor for coronary heart disease.

9. What are the major known risk factors for coronary heart disease?

10. What is the most significant food safety concern according to food scientists?

11. What are the keys to preventing illness from food-borne microorganisms?

12. *True or False:* Although perceived as a potential food safety threat, chemical residues in meat and dairy products are not a major problem in the United States.

13. What are some environmental issues facing livestock producers in the United States?

14. What are three specific water quality issues that involve animal agriculture?

15. What are some marketing issues facing livestock producers?

16. What are the major issues associated with concentration and integration of livestock operations?

17. What are two trade agreements that affect foreign markets available to U.S. animal products?

18. *True or False:* Value-base marketing has provided an effective incentive for livestock producers to improve carcass quality of livestock through genetic selection and nutrition.

19. What are some issues concerning use of public lands for livestock production?

SELECTED REFERENCES

Publications

Animal Industry Foundation. 1988. *Animal Agriculture: Myths and Facts.* Arlington, VA: Animal Industry Foundation.

Assessment of Marketing Strategies to Enhance Returns to Lamb Producers. 1991. College Station, TX: Texas Agricultural Market Research Center.

Bauman, D. E. 1992. Bovine somatotropin: Review of an emerging animal technology. *J. Dairy Sci.* 75:3432–3451.

Bjerklie, S. 1990. Poultry's decade of issues. *Meat and Poultry* 36:24.

Cheeke, P. R. 1993. *Impacts of Livestock Production on Society.* Diet/Health and the Environment. Danville, IL: Interstate Publishers, Inc.

Fox, M. W. 1986. *Agricide.* New York: Schocken Books.

Gleick, P. H. (ed.) 1993. *Water in Crisis.* New York: Oxford University Press.

Hafs, H. D. (ed.) 1993. Genetically Modified Livestock: Progress, Prospects and Issues. *J. Anim. Sci.* 71(suppl. 3).

Hallman, W. K. and Metcalfe, J. 1993. Public Perceptions of Agricultural Biotechnology: A survey of New Jersey Residents. Ecosystem Policy Research Center: Rutgers University.

Meat industry: A sampling of the issues and controversies affecting meat and poultry industries all over the planet. 1989. *Meat and Poultry* 35:12–14, 41, 43.

Moreng, R. E., and Avens, J. S. 1985. Agricultural Animal Welfare. Chap. 11 in *Poultry Science and Production.* Reston, VA: Reston Publishing Co.

National Cattlemen's Foundation. 1990. *Special Interest Group Profiles.* Washington, DC: Hill and Knowlton.

National Research Council, Committee on Diet and Health. 1989. Diet and Health. Implications for Reducing Chronic Disease Risk. Washington, D.C.: National Academy Press.

Pearson, A. M., and Dutson, T. R. 1990. *Meat and Health.* Amsterdam: Elsevier Science Publishers.

Regan, T. 1985. *The Case for Animal Rights.* Berkeley: University of California Press.

Rollin, B. E. 1990. Animal welfare, animal rights, and agriculture. *J. Anim. Sci.* 68:3456

Rollin, B. E. 1993. "Animal Production and the New Social Ethic for Animals," Food Animal Well-Being 1993 Conference Proceedings and Deliberations,

Purdue University Office of Agricultural Research Programs, West Lafayette, Indiana, p. 11.

Rollin, B. E. 1995. Farm Animal Welfare—Social, Bioethical and Research Issues. Ames: Iowa State University Press.

Roybal, J. 1990. A hassle over hot-iron branding. *BEEF* (Aug.).

Scientific Aspects of the Welfare of Food Animals. 1981. CAST Report No. 91. Ames, IA: Council for Agri. Sci. and Tech.

Singer, P. 1975. *Animal Liberation.* New York: New York Review of Books.

Smidt, D. 1983. *Indicators Relevant to Farm Animal Welfare.* Boston: Martinus Nijhoff Publishers.

Smith, G. C. 1993. Anti-meat propaganda: combatting myths with scientific facts. Presented to the International Meat Industry Convention. Chicago, IL.

Smith, G. C. 1993. Papers from AN 560 (Issues) class. Fort Collins, CO: Colorado State University.

Swinker, A. M. and Heird, J. C. 1994. Horse industry: trends, opportunities, and issues. *Encyclopedia of Agricultural Science.* San Diego: Academic Press Inc.

Tannenbaum, J. 1986. Animal rights: Some guideposts for the veterinarian. *JAVMA* 188:1258.

Thomas, J. A., and Myers, L. A. 1993. *Biotechnology and Safety Assessment.* New York: Raven Press.

U.S. Agriculture and the North American Free Trade Agreement. 1993 (July). Ames, IA: Council for Agricultural Science and Technology.

Water Quality. Agriculture's Role. 1992. Report 120. Ames, IA: Council of Agricultural Science and Technology.

Visuals

Animal Agriculture: Myths and Facts (videotape; 18 min.). Animal Industry Foundation, P.O. Box 9522, Arlington, VA 22209-0522.

The Animal Film (videotape; 136 min., pro-animal rights). 1981. The Cinema Guild, 1697 Broadway, New York, NY 10019.

Buffalo Lessons (videotape; 22 min.). 1991. AgriBase Inc., 7509 Tiffany Springs Parkway, Kansas City, MO 64190.

Cattlemen Care About Animal Welfare (videotape; 18 min.). 1991. Communication Dept., National Cattlemen's Association, Box 3469, Englewood, CO 80155.

Cattlemen Care About the Environment (videotape; 9 min.) and *Issues and Answers for the Nineties* (videotape; 7 min.). 1990. Communications Dept., National Cattlemen's Association, Box 3469, Englewood, CO 80155.

CHAPTER 36

Making Effective Management Decisions

Effective management of livestock operations implies that available resources are used to maximize net profit while the same resources are conserved or improved. Available resources include fixed resources (land, labor, capital, and management) and renewable biological resources (animals and plants). Effective management requires a manager who knows how to make timely decisions based on a careful assessment of management alternatives. Modern technology is providing useful tools to make more rapid and accurate management decisions.

Previous chapters have shown how biological principles determine the efficiency of animal production. It is important to identify other resources that, when combined with the efficiency of animal production, determine the profitability of an operation.

MANAGING FOR LOWER COSTS AND HIGHER RETURNS

Most livestock producers manage their operations with plans to make a profit. Simply stated, the profitability formula is:

$$\begin{matrix} \text{profit} \\ \text{or} \\ \text{<loss>} \end{matrix} = (\text{production} \times \text{price}) - \text{cost}$$

The formula can be expanded to make management decisions more focused:

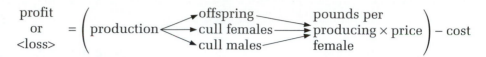

$$\begin{array}{c} \text{profit} \\ \text{or} \\ \text{<loss>} \end{array} = \left(\text{production} \begin{array}{c} \text{offspring} \\ \text{cull females} \\ \text{cull males} \end{array} \begin{array}{c} \text{pounds per} \\ \text{producing} \times \text{price} \\ \text{female} \end{array} \right) - \text{cost}$$

Note: Costs include feed, labor, veterinary, repairs, fuel, interest, other.

Obviously, profit occurs when output value exceeds input costs, and loss occurs when input costs exceed output value. Management decisions should focus on an optimum combination of output value and input costs to maximize profits while maintaining or improving resources.

The primary components of a long-term profitability formula include the following:

Costs. These are identified in earlier chapters (as enterprise budgets) for several species (see Chapter 23 for beef cattle costs, Chapter 25 for dairy cattle, Chapter 27 for swine, and Chapter 29 for sheep).

Production (output). Output is usually pounds and/or numbers sold. The outputs for the various species are shown in the enterprise budgets of the feeding and management chapters.

Price. The amount received per pound, per head, or per dozen (for eggs). Price is influenced primarily by supply and demand.

Maintaining or improving the resources. Refers to the land with the forage and crops produced from it. Maximizing short-term profits can easily deplete the land resource by overgrazing, erosion, and the like.

Each component of the profitability formula should initially be evaluated in terms of how it affects output value.

Price has a tremendous effect on output value, though individual producers have little influence over price. Producers accept the price the market offers at the time they sell their products (output). Therefore, the primary management focus for most livestock and poultry producers is to reduce costs while increasing production and maintaining or improving resources.

The enterprise budget analyses in the species management chapters (Chapters 23, 25, 27, and 29) contain the component parts of input costs. Managers can analyze these costs and make management decisions to reduce them. Throughout all of the chapters in the book, the biological principles affecting production are identified. Producers apply these principles in making cost-effective increases in production.

THE MANAGER

The manager is the individual responsible for planning and decision making. The management process in simple form is to plan, act, and evaluate. This process is described in more detail in Fig. 36.1.

It is imperative that livestock and poultry producers manage their operations as businesses. Current economic pressures associated with keen competition are forcing more producers to manage their operations as business enterprises.

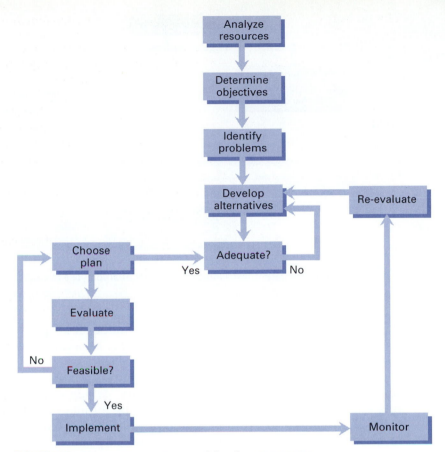

FIGURE 36.1 Major component parts of the planning process.

The manager may be an owner-operator with minimum additional labor, or the manager of a more complex organizational structure involving several other individuals or businesses (Fig. 36.2). An effective manager, whether involved in a one-person operation or a complex organizational structure, needs to (1) be profit-oriented; (2) identify objectives of the business and establish both short-term and long-range goals to achieve those objectives; (3) keep abreast of the current knowledge related to the operation; (4) know how to use time effectively; (5) attend to the physical, emotional, and financial needs of those employed in the operation; (6) incorporate incentive programs to motivate employees to perform at their full capacity each day; (7) have honest business dealings; (8) effectively communicate responsibilities to all employees and make employees feel they are part of the operation; (9) know what needs to be done and at what time; (10) be a self-starter; (11) set priorities and allocate resources accordingly; (12) remove or alleviate high risks; and (13) set a good example for others to follow.

FINANCIAL MANAGEMENT

Costs, returns, and profitability of a livestock operation can only be assessed critically with a meaningful set of records. Table 36.1 identifies the financial records needed by most livestock and poultry operations. Each record is

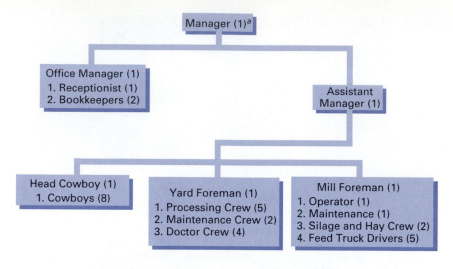

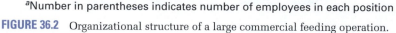

[a]Number in parentheses indicates number of employees in each position

FIGURE 36.2   Organizational structure of a large commercial feeding operation.

described and its purpose is given. Many producers keep records only sufficient to satisfy the IRS (Fig. 36.3). While this is important, additional records are necessary to make critical management decisions.

Profits can be determined by evaluating all cash costs of the operation, all inventory increases or decreases, and the value of opportunity costs. Opportunity costs represent the returns that would be forfeited if debt-free resources (such as owned land, livestock, and equipment) were used in their next-best level of employment; for example, the value of pastureland if it were leased, returns if all capital represented by equipment and livestock were invested in a certificate of deposit (or similar investment), and returns if all family labor and management were utilized in other employment. By calculating this, a price can be established for living on a farm or ranch, and a goal can be established for increasing profits.

Credit and money management become crucial during periods of inflation, high interest rates, and relatively low livestock prices. Prudent use of credit can enable a livestock operation to grow more rapidly than it could through the use of reinvested earnings and savings, so long as borrowed funds return more over time than they cost. Thus, farmers and ranchers have to look to credit as a financial tool and learn to use it effectively.

INCOME TAX CONSIDERATIONS

Frequently, managers feel that they are ineffective if income taxes are paid. Paying little or no income tax should not be a major goal of the operation. A common attitude is reflected in the frequently heard comment of livestock producers that "I've been producing livestock for 20 years and haven't made any money, but I haven't paid any taxes, either."

Well-managed livestock operations pay income taxes if maximizing profitability is a goal. Therefore, it is not poor management to pay income tax but to pay more than is owed.

TABLE 36.1 **Financial Records for Livestock and Poultry Operations**

Financial Record	Description and Purpose
Cash transactions	Recording of all cash receipts and expenditures is the simplest yet most time-consuming of all financial records. This provides most of the information needed for filing income-tax returns and in making loan applications.
Balance sheet	Provides a financial picture of the operation at one point in time—usually on the last day of the year. It reflects the net worth of the operation. Net worth = assets (what is owned) – liabilities (what is owed).
Income statement	A moving financial picture that describes most of the changes in net worth from one balance sheet to the next. Net income is calculated by subtracting the expenditures (cash, decrease in inventory, and depreciation) from income (cash receipts and increases in inventory).
Cash-flow statement	Shows cash generated and cash needed on a periodic basis (usually monthly) throughout the year. It assesses times when money must be borrowed and times when money is available for additional purchases or investment, or to retire existing debts. A cash-flow budget can be used to plan for the next calendar year. An active cash-flow statement tracks what is actually happening and evaluates the accuracy of the cash-flow budget.
Enterprise budget[a]	Identifies costs and returns associated with a specific product or enterprise. It can aid in making financial decisions by identifying specific problem areas where management changes can be made. Enterprise budgets are also useful where operations have more than one enterprise or where additional enterprises are being considered. Components include production and marketing assumptions, operating receipts, direct costs, net receipts, and break-even analysis. Estimates of market weight and price can be used in assessing risk in production decisions.
Partial budget	Involves only those income and expense items that would change when implementing a proposed management decision.
Income-tax forms	Form 1040 Schedule F is the primary income-tax form for sole proprietors and individual partners in a partnership. Producers who have completed Schedule F have basically completed an income statement. There are numerous other forms and 1040 schedules (e.g., Asset Sales, Asset Purchases, Self-Employment Tax, Farm Rental Income and Expenses, Tax Withholding, Depreciation) that are also completed, depending on individual operations and circumstances.

[a] Enterprise budget analyses are shown for beef cattle, dairy cattle, swine, and sheep in Chapters 23, 25, 27, and 29, respectively.

Tax laws are complex and constantly changing. Most cattle producers should consult a qualified tax adviser for tax reporting and longer-term tax management.

ESTATE AND GIFT-TAX PLANNING

Many livestock operations have a large amount of debt-free capital invested in land, livestock, buildings, and equipment. Adequate knowledge and proper planning are necessary for farmers and ranchers to pass on viable economic units to their heirs.

In recent years, significant changes in estate-tax policy have benefited family operations. For example, the unified credit allows each descendant to pass on a maximum of $600,000 of property to the next generation without paying federal estate taxes. Thus, a husband and wife could pass on $1.2 million as tax-free dollars. Producers can make substantial annual cash gifts to children and grandchildren without paying a federal gift tax.

SCHEDULE F	**Profit or Loss From Farming**	OMB No. 1545-0074
(Form 1040)		19**95**
Department of the Treasury Internal Revenue Service (M)	▶ Attach to Form 1040, Form 1041, or Form 1065.	Attachment Sequence No. **14**
	▶ See Instructions for Schedule F (Form 1040).	

Name of proprietor

Social security number (SSN)

A Principal product. Describe in one or two words your principal crop or activity for the current tax year.

B Enter principal agricultural activity code (from page 2) ▶

D Employer ID number (EIN), if any

C Accounting method: (1) ☐ Cash   (2) ☐ Accrual

E Did you "materially participate" in the operation of this business during 1995? If "No," see page F-2 for limit on passive losses. ☐ Yes ☐ No

Part I Farm Income—Cash Method. Complete Parts I and II (Accrual method taxpayers complete Parts II and III, and line 11 of Part I.)
Do not include sales of livestock held for draft, breeding, sport, or dairy purposes; report these sales on Form 4797.

1	Sales of livestock and other items you bought for resale	1	
2	Cost or other basis of livestock and other items reported on line 1	2	
3	Subtract line 2 from line 1	3	
4	Sales of livestock, produce, grains, and other products you raised	4	
5a	Total cooperative distributions (Form(s) 1099-PATR)	5a	**5b** Taxable amount	5b
6a	Agricultural program payments (see page F-2)	6a	**6b** Taxable amount	6b
7	Commodity Credit Corporation (CCC) loans (see page F-2):		
a	CCC loans reported under election	7a	
b	CCC loans forfeited or repaid with certificates	7b	**7c** Taxable amount	7c
8	Crop insurance proceeds and certain disaster payments (see page F-2):		
a	Amount received in 1995	8a	**8b** Taxable amount	8b
c	If election to defer to 1996 is attached, check here ▶ ☐	**8d** Amount deferred from 1994	8d
9	Custom hire (machine work) income	9	
10	Other income, including Federal and state gasoline or fuel tax credit or refund (see page F-3)	10	
11	**Gross income.** Add amounts in the right column for lines 3 through 10. If accrual method taxpayer, enter the amount from page 2, line 51. ▶	11	

Part II Farm Expenses—Cash and Accrual Method. Do not include personal or living expenses such as taxes, insurance, repairs, etc., on your home.

12	Car and truck expenses (see page F-3—also attach **Form 4562**)	12		25	Pension and profit-sharing plans	25
13	Chemicals	13		26	Rent or lease (see page F-4):	
14	Conservation expenses. Attach **Form 8645**	14		a	Vehicles, machinery, and equipment	26a
15	Custom hire (machine work)	15		b	Other (land, animals, etc.)	26b
16	Depreciation and section 179 expense deduction not claimed elsewhere (see page F-4)	16		27	Repairs and maintenance	27
				28	Seeds and plants purchased	28
				29	Storage and warehousing	29
17	Employee benefit programs other than on line 25	17		30	Supplies purchased	30
18	Feed purchased	18		31	Taxes	31
19	Fertilizers and lime	19		32	Utilities	32
20	Freight and trucking	20		33	Veterinary, breeding, and medicine	33
21	Gasoline, fuel, and oil	21		34	Other expenses (specify):	
22	Insurance (other than health)	22		a	....	34a
23	Interest:			b	....	34b
a	Mortgage (paid to banks, etc.)	23a		c	....	34c
b	Other	23b		d	....	34d
24	Labor hired (less employment credits)	24		e	....	34e
				f		34f

35	**Total expenses.** Add lines 12 through 34f ▶	35
36	**Net farm profit or (loss).** Subtract line 35 from line 11. If a profit, enter on **Form 1040, line 18,** and ALSO on **Schedule SE, line 1.** If a loss, you MUST go on to line 37 (estates, trusts, and partnerships, see page F-5).	36
37	If you have a loss, you MUST check the box that describes your investment in this activity (see page F-5). If you checked 37a, enter the loss on **Form 1040, line 18,** and ALSO on **Schedule SE, line 1.** If you checked 37b, you MUST attach **Form 6198.**	37a ☐ All investment is at risk. 37b ☐ Some investment is not at risk.

FIGURE 36.3 Internal Revenue Service form (Schedule F) showing income and expense items for which documented records must be kept. Source: Dept. of Treasury.

Livestock producers should review tax plans (estate, gift, and income) and consult with professionals who are knowledgeable about current tax laws.

COMPUTERS

Livestock operators need to keep a voluminous amount of information for financial, production, and inventory records in order to assess management alternatives critically. Some managers with conventional records systems have the ability to "use their heads" to make good management decisions with a high degree of accuracy. They are able to use their mind (the most complex computer) to assimilate, store, and recall useful information effectively. However, for most managers, the computer can be a valuable aid.

Small computers (microcomputers) are frequently used as a management tool by farmers and ranchers. Producers must recognize what computers can and cannot do, at the present time and in the future. The technology of computers is changing rapidly; therefore, producers must be ready to accept the frustration of change if they become involved with computers.

Every computer, regardless of size and shape, is basically a fast adding machine and an electronic filing cabinet. A microcomputer system consists primarily of a system of hardware and software. The hardware for microcomputers (Fig. 36.4) consists of a terminal (a typewriter keyboard and an optional television screen), a printer, data-storage devices (the filing cabinets), main memory, and a central processing unit (CPU—the adding machine). The data-storage devices can be magnetic (similar to cassette tapes), floppy disks (similar to 45-rpm records), or hard disks (large storage capacity in one unit). The devices that read and write the information on these disks are called *disk drives* or *drives.*

Software are the programs that make the hardware function. Each program typically consists of thousands of minutely detailed instructions that tell the computer exactly what data to manipulate mathematically, print, store, or display.

FIGURE 36.4 Microcomputers are frequently used in making management decisions with livestock and poultry. Courtesy of Colorado State University.

Good software programs are the most limiting factor in microcomputer utilization by livestock producers, although more useful programs are becoming available. Software developed for one microcomputer may not function on a different brand of machine. Computer utilization is expected to increase rapidly as more compatible programs are developed. The four ways producers can obtain software are by (1) purchasing a complete commercial package, (2) hiring someone to do the programming, (3) learning to program the computer themselves, or (4) obtaining it from their land grant university.

The three major areas in which microcomputers have the largest potential for livestock producers are (1) business accounting, (2) herd performance, and (3) financial management. A computer that would handle these three functions would need a relatively large memory.

Computers do not simplify a livestock business or cut overhead costs. They can even make a business more complex and more costly. If wisely acquired and properly utilized, a computer can increase profits, but usually not by reducing expenses. The computer will not solve the problem of a poorly organized operation. Well-planned and efficient office management practices must exist and function smoothly before the addition of a computer will do anything but aggravate the organizational problem of the business. The computer cannot assist the producer in managing the business effectively unless the computer receives all the required data in precisely the prescribed manner. Unless the computer can provide the information needed by producers at the appropriate time and in an easily understood form, it should not be part of a management program, regardless of price.

A producer should follow six steps before deciding to purchase a computer: (1) determine what important management decisions need to be made and what information is needed to make these decisions effectively, (2) see what programs are available to meet the management information needs, (3) determine the hardware needed to use the software, (4) contact local computer hardware dealers and evaluate the various alternatives and services, (5) estimate the cost-benefit ratio of the proposed management-information system, and (6) make the final decision after evaluating the experiences of other producers who are using computers in their operations.

MANAGEMENT SYSTEMS

Management-systems analysis provides a method of systematically organizing information needed to make valid management decisions. It permits variables to be more critically assessed and analyzed in terms of their contribution to the desired end point, usually net profit. Individuals who have been educated to think broadly in the framework of management systems can often make valid management decisions without the use of data-processing equipment. A pencil, hand calculator, and a well-trained mind are the primary components required for making competent management decisions. Without question, however, the use of the computer in synthesizing voluminous amounts of information enhances the management system.

Resources are different for each operation; no fixed recipe exists for successful livestock production. The uniqueness of each operation results from such variables as different levels of forage production, varying marketing alternatives, varying energy costs, different types of animals, environmental differences, feed nutrients at various costs, and varying levels of competence in labor and management. All of these variables and their interactions pose challenges to the producer seeking to combine them into sound management decisions for a specific operation.

It is a common practice to increase or maximize production of animals by using known biological relationships. Animal production typically has been maximized without careful consideration of cost-benefit ratios and of how increased productivity relates to land, feed, and management resources. However, recognition is now being given to the need for optimization rather than maximization of animal productivity. Thus, there is an increased interest in management systems that attempt to optimize production with net profit being the primary (and possibly the only) goal involved. As a result, some valid biological relationships will not be useful or applicable because an economic analysis may prevent their inclusion in sound management decisions. For example, it is a well-known biological fact that calves born earlier in the calving season will have heavier weaning weights. Because of this relationship, some ranchers move the calving season to a period earlier in the year without a careful economic assessment. An economic evaluation for one ranch demonstrated that by changing the calving season to match forage availability (in this case about 30 days to grazable forage), profits were increased significantly. Changing calving season alone was estimated to increase the ranch's carrying capacity by 30–40%, in terms of animal units. In addition, the annual cow cost was reduced by 18% per cow because of reduction in winter feed requirements.

In dairy production, there is often a difference between maximum milk production and optimum milk production if producers want to maximize profit. The extra feed and management needed to maximize milk production may cost more than the value of the added milk.

Some farmers and ranchers are limited in utilizing the computer in management systems because of the lack of software. However, more management-system software continues to be made available to producers.

To be competitive, livestock operations have to be operated like businesses. This means keeping accurate financial and production records for use in making management decisions. The computer is an important tool to improve the effectiveness of the management-decision process. However, the pencil, hand calculator, and a keen mind will continue to be important in making profitable management decisions.

CHAPTER SUMMARY

■ The profitability formula:

$$\begin{matrix} \text{profit} \\ \text{or} \\ \text{<loss>} \end{matrix} = (\text{production} \times \text{price}) - \text{cost}$$

identifies the components for animal producers to lower their costs and increase their revenue and profits.

■ Assessment of resources (human, financial, forage/feed, animals, etc.) and a decision-making process that integrates the resources is essential for the well-being of all resources.

■ Integrated resource management (management systems) is critical if future profits are to be sustained and resources are to be maintained or improved.

REVIEW QUESTIONS

1. What are the four components of the long-term profitability formula?

2. What is critical to assessing costs, returns, and profitability of livestock operations?

3. What are the three major areas in which microcomputers or personal computers have the largest potential for livestock producers?

SELECTED REFERENCES

Publications

Bourdon, R. 1992. *The Systems Concept of Beef Production.* Beef Improvement Federation Fact Sheet.

Harl, N. 1980. *Farm Estate and Business Planning.* 6th ed. Skokie, IL: Century Communications.

Killcreas, W. E., and Hickel, R. 1982. *A Set of Microcomputer Programs for Swine Record Keeping and Production Management.* Miss. Agric. and Forestry Expt. Sta. AEC Tech. Public. No. 35.

Klinefelter, D. A., and Hottel, B. 1981. *Farmers' and Ranchers' Guide to Borrowing Money.* Texas Agric. Expt. Sta. MP-1494.

Libbin, J. D., and Catlett, L. B. 1987. *Farm and Ranch Financial Records.* New York: Macmillan.

Luft, L. D. Sources of credit and cost of credit. In *Great Plains Beef Cattle Handbook,* GPE-4351 and 4352.

Maddux, J. 1981. Dec. *The Man in Management. Proceedings: The Range Beef Cow, a Symposium on Production, VII.* Rapid City, SD.

NCA-IRM-SPA™ Workbook for The Cow-Calf Enterprise. Texas A&M University, College Station, TX: James McGrann.

Ott, G. Planning for profit with partial budgeting. *Great Plains Beef Cattle Handbook,* GPE-4551.

Sobba, A. 1993. *Tax Tips for Cattlemen.* Beef Business Bulletin. Englewood, CO: National Cattlemen's Association.

Taylor, R. E. 1994. *Beef Production and Management Decisions.* New York: Macmillan Publishing Co.

Wilt, R. W., Jr., and Bell, S. D. 1977. *Use of Financial Cash Flow Statements as a Financial Management Tool.* Ala. Agric. Expt. Sta. Bull.

Visuals/Audio

Covey, S. R. 1991. *The Seven Habits of Highly Effective People* (audio cassette seminar). Covey Leadership Center, 3507 N. University Ave., Suite 100, Provo, UT 84604-4479.

Farm and Ranch Management; and *Agricultural Credit* (sound filmstrips). Vocational Education Productions, California Polytechnic State University, San Luis Obispo, CA 93407.

Finding the Profit in Agriculture (videotape). CEV, P.O. Box 65265, Lubbock, TX 79464-5265.

Computer Software

Swine Management Series One; and *Swine Record Keeping.* Vocational Education Productions, California Polytechnic State University, San Luis Obispo, CA 93407.

NCA-IRM-SPA™ Financial Calculations; Production Calculations. Brighton, IL: AGRO-Systems.

CHAPTER 37

Careers and Career Preparation in the Animal Sciences

Millions of domestic animals that provide food, fiber, and recreation for people create many and varied types of career opportunities. A placement survey of approximately 10,000 agriculture graduates in the United States is shown in Tables 37.1 and 37.2. Animal sciences placement data follow a similar pattern because they compose a high percentage of the graduates shown in these tables.

Beef cattle, dairy cattle, horses, poultry, sheep, and swine are the animals of primary importance in animal science curricula, whereas reproduction, nutrition, breeding (genetics), meats, and live animal appraisal are the specialized topics typically covered in animal science courses. A few animal science programs include studies of pet and companion animals. Most college and university curricula in the animal sciences are designed to assist students in the broad career areas of production, science, agribusiness, and the food industry. Some animal science departments combine production and agribusiness into an industry concentration. The major careers in these areas are shown in Table 37.3.

PRODUCTION

Many individuals are intrigued with animal production because of the possibility of being their own boss and working directly with animals (Fig. 37.1). These ambitions may be somewhat idealistic because the financial investment required in land is typically hundreds of thousands of dollars. Beef cattle, dairy, poultry, sheep, and swine production operations continue to become fewer in number, more specialized, and larger in terms of the amount of capital invested. Those who find a career in the production area usually have a family

TABLE 37.1   Placement of Bachelor Degree Recipients in Animal Science

Postgraduation Activity	Percent of Graduates
Agricultural production	35%
Graduate and professional study	30
Agribusiness and business management	20
Scientists and related specialists	8
Education, including extension	5
Miscellaneous	2

Source: Adapted from the Food and Agricultural Education Information System 1995 Placement Report.

operation to which they can return. A few others have the needed capital to invest.

Sometimes students who seek employment in animal production are disillusioned with the base salary, benefits, and working hours when production work is compared with agribusiness careers. However, some graduates are anxious and willing to obtain experience and prove they can work, and eventually locate permanent employment in production operations. Some producers of cattle, sheep, swine, and horse operations allow an employee to buy into an operation on a limited basis or own some of the animals after working a year to two.

Certain careers require education and experience in the production area even though the individual will not be producing animals directly. Career opportunities such as breed association and publication positions, fieldpersons, extension agents, and livestock marketing require an understanding of production, as one works directly with livestock producers.

SCIENCE

An animal science student who concentrates heavily in science courses is usually preparing for further academic work with a goal of achieving one or more advanced degrees. A minimum grade average of B is usually required for admission to graduate school or a professional veterinary medicine program.

TABLE 37.2   Average Starting Salaries from Graduates of Colleges of Agriculture in the United States

Graduate	Average Annual Starting Salary		
	1995	1990	1985
Bachelor of Science (B.S.)	$23,393	$20,000	$17,585
Master of Science (M.S.)	29,668	25,000	22,020
Doctor of Philosophy (Ph.D.)	38,667	35,000	29,448

Source: Adapted from the Food and Agricultural Education Information System 1995 Placement Report.

TABLE 37.3   Animal Sciences Careers in Production, Science, Agribusiness, and Food Industry

Animal Sciences			
Production	Science	Agribusiness	Food Industry
Feedlot positions	Graduate school for Master of Science (M.S.) and Doctor of Philosophy (Ph.D.) degrees	Sales and management positions with feed companies, packing companies, drug and pharmacy companies, equipment companies, and so on	Food-processing plants
Livestock production operations (beef, dairy, swine, sheep, and horses)	Research (university or industry) in nutrition, reproduction, breeding and genetics, products, and production management	Livestock publications	Food-ingredient plants
Ranch positions	University or college teaching	Advertising and promotion	Food-manufacturing plants
Breed associations	University extension and area extension	Finance (PCAs, banks, and so on)	Government—protection and regulatory agencies
AI studs and breeding	Management positions in industry	Public relations	Government—Department of Defense (food supply and food service)
Livestock buyers for feeders and packers	Government work	Meat grading (federal government)	Government—Department of Agriculture (research and information)
County extension agents	International opportunities	International opportunities	University research, teaching, and extension
Meat grading and handling distribution	Laboratory technicians	Graduate school for Master's in business Administration (M.B.A.)	Positions with food companies
Marketing (auctions, Cattle Fax, livestock sales management, and so on)	Veterinary school for Doctor of Veterinary Medicine (D.V.M.)	Foreign agriculture	Research and development with food companies
International opportunities	Private practice	Technical sales and service	
Livestock and meat market reporting (government)	Consulting	Positions in poultry-production units	
Riding instructors	University teaching and research	Companion animals	
Feed manufacturing	Meat inspection		
Companion animals	Industry/commercial		
Boarding	Companion animal research		
Breeding			
Training			
Humane Society			

Advanced degrees are usually obtained after entrance into graduate school or after being admitted to a professional veterinary medicine program where a Doctor of Veterinary Medicine (D.V.M.) degree is awarded. Table 37.4 shows where graduates of veterinary medical colleges are employed, their average starting salaries, and their income after several years of employment.

The Master of Science (M.S.) and Doctor of Philosophy (Ph.D.) degrees are the advanced degrees received by the science-oriented student. Although the 2-year college certificate or Bachelor of Science (B.S.) degree allows breadth of education, advanced degrees are generally directed to a specialization. These specialized animal science areas are typically in nutrition, reproduction, breeding (genetics and statistics), animal products, and, less frequently, management systems (Figs. 37.2–37.5).

FIGURE 37.1   There are numerous career opportunities in the production of the various species of farm animals. Some students interested in the production area will not be directly involved in animal production. These students will work for companies, associations, or groups closely associated with the animal producers. Courtesy of Institute of Agriculture and Natural Resources, University of Nebraska; lower left photo courtesy of University of Illinois.

TABLE 37.4 Professional Incomes and Employment for U.S. Veterinarians

Career Area	Avg. Income[a]	Avg. Starting Salary in 1995
Large animal exclusive	$63,547	$39,535
Large animal predominant	56,500	34,290
Mixed animal	50,862	31,892
Small animal predominant	55,179	30,970
Small animal exclusive	57,991	31,882
Equine predominant	64,240	31,924
Average (private practice)	$57,557	
College or university	$63,749	NA
Federal government	59,626	NA
State/local government	57,430	NA
Uniformed services	57,619	NA
Industry	94,046	NA
Advanced study	NA	NA
Average (public or corporate)	$67,043	

[a] Annual salary after several years of employment.
Source: JAVMA 208:206 (Jan. 15, 1995); 209:1712 (Nov. 15, 1996); 210:188 (Jan. 15, 1997).

AGRIBUSINESS

Although the number of individuals employed in livestock and poultry production has decreased, the number of individuals and businesses serving producers has greatly increased. Positions in sales, management, finance, advertising, public relations, and publications are prevalent in feed, drug, equipment, meat packing, and livestock organizations.

Agribusiness careers that relate to the livestock industry require a student to have a foundation of knowledge of livestock along with an excellent comprehension of business, economics, computer science, and effective communication. A graduate should understand people, know how to communicate with people, and enjoy working with people. Extracurricular activities that give students experience in leadership and working with people are invaluable for meaningful career preparation.

Some animal science students who concentrate heavily in business and economics courses will pursue an advanced degree. This is typically a Master's in Business Administration (M.B.A.). A combination of a B.S. degree in Animal Science with an M.B.A. is an excellent preparation for management positions in livestock-oriented businesses.

FOOD INDUSTRY

Red meats, poultry, milk, and eggs—the primary end products of animal production—provide basic nutrition and eating enjoyment for millions of consumers. Career opportunities are numerous in providing one of our basic needs—food. Processing, packaging, and distribution of food are important

FIGURE 37.2 Many students concentrating in the sciences are preparing themselves for entrance into graduate school. There they will complete M.S. and Ph.D. degrees, which are primarily research-oriented degrees. A few examples are depicted here. Courtesy of Institute of Agriculture and Natural Resources, University of Nebraska; lower right photo courtesy of West Virginia University.

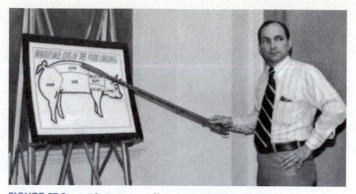

FIGURE 37.3 A Ph.D. is usually required for teaching in a college or university. Courtesy of Oklahoma State University.

components of the food-production chain. New food products are continually being produced, and new methods of manufacturing and fabricating food are being developed. Electrical stimulation of carcasses for increased tenderization, vacuum packaging of primal cuts for a longer shelf life, and preparation of convenience products are examples of recent innovations in meat processing. A vital and continuing challenge over the next several decades will be to provide a food supply that is nutritious, safe, convenient, attractive, and economical—which is also satisfying to the consumer (Fig. 37.6).

INTERNATIONAL OPPORTUNITIES

Much has been written during the past decade of the challenge to provide adequate nutrition to an ever-expanding world population. Many countries have

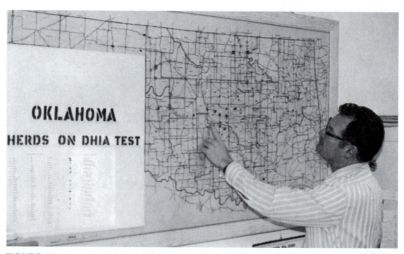

FIGURE 37.4 University extension personnel need a Ph.D. degree so they can interpret the current research and apply it in off-campus educational programs. Courtesy of Oklahoma State University.

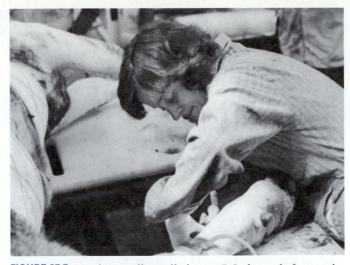

FIGURE 37.5 Students will usually have a B.S. degree before applying for admission to the College of Veterinary Medicine. Students with high academic performance in the science aspect of animal science are typically well prepared for completing the D.V.M. degree. Courtesy of Colorado State University.

tremendous natural resources for expanded food production but lack technical knowledge and adequate capital to develop these resources. Federal government programs, designed to help foreign countries help themselves, offer several career opportunities in organizations such as the Peace Corps, Vista, and USAID. Many of these opportunities of assisting people in agricultural production in developing countries are open to individuals educated and experienced in the animal sciences.

The Foreign Agricultural Service of the USDA employs people in animal economics, marketing, and administration as attachés and international secretaries. Animal science students take several courses in economics, marketing, foreign languages, and business administration if they wish to qualify for these positions.

Certain private industries offer opportunities in animal production and related businesses in foreign countries. Multinational firms that develop livestock feed companies, drug and pharmaceutical companies, companies that export and import animals and animal products, and consulting companies are some examples.

MANAGEMENT POSITIONS

Management and administrative positions exist in the production, science, agribusiness, and food industry areas of animal science. Because management sometimes implies fewer hours and more pay than other positions that require more physical labor, young men and women typically desire management positions. Nevertheless, although management and administrative salaries are usually higher, the work load and pressures are usually much greater.

FIGURE 37.6 The demand for animal products creates many industries that support the animals and process the products for human consumption. Courtesy of the University of Nebraska, Colorado State University, and Oklahoma State University.

Many college graduates have the impression that once the degree is in hand, it automatically qualifies them for management positions. In actuality, management positions are usually earned based on a person's proven ability to work with people, solve problems, and make effective decisions. After being employed for a few years, individuals having previously learned the academic principles must also experience the component parts of an operation or business. For example, presidents or vice presidents of feed companies have typically started their initial careers in feed sales. Most successful managers have had a continual series of learning experiences since the end of the formal education. Managers need to understand all aspects of the business they are managing, particularly the products that are produced and sold.

WOMEN IN ANIMAL SCIENCE CAREERS

Women are pursuing and finding career opportunities in all areas of the animal sciences. The number of women majoring in animal science has been growing steadily in most colleges and universities. In the early 1970s, the number of females majoring in animal science was 20–25% of the total majors. In many universities women now comprise more than half of the animal science and D.V.M. enrollments (Fig. 37.7). Many agricultural employers, especially in the agribusiness, are actively recruiting women with animal science degrees.

In recent years, women have found career opportunities that many felt were reserved for men only. Although some men expressed fear and had serious reservations, many of these women proved they could perform competently. It does, however, appear that some employers in private production agriculture are still reluctant to hire women on the same basis as men. Nevertheless, women interested in animal science careers should prepare themselves adequately in both education and experience. Many livestock agribusiness companies are actively recruiting women graduates.

Recent observations in a California study showed that women graduates in agriculture were at a disadvantage in salary and job status when compared with men graduating with similar degrees. However, a Colorado State University survey of its women graduates in animal sciences since 1979 revealed that many of them were in high-paying careers. This survey also demonstrated that women desiring employment in the production area had struggled in identifying meaningful employment. The agribusiness area (e.g., Farmer's Home Administration, feed companies, food-processing companies, and drug and pharmaceutical companies) had provided attractive careers for women graduates: several companies reported annual starting salaries were above $20,000.

FUTURE CAREER OPPORTUNITIES

A 1990 USDA national assessment report projected the employment opportunities for food and agricultural sciences graduates through 1995. The estimated number of annual job openings is 48,000, with approximately 43,500 qualified

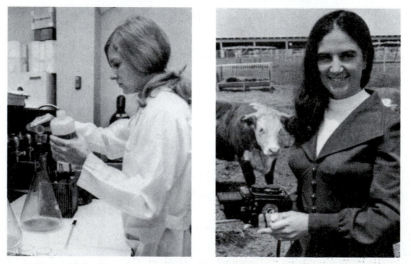

FIGURE 37.7 Women are seeking and finding career opportunities in the animal sciences. The examples shown are (A) teaching equitation, (B) veterinarian, (C) feedlot manager, (D) researcher, (E) photographer. Courtesy of Colorado State University; University of Illinois; Office of Agricultural Communications; *BEEF;* University of Wyoming; *The American Hereford Journal;* and Oklahoma State University.

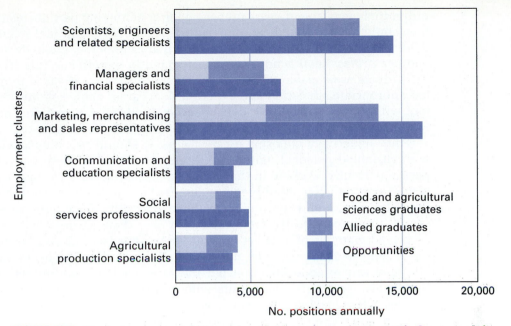

FIGURE 37.8 Employment opportunities and available graduates (1990–1995). Courtesy of the USDA.

college graduates available each year. The supply of graduates is expected to be approximately 11% less than the available job openings.

Figure 37.8 shows the employment opportunities and the available graduates for six different career areas. The USDA report has the following summary:

■ In summary, basic plant and animal research, food and fiber processing, and agribusiness management and marketing are expected to provide the most significant employment opportunities for college graduates with expertise in agriculture, natural resources, and veterinary medicine through 1995. In contrast, college graduates seeking positions in production agriculture, education, and communications are encountering strong competition for somewhat limited employment opportunities within the U.S. food and agricultural system.

CAREER PREPARATION

An education should provide opportunities for people to earn a living, continue learning, and live a full, productive life even beyond the typical retirement years. A broad-based education is important because a person may be preparing for a career that does not presently exist or a career not recognized at the present time. Change has occurred and will no doubt continue to occur, so a person needs to be flexible and adaptable and willing to continue learning throughout life.

Occasionally someone chooses a career for which he or she has little real talent or for which she or he is not qualified or properly motivated. These individuals may spend years preparing for or achieving partial success when they could have been outstanding in another area. In the animal sciences, students preparing for acceptance into veterinary medicine, graduate school, or another professional school seem to frequently experience this frustration. Individuals are capable of success in several different careers if they receive the appropriate education and experience. A career choice should be consistent with a person's interest, desire, and motivation. A career goal should be established with flexibility for change if it proves to be unrealistic.

After tentatively choosing a career area, it should be pursued vigorously with sustained personal motivation. Students should write to and visit with people currently working in the student's chosen career area. Every personal visit or written letter should request another lead or source of further information. Students should continually evaluate the reality of a chosen career. They should think realistically about the facts rather than getting caught up in the perceived glamour of a career choice. Career-interest tests can be helpful in identifying areas to pursue in more detail. However, these tests are not infallible; they can provide direction but they are not the final answer. Students should always temper the results with additional information from several sources before making final decisions.

Employers hire people, not a college major or area of study. The classes a student takes in college are only part of the necessary preparation. Therefore, in addition to academic preparation, students should develop their marketable personal assets, emphasized by employers as the ability to lead, solve problems, communicate and work with people, and consistently work hard. A person develops many of these personal characteristics through experience, part-time and summer jobs, and extracurricular activities. Internships are especially valuable in obtaining exposure to a career area and developing personal abilities and skills through practical experience. A combination of useful courses, practical experience, effective communication, and excellent work skills is the best assurance of finding and keeping a meaningful career in the animal sciences.

CHAPTER SUMMARY

■ Livestock production and the associated organizations and agribusinesses provide numerous and varied career opportunities for students majoring in the animal sciences.

■ Students concentrating on science courses are usually preparing for advanced degrees, such as, doctor of veterinary medicine (D.V.M.) or the research-oriented degrees of master of science (M.S.) and doctor of philosophy (Ph.D.).

■ Students concentrating on economics, agricultural business, and business courses are usually preparing for careers in livestock production and the agribusiness careers in the meat, animal health, and feed industries.

■ In addition to an excellent course preparation, employers emphasize skills in effective communication, computers, and work ethic.

■ Meaningful internships make a significant contribution to career preparation.

SELECTED REFERENCES

Publications

American Meat Science Association, *Careers in the Meat Industry.* Chicago, IL: American Meat Science Association.

Coulter, K. J., Stanton, M., and Goecker, A. D. 1990. *Employment Opportunities for College Graduates in the Food and Agricultural Sciences.* Washington, DC: USDA, Higher Education Program.

Employment, starting salaries, and educational indebtedness of 1989 graduates of U.S. veterinary medical colleges. 1989. *JAVMA* 195:1407.

Food and Agricultural Education Information System. Degrees Awarded and Placement for Agricultural and Natural Resources. May 1993.

National Association of State Universities and Land Grant Colleges. July 1990. *Survey of Agricultural Degrees Granted and Post-Graduation Activities of Graduates.*

Wood, J. B., Dupre, D. H., and Thompson, O. E. 1981. *Women in the agricultural labor market. California Agriculture* (Sept.–Oct., 16).

Visuals

Careers: Marketing Specialists; Social Service Professionals; Agricultural Producers; Agricultural Scientists; and *Financial Specialists and Managers* (five videotapes). CEV, P.O. Box 65265, Lubbock, TX 79464-5265.

Careers: Marketing Yourself; the Right Fit; Resumes; Interviews; Getting the Interview; Getting in the Door; Making the Cut; Practice Interviewing; Getting the Offer (eight videotapes). CEV, P.O. Box 65265, Lubbock, TX 79464-5265.

Job Hunting Skills; Working Toward a Career; Finding and Landing the Right Job; How to Interview for a Job; Interviewing; Resume Writing; The Job Application (six videotapes). CEV, P.O. Box 65265, Lubbock, TX 79464-5265.

Unbridled Opportunities—Careers in the Horse Industry. American Horse Council, 1700 K Street N.W., Suite 300, Washington, DC 20006.

Veterinary Medicine: The Challenge That Lasts a Lifetime (videotape; 13 min.). College of Veterinary Medicine and Biomedical Sciences, Colorado State University, Fort Collins, CO 80523.

Glossary

abomasum   The fourth stomach compartment of ruminant animals that corresponds to the true stomach of monogastric animals.

abortion   Delivery of fetus between conception and several days before normal parturition.

abscess   Localized collection of pus in a cavity formed by disintegration of tissues.

absorption   The passage of liquid and digested (soluble) food across the gut wall.

accessory organs   The seminal vesicles, prostate, and Cowper's glands in the male. These glands add their secretions to the sperm to form semen.

accuracy (ACC) of selection   Numerical value, ranging from 0–1.0, denoting the confidence that can be placed in the EPD (expected progeny difference); e.g., high ($\geq$0.70), medium (0.40–0.69), low ($\leq$0.40).

adipose   Fat cells or fat tissue.

ad libitum   Free choice; allowing animals to eat all they want.

afterbirth   The membranes attached to the fetus that are expelled after parturition.

AI   Abbreviation for *artificial insemination.*

air dry   Refers to feeds in equilibrium with air; they would contain approximately 10% water or 90% dry matter.

albumen   The white of an egg.

alimentary tract   Passageway for food and waste products through the body.

alleles Genes occupying corresponding loci in homologous chromosomes that affect the same hereditary trait but in different ways.

allelomimetic behavior Doing the same thing. Animals tend to follow the actions of other animals.

alveolus (plural, *alveoli*) A hollow cluster of cells. In the mammary gland, these cells secret milk.

amino acid An organic acid in which one or more of the hydrogen atoms has been replaced by the amino group ($-NH_2$). Amino acids are the building blocks in the formation of proteins.

amnion A fluid-filled membrane located next to the fetus.

ampulla The dilated or enlarged upper portion of the vas deferens in bulls, bucks, and rams, where sperm are stored for sudden release at ejaculation.

anabolic A constructive, or "building up," process.

anatomy Science of animal body structure and the relation of the body parts.

androgen A male sex hormone, such as testosterone.

anemia Deficiency of hemoglobin, often accompanied by a reduced number of red blood cells. Usually results from an iron deficiency.

anestrous Period of time when female is not in estrus; the nonbreeding season.

antemortem Before death.

anterior Situated in front of, or toward the front part of, a point of reference. Toward the head of an animal.

anterior pituitary (AP) The part of the pituitary gland, located at the base of the brain, that produces several hormones.

anthelmintic A drug or chemical agent used to kill or remove internal parasites.

antibiotic A product produced by living organisms, such as yeast, which destroys or inhibits the growth of other microorganisms, especially bacteria.

antibody A specific protein molecule that is produced in response to a foreign protein (antigen) that has been introduced into the body.

antigen A foreign substance that, when introduced into the blood or tissues, causes the formation of antibodies. Antigens may be toxins or native proteins.

antiseptic A chemical agent used on living tissue to control the growth and development of microorganisms.

AP See *anterior pituitary.*

APCS A popular milk testing system where weights are recorded at two milkings.

artery Vessel through which blood passes from the heart to all parts of the body.

arteriosclerosis A disease resulting in the thickening and hardening of the artery walls.

artificial insemination (AI)   The introduction of semen into the female reproductive tract (usually the cervix or uterus) by a technique other than natural service.

artificial vagina   A device used to collect semen from a male when he mounts in a normal manner to copulate. The male ejaculates into this device, which simulates the vagina of the female in pressure, temperature, and sensation to the penis.

ascaris   Any of the genus (*Ascaris*) of parasitic roundworms.

as fed   Refers to feeding feeds that contain their normal amount of moisture.

assimilation   The process of transforming food into living tissue.

atherosclerosis   A form of arteriosclerosis involving fatty deposits in the inner walls of the arteries.

atrophy   Shrinking or wasting away of a tissue or organ.

auction   A market facility where an auctioneer sells animals to the highest bidder.

autopsy   A postmortem examination in which the body is dissected to determine the cause of death.

avian   Refers to birds, including poultry.

balance sheet   A statement of assets owned and liabilities owed in dollar terms that shows the equity or net worth at a specific point in time (e.g., net worth statement).

band   (1) A relatively large group of range sheep. (2) Method of identification—e.g., to put a band around the leg of a chicken.

Bang's disease   See *brucellosis*.

barren   Not capable of producing offspring.

barrow   A male swine that was castrated before reaching puberty.

basal metabolism   The chemical changes that occur in an animal's body when the animal is in a thermoneutral environment, resting, and in a postabsorptive state. It is usually determined by measuring oxygen consumption and carbon dioxide production.

beef   The meat from cattle (bovine species) other than calves (the meat from calves is called *veal*).

beri-beri   A disease caused by a deficiency of vitamin B_1.

biologicals   Pharmaceutical products used especially to diagnose and treat diseases.

biotechnology   The use of microorganisms, plant cells, animal cells, or parts of cells (such as enzymes) to produce industrially important products or processes.

birth weight expected progeny difference (EPD)   The expected average increase or decrease in the birth weight of a beef bull's calves when compared to the other bulls in the sire summary.

blemish   Any defect or injury that mars the appearance of, but does not impair the usefulness of, an animal.

bloat   An abnormal condition in ruminants characterized by a distention of the rumen, usually seen on an animal's upper left side, owing to an accumulation of gases.

blood spots   Spots in the egg caused by a rupture of one or more blood vessels in the yolk follicle at the time of ovulation.

BLUP   Best linear unbiased prediction, method for estimating breeding values of breeding animals.

boar   (1) A male swine of breeding age. (2) Denotes a male pig, which is called a *boar pig.*

bog spavin   A soft enlargement of the anterior, inner aspect of the hock.

bolus   (1) Regurgitated food. (2) A large pill for dosing animals.

bone spavin   A bony (hard) enlargement of the inner aspect of the hock.

bots   Any of a number of related flies whose larvae are parasitic in horses and sheep.

bovine   A general family grouping of cattle.

boxed beef   Cuts of beef put in boxes for shipping from packing plant to retailers. These primal and subprimal cuts are intermediate cuts between the carcass and retail cuts.

boxed lamb   See *boxed beef.* Similar process except lamb instead of beef.

break joint   Denotes the point on a lamb carcass where the foot and pastern are removed at the cartilaginous junction of the front leg.

bred   Female has been mated to the male. Usually implies the female is pregnant.

breech   The buttocks. A breech presentation at birth is where the rear portion of the fetus is presented first.

breed   Animals of common origin with characteristics that distinguish them from other groups within the same species.

breeding value   A genetic measure for one trait of an animal, calculated by combining into one number several performance values that have been accumulated on the animal and the animal's relatives.

brisket disease   A noninfectious disease of cattle characterized by congestive right heart failure. It affects animals residing at high altitudes (usually above 7,000 ft).

British breeds   Breeds of beef cattle originating in England. Examples are Angus, Hereford, and Shorthorn.

British thermal unit (Btu)   The quantity of heat required to raise the temperature of 1 lb of water 1°F at or near 39.2°F.

brockle-faced   White-faced with other colors splotched on the face and head.

broiler   A young meat-type chicken of either sex (usually up to 6–8 weeks of age) weighing 3–5 lb. Also referred to as a *fryer* or *young chicken.*

broken-mouth   Some teeth are missing or broken.

broodiness   The desire of a female bird to sit on eggs (incubate).

browse Woody or brushy plants. Livestock feed on tender shoots or twigs.

brucellosis A contagious bacterial disease that results in abortions; also called *Bang's disease.*

buck A male sheep or goat. This term usually denotes animals of breeding age.

bulbourethral (Cowper's) gland An accessory gland of the male that secretes a fluid which constitutes a portion of the semen.

bull A bovine male. The term usually denotes animals of breeding age.

buller-steer syndrome A behavior problem where a steer has a sexual attraction to other steers in the pen. The steer is ridden by the other steers, resulting in poor performance and injury.

bullock A young bull, typically less than 20 months of age.

buttermilk The fluid remaining after butter has been made from cream. By use of bacteria, cultured buttermilk is also produced from milk.

buttons May refer to cartilage or dorsal processes of the thoracic vertebrae. Also see *cotyledons.*

by-product A product of considerably less value than the major product. For example in U.S. meat animals, the hide, pelt, and offal are by-products, whereas meat is the major product.

C-section See *cesarean section.*

calf A young male or female bovine animal under a year of age.

calorie The amount of heat required to raise the temperature of 1 g of water from 15°C to 16°C.

calve Giving birth to a calf. Same as parturition.

calving interval The amount of time (days or months) between the birth of a calf and the birth of a subsequent calf, both from the same cow.

candling The shining of a bright light through an egg to see if it contains a live embryo.

canter A slow, easy gallop.

capon Castrated male chicken. Castration usually occurs between 3 and 4 weeks of age.

capped hocks Hocks that have hard growths that cover, or "cap," their points.

carbohydrates Any foods, including starches, sugars, celluloses, and gums, that are broken down to simple sugars through digestion.

carcass merit The value of a carcass for consumption.

carnivorous Subsisting or feeding on animal tissues.

carotene The orange pigment found in carrots, leafy plants, yellow corn, and other feeds, which can be broken down to form two molecules of vitamin A.

caruncle (1) The red and blue fleshy, unfeathered area of skin on the upper region of the turkey's neck. (2) The "buttons" on the ruminant uterus where the cotyledons on the fetal membranes attach.

casein The major protein of milk.

cash-flow statement   A financial statement summarizing all cash receipts and disbursements over the period of time covered by the statement.

castrate   (1) To remove the testicles. (2) An animal that has had its testicles removed.

cattalo   A cross between domestic cattle and bison.

cecum (ceca)   A blind pouch at the junction of the small and large intestine. Poultry have two ceca.

cervix   The portion of the female reproductive tract between the vagina and the uterus. It is usually sealed by thick mucus except when the female is in estrus or delivering young.

cesarean section   Delivery of fetus through an incision in abdominal and uterine walls.

chalaza   A spiral band of thick albumen that helps hold the yolk of an egg in place.

chemotherapeutics   Chemical agents used to prevent or treat diseases.

chevon   Meat from goats.

chick   A young chicken that has recently been hatched.

chorion   The outermost layer of fetal membranes.

chromosome   A rodlike or stringlike body found in the nucleus of the cell that contains genetic information. The chromosome contains the genes.

chyme   The thick, liquid mixture of food that passes from the stomach to the small intestine.

chymotrypsin   A milk-digesting enzyme secreted by the pancreas into the small intestine.

class   A group of animals categorized primarily by sex and age.

clip   One season's yield of wool.

clitoris   The ventral part of the vulva of the female reproductive tract that is homologous to the penis in the male. It is highly sensory.

cloaca   Portion of the lower end of the avian digestive tract that provides a passageway for products of the urinary, digestive, and reproductive tracts.

closed face   A condition in which sheep cannot see because wool covers their eyes.

clutch   Eggs laid by a hen on consecutive days.

coccidia   A protozoan organism that causes an intestinal disease called *coccidiosis*.

coccidiosis   A morbid state caused by the presence of organisms called coccidia, which belong to a class of sporozoans.

cock   A male chicken; also called a *rooster*.

cockerel   Immature male chicken.

cod   Scrotal area of steer remaining after castration.

coefficient of determination Percentage of variation in one trait that is accounted for by variation in another trait.

colic A nonspecific pain of the digestive tract.

colon The large intestine from the end of the ileum and beginning with the cecum to the anus.

colostrum The first milk given by a female after delivery of her young. It is high in antibodies that protect young animals from invading microorganisms.

colt A young male of the horse or donkey species.

comb The fleshy outgrowth on the top of a chicken's head, usually red in color, with varying sizes and shapes.

commercial (1) A carcass grade of cattle. (2) Livestock that are not registered or pedigreed by a registry (e.g., breed) association.

compensatory gain A faster-than-normal rate of gain after a period of restricted gain.

compensatory growth See *compensatory gain.*

composite breed A breed that has been formed by crossing two or more breeds.

concentrate A feed that is high in energy, low in fiber content, and highly digestible.

conception Fertilization of the ovum (egg).

conditioning The treatment of animals by vaccination and other means before putting them in the feedlot.

conformation The physical form of an animal; its shape and arrangement of parts.

contagious disease Infectious disease; a disease that is transmitted from one animal to another.

contemporaries A group of animals of the same sex and breed (or similar breeding) that have been raised under similar environmental conditions (same management group).

continental breeds Breeds of beef cattle originating in countries other than England. Sometimes called European or exotic breeds. Examples are Charolais, Limousin, and Simmental.

contracted heels A condition in which the heels of a horse are pulled in so that expansion of the heel cannot occur when the foot strikes the ground.

core samples Samples (wool, feed, or meat) taken by a coring device to determine the composition of the sample.

corpus luteum A yellowish body in the mammalian ovary. The cells that were follicular cells develop into the corpus luteum, which secretes progesterone. It becomes yellow in color from the yellow lipids that are in the cells.

correlation coefficient A measure of the association of one trait with another.

cost of gain The total cost divided by the total pound gained; usually expressed on a per-pound basis.

cotyledon An area of the placenta that contacts the uterine lining to allow nutrients and wastes to pass between the mother and the developing young. Sometimes referred to as *button.*

cow A sexually mature female bovine animal—usually one that has produced a calf.

cow–calf operation A management unit that maintains a breeding herd and produces weaned calves.

cow hocked A condition in which the hocks are close together but the feet stand apart.

creep An enclosure in which young can enter to obtain feed but larger animals cannot enter. This process is called *creep feeding.*

crimp The waves, or kinks, in a wool fiber.

crossbred An animal produced by crossing two or more breeds.

crossbreeding Mating animals from genetically diverse groups (e.g., breeds) within a species.

crutching See *tagging.*

cryptorchidism The retention of one or both testicles in the abdominal cavity in animals that typically have the testicles hanging in a scrotal sac.

cud Bolus of feed a ruminant animal regurgitates for further chewing.

cull To eliminate one or more animals from the breeding herd or flock.

curb A hard swelling that occurs just below the point of the hock.

curd Coagulated milk.

cutability Fat, lean, and bone composition of meat animals. Used interchangeably with yield grade. (See also *yield grade.*)

cutting chute A narrow chute where animals proceed through gates in single file such that animals can be directed into pens along the side of the chute.

cwt An abbreviation for hundredweight (100 lb).

cycling Infers that nonpregnant females have active estrous cycles.

dam Female parent.

dark cutter Color of the lean (muscle) in the carcass has a dark appearance, usually caused by stress (excitement, etc.) to the animal before slaughter.

debeaking To remove the tip of the beak of chickens.

dehorn To remove the horns from an animal.

deoxyribonucleic acid (DNA) A complex molecule consisting of deoxyribose (a sugar), phosphoric acid, and four nitrogen bases (a gene is a piece of DNA).

depreciation An accounting procedure by which the purchase price of an asset with a useful life of more than 1 year is prorated over time.

dewclaws   Hard horny structures above the hoof on the rear surface of the legs of cattle, swine, and sheep.

dewlap   Loose skin under the chin and neck of cattle.

DHIA   Dairy Herd Improvement Association, an association which dairy producers participate in keeping dairy records. Sanctioned by the National Cooperative Dairy Herd Improvement Program.

DHIR   Dairy Herd Improvement Registry, a dairy record-keeping plan sponsored by the breed associations.

diet   Feed ingredients or mixture of ingredients (including water) which are consumed by animals.

digestibility   The quality of being digestible. If a high percentage of a given food taken into the digestive tract is absorbed into the body, that food is said to have *high digestibility*.

digestion   The reduction in particle size of feed so that the feed becomes soluble and can pass across the gut wall into the vascular or lymph system.

diploid   Having the normal, paired chromosomes of somatic tissue as produced by the doubling of the primary chromosomes of the germ cells at fertilization.

disease   Any deviation from a normal state of health.

disinfect   To kill, or render ineffective, harmful microorganisms and parasites.

disinfectant   A chemical that destroys disease-producing microorganisms or parasites.

distal   Position that is distant from the point of attachment of an organ.

DM   See *dry matter.*

DNA   See *deoxyribonucleic acid.*

dock   (1) To cut off the tail. (2) The remaining portion of the tail of a sheep that has been docked. (3) To reduce or lower in value.

doe   A female goat or rabbit.

dominance   (1) A situation in which one gene of an allelic pair prevents the phenotypic expression of the other member of the allelic pair. (2) A type of social behavior in which an animal exerts influence over one or more other animals.

dominant gene   A gene that overpowers and prevents the expression of its recessive allele when the two alleles are present in a heterozygous individual.

dorsal   Of, on, or near the back of an animal.

double muscling   A genetic trait in cattle where muscles are greatly enlarged rather than duplicate muscles.

down   Soft, fluffy type of feather located under the contour feathers. Serves as insulating material.

drake   Mature male duck.

drench   To give fluid by mouth (e.g., medicated fluid is given to sheep for parasite control).

dressing percentage   The percentage of the live animal weight that becomes the carcass weight at slaughter. It is determined by dividing the carcass weight by the live weight, then multiplying by 100.

drop   Body parts removed at slaughter—primarily hide (pelt), head, shanks, and offal.

drop credit   Value of the drop.

dry (cow, ewe, sow, mare)   Refers to a nonlactating female.

dry matter   Feed after water (moisture) has been removed (100% dry).

dubbing   The removal of part or all of the soft tissues (comb and wattles) of chickens.

dung   The feces (manure) of farm animals.

dwarfism   The state of being abnormally undersized. Two kinds of dwarfs are recognized; one is proportionate and the other is disproportionate.

dysentery   Severe diarrhea.

dystocia   Difficult birth.

ectoderm   The outermost layer of the three layers of the primitive embryo.

edema   Abnormal fluid accumulation in the intercellular tissue spaces of the body.

ejaculation   Discharge of semen from the male.

emaciation   Thinness; loss of flesh where bony structures (hips, ribs, and vertebrae) become prominent.

embryo   Very early stage of individual development within the uterus. The embryo grows and develops into a fetus. In poultry, the embryo develops within the eggshell.

embryo transfer   The transfer of fertilized eggs from a donor female to one or more recipient females.

endocrine gland   A ductless gland that secretes a hormone into the bloodstream.

endoderm   The innermost layer of the three layers of the primitive embryo.

enterotoxemia   A disease of the intestinal tract caused by bacterial secretion of toxins. Its symptoms are characteristic of food poisoning.

entropion   Turned-in eyelids.

environment   The sum total of all external conditions that affect the well-being and performance of an animal.

enzyme   A complex protein produced by living cells that causes changes in other substances in the cells without being changed itself and without becoming a part of the product.

EPD   See *expected progeny difference.*

epididymis   The long, coiled tubule leading from the testis to the vas deferens.

epididymitis   An inflammation of the epididymis.

epiphysis A piece of bone separated from a long bone in early life by cartilage, which later becomes part of the larger bone.

epistasis A situation in which a gene or gene pair masks (or controls) the expression of another nonallelic pair of genes.

equine Refers to horses.

equine encephalomyelitis An inflammation of the brain of horses.

eruction (or eructation) The elimination of gas by belching.

esophageal groove A groove in the reticulum between the esophagus and omasum. Directs milk in the nursing young ruminant directly from the esophagus to the omasum.

essential nutrient A nutrient that cannot be synthesized by the body and must be supplied in the diet.

estrogen Any hormone (including estradiol, estriol, and estrone) that causes the female to come physiologically into heat and to be receptive to the male. Estrogens are produced by the follicle of the ovary and by the placenta.

estrous An adjective meaning "heat," which modifies such words as "cycle." The estrous cycle is the heat cycle, or time from one heat to the next.

estrous synchronization Controlling the estrous cycle so that a high percentage of the females in the herd express estrus at approximately the same time.

estrus The period of mating activity in the female mammal. Same as heat.

ET Abbreviation for *embryo transfer*.

ethology Study of animal behavior in the animal's natural environment.

European breeds See *continental breeds*.

eviscerate The removal of the internal organs during the slaughtering process.

ewe A sexually mature female sheep. A ewe lamb is a female sheep before attaining sexual maturity.

exocrine gland Gland that secretes fluid into a duct.

exotic breeds See *continental breeds*.

expected progeny difference (EPD) One-half of the breeding value; the difference in performance to be expected from future progeny of a sire, compared with that expected from future progeny of an average bull in the same test.

family selection Selection based on performance of a family.

farrow To deliver, or give birth to, pigs.

fat Adipose tissue.

FDA See *Food and Drug Administration*.

feather picking The picking of feathers from one bird by another.

feces Bowel movements, excrement from the intestinal tract.

feed additive Ingredient (such as an antibiotic or hormone-like substance) added to a diet to perform a specific role (e.g., to improve gain or feed efficiency).

feed bunk A trough or container used to feed farm animals.

feed efficiency   (1) The amount of feed required to produce a unit of weight gain or milk; for poultry, this term can also denote the amount of feed required to produce a given quantity of eggs. (2) The amount of gain made per unit of feed.

feeder   Animals (e.g., cattle, lambs, pigs) that need further feeding prior to slaughter.

feeder grades   Visual classifications (descriptive and/or numerical) of feeder animals. Most of these grades have been established by the USDA.

felting   The process of pressing wool fibers together in conjunction with heat and moisture to produce a fabric.

femininity   Well-developed secondary female sex characteristics, udder development and refinement in head and neck.

feral   Domesticated animals that return to nature to survive and reproduce.

fertility   The capacity to initiate, sustain, and support reproduction. With reference to poultry, the term typically refers to the percentage of eggs that, when incubated, show some degree of embryonic development.

fertilization   The process in which a sperm unites with an egg to produce a zygote.

fetus   Later stage of individual development within the uterus. Generally, the new individual is regarded as an embryo during the first half of pregnancy, and as a fetus during the last half.

fill   The contents of the digestive tract.

filly   A young female horse.

fineness   A term used to describe the diameter of wool fibers.

finish   The degree of fatness of an animal.

Finnsheep   A highly prolific breed of sheep introduced into the United States in 1968.

fistula   A running sore at the top of the withers of a horse, resulting from a bruise followed by invasion of microorganisms.

flank firmness   Firmness of the flank muscle in lamb carcass evaluation.

flank streaking   Streaks of fat in the flank muscle of lamb carcasses.

fleece   The wool shorn at one time from all parts of the sheep.

flehmen   A pattern of behavior expressed in some male animals (e.g., bull, ram, stallion) during sexual activity. The upper lip curls up and the animal inhales in the vicinity of the vulva or urine.

Food and Drug Administration (FDA)   A U.S. government agency responsible for protecting the public against impure and unsafe foods, drugs, veterinary products, and other products.

flock   A group of sheep or poultry.

flushing   Placing females (typically sheep and swine) on a gaining level of nutrition before breeding to stimulate greater ovulation rates.

fly strike   An infestation with large numbers of blowfly maggots.

foal A young male or female horse (noun) or the act of giving birth (verb).

follicle A blisterlike, fluid-filled structure in the ovary that contains the egg.

follicle-stimulating hormone (FSH) A hormone produced and released by the anterior pituitary that stimulates the development of the follicle in the ovary.

footrot A disease of the foot in sheep and cattle. In sheep it causes rotting of tissue between the horny part of the foot and the soft tissue underneath.

forb Weedy or broad-leaf plants, as contrasted to grasses, that serve as pasture for animals.

forging The striking of the heel of the front foot with the toe of the hind foot by a horse in action.

founder Nutritional ailment resulting from overeating. Lameness in front feet, with excessive hoof growth, usually occurs.

frame score A numerical rating of frame size.

frame size A measure of skeletal size. It can be visual or by measurement (usually taken at the hips).

freemartin Female born twin to a bull (approximately 9 of 10 will not conceive).

freshen To give birth to young and initiate milk production. This term is usually used in reference to dairy cattle.

fryer See *broiler.*

FSH See *follicle-stimulating hormone.*

full-mouth Animal has all permanent teeth fully exposed.

full sibs Animals having the same sire and dam.

gallop A four-beat gait in which each of the two front feet and both of the hind feet strike the ground at different times.

gametes Male and female reproductive cells. The sperm and the egg.

gametogenesis The process by which sperm and eggs are produced.

gander Mature male goose.

gelding A male horse that has been castrated.

gene A small segment of chromosome (DNA) that is a genetic code for a trait and determines how the trait will develop.

general combining ability The ability of individuals of one line or population to combine favorably or unfavorably with individuals of several other lines or populations.

generation interval Average age of the parents when offspring are born.

generation turnover Length of time from one generation of animals to the next generation.

genetic engineering The technique of removing, modifying, or adding genes to a DNA molecule.

genome All of the genes of a species.

genotype The genetic constitution, or makeup, of an individual. For any pair of alleles, three genotypes (e.g., *AA, Aa,* and *aa*) are possible.

gestation The time from breeding or conception of a female until she gives birth to her young.

gilt A young female swine prior to the time that she has produced her first litter.

goiter Enlargement of the thyroid gland, usually caused by iodine-deficient diets.

gonad The testis of the male; the ovary of the female.

gonadotrophin Hormone that stimulates the gonads.

gossypol A toxic product contained in cottonseed.

grade (1) A designation of live or carcass merit—e.g., choice grade. (2) Livestock not registered with registry (e.g., breed) association.

grading up The continued use of purebred sires of the same breed in a grade herd or flock.

grass tetany A disease of cattle and sheep marked by staggering, convulsions, coma, and frequently death, caused by a mineral imbalance (magnesium) while grazing lush pasture.

grease wool Wool as it comes from the sheep and prior to cleaning. It contains the natural oils from the sheep.

gross energy The amount of heat, measured in calories, produced when a substance is completely oxidized. It does not reveal the amount of energy that an animal could derive from eating the substance.

growth The increase in protein over its loss in the animal body. Growth occurs by increases in cell numbers, cell size, or both.

habituation The gradual adaptation to a stimulus or to the environment.

half sib Animals having one common parent.

hand mating Same as hand breeding—bringing a female to a male for service (breeding), after which she is removed from the area where the male is located.

hank A measurement of the fineness of wool. A hank is 560 yards of yarn. More hanks of yarn are produced from fine wools than coarse wools.

haploid One-half of the diploid number of chromosomes for a given species, as found in the germ cells.

hatchability A term that indicates the percentage of a given number of eggs set from which viable young hatch, sometimes calculated specifically from the number of fertile eggs set.

hay Harvested forage such as alfalfa hay.

heat See *estrus.*

heat increment The increase in heat production after consumption of feed when an animal is in a thermoneutral environment. It includes additional heat generated in fermentation, digestion, and nutrient metabolism.

heaves   A respiratory defect in horses during which the animal has difficulty completing the exhalation of inhaled air.

heifer   A young female bovine cow before the time that she has produced her first calf.

heiferette   A heifer that has calved once, after which the heifer is fed for slaughter; the calf has usually died or been weaned at an early age.

hemoglobin   The iron-containing pigment of the red blood cells. It carries oxygen from the lungs to the tissues.

hen   An adult female domestic fowl, such as a chicken or turkey.

herbivorous   Subsisting or feeding on plants.

herd   A group of animals. Used with beef, dairy, or swine.

heritability   The portion of the total variation or phenotypic differences among animals that is due to heredity.

hernia   The protrusion of some of the intestine through an opening in the body wall (also commonly called *rupture*). Two types of hernias, umbilical and scrotal, occur in farm animals.

heterosis   Performance of offspring that is greater than the average of the parents. Usually the amount of superiority of the crossbred over the average of the parental breeds. Also referred to as *hybrid vigor.*

heterozygous   A term designating an individual that possesses unlike genes for a particular trait.

hides   Skins from animals such as cattle, horses, and pigs; beef hides weigh more than 30 lb each as contrasted to calf skins, which weigh less.

hinny   The offspring that results from crossing a stallion with a female donkey (jenny).

hobble   To tie two of an animal's legs together. An animal is hobbled to prevent it from kicking or moving a long distance.

homeotherm   A warm-blooded animal. An animal that maintains its characteristic body temperature even though environmental temperature varies.

homogenized   Milk that has had the fat droplets broken into very small particles so that the milk fat stays in suspension in the milk fluids.

homologous   Corresponding in type of structure and derived from a common primitive origin.

homologous chromosomes   Chromosomes having the same size and shape that contain genes affecting the same characters. Homologous chromosomes occur in pairs in typical diploid cells.

homozygous   A term designating an individual whose genes for a particular trait are alike.

hormone   A chemical substance secreted by a ductless gland. Usually carried by the bloodstream to other places in the body where it has its specific effect on another organ.

hybrid vigor   See *heterosis.*

hydrocephalus   A condition characterized by an abnormal increase in the amount of cerebral fluid, accompanied by dilation of the cerebral ventricles.

hypertension   High blood pressure.

hypothalamus   A portion of the brain found in the floor of the third ventricle. It regulates reproduction, hunger, and body temperature and has other functions.

hypoxia   A condition resulting from deficient oxygenation of the blood.

immunity   The ability of an animal to resist or overcome an infection.

implant   To graft or insert material to intact tissues.

implantation   The attachment of the fertilized egg to the uterine wall.

imprinting   Learning associated with maturational readiness.

inbreeding   The mating of individuals who are more closely related than the average individuals in a population. Inbreeding increases homozygosity in the population but it does not change gene frequency.

incisor   A front tooth.

incubation period   The time that elapses from the time an egg is placed into an incubator until the young is hatched.

independent culling level   Selection method in which minimum acceptable phenotypic levels are assigned to several traits.

index   (1) An overall merit rating of an animal. (2) A method of predicting the milk-producing ability that a bull will transmit to his daughters.

infection   Invasion of the body tissues by microbial agents or parasites other than insects.

infectious   Capable of invading and growing in living tissues. Used to describe various pathogenic microorganisms such as viruses, bacteria, protozoa, and fungi.

influenza   A virus disease characterized by inflammation of the respiratory tract, high fever, and muscular pain.

ingest   Anything taken into the stomach.

inheritance   The transmission of genes from parents to offspring.

insemination   Deposition of semen in the female reproductive tract.

instinct   Inborn behavior.

integration   The bringing together of all segments of a livestock or poultry production program under one centrally organized unit.

intelligence   The ability to learn to adjust successfully to situations.

interference   The striking of the supporting leg by the foot of the striding leg by a horse in action.

interstitial cells   The cells between the seminiferous tubules of the testicle that produce testosterone.

intravenous   Within the vein. An intravenous injection is an injection into a vein.

in vitro   Outside the living body; in a test tube or other artificial environment.

jack   A male donkey.

jackass   See *jack.*

jennet   A female donkey.

jenny   A female donkey.

Karakul   A breed of fat-tailed sheep having coarse, wiry furlike hair. Used to produce Persian lambskins.

ked   An external parasite that affects sheep. Although commonly called *sheep tick,* it is actually a wingless fly.

kemp   Coarse, opaque, hairlike fibers in wool.

ketosis   A condition (also called *acetonemia*) that is characterized by a high concentration of ketone bodies in the body tissues and fluids.

kid   Young goat.

kilocalorie (kcal, Kcal)   An amount of heat equal to 1,000 calories. (See also *calorie.*)

kosher meat   Meat from ruminant animals with split hooves where the animals have been slaughtered according to Jewish law.

lactalbumin   A nutritive protein of milk.

lactation   The secretion and production of milk.

lactose   Milk sugar. When digested, it is broken down into one molecule of glucose and one of galactose.

lamb   (1) A young male or female sheep, usually less than a year of age. (2) To deliver, or give birth to, a lamb.

lamb dysentery   See *dysentery.*

lambing   Act of giving birth. Same as parturition.

lambing jug   A small pen in which a ewe is put for lambing. It is also used for containing the ewe and her lamb until the lamb is strong enough to run with other ewes and lambs.

laminitis   Lameness associated with inflammation of the laminae that attaches the horse's hoof wall to the fleshy part of the foot. Typically related to founder.

lard   The fat from pigs that has been produced through a rendering process.

layer   A hen that is kept for egg production.

legume   Any plant of the family *leguminosae,* such as pea, bean, alfalfa, and clover.

leucocytes   White blood cells.

LH   See *luteinizing hormone.*

libido   Sex drive or the desire to mate on the part of the male.

lice   Small, flat, wingless insect with sucking mouth parts that is parasitic on the skin of animals.

linebreeding   A mild form of inbreeding that maintains a high genetic relationship to an outstanding ancestor.

line crossing   The crossing of inbred lines.

lipid   An organic substance that is soluble in alcohol or ether but insoluble in water; used interchangeably with the term *fat.*

litter   The young produced by multiparous females such as swine. The young in a litter are called *littermates.*

liver flukes   A parasitic flatworm found in the liver.

locus   The place on a chromosome where a gene is located.

longevity   Life span of an animal. Usually refers to a long life span.

luteinizing hormone (LH)   A protein hormone, produced and released by the anterior pituitary, which stimulates the formation and retention of the corpus luteum. It also initiates ovulation.

lymph   Clear yellowish, slightly alkaline fluid contained in lymphatic vessels.

macroclimate   The large, general climate in which an animal exists.

macromineral   A mineral that is needed in the diet in relatively large amounts.

maintenance   A condition in which the body is maintained without an increase or decrease in body weight and with no production or work being done.

mammal   Warm-blooded animals that suckle their young.

mammary gland   Gland that secretes milk.

management   The act, art, or manner of managing, handling, controlling, or directing a resource or integrating several resources.

marbling   The distribution of fat in muscular tissue; intramuscular fat.

mare   A sexually developed female horse.

market class   Animals grouped according to the use to which they will be put, such as slaughter, feeder, or stocker.

market grade   Animals grouped within a market class according to their value.

masticate   To chew food.

mastitis   Inflammation of the mammary gland.

masturbation   Ejaculation by a male by some process other than sexual intercourse.

mean   (1) Statistical term for average. (2) Term used to describe animals having bad behavior.

meat   The tissues of the animal body that are used for food.

meat spots   Spots in the egg that are blood spots which have changed color or tissue sloughed off from the reproductive organs of the hen.

meiosis   A special type of cell nuclear division that is undergone in the production of gametes (sperm in the male, ova in the female). As a result of meiosis, each gamete carries half the number of chromosomes of a typical body cell in that species.

melengestrol acetate (MGA)   A feed additive that suppresses estrus in heifer and is widely used in the feedlot industry.

mesoderm   The middle layer of the three layers of the primitive embryo.

messenger RNA   The ribonucleic acid that is the carrier of genetic information from nuclear DNA. It is important in protein synthesis.

metabolism   (1) The sum total of chemical changes in the body, including the "building up" and "breaking down" processes. (2) The transformation by which energy is made available for body uses.

metabolizable energy   Gross energy in the feed minus the sum of energy in feces, gaseous products of digestion, and energy in urine. Energy that is available for metabolism by the body.

metritis   Inflammation (infection) of the uterus.

MGA   See *melengestrol acetate.*

microclimate   A small, special climate within a macroclimate created by the use of such devices as shelters, heat lamps, and bedding.

microcomputer   A small computer that has a smaller memory capacity than a larger or mainframe computer.

micromineral   A mineral that is needed in the diet in relatively small amounts. The quantity needed is so small that such a mineral is often called a *trace mineral.*

Milk EPD   A genetic estimate of the milking ability of a beef bull's daughters when compared to the daughters of other bulls.

milk fat   The fat in milk; synonymous with butterfat.

milk fever   See *parturient paresis.*

milk letdown   The release of milk into the teat cisterns.

Milk Only records   Dairy record system similar to DHI except no milk fat samples are taken.

minimum culling level   A selection method in which an animal must meet minimum standards for each trait desired in order to qualify for being retained for breeding purposes.

mites   Very small arachnids that are often parasitic upon animals.

mitosis   A process in which a cell divides to produce two daughter cells, each of which contains the same chromosome complement as the mother cell from which they came.

modifying genes   Genes that modify the expression of other genes.

mohair   Fleece of the Angora goat.

monogastric   Having only one stomach or only one compartment in the stomach. Examples are swine and poultry.

monoparous   A term designating animals that usually produce only one offspring at each pregnancy. Horses and cattle are monoparous.

moon blindness   Periodic blindness that occurs in horses.

morbidity   Measurement of illness; morbidity rate is the number of individuals in a group that become ill during a specified time.

mortality rate   Number of individuals that die from a disease during a specified time, usually 1 year.

mouthed   The examination of an animal's teeth.

mule   The hybrid that is produced by mating a male donkey with a female horse. They are usually sterile.

mulefoot   Having one instead of the expected two toes, on one or more of the feet.

multiparous   A term that designates animals that usually produce several young at each pregnancy. Swine are multiparous.

mutation   A change in a gene.

mutton   The meat from a sheep that is over 1 year old.

muzzle   The nose of horse, cattle, or sheep.

myofibrils   The primary component part of muscle fibers.

navel   The area where the umbilical cord was formerly attached to the body of the offspring.

necropsy   Perform a postmortem examination.

net energy   Metabolizable energy minus heat increments. The energy available to the animal for maintenance and production.

nicking   The way in which certain lines, strains, or breeds perform when mated together. When outstanding offspring result, the parents are said to have *nicked* well.

nipple   See *teat*.

nodular worm   An internal parasitic worm that causes the formation of nodules in the intestines.

nonruminant   Simple-stomached or monogastric animal.

NPN (nonprotein nitrogen)   Nitrogen in feeds from substances such as urea and amino acids, but not from preformed proteins.

nucleotide   Compound composed of phosphoric acid, sugar, and a base (purine or pyrimadine), all of which constitute a structural unit of nucleic acid.

nutrient   (1) A substance that nourishes the metabolic processes of the body. (2) The end product of digestion.

nutrient density   Amount of essential nutrients relative to the number of calories in a given amount of food.

obesity   An excessive accumulation of body fat.

offal   All organs and tissues removed from inside the animal during the slaughtering process.

omasum   One of the stomach components of ruminant animals that has many folds.

omnivorous   Feeding on both animal and vegetable substances.

on full feed A term that refers to animals that are receiving all the feed they will consume. See also *ad libitum.*

oogenesis The process by which eggs, or ova, are produced.

open Refers to nonpregnant females.

open-faced Face of sheep that is free from wool, particularly around the eyes.

opportunity costs Returns given up if debt-free resources (e.g., land, livestock, equipment) are used in their next-best level of employment.

optimum level of performance The level at which a trait or traits maximize net profit. Resources are managed to achieve a combined balance of traits that sustains high levels of profitability.

osteopetrosis Abnormal thickening, hardening, and fragility of bones, making them weaker.

osteoporosis An abnormal decrease in bone mass with an increased fragility of the bones.

outbreeding The process of continuously mating females of the herd to unrelated males of the same breed.

outcrossing The mating of an individual to another in the same breed that is not related to it. Outcrossing is a specific type of outbreeding system.

ova Plural of ovum, meaning *eggs.*

ovary The female reproductive gland in which the eggs are formed and progesterone and estrogenic hormones are produced.

overeating disease A toxic condition caused by the presence of undigested carbohydrates in the intestine, which stimulates harmful bacteria to multiply. When the bacteria die, they release toxins. Called *enterotoxemia* in some animals.

overshot jaw Upper jaw is longer than lower jaw. Also called *parrot mouth.*

oviduct A duct leading from the ovary to the horn of the uterus.

ovine Refers to sheep.

ovulation The shedding, or release, of the egg from the follicle of the ovary.

ovum The egg produced by a female.

Owner-sampler record Dairy record system similar to DHI except milk weights and samples are taken by the dairy producer instead of a DHIA supervisor.

pace A lateral two-beat gain in which the right rear and front feet hit the ground at one time and the left rear and front feet strike the ground at another time.

paddling The outward swinging of the front feet of a horse that toes in.

pale, soft, exudative (PSE) A genetically predisposed condition in swine in which the pork is very light colored, soft, and watery.

palpation Feeling by hand.

parasite An organism that lives a part of its life cycle in or on, and at the expense of, another organism. Parasites of farm animals live at the expense of the farm animals.

parity   Number of different times a female has had offspring.

parrot mouth   Upper jaw is longer than lower jaw.

parturient paresis   Partial paralysis that occurs at or near time of giving birth to young and beginning lactation. The mother mobilizes large amounts of calcium to produce milk to feed newborn, and blood calcium levels drop below the point necessary for impulse transmission along the nerve tracks. Commonly called *milk fever.*

parturition   The process of giving birth.

pasteurization   The process of heating milk to 161°F and holding it at that temperature for 15 seconds to destroy pathogenic microorganisms.

pasture rotation   The rotation of animals from one pasture to another so that some pasture areas have no livestock on them in certain periods.

pathogen   Biologic agent (e.g., bacteria, virus, protozoa, nematode) that may produce disease or illness.

paunch   Another name for rumen.

pay weight   The actual weight for which payment is made. In many cases it is the shrunk weight (actual weight minus pencil shrink).

pedigree   The record of the ancestry of an animal.

pelt   The natural, whole skin covering, including the wool, hair, or fur; e.g., a sheep pelt has the wool left on.

pencil shrink   An arithmetic deduction (percent of liveweight) from an animal's weight to account for fill.

pendulous   Hanging loosely.

penis   The male organ of copulation. It serves both as a channel for passage of urine from the bladder as an extension of the urethra and as a copulatory organ through which sperm are deposited into the female reproductive tract.

per capita   Per person.

performance test   The evaluation of an animal according to it performance.

pernicious anemia   A chronic type of mycrocitic anemia caused by a deficiency of vitamin B_{12} or a failure of intestinal absorption of vitamin B_{12}.

pharmaceutical   A medicinal drug.

phenotype   The characteristics of an animal that can be seen and/or measured (e.g., the presence or absence of horns, the color, or the weight of an animal).

pheromones   Chemical substances that attract the opposite sex.

photoperiod   Time period when light is present.

physiology   The science that pertains to the functions of organs, organ systems, or the entire animal.

picking   The removal of feathers in dressing poultry.

pigeon-toed   See *toeing in.*

pin bones   In cattle, the posterior ends of the pelvic bones that appear as two raised areas on either side of the tail head.

pink tooth   Congenital porphyria, teeth are pink gray and the animals tend to sunburn easily.

pinworms   A small nematode worm with unsegmented body found as a parasite in the rectum and large intestine of animals.

pituitary   Small endocrine gland located at the base of the brain.

placenta   The vascular organ that unites the fetus to the uterus.

pneumonia   Inflammation or infection of alveoli of the lungs caused by either bacteria or viruses.

poikilotherm   A cold-blooded animal; one whose body temperature varies with that of the environment.

polled   Naturally or genetically hornless.

poll evil   An abscess behind the ears of a horse.

Polypay   A synthetic breed of sheep developed in the United States by combining the Dorset, Targhee, Rambouillet, and Finnsheep breeds.

Porcine stress syndrome (PSS)   A genetic defect in swine inherited as a simple recessive. It is associated with heavily muscled animals that may suddenly die when exposed to stressful conditions. Their muscle is usually pale, soft, and exudative (PSE).

pork   The meat from swine.

posterior   Toward the rear end of an animal.

postgastric fermentation   The fermentation of feed that occurs in the cecum, behind the area where digestion has occurred.

postnatal   See *postpartum.*

postpartum   After birth.

postpartum interval   The length of time from parturition until the dam is pregnant again.

poult   A young turkey of either sex, from hatching to approximately 10 weeks of age.

poultry   This term includes chickens, turkeys, geese, pigeons, peafowls, guineas, and game birds.

predicted difference   Dairy bull record based on superiority or inferiority of the bull's daughters compared to their herd mates.

predicted transmitting ability (PTA)   Estimate of genetic transmitting ability (e.g., one-half of the breeding value) of dairy bulls. Estimated amount by which daughters of a bull will differ from the breed average.

pregastric fermentation   Fermentation that occurs in the rumen of ruminant animals. It occurs before feed passes into the portion of the digestive tract in which digestion actually occurs.

pregnancy disease   A metabolic disease in late pregnancy affecting primarily ewes carrying twins or triplets. A form of ketosis. Also called *pregnancy toxemia.*

pregnancy testing   Evaluation of females for pregnancy through palpation or using an ultrasound machine.

prenatal   Prior to being born. Before birth.

probe   A device used to measure backfat thickness in pigs and cattle.

production testing   An evaluation of an animal based on its production record.

progeny testing   An evaluation of an animal on the basis of performance of its offspring.

progesterone   A hormone produced by the corpus luteum that stimulates pro-gestational proliferation in the uterus of the female.

prolapse   Abnormal protrusion of part of an organ, such as uterus or anus.

prostaglandins   Chemical mediators that control many physiological and bio-chemical functions in the body. One prostaglandin ($PGF_{2\alpha}$) can be used to synchronize estrus.

prostate   A gland of the male reproductive tract that is located just back of the bladder. It secretes a fluid that becomes part of semen at ejaculation.

protein   A substance made up of amino acids that contains approximately 16% nitrogen (based on molecular weight).

protein supplement   Any dietary component containing a high concentration (at least 25%) of protein.

proximal   Nearest; situated toward the point of origin (e.g., limb or bone).

PSE   See *pale, soft, exudative.*

PSS   See *porcine stress syndrome.*

PTA   See *predicted transmitting ability.*

puberty   The age at which the reproductive organs become functionally oper-ative.

pullet   Young female chicken from day of hatch through onset of egg produc-tion; sometimes the term is used through the first laying year.

purebred   An animal eligible for registry with a recognized breed association.

quality grades   Animals grouped according to value as Prime, Choice, etc., based on conformation and fatness of the animals.

rack   (1) A rapid four-beat gait of a horse. (2) A wholesale cut of lamb located between the shoulder and loin.

ram   A male sheep that is sexually mature.

ration   The amount of total feed fed to an animal over a 24-hour period.

reach   See *selection differential.*

realized heritability   The portion obtained of what is reached for in selection.

realizer   A feeder animal (usually cattle) that has serious health problems or injury. Economics dictate the animal be sold rather than continue the dura-tion of the feeding program.

reasoning   The ability of an animal to respond correctly to a stimulus the first time the animal encounters a new situation.

recessive gene   A gene that has its phenotype masked by its dominant allele when the two genes are present together in an individual.

reciprocal recurrent selection   The selection of breeding animals in two populations based on the performance of their offspring after animals from two populations are crossed.

recombinant DNA (rDNA)   Isolated DNA molecules that can be inserted into the DNA of another cell. rDNA is used in the genetic-engineering process.

rectal prolapse   Protrusion of part of large intestine through the anus.

recurrent selection   Selection for general combining ability by selecting males that sire outstanding offspring when mated to females from varying genetic backgrounds.

red meat   Meat from cattle, sheep, swine, and goats, as contrasted to the white meat of poultry.

registered   Recorded in the herdbook of a breed.

regurgitate   To cast up digested food to the mouth as is done by ruminants.

reinforcement   A reward for making the proper response to a stimulus or condition.

replicate   To duplicate, or make another exactly like, the original.

reproduction   The production of live, normal offspring.

retained placenta   Placenta remains within the reproductive tract after parturition has occurred.

reticulum   One of the stomach components of ruminant animals that is lined with small compartments, giving a honeycomb appearance.

rhinitis   Inflammation of the mucous membranes lining the nasal passages.

ribonucleic acid (RNA)   An essential component of living cells, composed of long chains of phosphate, ribose sugar, and several bases.

rhinopneumonitis   Equine herpes virus-1. It produces acute catarrh upon primary infection.

rickets   A disease of disturbed ossification of the bones caused by a lack of vitamin D or unbalanced calcium/phosphorus ratio.

ridgling   Another term for *cryptorchid*.

ringbone   An ossification of the lateral cartilage of the foot of a horse all around the foot.

RNA   See *ribonucleic acid*.

riparian   An area next to water (stream, river, or lake) where more vegetation grows (compared to a greater distance from the water source) because of the added moisture from the water. Grazing animals usually inhabit this area more frequently than others, thus increasing the possibility of overgrazing.

roman nose   A nose having a prominent bridge; e.g., a roman-nosed horse.

roughage   A feed that is high in fiber, low in digestible nutrients, and low in energy. Such feeds as hay, straw, silage, and pasture are examples.

rumen   The large fermentation pouch of the ruminant animal in which bacteria and protozoa break down fibrous plant material that is swallowed by the animal, sometimes referred to as the *paunch*.

ruminant   A mammal whose stomach has four parts (rumen, reticulum, omasum, and abomasum). Cattle, sheep, goats, deer, and elk are ruminants.

rumination   The regurgitation of undigested food and chewing it a second time, after which it is again swallowed.

salmonella   Gram-positive, rod-shaped bacteria that cause various diseases such as food poisoning in animals.

scale   (1) Size. (2) Equipment on which an animal is weighed.

scoured wool   Wool that has been cleaned of grease and other foreign material.

scours   Diarrhea; a profuse watery discharge from the intestines.

screwworms   Larvae of several American flies that infest wounds of animals.

scrotal circumference   A measurement (usually cm or in.) of the circumference of both testicles and the scrotal sac that surrounds them.

scrotum   A pouch that contains the testes. It is also a thermoregulatory organ that contracts when cold and relaxes when warm, thus tending to keep the testes at a lower temperature than that of the body.

scurs   Small growths of hornlike tissue attached to the skin of polled or dehorned animals.

scurvy   A deficiency disease in humans that causes spongy gums and loose teeth. It is caused by a lack of vitamin C (ascorbic acid).

seed stock   Breeding animals; sometimes used interchangeably with *purebred*.

selection   Differentially reproducing what one wants in a herd or flock.

selection differential   The difference between the average for a trait in selected animals and the average of the group from which they come; also called *reach*.

selection index   Selection method in which several traits are evaluated and expressed as one total score.

semen   The fluid containing the sperm that is ejaculated by the male. Secretions from the seminal vesicles, the prostate gland, the bulbourethral glands, and the urethral glands provide most of the fluid.

seminal vesicles   Accessory sex glands of the male that provide a portion of the fluid of semen.

seminiferous tubules   Minute tubules in the testicles in which sperm are produced. They comprise about 90% of the mass of the testes.

service   To breed or mate.

settle   To become pregnant.

sex-limited   Existing in only one sex, such as milk production in dairy cattle.

shearing   The process of removing the fleece (wool) from a sheep.

sheath rot   Inflammation of the prepuce in male sheep.

sheep bot   Any of a number of related flies whose larvae are parasitic in sheep. They usually are found in the sinuses.

shipping fever   A widespread respiratory disease of cattle and sheep.

shoat   A young pig of either sex.

shoe boil   Blemish of the horse caused by the horseshoe putting pressure on the elbow when the horse lies down.

shrink   Loss of weight—commonly used in the loss in liveweight when animals are marketed or loss in weight from grease wool to clean wool.

sib   A brother or sister.

sickle hocks   Hocks that have too much set, causing the hind feet to be too far forward and too far under the animal.

side bones   Ossification of the lateral cartilages of the foot of a horse.

sigmoid flexure   The S-curve in the penis of boars, rams, bucks, and bulls.

silage   Forage, corn fodder, or sorghum preserved by fermentation that produces acids similar to the acids that are used to make pickled foods for people.

sire   Male parent.

sire index   A dairy bull test record obtained by comparing a bull's daughters with their contemporary herd mates.

skins   Skins come from smaller animals such as pigs, sheep, goats, and wild animals. A beef hide weighing less than 30 lb is called a skin.

sleeping sickness   An infectious disease common in tropical Africa and transmitted by the bite of a tsetse fly.

slotted floor   Floor having any kind of openings through which excreta may fall.

SNF   See *solids-nonfat*.

snood   The relatively long, fleshy extension at the base of the turkey's beak.

software   Program instructions to make computer hardware function.

soilage   Green forage that is cut and brought to animals as food.

solids-nonfat   Total milk solids minus fat. It includes protein, lactose, and minerals.

somatotropin   The growth hormone from the anterior pituitary that stimulates nitrogen retention and growth.

sore mouth   A virus-caused disease affecting primarily lambs.

sow   A female swine that has farrowed one litter or has reached 12 months of age.

spay   To remove the ovaries.

specific combining ability   The ability of a line or population to exhibit superiority or inferiority when combined with other lines or populations.

spermatid   The haploid germ cell prior to spermiogenesis.

spermatogenesis   The process by which spermatozoa are formed.

spermiogenesis   The process by which the spermatid loses most of its cytoplasm and develops a tail to become a mature sperm.

spider syndrome   A recessive genetic abnormality common to black-faced sheep. The front legs are usually bent out from the knees and the hind legs typically show some deformities.

spinning count   The number of hanks of yarn that can be spun from a pound of clean wool. One method of evaluating fineness of wool.

splay footed   See *toeing out.*

spool joint   The joint where the foot and pastern are removed from the front leg. Used to identify a mutton carcass.

spur   A sharp projection on the back of a male bird's shank.

stags   Castrated male sheep, cattle, goats, or swine that have reached sexual maturity prior to castration.

stallion   A sexually mature male horse.

staple length   Length of wool fibers.

steer   A castrated bovine male that was castrated early in life before puberty.

sterility   Inability to produce offspring.

stifle   Joint of the hindleg between the femur and tibia.

stifled   Injury of the stifle joint.

stillborn   Offspring born dead.

stocker   Weaned cattle that are fed high-roughage diets (including grazing) before going into the feedlot.

stomach worms   *Haemonchus contortus,* or worms of the stomach of cattle, swine, sheep, and goats.

strangles   An infectious disease of horses, characterized by inflammation of the mucous membranes of the respiratory tract.

streptococcus   Sperical, gram-positive bacteria that divide in only one plane and occur in chains. Some species cause serious disease.

stress   An unusual or abnormal influence causing a change in an animal's function, structure, or behavior.

stringhalt   A sudden and extreme flexion of the back of a horse, producing a jerking motion of the hindleg in walking.

strongyles   Any of various roundworms living as parasites, especially in domestic animals.

stud   Usually the same as stallion. Also a place where male animals are maintained (e.g., *bull stud*).

suckling gain   The gain that a young animal makes from birth until it is weaned.

subcutaneous   Situated beneath, or occurring beneath, the skin. A subcutaneous injection is an injection made under the skin.

sulfonamides   A sulfa drug capable of killing bacteria.

superovulation   The hormonally induced ovulation of a greater than normal number of eggs.

supplement   A feed used with another to improve the nutritive balance of the total feed.

sweeny   Atrophy of muscle (typically shoulder) in horses.

sweetbread   An edible by-product also known as the *pancreas.*

switch   The tuft of long hair at the end of tail (cattle and horses).

syndactyly   Union of two or more digits (e.g., in cattle, the two toes would be a solid hoof).

synthetic breeds   See *composite breed.*

tagging   Clipping wool from the dock, udder, and vulva regions of the ewe prior to breeding and lambing.

tags   (1) Wool covered with manure. (2) Abbreviated form of ear tags, used for identification.

tallow   The fat of cattle and sheep.

tandem selection   Selection for one trait for a given period of time followed by selection for a second trait and continuing in this way until all important traits are selected.

TDN   Total digestible nutrients; includes the total amounts of digestible protein, nitrogen-free extract, fiber, and fat (multiplied by 2.25), all summed together.

teaser ram   A ram made incapable of impregnating a ewe by vasectomy or by use of an apron to prevent copulation, which is used to find ewes in heat.

teasing   The stallion in the presence of the mare to see if she will mate.

teat   The protuberance of the udder through which milk is drawn.

tendon   Tough, fibrous connective tissue at ends of muscle bundles that attach muscle to bones or cartilage structures.

terminal sire   The sire used in a terminal crossbreeding program. It is intended that all offspring from a terminal sire be sold as market animals.

testicle   The male sex gland that produces sperm and testosterone.

testosterone   The male sex hormone that stimulates the accessory sex glands, causes the male sex drive, and causes the development of masculine characteristics.

tetanus   An acute infectious disease caused by toxin elaborated by the bacterium *Clostridium tetani,* in which toxic spasms of some of the voluntary muscles occur.

tetrad   A group of four similar chromotids formed by the splitting longitudinally of a pair of homologous chromosomes during meiotic prophase.

thermoneutral zone (TNZ)   Range in temperature where rate and efficiency of gain is maximized. Comfort zone.

thoroughpin   A hard swelling that is located between the Achilles tendon and the bone of the hock joint.

toeing in   Toes of front feet turn in. Also called *pigeon-toed.*

toeing out   Toes of front feet turn out. Also called *splay-footed.*

tom   A male turkey.

TPI   Total prediction index used in dairy cattle breeding. It includes the predicted differences for milk production, fat percentage, and type into one figure in a ratio of milk production × 3:fat percentage × 1:type × 1.

transcription   The synthesis of RNA from DNA in the nucleus by matching the sequences of the bases.

transgenic animals   Animals that contain genes transferred from other animals, usually from a different species.

transmissable gastroenteritis (TGE)   A serious, contagious diarrhea disease in baby pigs.

tripe   Edible product from walls of ruminant stomach.

trot   A diagonal two-beat gait in which the right front and left rear feet strike the ground in unison, and the left front and right rear feet strike the ground in unison.

twist   Vertical measurement from top of the rump to point where hindlegs separate.

twitch   To squeeze tightly the upper lip of a horse by means of a small rope that is twisted.

type   (1) The physical conformation of an animal. (2) All those physical attributes that contribute to the value of an animal for a specific purpose.

udder   The encased group of mammary glands of mammals.

ultrasound   A process used to measure fat thickness and ribeye area in swine and cattle. The machine sends sound waves into the back of the animal and records these waves as they bounce off the tissues. Different wavelengths are recorded for fat than for lean. Also used to diagnose pregnancy.

umbilical cord   A cord through which arteries and veins travel from the fetus to and from the placenta, respectively. This cord is broken when the young are born.

undershot jaw   Lower jaw is longer than upper jaw.

unsoundness   Any defect or injury that interferes with the usefulness of an animal.

urinary calculi   Disease where mineral deposits crystallize in the urinary tract. The deposits may block the tract, causing difficulty in urination.

uterus   That portion of the female reproductive tract where the young develop during pregnancy.

vaccination   The act of administering a vaccine or antigens.

vaccine   Suspension of attenuated or killed microbes or toxins administered to induce active immunity.

vagina   The copulatory portion of the female's reproductive tract. The vestibule portion of the vagina also serves for passage of urine during urination. The vagina also serves as a canal through which young pass when born.

variety meats   Edible organ by-products (e.g., liver, heart, tongue, tripe).

vas deferens   Ducts that carry sperm from the epididymis to the urethra.

vasectomy   The removal of a portion of the vas deferens. As a result, sperm are prevented from traveling from the testicles to become part of the semen.

veal   The meat from very young cattle, under 3 months of age.

vein   Vessel through which blood passes to the heart from various organs or body parts.

vermifuge   A chemical substance given to the animals to kill internal parasitic worms.

VFA   See *volatile fatty acids.*

villi   Projections of the inner lining of the small intestine.

virus   Ultramicroscopic bundle of genetic material capable of multiplying only in living cells. Viruses cause a wide range of disease in plants, animals, and humans, such as rabies and measles.

viscera   Internal organs and glands contained in the thoracic and abdominal cavities.

vitamin   An organic catalyst, or component thereof, that facilitates specific metabolic functions.

volatile fatty acids (VFA)   A group of fatty acids produced from microbial action in the rumen; examples are acetic, propionic, and butyric acids.

vulva   The external genitalia of a female mammal.

walk   A four-beat gait of a horse in which each foot strikes the ground at a time different from each of the other three feet.

warble   The larval stage of the heel fly that burrows out through the hide of cattle in springtime.

wattle   Method of identification in cattle where strips of skin (3–6 inches long) are usually cut on the nose, jaw, throat, or brisket.

weaner   An animal that has been weaned or is nearing weaning age.

weaning   Separating young animals from their dams so that the offspring can no longer suckle.

weaning weight EPD   A genetic estimate of the weaning weight of a beef bull's calves when compared to other bulls in the sire summary.

weanling   An animal of weaning age.

wet   Used to describe a milking female—e.g., wet cow or wet ewe.

wether   A male sheep castrated before reaching puberty.

white muscle disease   A muscular disease caused by a deficiency of selenium or vitamin E.

winking   Indication of estrus in the mare where the vulva opens and closes.

withdrawal time   The time before slaughter that a drug should not be given to an animal.

withers   Top of the shoulders.

wool   The fibers that grow from the skin of sheep.

wool blindness   Sheep cannot see owing to wool covering their eyes.

wool top   A continuous untwisted strand of combed wool in which the fibers lie parallel and the short fibers have been combed out.

woolens   Cloth made from short wool fibers that are intermingled in the making of the cloth by carding.

worsteds   Cloth made from wool that is long enough to comb and spin into yarn. The finish of worsteds is harder than woolens, and worsted clothes hold a press better.

yearling   Animals that are approximately 1 year old.

yearling weight EPD   A genetic estimate of yearling weight of a bull's progeny compared to other bulls in the same sire summary.

yield   Used interchangeably with *dressing percentage.*

yield grades   The grouping of animals according to the estimated trimmed lean meat that their carcass would provide; cutability.

yolk   (1) The yellow part of the egg. (2) The natural grease (lanolin) of wool.

yolk sac   Layer of tissue encompassing the yolk of an egg.

zone of thermoneutrality   The environmental temperature (about 65°F) at which heat production and heat elimination are approximately equal for most farm animals.

zygote   (1) The cell formed by the union of two gametes. (2) An individual from the time of fertilization until death.

Index